World Regional Geography

WORLD REGIONAL GEOGRAPHY

Robert E. Norris
Oklahoma State University

West Publishing Company
St. Paul   New York   Los Angeles   San Francisco

Copyediting   Peggy Hoover
Interior and Cover Designs   Diane Beasley Design
Cartographic Design and Production   Cartography Services, Oklahoma State University
Composition   Carlisle Communications
Cover Photograph   Comstock, Inc., R. Michael Stuckey

About the Cover Photograph

The Chang Jiang (or Ch'ang-Chiang) is the longest river in China, and among world rivers it is exceeded in length only by the Amazon and the Nile. Chang Jiang means "long" river. An ancient, but possibly more familiar, name for the river is Yangtze or Yang-tzu. In the southwest plateau section, between the cities of Chengdu and Chongoing, the river is known as the Chin-sha or "golden sand". This golden tone can be seen in the cover photograph.

COPYRIGHT ©1990   By WEST PUBLISHING COMPANY
50 W. Kellogg Boulevard
P.O. Box 64526
St. Paul, MN 55164-1003

All rights reserved

Printed in the United States of America

97 96 95 94 93 92 91 90   8 7 6 5 4 3 2 1 0

LIBRARY OF CONGRESS CATALOGING-IN-PUBLICATION DATA

Norris, Robert E.
   World regional geography / Robert E. Norris.
       p.   cm.
   Bibliography: p.
   Includes index.
   ISBN 0-314-48133-8
   1. Geography.   I. Title.
G128.N673   1990
910—dc20

89-8901
CIP

Chapter-opening Photo Credits

2–3 National Aeronautics and Space Administration; **6–7** Art Bilsten, Photo Researchers, Inc.; **26–27** Cary Wolinsky, Stock, Boston; **56–57** Harry Gruyaert, Magnum; **80–81** T. Omas D.W. Friedmann, Photo Researchers, Inc.; **122–23** Mike Mazzaschi, Stock, Boston; **164–65** Maggie Steber, Stock, Boston; **202–3** Allen Green, Photo Researchers, Inc.; **230–31** Ulrike Welsch, Photo Researchers, Inc.; **258–59** Abbas, Magnum; **302–3** George Holton, Photo Researchers, Inc.; **338–39** Toni Angermayer, Photo Researchers, Inc.; **394–95** Louis Goldman, Rapho/Photo Researchers, Inc.; **422–23** George Holton, Photo Researchers, Inc.; **452–53** Henri Cartier-Bresson, Magnum; **482–83** Rene Burri, Magnum; **514–15** Chad Ehlers, International Stock Photography; **536–37** Ira Kirschenbaum, Stock, Boston; **556–57** Chris Steele Perkins, Magnum.

Color Insert Photo Credits

World Levels of Economic Development
First page: John Garrett, Tony Stone Worldwide.
Third page: (top) Julie Houck, Stock, Boston; *(bottom)* Rio Branco, Magnum.
Fourth page: (top) Diane M. Lowe, Stock, Boston; *(bottom)* Sebastiao Salgado, Magnum.

World Levels of Income
First page: Ian Berry, Magnum.
Third page: (top) Richard Hutchings, Photo Researchers, Inc.; *(bottom)* Paul Fusco, Magnum.
Fourth page: (top) Peter Menzel, Stock, Boston; *(bottom)* Cary Wolinsky, Stock, Boston.

World Levels of Agriculture
First page: George Hunter, Tony Stone Worldwide.
Third page: (top) Noboru Komine, Photo Researchers, Inc.; *(bottom)* Randall Hyman, Stock, Boston.
Fourth page: (top) Hugh Sitton, Tony Stone Worldwide; *(bottom)* Victor Englebert, Photo Researchers, Inc.

World Levels of Energy Use
First page: Peter Menzel, Stock, Boston.
Third page: (top) D.C. Lowe, Tony Stone Worldwide; *(bottom)* Elliott Erwitt, Magnum.
Fourth page: (top) Sebastiao Salgado, Magnum; *(bottom)* David R. Austen, Stock, Boston.

This book is dedicated to the memory of L. Lloyd Haring (1922–82). Lloyd was my first college geography teacher and later became a close friend and co-author. Lloyd was always a source of encouragement for me and has been an inspiration even after his untimely death. Without Lloyd's help and guidance in the early part of my career, life would have been much tougher than it was.

Contents in Brief

Chapter 1
INTRODUCTION TO WORLD GEOGRAPHY 2

Chapter 2
THE PHYSICAL SETTING 6

Chapter 3
THE HUMAN DIMENSION 26

Chapter 4
CONCEPTS AND TOOLS 56

Chapter 5
THE UNITED STATES AND CANADA 80

Chapter 6
EUROPE 122

Chapter 7
THE SOVIET UNION 164

Chapter 8
JAPAN 202

Chapter 9
AUSTRALIA AND NEW ZEALAND 230

Chapter 10
MIDDLE AMERICA 258

Chapter 11
SOUTH AMERICA 302

Chapter 12
AFRICA 338

Chapter 13
THE MIDDLE EAST 394

Chapter 14
SOUTH ASIA 422

Chapter 15
SOUTHEAST ASIA 452

Chapter 16
EAST ASIA 482

Chapter 17
THE PACIFIC ISLANDS 514

Chapter 18
THE POLAR REGIONS 536

Chapter 19
WORLD PROBLEMS AND PROSPECTS 556

Contents

Preface xxv

Chapter 1
INTRODUCTION TO WORLD GEOGRAPHY 2
Key Words 5
References and Readings 5

Chapter 2
THE PHYSICAL SETTING 6
The Earth in Space 7
The Composition, Size, and Shape of the Earth 8
The Earth–Sun Relationship 9
The Continents 11
Landforms 14
Climate 18
Vegetation and Soils 22
Key Words 25
References and Readings 25

Chapter 3
THE HUMAN DIMENSION 26
Introduction 27
Current Population 27
Demographic Terms 30
Demographic Transition 31
Population Pyramids 32
Racial Types 35
Cultural Traits 38
Language 39
Religion 42
Economic Geography 48
  Agriculture 48
  Mineral Production 50
  Industry 50
  Trade, Transportation, and Communication 51
Summary 53
Key Words 53
References and Readings 54

Chapter 4
CONCEPTS AND TOOLS 56
Spatial Distributions 57
Spatial Interaction 59
Spatial Diffusion 60
Spatial Models 63
Spatial Interaction Model 67
Maps and Mapping 69
Map Scales and Projections 73
Computer Maps 76
Remote Sensing 76
Key Words 78
References and Readings 78

Chapter 5
THE UNITED STATES AND CANADA 80
Introduction 81
Physical Geography of the United States and Canada 81
  Landforms 81
  Climate 89
  Vegetation 91
  Soils 93
Human Geography of the United States and Canada 95
  Population 95
  Cultural Influences 100
  Religion 100
  Language 101
  Ethnicity 102

Economic Geography of the United States and Canada 104
   Agriculture 104
   Mineral Production 108
   Industry 113
   Transportation 115
Political Geography of the United States and Canada 120
Key Words 120
References and Readings 120

Chapter 6
EUROPE 122
Introduction 123
Physical Geography of Europe 126
   Landforms 126
   Climate and Vegetation 133
Human Geography of Europe 135
   Population 136
   Religion 142
   Language 145
   Ethnicity 147
Economic Geography of Europe 148
   Agriculture 148
   Mineral Production 152
   Trade and Transportation 154
Political Geography of Europe 161
Key Words 162
References and Readings 162

Chapter 7
THE SOVIET UNION 164
Introduction 165
Physical Geography of the Soviet Union 167
   Landforms 167
   Climate and Vegetation 170
Human Geography of the Soviet Union 174
   Growth of the Soviet Union 174
   Population 175
   Urbanization 176
   Religion 180
   Language and Ethnicity 182
   Education and Health 183
Economic Geography of the Soviet Union 187
   Agriculture 187
   Mineral Production 190
   Industry 191
   Transportation 196
   Communication 198
Political Geography 199
Key Words 200
References and Readings 200

Chapter 8
JAPAN 202
Introduction 203
Physical Geography of Japan 204
   Landforms 204
   Climate and Vegetation 207
Human Geography of Japan 211
   Population 211
   Urbanization 211
   Race and Culture 213
   Language 214
   Religion 215
Economic Geography of Japan 217
   Agriculture 217
   The Fishing Industry 220
   Mineral Production 222
   Manufacturing 223
   Trade and Transportation 225
   Standard of Living 225
Political Geography of Japan 228
Key Words 229
References and Readings 229

Chapter 9
AUSTRALIA AND NEW ZEALAND 230
Introduction 231
Physical Geography of Australia and New Zealand 233
   Landforms of Australia 233
   Landforms of New Zealand 235
   Climate and Vegetation of Australia 236
   Climate and Vegetation of New Zealand 239
Human Geography of Australia and New Zealand 241
   Settlement 241
   Population 243
   Language and Religion 246
   Minorities 246
   Health and Welfare 246

Economic Geography of Australia and
New Zealand   247
   Agriculture   247
   Mineral Production   249
   Industry   250
   Trade   251
   Transportation   252
Political Geography of Australia and
New Zealand   255
Key Words   255
References and Readings   255

Chapter 10
MIDDLE AMERICA   258

Introduction   259
   Mexico   260
   Central America   261
   The Caribbean Islands   262
Physical Geography of Middle America   263
   Landforms of Mexico   263
   Climate and Vegetation of Mexico   265
   Landforms of Central America   269
   Climate and Vegetation of
   Central America   270
   Landforms of the Caribbean Islands   271
   Climate of the Caribbean Islands   273
Human Geography of Middle America   275
   Population Patterns   275
   Population Distribution   276
   Population Structure   278
   Language and Religion   279
   Ethnicity   280
Economic Geography of Middle America   283
   Agriculture   283
   Mineral Production   288
   Industry   289
   Trade and Transportation   296
Political Geography of Middle America   297
Key Words   298
References and Readings   298

Chapter 11
SOUTH AMERICA   302

Introduction   303
Physical Geography of South America   304
   Landforms   304
   Climate and Vegetation   309
Human Geography of South America   312
   Population   312
   Urban Centers, Towns, and Villages   313
   Language and Religion   317
Economic Geography of South America   318
   Agriculture   318
   Mineral Production   326
   Industry   328
   Trade and Transportation   329
Political Geography of South America   333
Key Words   335
References and Readings   335

Chapter 12
AFRICA   338

Introduction   339
   Physical Geography of Africa   341
   Cultural Geography of Africa   343
NORTHERN AFRICA   344
Introduction   344
Physical Geography of Northern Africa   347
Human Geography of Northern Africa   349
   Population   349
   Income and Life Expectancy   351
   Culture, Language, and Religion   353
Economic Geography of Northern Africa   354
   Agriculture   354
   Mineral Production and Industry   355
   Transportation   356
Political Geography of Northern Africa   358
TROPICAL AFRICA   359
Introduction   359
Physical Geography of Tropical Africa   360
Human Geography of Tropical Africa   363
   Population   363
   Ethnicity   368
   Poverty   368
   Health   370
   Religion   371
   Language   371
Economic Geography of Tropical Africa   372
   Agriculture   372
   Mineral Production and Industry   374
   Transportation and Communications   375
   Political Geography of Tropical Africa   377

SOUTHERN AFRICA   378
Introduction   378
Physical Geography of Southern Africa   380
Human Geography of Southern Africa   382
   Population and Urbanization   382
   Ethnicity   383
   Religion and Language   384
Economic Geography of Southern Africa   385
   Economic Conditions   385
   Agriculture   385
   Mineral Production and Industry   388
   Transportation and Communications   389
Political Geography of Southern Africa   389
Key Words   392
References and Readings   392

Chapter 13
THE MIDDLE EAST   394

Introduction   395
Physical Geography of the Middle East   398
Human Geography of the Middle East   401
   Population and Urbanization   401
   Language   406
   Religion   406
Economic Geography of the Middle East   407
   Crude-Oil Production   407
   Agriculture   409
   Mineral Production and Industry   411
   Transportation and Communications   415
Political Geography of the Middle East   417
Key Words   420
References and Readings   420

Chapter 14
SOUTH ASIA   422

Introduction   423
Physical Geography of South Asia   426
Human Geography of South Asia   431
   Population   431
   Urbanization   433
   Religion   435
   Language   440
Economic Geography of South Asia   440
   Agriculture   440
   Mineral Production and Industry   442
   Transportation and Communications   444
Political Geography of South Asia   446
Key Words   449
References and Readings   449

Chapter 15
SOUTHEAST ASIA   452

Introduction   453
Physical Geography of Southeast Asia   456
Human Geography of Southeast Asia   460
   History   460
   Population   465
   Ethnicity   468
   Religion and Language   470
Economic Geography of Southeast Asia   471
   Agriculture   471
   Mineral Production and Industry   474
   Transportation and Communications   476
Political Geography of Southeast Asia   480
Key Words   480
References and Readings   480

Chapter 16
EAST ASIA   482

Introduction   483
Physical Geography of East Asia   484
   Landforms   484
   Rivers   484
   Climate   489
Human Geography of East Asia   491
   History   491
   Population   493
   Urbanization   495
   Religion   496
   Language   498
Economic Geography of East Asia   501
   Agriculture   501
   Industry   503
   Transportation and Communications   507
Political Geography of East Asia   510
Key Words   511
References and Readings   511

Chapter 17
THE PACIFIC ISLANDS 514

Introduction 515
Physical Geography of the Pacific Islands 516
   Landforms 516
   Climate 520
   Vegetation 523
Human Geography of the Pacific Islands 525
   History 525
   Population 526
   Ethnicity 528
   Languages 528
   Religion 529
Economic Geography of the Pacific Islands 529
   Agriculture 529
   Mineral Production and Industry 529
   Transportation and Communications 530
Political Geography of the Pacific Islands 531
Key Words 533
References and Readings 533

Chapter 18
THE POLAR REGIONS 536

Introduction 537
Physical Geography of the Polar Regions 539
   Landforms 539
   Climate 544
   Vegetation and Wildlife 545
Human Geography of the Polar Regions 549
Economic Geography of the Polar Regions 552
Political Geography of the Polar Regions 553
Key Words 555
References and Readings 555

Chapter 19
WORLD PROBLEMS AND PROSPECTS 556

Introduction 557
Environmental Problems 558
   Air Pollution 558
   The Greenhouse Effect 559
   Water Pollution 559
   Destruction of Farm Land 560
   Depletion of Resources 561
Population Problems 562
   Rate of Increase 562
   Population Movement 562
Economic Problems 567
   Distribution of Income 567
   Lack of Income Diversity 567
   Debtor Nations 567
Political Problems 567
   Current Conflicts 567
   Nuclear Weapons 569
   Military Governments 569
   Territorial Expansion 569
Prospects for the Future 570
Key Words 571
References and Readings 571

Appendix A Conversions for Units of Measurement A-1
Appendix B Climatic Classification A-3
Appendix C Selected Data on World Regions A-5
Bibliography B-1
Glossary G-1
Index I-1

LIST OF FIGURES

Figure 2.1	Internal Structure of the Earth 9		Figure 4.8	Christaller's Market Area Development 66
Figure 2.2	The Solar System 10		Figure 4.9	Christaller's Hexagonal Market Areas 67
Figure 2.3	The Change of Seasons 12		Figure 4.10	Distance-Decay Curve 68
Figure 2.4	Historical Change in Earth's Geology 14		Figure 4.11	An Area Map 69
Figure 2.5	Mid-ocean Ridges and Trenches 15		Figure 4.A	Cartesian Coordinate System Applied to a Sphere 70
Figure 2.6	Crustal Plates 16		Figure 4.12	Schematic of Statistical Map 71
Figure 2.7	World Climate Regions 20		Figure 4.13	Schematic of a Dot Map 72
Figure 2.8	The General Circulation of the Atmosphere 21		Figure 4.14	Schematic of a Choropleth Map 72
Figure 2.9	Natural Vegetation Regions of the World 23		Figure 4.15	Schematic of an Isoline Map 73
Figure 2.10	Major Soil Regions of the World 24		Figure 4.16	Map Scale 74
Figure 3.1	World Birth Rates, 1986 31		Figure 4.17	Map Projections 75
Figure 3.2	World Death Rates, 1986 33		Figure 5.1	World Location of the United States and Canada 82
Figure 3.3	Population Age Structures for the United States and Egypt 34		Figure 5.2	Land Elevations (A) and Physiographic Regions (B) of the United States and Canada 84
Figure 3.4	World Racial Group Regions 38			
Figure 3.5	World Language Regions 40		Figure 5.3	Patterns of Annual Precipitation in the United States 91
Figure 3.6	World Religions 44			
Figure 3.7	World Areas of Cultivated Land 48		Figure 5.4	Köppen Climate Regions of the United States and Canada 92
Figure 3.8	Coal and Iron Ore Deposits of the World 51		Figure 5.5	Natural Vegetation Regions of the United States and Canada 94
Figure 3.9	Manufacturing Areas of the World 52		Figure 5.6	Soil Regions of the United States and Canada 96
Figure 3.10	World Shipping Routes 53		Figure 5.7	Population Distribution of the United States and Canada 97
Figure 4.1	Pattern, Density, and Dispersion 58		Figure 5.8	Religious Affiliations in the United States 102
Figure 4.2	Redistribution of Population 59			
Figure 4.3	Diffusion of FM Radio Stations in the United States 61		Figure 5.9	Agricultural Regions of the United States and Canada 105
Figure 4.4	Types of Spatial Diffusion 62		Figure 5.10	Major Coalfields of the United States and Canada 109
Figure 4.5	The Logistic Curve 63			
Figure 4.6	Von Thünen's Landscape 64		Figure 5.11	Petroleum-Producing Areas of the United States and Canada 111
Figure 4.7	Distorted von Thünen Landscape 65			

Figure 5.12	Major Nonfuel Mineral Deposits of the United States and Canada 112	Figure 7.4	Climate Regions of the Soviet Union 172
Figure 5.13	Industrial Regions of the United States and Canada 114	Figure 7.5	Natural Vegetation and Agricultural Zones of the Soviet Union 173
Figure 5.14	Major Railroads of the United States and Canada 117	Figure 7.6	Historical Expansion of the Territory of the Soviet Union 175
Figure 5.15	Inland Waterway Traffic in the United States 118	Figure 7.7	Population Density of the Soviet Union 176
Figure 5.16	Major Pipelines of the United States and Canada 119	Figure 7.8	Major Cities of the Soviet Union 178
Figure 6.1	World Location of Europe 124	Figure 7.9	Ethnic Regions of the Soviet Union 184
Figure 6.2	Divisions of Europe 125	Figure 7.10	Agricultural Areas of the Soviet Union 189
Figure 6.3	Land Elevations and Water Bodies of Europe 127	Figure 7.11	Mineral Deposits of the Soviet Union 192
Figure 6.4	Physiographic Regions of France 129	Figure 7.12	Coal and Petroleum Deposits of the Soviet Union 194
Figure 6.5	The Upland Regions of Spain and France 130	Figure 7.13	Manufacturing Regions of the Soviet Union 196
Figure 6.6	Mountains and Rivers of Europe 132	Figure 7.14	Railroads of the Soviet Union 198
Figure 6.7	Köppen Climate Regions of Europe 134	Figure 8.1	World Location of Japan 204
Figure 6.8	Summer (A) and Winter (B) Precipitation Patterns of Europe 136	Figure 8.2	Land Elevations of Japan 206
		Figure 8.3	The Kanto Plain 208
Figure 6.9	Population Density of Europe 138	Figure 8.4	Climate Regions of Japan 209
Figure 6.10	Capitals and Other Major Cities of Europe 140	Figure 8.5	The Summer and Winter Winds over Japan 210
Figure 6.11	Population Increase in Europe for 1986 143	Figure 8.6	Population Patterns of Japan 212
		Figure 8.7	Major Cities of Japan 213
Figure 6.12	Religions of Europe 144	Figure 8.8	Land Use in Japan 221
Figure 6.13	Languages of Europe 146	Figure 8.9	Wheat, Barley, and Oats Producing Areas of Japan 222
Figure 6.14	Agricultural Regions of Europe 150	Figure 8.10	Manufacturing Regions of Japan 223
Figure 6.15	Major Coalfields of Europe 153		
Figure 6.16	Industrial Regions of Europe 155	Figure 8.11	Railroads of Japan 227
		Figure 9.1	World Location of Australia and New Zealand 232
Figure 6.17	Petroleum-Producing Areas and Pipelines of Europe 157	Figure 9.2	Landforms of Australia and New Zealand 234
Figure 6.18	Transportation Links of Europe 158	Figure 9.3	Physiographic Regions of Australia 236
Figure 7.1	World Location of the Soviet Union 166	Figure 9.4	Climate Regions of Australia and New Zealand 238
Figure 7.2	Land Elevations and Water Bodies of the Soviet Union 168	Figure 9.5	Summer and Winter Temperature and Precipitation Patterns of Australia and New Zealand 240
Figure 7.3	Locations and Names of Landforms of the Soviet Union 171		

Figure	Title	Figure	Title
Figure 9.6	Population Patterns of Australia and New Zealand 244	Figure 10.23	Sugar Cane Production Areas of Cuba 289
Figure 9.7	Agricultural Areas of Australia and New Zealand 248	Figure 10.24	Mineral Deposits of Middle America 290
Figure 9.8	Mineral Deposits of Australia and New Zealand 250	Figure 10.A	The Five Nations of Mexico 292
Figure 9.9	Roads and Railroads of Australia and New Zealand 253	Figure 10.25	Industrial Areas of Mexico 294
Figure 10.1	World Location of Middle America 260	Figure 10.26	The Panama Canal 295
Figure 10.2	The Countries of Middle America 261	Figure 10.27	Roads and Railroads of Mainland Middle America 298
Figure 10.3	The States of Mexico 262	Figure 11.1	World Location of South America 305
Figure 10.4	The Countries of Central America 263	Figure 11.2	The Countries of South America 306
Figure 10.5	The Greater Antilles and Lesser Antilles 264	Figure 11.3	Land Elevations and Water Bodies of South America 307
Figure 10.6	Land Elevations and Water Bodies of Mexico 266	Figure 11.4	Climate Regions of South America 311
Figure 10.7	Climate Regions of Mexico 267	Figure 11.5	Population Patterns of South America 314
Figure 10.8	Altitudinal Zones of Middle America 268	Figure 11.6	Industrial Areas and Major Cities of South America 316
Figure 10.9	Land Elevations and Water Bodies of Central America 270	Figure 11.7	The Latin American City 317
Figure 10.10	Climate Regions of Central America 271	Figure 11.8	Agricultural Regions and Land Use of South America 320
Figure 10.11	Annual Precipitation of Central America 272	Figure 11.9	Forest Sequence after Slash-and-Burn Agriculture 321
Figure 10.12	Geology of the Caribbean 273	Figure 11.A	Coca Cultivation Areas of South America 323
Figure 10.13	Guadeloupe, Dominica, and Martinique 276	Figure 11.10	Agricultural Regions of Argentine Pampa 325
Figure 10.14	Tropical Storm (Hurricane) Tracks of the Caribbean 277	Figure 11.11	Mineral Deposits of South America 326
Figure 10.15	Precipitation Patterns of a Typical Caribbean Island 278	Figure 11.12	Railroads of South America 332
Figure 10.16	Profile of a Hurricane 279	Figure 11.13	International Highways of South America 334
Figure 10.17	Population Patterns of Middle America 280	Figure 12.1	World Location of Africa 340
Figure 10.18	Major Cities of Middle America 281	Figure 12.2	Land Elevations and Water Bodies of Africa 342
Figure 10.19	Population Pyramids for the United States and Mexico 282	Figure 12.3	Regional Divisions of Africa 345
		Figure 12.4	The Countries of Northern Africa 346
Figure 10.20	Racial Composition of Middle America 284	Figure 12.5	Land Elevations and Water Bodies of Northern Africa 348
Figure 10.21	Agricultural Regions of Central America 286	Figure 12.6	Precipitation Patterns of Northern Africa 349
Figure 10.22	Banana Production Areas of Central America 286	Figure 12.7	Population Distribution of Africa 350

List of Figures **xvii**

Figure 12.8	Life Expectancy (A) and Per Capita Gross National Product Rates (B) for Africa 352	Figure 13.2	The Countries of the Middle East 397
Figure 12.9	The Arab World 353	Figure 13.3	Locations of Important Mountain Ranges of the Middle East 399
Figure 12.10	Agriculture of Northern Africa 355	Figure 13.4	Land Elevations and Water Bodies of the Middle East 400
Figure 12.11	Mineral Production Areas of Northern Africa 356	Figure 13.5	The Jordan River Valley 401
Figure 12.12	Railroads of Northern Africa 358	Figure 13.6	Precipitation Patterns of the Middle East 402
Figure 12.13	Countries of Tropical Africa 360	Figure 13.7	Population Patterns of the Middle East 403
Figure 12.14	Land Elevations and Water Bodies of Tropical Africa 361	Figure 13.8	Street Plans of Istanbul (A) and Ankara (B) 404
Figure 12.15	Annual Precipitation of Africa 364	Figure 13.9	The Evolution of Israel 408
Figure 12.16	Natural Vegetation Regions of Africa 365	Figure 13.10	The Holy Land of Biblical Times 410
Figure 12.A	Burkina Faso 367	Figure 13.11	Oilfields and Major Pipelines of the Middle East 411
Figure 12.17	Infant Mortality (A), Birth Rates (B), and Rate of Population Increase (C) for Africa 369	Figure 13.12	Cultivated Areas of the Middle East 411
Figure 12.18	AIDS in Tropical Africa 371	Figure 13.13	Roads and Railroads of the Middle East 418
Figure 12.19	Language Regions of Africa 373	Figure 14.1	Three Regions of Asia 424
Figure 12.20	Agricultural Regions of Tropical Africa 374	Figure 14.2	World Location of South Asia 425
Figure 12.21	Mineral Deposits of Tropical Africa 376	Figure 14.3	The Countries of South Asia 426
Figure 12.22	Roads, Railroads, and Airports of Tropical Africa 377	Figure 14.4	Land Elevations and Water Bodies of South Asia 428
Figure 12.23	Tribal and Modern National Boundaries of Tropical Africa 378	Figure 14.5	July (A) and January (B) Prevailing Wind Patterns of South Asia 429
Figure 12.24	Countries of Southern Africa 380	Figure 14.6	July (A) and January (B) Temperature Averages for South Asia 431
Figure 12.25	Land Elevations and Water Bodies of Southern Africa 381	Figure 14.7	Population Patterns of South Asia 432
Figure 12.26	Climate Regions of Southern Africa 383	Figure 14.8	The Ganges River Valley 433
Figure 12.27	Johannesburg and Pretoria 384	Figure 14.9	Major Cities of South Asia 435
Figure 12.B	Madagascar 386	Figure 14.10	City Plan of Kabul 436
Figure 12.28	Agricultural Regions of Southern Africa 387	Figure 14.A	Locations of Major Afghan Refugee Camps in Pakistan 437
Figure 12.29	Mineral Deposits of Southern Africa 389	Figure 14.11	Source Areas and Diffusion Routes of the Major Religions of South Asia 438
Figure 12.30	Roads and Railroads of Southern Africa 390		
Figure 13.1	World Location of the Middle East 396	Figure 14.12	Language Regions of South Asia 442

List of Figures

Figure 14.13	Agricultural Regions of South Asia 443		Figure 16.9	Agricultural Regions of China 504
Figure 14.14	Coal Fields, Oil and Gas Fields, and Iron Ore Deposits of India 444		Figure 16.10	Mineral Deposits and Industrial Areas of China 506
Figure 14.15	Railroads and Industrial Areas of India 445		Figure 16.11	Major Roads of China 508
Figure 14.B	Bangladesh 447		Figure 16.12	Railroads of China 509
Figure 15.1	World Location of Southeast Asia 454		Figure 17.1	Ethnic Regions of the Pacific Islands 516
Figure 15.2	The Countries of Southeast Asia 455		Figure 17.2	Mountains of the Hawaiian Islands 518
Figure 15.3	Land Elevations and Water Bodies of Southeast Asia 457		Figure 17.3	Geology of the Pacific 519
			Figure 17.4	Formation of Atolls 520
Figure 15.4	Part of Indonesia 458		Figure 17.5	Prevailing Winds of the Pacific Region 521
Figure 15.5	Climate Regions of Southeast Asia 459		Figure 17.6	Wallace's Line 523
Figure 15.6	Wind Changes and Typhoon Tracks in Southeast Asia 460		Figure 17.7	Typical Pacific Island 524
			Figure 17.8	Exploration of the Pacific Region 527
Figure 15.7	Cultural Cores and Historical Invasion Routes of Southeast Asia 462		Figure 17.9	Agriculture on Guam 530
			Figure 18.1	Polar Projections of the Arctic and the Antarctic 538
Figure 15.8	European Colonial Holdings in Southeast Asia 465		Figure 18.2	Land Elevations and Water Bodies of the Polar Region of North America 540
Figure 15.9	Population Patterns of Southeast Asia 466		Figure 18.3	Land Elevations and Water Bodies of the Polar Region of Europe and Asia 541
Figure 15.10	Singapore 468			
Figure 15.11	Agricultural Regions of Mainland Southeast Asia 472		Figure 18.4	Land and Ice Elevations of the Antarctic 543
Figure 15.12	Mineral Deposits of Southeast Asia 475		Figure 18.5	Atmospheric Circulation 546
			Figure 18.6	Scientific Stations in Antarctica 550
Figure 15.13	Roads and Railroads of Mainland Southeast Asia 478		Figure 18.7	Native Peoples of the Arctic 550
Figure 16.1	World Location of East Asia 485		Figure 18.8	Cities, Permanent Bases, and Towns in the Arctic 552
Figure 16.2	The Countries of East Asia 486			
Figure 16.3	Land Elevations and Water Bodies of East Asia 487		Figure 18.9	Mineral Deposits of the Arctic 553
Figure 16.4	Physiographic Features and Rivers of East Asia 488		Figure 18.10	Political Divisions of Antarctica 554
Figure 16.5	Climate Regions of East Asia 490		Figure 19.1	World Pollution 558
			Figure 19.2	Greenhouse Effect 560
Figure 16.6	Evolution of China 492		Figure 19.3	Major Hazardous Waste Sites in the United States 561
Figure 16.7	Population Patterns of East Asia 494		Figure 19.4	World Population Explosion 563
Figure 16.8	Major Cities of China 497		Figure 19.5	World Conflict Areas 568
Figure 16.A	Agricultural Regions of North Korea and South Korea 502		Figure 19.6	World Military Regimes 570

LIST OF FOCUS BOXES

Chapter 3	Thomas Malthus on Population 28		*Chapter 11*	Cocaine 322
Chapter 4	Latitude and Longitude 70 The Metric System 77		*Chapter 12*	Burkina Faso 367 Madagascar 386 Apartheid in South Africa 391
Chapter 5	Sports in America 98		*Chapter 13*	Jerusalem 405 The Jews 412
Chapter 6	Working Mothers and Child Care 142 The Balkans 147		*Chapter 14*	Migration from Afghanistan 437 Storms and Crowding in Bangladesh 447
Chapter 7	Christianity in the Soviet Union 182 Pronunciation of Russian Words 186 Potatoes and Vodka 187		*Chapter 15*	The World of Rice 474
			Chapter 16	Chinese Place Names and the Pin-yin System 498 Contrasting Koreas 502 China's Fertility Debate 505
Chapter 8	Sports in Japan 218 Japan's American Investments 226		*Chapter 17*	The Island World 517
Chapter 9	Australian Immigration 245		*Chapter 18*	Ozone Depletion 548
Chapter 10	The United Fruit Company 287 The Five Nations of Mexico 291		*Chapter 19*	International Refugees 564

Color Inserts

World Levels of Economic Development (after page 38)

World Levels of Agriculture (after page 134)

World Levels of Energy Use (after page 326)

World Levels of Income (after page 486)

PREFACE

This text was designed for a one-semester (or one-quarter) college course in world regional geography. Such a course is generally offered at the freshman or sophomore level, so little expertise in geography is expected. The information is presented with that in mind. The reading level of the book is for freshmen in college, for students with no geography training, and for students who may take only one course in geography.

From the human perspective, the world is a very large place. A course of study that covers the world could be a lifelong experience, and a book that attempts to cover all the elements of the world would take up many volumes. Authors of a world geography book therefore are confronted with the problem of reducing the vast array of information about the world to a form that is manageable and organizing the material so that it can be easily comprehended. The process is similar to threshing grain. The extraneous information (the straw) must be gleaned from the kernals (the seeds), then the seeds must be put into useful order (the grain bin). The study of geography, then, is a search for order.

Geographers use regions to help organize the pertinent information. This book is organized on the basis of 14 large world regions. Regions other than these could have been selected, but these 14 are useful and homogeneous enough to promote understanding of the information about them. The overall organization of the book follows economic lines. The regional discussion begins with what are called the "developed" areas of the world—the countries where advanced economic development is a common feature. The discussion then turns to the less developed areas and ends with the least developed.

Since about 1850 the study of geography has been divided into *physical* topics and *human* topics, but in regional courses both these elements are considered. The physical setting, such as the home for humans, is an important factor in the study of any region. Each chapter on regions, then, begins with a discussion of the landforms, climates, and vegetation found in that region. The discussion on the human conditions of the region follows the physical aspects. The human geography section of each chapter is presented according to the major themes in human geography: population (including distribution and density, as well as racial and ethnic makeup), culture (including language and religion), economics (agriculture, mineral production and industry, trade and transportation), and politics. Other topics in human geography could have been selected, but these seem to be the most important for understanding the basic human aspects of any region.

The straightforward physical and human coverage in this book is designed to correspond to the "back to basics" movement in geography. A world geography textbook should contain as much information as possible about the physical and human parts of the world, so a balance between these important elements of geography was sought. History is important, but a world geography textbook should be a geography book first. Thus, history is not stressed in this book. The information contained here is as current as possible, but the book was not designed as a "current events" textbook. Current events, like history, also are important, but they should be stressed by the instructor as they occur. In fact, teachers are encouraged to emphasize their own interests, because the book was designed to be a reference, or guide, to information about the physical and human conditions of world regions. It is a supplement to lectures, not a replacement for them.

This book is largely descriptive. It was designed to provide information (facts) about the world. Many students who use this book will take only one college course in geography. Therefore, one chapter, Chapter 4, is devoted to some theories, concepts, and tools geographers have found useful, so the analytical part of geography appears in a single chapter instead of scattered throughout the text. The theories apply to many regions, not just one, and the tools can be used to advantage in understanding each region. In order to get the most benefit from Chapter 4, the teacher should refer to it, and the students should reread it, occasionally throughout the course.

This text is accompanied by a comprehensive instructor's manual with test bank prepared by Lou Seig of Oklahoma State University, which has been developed to correspond closely with each chapter in the textbook. An extensive test bank portion of the manual provides a substantive review of all major themes and concepts discussed within each chapter. The test bank is also available on *Westest,* a computerized testing system.

The final product is the result of many difficult choices concerning what to cover and what to leave out. The reviewers and editors helped with some of these decisions, while other decisions reflect the opinions of the author. Any errors are the responsibility of the author. Comments from readers on either the coverage or about errors are welcome.

The reviewers deserve special thanks for taking time to read the many typed pages of early manuscript carefully. Without their help, this book would be much less than what it is. These people are:

John Anderson, University of Louisville
Marvin Baker, University of Oklahoma
Thomas Detwyler, University of Wisconsin at Stevens Point
Richard Gelpke, Boston University
Linda Greenow, State University of New York at New Paltz
Merrill Johnson, University of New Orleans
Tom Love, Linfield College, McMinn, Oregon
Robert Peplies, East Tennessee State University
Martin Petit, Illinois Central College
Gregory Rose, Ohio State University at Marion
Richard Ulack, University of Kentucky
Roland Williams, West Liberty (W.Va.) State College

Many other people should be thanked for the part they played in the development of this book. All the people at West Publishing Company were helpful and encouraging. It was a pleasure to work with them. Special thanks goes to Tom LaMarre, the West editor who originally proposed the idea for such a project and then followed through with suggestions and encouragement. Without Tom's patience and persistence, the final product would never have materialized. Special thanks also goes to West production editor Mark Jacobsen, who brought the scattered parts together and made them into a book.

Much indebtedness is owed to Gayle Maxwell (Coordinator, Cartography Services, Oklahoma State University) and Chris Head (Assistant to the Cartographer). Gayle took rough sketches of maps, designed the layouts for the finished maps, and oversaw the construction of each map. The effort involved in this work was great. Thanks, then, to all the cartographers who worked on the project, most of whom are geography students at Oklahoma State University.

Many other people should be recognized and receive my sincere thanks for their help while the writing was in progress. My family always gave me encouragement and provided the time to write by doing jobs that I should have done. Thanks to my wife Edith, daughter Jennifer, and long-distance cheerleader, daughter Liz McKinley.

The many students that have taken my courses over the years should be mentioned also. They have shown where some ideas work and others do not. They have pointed out the things they like and dislike about textbooks. This book was written with them in mind.

Robert E. Norris
Stillwater, Oklahoma

World Regional Geography

Chapter 1

INTRODUCTION TO WORLD GEOGRAPHY

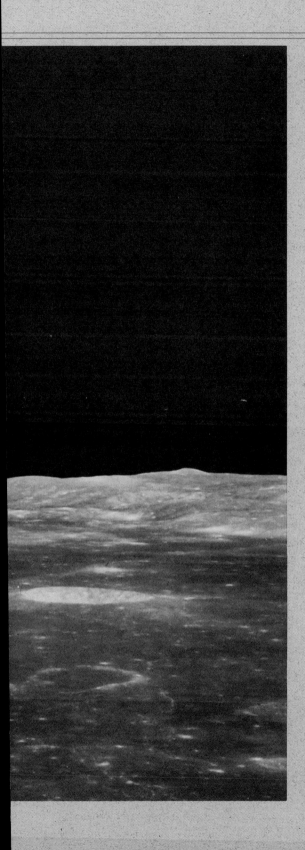

The study of geography is divided into two broad categories: human geography and physical geography. **Human geography** is the study of people, including their cultures, religions, politics, occupations, and distribution on the earth. It also includes the study of the activities of human beings, such as migration and other population movements, the growing of crops, the production and buying and selling of goods, and such divisive human activities as wars. Because human geography is so broad, it usually is studied by using subdivisions such as economic geography, political geography, and other appropriate segments. **Physical geography** is the study of the natural setting in which humans live. It includes the study of landforms, water, weather and climate, vegetation and soils, and the processes that create variations in these physical elements.

Geography is the science that seeks to explain both spatial distributions and place-to-place differences in the human and physical aspects of the world. A major goal in the study of geography is to learn where things are located, but it is also important to discover why things are located where they are. Maps are important to geographers because they help answer the "where" question, but the "why" question takes some effort beyond merely looking at a map. Sometimes explanations of "why" can be gained through comparing one map with another; at other times, the history of the phenomenon must be considered. There have been many types of explanations, and research in geography seeks to find better means of explaining the world in which we live.

Regional geography is the study of both the human elements and the physical elements within a specific area. A *region* is an area of the earth's surface that has internal factors that set it apart from surrounding areas. Anyone could design a region merely by drawing an enclosed line on a map, but geographers have decided that regions should have some internal cohesion as well. Regions should have some attributes that are common or similar throughout the area, and different from those of surrounding areas. Part of the geographer's work is to identify the attributes that can be used to designate a region. This seemingly easy task can become very complicated, especially when both the physical factors and the human factors must be considered.

Geographers have defined other types of regions. The Rocky Mountains, for example, are located within a region that is based on one physical attribute—the uplifted and dissected mountain system itself. The Muslim World, on the other hand, is an area in Northern Africa and the Middle East where the Islamic religion predominates. It is a region based on religion, or one human factor. These types of regions are called **single-factor** regions because they are based on one measurable attribute. Many single-factor regions can be identified—for example, the Corn Belt or the Wheat Belt in the United States.

Nodal regions are another type of single-factor region. They are based on movement rather than on something observable, such as mountains or people. For example, the distribution of a daily newspaper from the newspaper office to readers' houses defines a nodal region. The newspaper office is the node, and all the houses receiving daily delivery are distribution points. When maps are made with lines drawn from the node to the peripheral points, the flows can be seen and a region appears. Other nodal regions might be a school district where movement is from schoolchildren's homes to the node (the school), commuters to a city, or goods distributed from a

---

human geography The branch of geography dealing with the place-to-place differences in human activities, such as economics, politics, and cultures (including languages and religions).

physical geography The branch of geography that focuses on the physical environment, including landforms, weather, climate, vegetation, and soils.

regional geography An approach to the study of geography in which certain demarcated areas (regions) are important. The focus is on the human and physical aspects of a region.

single-factor region An area demarcated by only one factor, such as where a particular crop is grown or where certain types of languages are spoken.

nodal region An area demarcated by lines of movement from a node to peripheral points.

multifactor region An area demarcated by numerous factors, such as both the physical environment and various human activities.

functional region An area demarcated for special purposes (functions), such as administration, distribution, and collection or other political or economic activities.

---

factory. Nodal regions, although based on one phenomenon, are dynamic because they indicate that something is moving or changing.

More complicated regions are called **multifactor regions.** For example, most of the Muslim World, is located in an area of extreme dryness, so, a climatic or physical factor might be included in the definition, along with the religious or human factor. The Muslim World is sometimes referred to as the "Arab World," which takes into consideration the type of people most commonly found there and includes both the cultural characteristic and a common language that many of the people use. Numerous factors, then, go into making up a multifactor region.

In this book, the countries of the world are divided into groups that form what geographers call **functional regions.** The economically developed world will be studied first, then the developing world, and finally we will look at the underdeveloped world. The word "developed" refers primarily to the economic structures of the countries, but it also refers to such factors as the health

spatial models Methods that help us understand why things are located in certain places. Models can be verbal (descriptions), visual (maps and graphs), or symbolic (mathematical expressions).

and living conditions of the people. Besides the economic divisions, each chapter is devoted either to one country or to a group of countries that are located within close proximity of one another. These groups of countries make up what are called *world regions.*

In order to study both the physical and the human dimensions of the world, it is necessary to separate the elements. We shall look at the physical makeup of the world first (Chapter 2) and then study some common elements of the human dimension (Chapter 3). A student of geography should be able to tell something about the people of a region by merely knowing where that region is located. For example, people that live in the desert areas of the world all have a similar battle to fight: They must all adapt to the severe dryness. Deserts are not distributed randomly; they are found only in particular locations and are created under certain conditions. By knowing the location of an area, a student should be able to tell whether the conditions exist or not. A similar kind of understanding can be used for all places on the earth. In this manner, geography becomes a logical science, not merely an exercise in rote learning.

Chapter 4 covers some concepts and tools that geographers have designed to help our understanding of the vast amount of information we have about our world and its people. The concepts help us make sense out of why things are located where they are. The **spatial models** discussed in Chapter 4 indicate the order that exists in the human world. For example, a model that tells us something about the internal structure of cities helps us to understand cities no matter where they are found. Every city is unique, but all cities have some similar characteristics—each has a central market area, where land values are higher than anywhere else in the city, and as land values decrease with distance from the central area, so do population densities. Cities worldwide have these things in common, and spatial models help us identify these common characteristics. Besides models, geographers' tools include such things as maps, computers, and the developing technologies, such as satellite surveillance.

Once we have an understanding of the physical and human worlds and how geographers make sense out of them, we will begin a mental tour around the planet in order to study the people and environments of world regions. It would be a wonderful learning experience if we could actually take the journey, but we will have to settle for a vicarious look. The knowledge gained from the next three chapters will give us a basic understanding of how to study the regions of the world.

■ *Key Words*

functional regions
human geography
multifactor regions
nodal regions
physical geography
regional geography
single-factor regions
spatial models

■ *References and Readings*

Abler, Ronald, et al. *Spatial Organization: The Geographer's View of the World.* Englewood Cliffs, N.J.: Prentice-Hall, 1971.

Haggett, Peter. *Geography: A Modern Synthesis.* New York: Harper & Row, 1972.

Meyer, Alfred H., and Strietelmeier, John H. *Geography in World Society: A Conceptual Approach.* New York: Lippincott, 1963.

Taaffe, Edward J. (ed.). *Geography.* Englewood Cliffs, N.J.: Prentice-Hall, 1970.

Whittlesey, Derwent. "The Regional Concept and the Regional Method." In *American Geography: Inventory and Prospect,* ed. Preston E. James and Clarence F. Jones. Washington, D.C.: Association of American Geographers, 1959.

Chapter 2

THE PHYSICAL SETTING

■ *The Earth in Space*

Faced with the problem of dating the origin of the earth, Archbishop James Ussher decided on Sunday, October 23, 4004 B.C. He made his calculation in 1658, and his date of origin stood for more than 200 years. During the last half of the nineteenth century, various geologists each moved the date of the earth's origin back in time by huge leaps. Finally, when it became possible to measure time using the decay rates of certain isotopes, the earth was estimated to be nearly 5 *billion* years old. Each time the earth was determined to be older than previously thought, theories about the age of the universe had to be revised. The generally accepted age of the earth is now 4.75 billion years, plus or minus 50 million years. This age has been confirmed through the study of rocks brought back from the moon by the astronauts.

Most scientists agree that the universe was created at a definite point in time and that until that time all matter and energy were combined into one relatively small ball, or mass. Sometime between 15 billion and 20 billion years ago, the ball exploded. This event is referred to as **"the big bang,"** and according to the "big bang" theory the explosion threw atomic particles, energy, and matter outward in all directions from the original mass. As the concentrated energy and matter expanded, new elements were formed from the existing atomic particles. The fusion of atomic protons, electrons, and neutrons created the elements that are now found in the universe.

Our solar system apparently began as a rapidly rotating cloud of particles. The bulk of this disk-shaped cloud condensed near the center and became the sun. The remaining outlying particles, called *planetesimals,* also rotated around central bodies that became the planets. Part of the planetesimal material collected to become the moons of the planets. All these whirling collections of material produced an interesting order in the solar system. The sun rotates from west to east, and all nine planets revolve around the sun in that same direction. In addition, the orbits the planets follow are all in the same plane in space, and this plane, called "the ecliptic," bisects the equatorial region of the sun. Furthermore, all the planets except Venus and Uranus rotate on their own axes from west to east. As we shall see, the orderly movement of the earth around the sun is significant for all life on the earth.

■ The Composition, Size, and Shape of the Earth

The internal composition of the earth has been theorized largely from the study of **seismic waves**—shock waves generated by earthquakes. These shock waves travel downward from the surface of the earth toward the center, then return to the surface at various points. At locations around the world, geologists monitor the waves on a daily basis. The shock waves travel at different speeds, according to the density of the material through which they pass. This study of seismic waves has revealed that the earth is made up of a core surrounded by internal divisions consisting of a series of concentric shells (Figure 2.1).

The core of the earth is divided into a relatively small **inner core** and a much larger **outer core.** The inner core has a radius of about 780 miles

"big bang" theory The idea that the universe started with a tremendous explosion (a big bang) that threw particles outward in all directions from a central mass or core. The galaxies are currently moving away from each other at great speeds, which lends support to the theory.

seismic waves A group of elastic waves generated within the earth whenever a sudden displacement of rock material occurs. Also, long-period ocean waves generated by a seismic disturbance.

inner core The innermost portion of the earth, assumed to be a rigid mass about 1,200 miles (1,920 km) in diameter and very hot (8,000 to 10,000° F).

outer core The portion of the earth located between the mantle and inner core, assumed to be molten or in a liquid state, with a radius of about 1,500 miles (2,400 km).

(1,255 km); the outer core is about 1,380 miles (2,220 km) thick. The inner core is composed primarily of iron molecules, with some nickel and possibly some silicon and sulfur. At a little more than 3,000 miles (4,827 km) below the surface of the earth, heat and pressure are extreme. The temperature is about 8,000° F, or the same as on the surface of the sun. The pressure is 3.5 million times atmospheric pressure (15 pounds per square inch). The heat and pressure create unusual conditions in the iron molecules. Although the inner core is in a solid state, the iron material is not like iron material found on the surface. In the inner core the molecules are squeezed and stretched, and the core material is about 12 times the density of surface water.

The outer part of the core is also made up of iron and nickel, but the material is in a liquid state. The heat and pressure decrease with distance from the inner core. The temperatures in the outer core range from nearly 8,000° F near the inner core to less than half that on the outer edges. The pressure drops in a similar manner, or to about half that of the inner core, and the density of the material also decreases with distance from the inner core.

> **mantle** The portion of the earth's interior located between the outer shell (the crust) and the outer core. The mantle is thought to be somewhat solid, but fluid enough that convectional currents move material laterally between the core and the crust. The mantle is about 1,800 miles (2,900 km) thick.

The next shell outward from the inner and outer cores is the **mantle.** The mantle is said to be "solid," but the material is capable of flowing, somewhat like cold molasses. The mantle is about 1,800 miles (2,896 km) thick and is made of minerals that are silicate compounds. A silicate is a silicon (a nonmetallic chemical element) in combination with oxygen. Iron is again the primary metal, but there is also much magnesium. The density of the rock material in the mantle is about half the density of the core.

Surrounding the mantle is a very thin *crust*. The crust varies in thickness between about 10 and 30 miles (16 and 48 km), less than one percent of the total radius of the earth. The boundary between the crust and the mantle is a sharply defined transition called the "Mohorovičić discontinuity." Named after its discoverer, it is usually referred to simply as "the Moho." The crust contains the least dense rock material of all the earth's concentric shells. Indeed, this material is said to be "floating" on the denser material underneath. The crust contains all the mountains, plains, and plateaus of the earth's surface, as well as all the bodies of water. The surface of the crust is in fact the surface of the earth.

Compared with the size of the universe, the earth is but a very tiny speck orbiting an insignificant star within a medium-sized galaxy. From the human standpoint, however, the earth is large, even in this time of rapid transportation and space flight. The distance from the center of the earth to the surface is about 3,950 miles (6,356 km). The exact radius varies, depending on where it is measured, because the spinning of the earth causes it to bulge at the equator and to flatten some at the poles. The polar diameter is about 7,900 miles (12,714 km), compared with about 7,927 miles (12,757 km) for the equatorial diameter. The bulging and flattening has created what is called an *oblate spheroid* rather than a true spheroid. The circumference of the earth is also affected by the oblateness, but the average figure is about 24,900 miles (40,000 km).

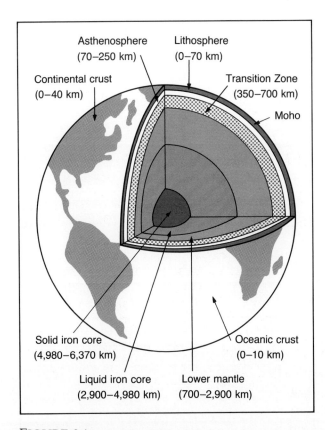

Figure 2.1
Internal Structure of the Earth. The vertical zones of the earth's interior include the inner core, the outer core, the mantle, and the crust.

The Earth-Sun Relationship

Because the sun is the primary source of energy for the earth, the spatial relationship of the earth to the sun (Figure 2.2) is of vital importance to life on the planet. The average distance between the earth and the sun is about 92,955,800 miles (149,565,589 km). Astronomers use this distance, called an *astronomical unit* (AU), as a measure much like a "light year." (For example, Pluto is about 40 AU from the sun.) Because the earth's orbit is elliptical rather than circular, the actual distance between the earth and the sun varies between about 94,555,000 miles (152,138,990 km) and 91,445,000 miles (147,135,000 km). On January 3 the earth is about 3 million miles closer to the sun than it is six months later, on July 4. The January position is called **perihelion;** the July position is called **aphelion.**

But the variation in the distance between the earth and the sun has little to do with the temperature fluctuation on the earth. The variation in the speed of the earth's revolution, or movement through space, compensates somewhat for the

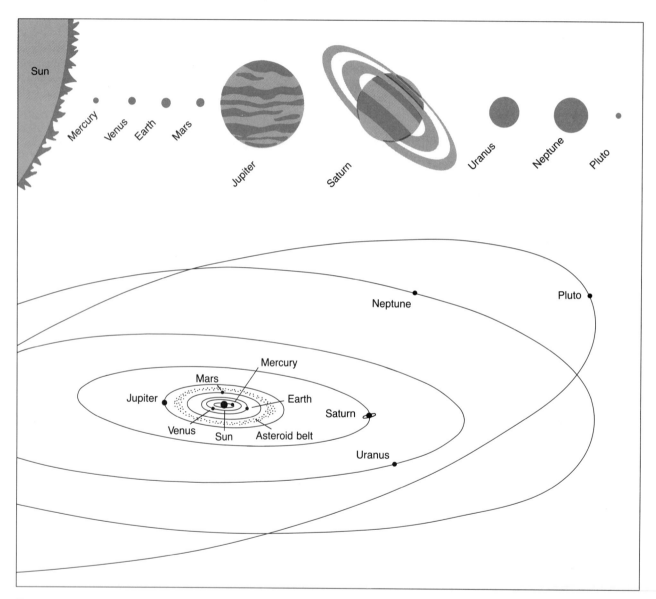

FIGURE 2.2

The Solar System. The planets all revolve around the sun in the same direction. Except for Venus and Uranus they rotate in the same direction, and except for Pluto they lie in the same plane (the ecliptic).

perihelion The point in the earth's orbit when the distance between the earth and the sun is at its minimum, as opposed to aphelion. (*Peri* comes from Greek, meaning "around", and *helios* is Greek for "the sun".)

aphelion The point in the earth's orbit when the distance between the earth and the sun is at its maximum, as opposed to perihelion. (*Apo* comes from Greek, meaning "away from", and *helios* is Greek for "the sun".)

difference in distance. The revolution is slightly faster during perihelion than during aphelion, so any particular place on the earth's surface receives about the same amount of solar heating during perihelion as during aphelion (see Figure 2.3).

The most significant factor related to heating the earth's surface is the tilt of the earth on its axis. The axis is the imaginary line running from pole to pole around which the earth rotates. The axes of the sun and the earth are not parallel. Because the earth is tilted from the *plane of the ecliptic* by 23.5 degrees, the North Pole points toward the sun during the northern hemisphere summer. The plane of the ecliptic corresponds to the most direct rays of the sun. These more direct rays heat more than the oblique rays because they are concentrated on less surface area, pass through less atmosphere, and are longer in duration. The tilt angle remains constant, so the earth's axis is always parallel to itself no matter where it is in the orbit. As the earth moves through its orbit, then, the most direct rays of the sun migrate from one latitude to another.

The yearly latitudinal migration of the sun's most direct rays causes the seasonal changes in temperature, and this in turn sets up the numerous cyclical rhythms in nature—in other words, it causes our seasons to change from summer to fall to winter to spring and back to summer. The rhythm is regular and predictable. On June 21 the most direct rays of the sun strike the earth at 23.5 degrees north latitude. In the northern hemisphere this time is called the *summer solstice;* it is the first day of summer and the longest day of the year in the northern hemisphere. However, for people in the southern hemisphere it is the first day of winter and the shortest day of the year. Six months later, on December 21, the sun's direct rays have migrated to 23.5 degrees south latitude. In the northern hemisphere this is called the *winter solstice,* and it is the shortest day of the year and the first day of winter there—and the opposite in the southern hemisphere. During the summer solstice, when the most direct rays of the sun are in the northern hemisphere, oblique rays strike the southern hemisphere and cause it to be winter there. Thus, the seasons of the two hemispheres are reversed. On March 21 and September 22 the most direct rays of the sun strike the earth at the equator. These times are called the *vernal equinox* and the *autumnal equinox,* respectively.

The tropical areas of the earth lie between 23.5 degrees north latitude and 23.5 degrees south latitude, because that area receives the most heating from the more direct rays of the sun. The polar regions are cold because they always receive only oblique rays. The seasonal changes are most pronounced in the middle latitude regions—the areas midway between the tropics and the poles.

These latitudinal differences in the amount of heat energy received also affect the circulation of the atmospheric winds. Our daily weather, the seasonal changes, and world climates are all affected by the tilt of the earth with respect to the plane of the ecliptic. This concept is basic to understanding the differential heating of the earth's surface.

■ *The Continents*

For many years most scientists believed that the earth's crust was rigid—that the continents were fixed in place and that the oceans filled the low places between them. The idea that the continents have moved has been discussed since the seventeenth century, but scientists ignored the idea. During the 1920s, however, Alfred Wegener, a German meteorologist, put together a comprehensive theory about the movement of land masses. The theory included the idea that all the earth's continents were once one large land mass that since has separated into the present continents. According to Wegener, the huge continent, which he called **"Pangaea,"** came into existence about 200 million years ago. He supported his idea with evidence he collected, especially from rocks and fossils from both sides of the Atlantic Ocean where South America and Africa seem to fit together. The major flaw in the theory, and the reason it was not accepted by most scientists, was that Wegener

Pangaea The name given to the supercontinent that is assumed to have existed on earth about 200 million years ago. It separated into several large sections that became the current continents.

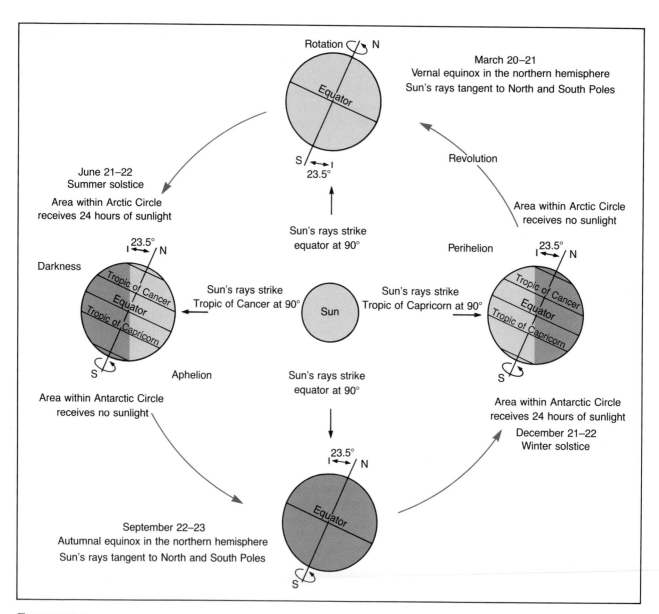

FIGURE 2.3

The Change of Seasons. The position of the earth, due to its tilt, changes with respect to the sun during a year. This means that the sun's rays strike the earth at different angles during the different seasons.

magma Molten, fluid rock occurring naturally within the earth. Igneous rocks are formed through the cooling and solidification of magma.

vulcanism The releasing of magma from the mantle either into vents in the earth's crust (intrusive) or out onto the earth's surface (extrusive).

subduction The lowering of sections of the earth's crust at plate boundaries so that the rock melts and combines with the magma of the mantle.

ring of fire The volcanic activity surrounding the Pacific Ocean where about 80 percent of the world's active volcanoes are located.

plate tectonics The theory that the earth's crust consists of a mosaic of rigid plates that move and generate landforms (especially mountains) where the plates meet.

sima The lower layer of the earth's crust that is dense, dark in color, and made primarily of basaltic rocks. The word is a combination of "si" and "ma" from the first two letters of the most prominent minerals in the rocks: silica and magnesium.

sial The upper layer of the earth's crust that is lighter in color and less dense than the sima (lower layer). The word is a combination of "si" and "al" from the first two letters of the most prominent minerals in the rocks: silica and aluminum.

lithosphere The part of the earth composed primarily of rock, as opposed to the hydrosphere (water) and atmosphere (air).

---

could not identify a mechanism that would have caused Pangaea to separate.

During World War II it was discovered that the ocean floors are divided by continuous linear ridges running lengthwise near mid-ocean and that there are deep trenches along the margins of the ocean basins. This discovery gave the clue to the explanation Wegener had missed. In 1944 the geologist Arthur Holmes put the pieces together. Holmes suggested that the sea floors are spreading—that new rock material rises from the earth's mantle and breaks through the crust along the ridges, and that the material solidifies and moves outward away from the ridges and then disappears again into the crust at the trenches. The rising material is molten rock, or **magma,** and comes up through the crust by a process called **vulcanism.** On reaching the surface of the crust, but under the ocean water, the magma turns into solid rock. This crustal material eventually returns to the mantle in a process called **subduction,** in which the rock melts and turns once again into magma. This continuous movement is driven by a convectional, or boiling, process within the earth's mantle. The boiling causes a horizontal movement in the upper part of the mantle that drags the crust along with it. As the crust moves, it pulls the continents along (see Figure 2.4).

In the early 1960s the geophysicist Harry Hess completed the theory of crustal movement by suggesting that the earth's crust is made up of numerous large plates. According to Hess, the plates become detached from one another and the spreading of the sea floor causes them to move (see Figure 2.5). This idea explains the location of volcanoes. Most volcanoes occur where two or more plates meet because the crust is weak where the plates converge. The **"ring of fire"** surrounding the Pacific Ocean contains about 80 percent of all the volcanoes in the world, all of which are associated with converging plates.

The movement of crustal plates, called **plate tectonics,** also explains how the continents move (see Figure 2.6). The continents are thought to ride along on top of the crustal plates much like passengers on a moving sidewalk. The continents cannot be subducted back into the crust because the material they are made of is less dense than the underlying plates. As a result, mountain-building occurs where certain plates are colliding.

Also related to the crustal movement theory is the fact that the crust is composed of two layers. The lower layer is primarily a continuous cover over the mantle, broken only at the oceanic ridges and in subduction zones. It is thicker under the land masses than under the oceans. Composed largely of basalt, this lower layer of the crust is a relatively dense, dark-colored rock. It is called the **sima**—from "*si*lica" and "*ma*gnesium," the most common minerals in its structure. The top layer of the crust contains the land masses we call continents. The most common rock in this layer is granite, which is lighter in color than basalt, and less dense. This layer also is named for the two most common minerals found in it, "*si*lica" and "*al*uminum," which produces the term **sial.** Together the sima and the sial make up the **lithosphere.**

The Physical Setting

So both the locations of the continents and the locations of relief features—such as mountain chains, plateaus, and valleys—can be explained by the dynamics of the plate tectonics. The surface of the earth is constantly being sculptured and rearranged by the movements of the plates. Through vulcanism, new rocks appear at the mid-oceanic ridges and then disappear through subduction at the trenches. The crust is being bent and broken and shifted around. New land rises up and then is eroded down. All these activities have been going on since the crust first solidified from a molten state, and they are still occurring. In some areas the sea floors are spreading at a rate of about 4 inches (10 cm) a year, which means the continents are moving apart at about the same rate. Other activities, such as earthquakes and volcanoes, cause more rapid changes in the crust. In addition to the convectional currents in the mantle, another underlying cause of all this activity is the gradual cooling and shrinking of the earth.

■ Landforms

The study of landforms is called *geomorphology,*— literally, the study of the shape of the land. Geomorphologists study the current characteristics of

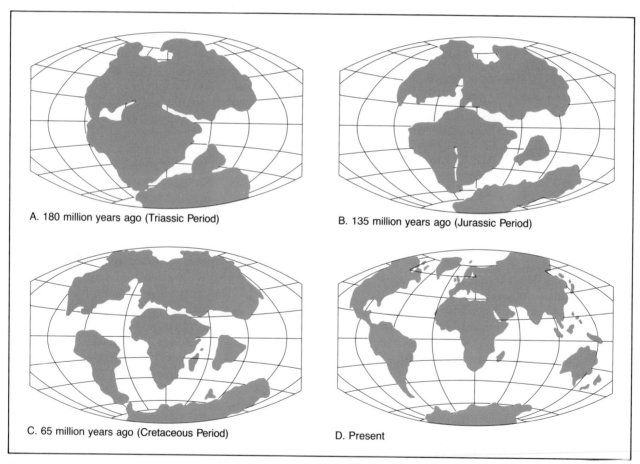

FIGURE 2.4

Historical Change in Earth's Geology. The sequence of diagrams illustrates the changes that apparently have occurred in the locations of the continents according to the floating continent theory. All of the continents originally were one large land mass known as Pangaea.

landforms as well as the processes of development. The landscape of a region is known as its *topography,* and topographical analysis usually includes three variables: (1) the structure of a region, which includes the type of underlying rocks, their arrangement, and the surface materials that result from these rocks; (2) the forces that created the shape of the landforms, including such internal crustal movements as plate tectonics as well as external factors, such as erosion; and (3) the stage—that is, how long the processes have been working to shape the landscape.

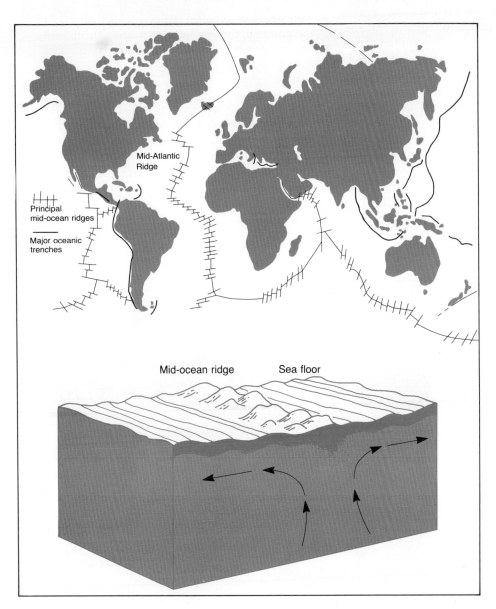

FIGURE 2.5

Mid-ocean Ridges and Trenches. Sea-floor spreading occurs when mid-ocean ridges rise due to the rock material from inside the earth pouring through the crust. The ocean trenches mark the margins of the continents.

The geomorphic processes are of two basic types, internal and external. The internal processes work from within the earth to build new land; the external processes work on the exposed surface to wear it down. This building up and wearing down of the landscape has been going on for billions of years and continues even now. The internal processes are of three main types. (1) *Plate movement* is the term for the large crustal movements, such as those associated with plate tectonics and sea-floor spreading. (2) *Vulcanism* is the movement of molten rock in the mantle outward toward the surface of the earth. "Extrusive vulcanism" occurs when magma, rock, and ash are "extruded" out onto the surface of the earth; "intrusive volcanism" occurs when the molten rock is deposited in odd-shaped masses or in cracks and veins just under the surface. (3) **Diastrophism** is crustal movement on a smaller scale than plate movement. It includes local or regional warping, folding, and faulting of the crust. Warping describes the bending of the earth's crust—for example, when rivers deposit their loads of sediment at deltas the weight of the deposited material can cause the underlying rock structure to bend downward. Folds can be of many types, from minor rises in the crust to a complicated doubling over of the crust, as in mountain-building. Faults are breaks in the crust that are usually linear and are marked on the surface by **fault lines.** Slippage along the faults causes earthquakes.

diastrophism All processes that change the shape of the earth's surface through crustal movements.

fault lines A fault is a fracture in the earth's crust along which movement causing displacement of one side in relation to the other has occurred. A fault line, then, is the line followed by a fault.

weathering The breakdown of rock material by exposure to the atmosphere, creating small, easily moved (eroded) fragments from the larger rocks.

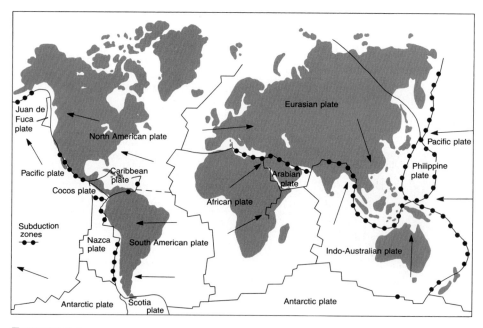

FIGURE 2.6

Crustal Plates. The edges of the major crustal plates tend to correspond to the locations of the mid-ocean ridges and trenches (see Figure 2.5). The generalized direction of movement of the plates is for the western hemisphere to move west, while the continents of the eastern hemisphere move toward each other.

> **mass wasting** All processes where soil and rock debris move downslope.

The external processes that wear down the landscape are weathering, mass wasting, and erosion. **Weathering** is the breaking up of rocks through physical and chemical processes. Large rocks are broken into smaller ones, which in turn are broken down further until eventually there is only sand or rock powder, and even the powder may disappear from chemical weathering. Soil is formed when organic material, air, and water are mixed with the fine rock particles. **Mass wasting** is the movement of large amounts of surface material, usually by slipping down slopes by gravity. A landslide is an example of mass wasting. *Erosion* is the gradual wearing away of surface material and can occur in many ways.

The most common form of erosion is the removal of soil by running water. This is called "fluvial erosion." Large streams can carry huge amounts of material that they deposit as sediment. Erosion includes (1) the wearing away and removal of soil and rocks, (2) the transportation of the material, and (3) the depositing of the sediments. Other forces that can cause erosion are wind ("*aeolian erosion*"), moving ice ("*glacial erosion*"), and waves along a shoreline. Both erosion and deposition create landforms.

Landforms result from a myriad of factors, but the primary things to remember are structure, process, and stage. *Structure* refers to the type of rocks, the *process* is how the rocks have been changed, and the *stage* is how long the process has been working on the rocks to change them. All landforms can be described in terms of these three variables. The great variety and complexity of the earth's landscape is a result of the interaction between the building processes and the erosional processes. The building processes result from the earth's internal structure and activity. The external processes are related to the conditions on the surface of the earth, which in turn are a response to the relationship between the earth and the sun.

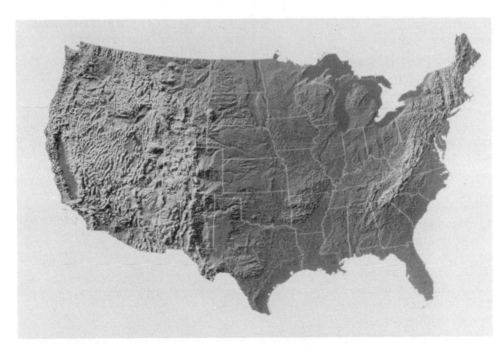

The topography of the United States can be seen in this computer-generated view. The map was produced by Larry Ostresh of the University of Wyoming.

The Grand Canyon in Arizona is a spectacular example of landforms created by fluvial erosion. The waters of the Colorado River have worn away and transported massive amounts of soil and rock material, leaving this mile-deep canyon that is a major tourist attraction.

Climate

A region's climate is the average of its daily weather conditions over a period of years. The most important components of climate are temperature and precipitation, but other elements, such as wind, evaporation rates, and amount of cloudiness, also are important. A region's climate is the primary factor in the development of soil types and the growth of natural vegetation. For this reason, understanding the global distribution of climates is the key to understanding the various environments in which people live and carry on activities.

Most climatic classification schemes use precipitation and temperature data. Averages are calculated and then mapped, and world climate regions are based on these mapped weather averages. There are many classification schemes, some of which use more than the temperature and precipitation variables. The oldest and probably the most commonly used classification system was designed by a German botanist, Wladimir Köppen, in 1918. The system is sophisticated enough to have withstood the test of time, yet simple enough for the average person to comprehend.

Köppen used five major climate categories, which he subdivided according to yearly temperature ranges and yearly variations in the amount of precipitation. He used letters to designate the major climate categories (see Table 2.1).

Notice that the five categories range from A, the tropical regions, to E, the polar regions—which corresponds to the relationship between latitude and climate. Köppen used a sixth category, H, to indicate highlands, or complex mountain climates, which can be found at any latitude. For example, polar conditions may exist on the tops of high mountains located in the low latitudes.

Each letter in the Köppen system stands for a climatic condition based on temperature and precipitation. Originally, however, Köppen studied

Glacial erosion created the nearly vertical cliffs found in Yosemite National Park in California.

isotherm An isoline drawn on a map that connects points where actual or average temperatures for a specified time are the same.

where types of vegetation were found. For example, the A designation is given only to places where palm trees grow, or where the average temperature of each month is above 64.4° F (18° C). An **isotherm** that indicated the latitudinal limits of the growth of palm trees was drawn on a map of the world. The E designation is for places where the average monthly temperature is never above 50° F (10° C), climates where trees cannot exist. The climates between the As and the Es are designated in a similar manner. (For the complete Köppen system, see the Appendix.) Köppen also classified subtypes of the five major climatic types using one or two more letters. For example, a "Cfa" climate is moist with a mild winter (C), with precipitation spread evenly over the year, or non-seasonal (f), and has hot summers (a). The second and third letters stand for particular conditions that can be calculated using monthly temperature and precipitation data.

The Köppen system is not perfect, but it does allow a quick and fairly accurate description of average weather conditions in a region. A mere two or three letters can provide a good deal of information about a region's climatic condition. Temperature and precipitation are the basic elements of the Köppen scheme, but knowing what controls these elements is important to understanding why the conditions exist. (See Figure 2.7 for location of climates.)

The landforms of Arizona can be seen in this computer-generated map. Notice the irregular gash of the Grand Canyon running northeast to southwest in the upper left corner of the state, the Basin and Range country of the southern part of the state, and the plateau on the north. The map was produced by Larry Ostresh of the University of Wyoming.

TABLE 2.1

Köppen's Five Major Climate Categories

LETTER	TYPE OF CLIMATE	DESCRIPTION OF CLIMATE	LOCATION OF CLIMATE
A	Tropical moist	Hot all year	Low latitudes
B	Dry	Desert/steppe	Low-mid latitude
C	Moist, mild winter	Subtropical	Mid-latitudes
D	Moist, severe winter	Continental	Mid-high latitude
E	Polar	Always cold	High latitudes

The Physical Setting

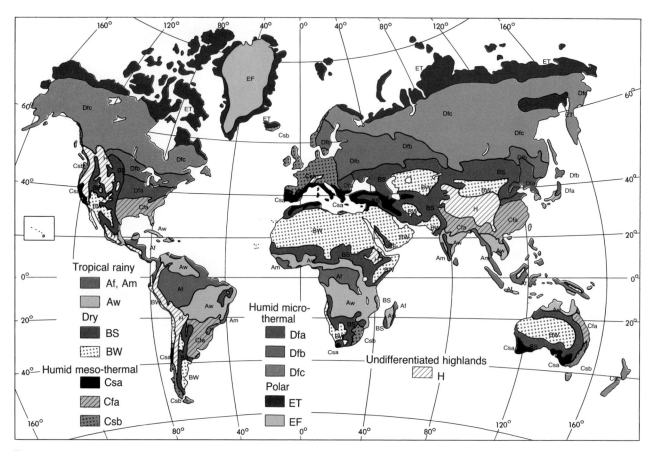

FIGURE 2.7

World Climate Regions. The global distribution of climates is shown on this map. The precise boundary lines between the climate regions are actually only transition zones. For a complete listing of all the climates see Appendix B.

The major control of climate is *latitude*. Differences in latitude produce differences in the amount of heat energy received from the sun, as well as seasonal differences in the length of days. Some factors, however, tend to disrupt the association between latitude and climate. The continental *location* of a place also affects the climate. For example, east coast locations are affected by warm offshore ocean currents that produce moisture for precipitation. Cold ocean currents off west coasts of continents allow little evaporation and therefore can produce dry climates. Continental interior locations often are dry areas because they are far from a source of moisture. In addition, the oceans tend to modify coastal temperatures, keeping summer temperatures lower and winter temperatures higher than what would be expected for a particular latitude. On the other hand, places located in the deep interior of continents experience the most extreme ranges of temperature because they do not benefit from the modifying influence of the oceans. Thus, location with respect to oceans is an important control for both temperature and precipitation, and the general oceanic circulation of water—currents—is influential as an agent of this control.

The general circulation of the atmosphere (Figure 2.8) affects the climate too. The constant movement of air can bring warm air masses to cold regions, and cold air masses to warm regions. The air movement disrupts the latitudinal and/or continental influences on climates. It also produces rain patterns that would not otherwise exist. When warm air meets cold air, the warm air is lifted over

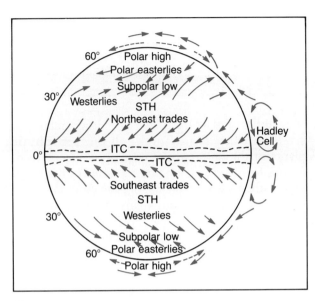

FIGURE 2.8
The General Circulation of the Atmosphere. The general circulation is disrupted by the continents, and the entire system shifts north–south with the changing seasons.

the cool air. The rising warm air is cooled, and if it contains moisture, precipitation is likely to occur. This is called "frontal precipitation" because the zone between two air masses is called a *front*. The mid-latitude regions get most of their precipitation through this type of convergence of air masses. In the tropical regions, where the "trade winds" meet, a constant zone of uplifting air is created. This *intertropical convergence zone* also causes precipitation. The front and the intertropical convergence zone are the two major rain-producing mechanisms, and they both depend on the general circulation of the atmosphere.

The landforms of a region can disrupt the general atmospheric circulation. Thus, *topography* is another climatic control. For example, a mountain range can cause prevailing winds to rise, and this lifting may cause precipitation. As a result, the windward sides of mountain ranges can be extremely wet. Mountain ranges also can act as barriers that prevent moisture-laden winds from penetrating the continental interior regions. An example of both these conditions is provided by

The Rocky Mountains are examples of topographic barriers that act as climate controls. Notice that the peaks are shrouded with clouds. Also notice that the land at the higher elevations is covered with snow while the lower, warmer levels are snow-free. The photograph was taken near Loveland Pass (11,988 feet, or 3,654 m), west of Denver, Colorado.

The Physical Setting 21

the Himalaya Mountains. In the Himalayas, uplifting on the south side of the mountain range, in conjunction with the monsoon winds, produces some of the rainiest conditions anywhere on earth. To the north of the Himalayas, however, are the extremely dry desert regions of central Asia—the mountains keep moisture from getting to the deserts.

■ Vegetation and Soils

The close association between a region's climate and the types of vegetation and soil found there has been noted above. Some of the Köppen climate names show this association. For example, the A-type climates are subdivided into Af, Am, and Aw. The "Af" climates are called "Tropical Rainforest," indicating that the climates are located in the tropics and that the most prominent vegetation found there is rainforest. A *rainforest* is a woodland that receives at least 81 inches (203.2 cm) of rainfall a year. The trees are tall, broadleaf evergreen types that form a continuous canopy. The "Am" climate is called "Tropical Monsoon," which indicates a seasonal rainfall distribution with huge amounts of rain during the summer months. The *monsoon forest* is a forest that has adapted to dry winters and overly wet summers. The trees are deciduous, which means they lose their leaves during the dry season; they also are shorter than the trees in the rainforest and are more widely spaced. The "Aw" climates are called "Tropical Savanna," which again indicates tropical conditions, but not enough rainfall to support a continuous tree cover. A *savanna* is a grassland with scattered clumps of trees.

The other climate regions of the world also contain specialized natural vegetation (see Figure 2.9). The most important factor relating to natural vegetation is the amount of precipitation the region receives and the period during the year when the precipitation comes. Places that get a lot of moisture throughout the year will have a natural covering of trees. As less moisture is received, or if moistures comes during pronounced seasons, the trees become smaller and more scattered. With less moisture still, trees give way to shrubs, which in turn give way to grass. Finally, in regions with very little rainfall the plant life either adapts or ceases to exist. For example, all the **xerophytic** plants of the deserts have developed special techniques for surviving on scant amounts of precipitation; they store water in some manner and have small, waxy leaves, and many have stickers for protection (cacti).

xerophytic Describes all vegetation that has adapted to arid conditions (*xero* means dry, and *phytic* refers to plants).

Temperature and precipitation, the primary determinants of climate and vegetation, also play a major role in the type of soil that is found in a region. Soil is composed of weathered rock particles mixed with organic material, water, and oxygen. Four main factors interact to produce soil: (1) the composition of the rock that provides the material in which the soil forms, (2) the temperature and precipitation in the region, (3) the natural

In the western foothills of the Sierra Mountains in California, the temperature and precipitation amounts are perfect for growing large trees. The tree whose base can be seen in this photograph is a sequoia known as "General Sherman."

parent material The rock material from which soils are formed.

vegetation, and (4) the length of time the soil has to form. The type of soil produced will depend on the underlying rock found in a region. The fact that this bedrock is called the **parent material** of the soil indicates its importance. Different environmental conditions also produce variation in soil types. Large amounts of moisture, for example, will create deep soils but tend to remove nutrients, which makes the soils infertile. Desert soils, on the other hand, are shallow but fertile because the lack of moisture leaves most of the nutrients in place.

Finally, it takes a very long time for soil to form. The longer the environmental factors exist without any change in conditions, the more well developed the soil will be.

Temperatures are important in the soil-forming process too. For example, high temperatures accompanied by a lot of moisture promote rapid decaying of the dead vegetation that covers the soil. Bacterial decomposition and the rotting away of the vegetative cover is important because the material formed becomes soil humus that provides the soil with nutrients. On the other hand, high temperatures along with low amounts of rain create high soil moisture evaporation rates, which slows down the decaying of organic matter. Thus, we might expect that the more vegetation there is, the better the soil, but this is not neces-

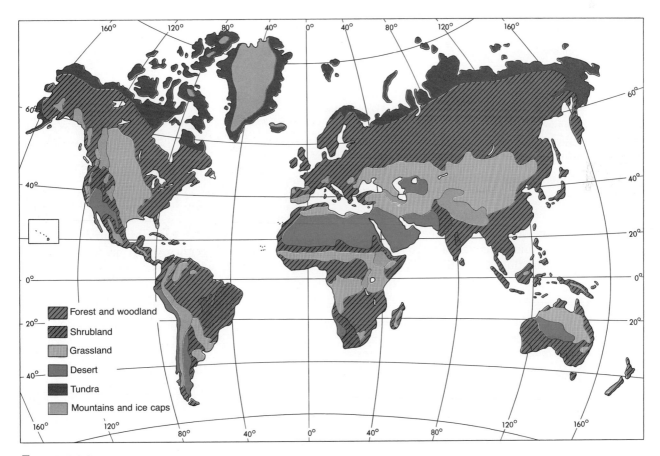

FIGURE 2.9

Natural Vegetation Regions of the World. Although much of the world's natural vegetation has been removed for agriculture, certain regions are distinguishable. These regions correspond closely to climate regions (see Figure 2.7).

The Physical Setting **23**

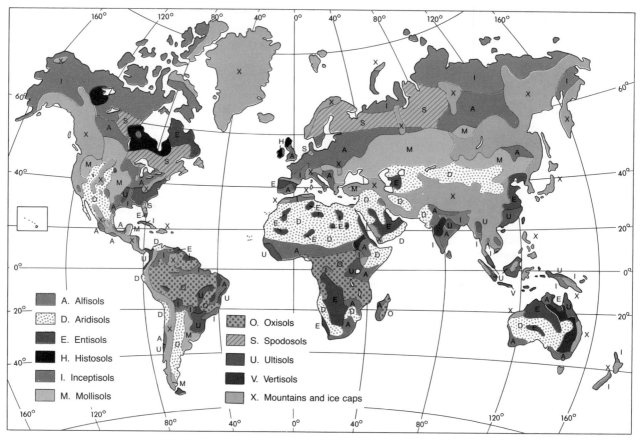

FIGURE 2.10

Major Soil Regions of the World. Because both climate and natural vegetation are elements in the formation of soil, maps of the three world features—climate, vegetation, and soils—are similar.

sarily true. Large amounts of precipitation are required for a continuous cover of large trees, and the excess moisture flushes out the soil nutrients.

The best soils for agriculture are in the areas where annual precipitation is about 35 to 45 inches—that is, where it is too dry for complete

TABLE 2.2

Types of Processes Producing the Five Major Soil Types

PROCESS	PRECIPITATION	TEMPERATURE	VEGETATION
Lateritic	Great amounts	High	Rainforest
Gleization	Large amounts	Low	Forest
Podzolization	Moderate amounts	Moderate	Forest
Calcification	Low to variable	High summer	Grassland
Salinization	Low amounts	High	Xerophytic

24   Chapter 2

tree cover but wet enough to provide sufficient soil moisture. These areas are the mid-latitude grasslands, where long, hot summers, combined with adequate moisture provided by the continuous stream of cyclonic storms, and the decayed grass work together to form very dark, fertile soils.

The numerous factors affecting soil formation can be combined in so many ways, depending on local conditions, that there are literally thousands of soil types (see Figure 2.10). For our purposes here, only the primary factors of climate and vegetation will be considered. And although there are numerous soil classifications, some of which contain hundreds of categories, only the five major soil types most useful at the global scale are classified here. The types of processes that produce the five soils are shown in Table 2.2.

The association among climate, vegetation, and soils from global area to global area can be seen by comparing three world maps (see Figures 2.7, 2.9, and 2.10). But the generalization at the world level does not mean that the association always holds true at the local level. At the local level, factors other than climate and vegetation can be important. For example, soil types can vary according to the texture (the size of the individual mineral particles) and the acidity of the soil. In general, however, the brief descriptions given here are sufficient for global analysis.

The elements of the physical environment covered in this chapter will be discussed for individual regions in the chapters that follow. The ideas, concepts, and terms introduced here should be referred to again when they appear in later chapters.

Key Words

- aphelion
- "big bang"
- diastrophism
- fault line
- inner core
- isotherm
- lithosphere
- magma
- mantle
- mass wasting
- outer core
- Pangaea
- parent material
- perihelion
- plate tectonics
- "ring of fire"
- seismic waves
- sial
- sima
- subduction
- vulcanism
- weathering
- xerophytic

References and Readings

Goudie, Andrew S., et al. (eds.). *The Encyclopedic Dictionary of Physical Geography.* New York: Basil Blackwell, 1985.

Hole, Francis D., and Campbell, James B. *Soil Landscape Analysis.* Totowa, N.J.: Rowman & Allanheld, 1985.

McKnight, Tom. *Physical Geography: A Landscape Appreciation.* Englewood Cliffs, N.J.: Prentice-Hall, 1987.

Strahler, Arthur N., and Strahler, Alan H. *Modern Physical Geography.* New York: Wiley, 1987.

Chapter 3

THE HUMAN DIMENSION

■ Introduction

Human geographers look at the patterns of human distribution over the surface of the earth, the place-to-place differences in people's use of the earth's resources, and the economic and political interactions between human groups. They are concerned primarily with how humans have adapted to the physical environment. Some things about human societies are much easier to measure than others. For example, human geography usually begins with an analysis of the number and type of people in an area, or the dynamics of the population. Figures on population for most parts of the world are easy to obtain and are quite accurate. Other factors are not so easy to assess. Rough approximations and subjective or intuitive judgments must be used to establish such things as the bases of political power. The purpose of this chapter, then, is to introduce the quantitative topics that the regional chapters will cover and to show how geographers use the information.

■ Current Population

The human population of the world today is about 5 billion and rising rapidly. The "teeming millions," a concern of the nineteenth century, is now the teeming billions. The rate of natural increase is 1.7 percent annually, which means a doubling time of 41 years. The impact of this becomes more apparent when we consider that the world's population did not reach one billion until about 1830. It took 100 years for the population to double from 1 to 2 billion, then only about 50 years for it to double again. At the current rate of increase, the world's population will reach 10 billion by the year 2025. This rapid growth since 1950 is what has been called the "population explosion."

Perhaps more important than the rapid increase in the world's population is the distribution of the population. Both population and population growth rates are unevenly distributed. There are four densely populated clusters: East Asia, South Asia, Europe, and eastern North America. About 1.26 billion people live in China, Japan, the Koreas, and the smaller enclaves, such as Hong Kong, Taiwan, and Macao. Slightly more than one billion people live south of the Himalayas, primarily in India, Pakistan, and Bangladesh. Europe, from the Atlantic Ocean to the Ural Mountains, is inhabited

---

Focus Box

Thomas Malthus on Population

The present rate of population increase has convinced some scientists that the earth simply cannot support many more people without some catastrophe occurring. Yet this is not a new idea. Thomas R. Malthus (1766–1834), a British economist, predicted that warfare, famine, and disease are inevitable conditions aggravated by unchecked population growth.

Malthus was born in Surrey, England, and studied theology at Cambridge University. Although he became an ordained minister, he started to write on problems related to population. Writing and lecturing on these topics gradually became his main interest, and in 1805 he was appointed professor of modern history and political economy at Haileybury College. The appointment was based partly on the publication of his *Essay on the Principle of Population* in 1798. One of the first writers to be concerned with overpopulation, Malthus wrote:

> Population, when unchecked, increases in a geometric ratio. Subsistence only increases in an arithmetical ratio. A slight acquaintance with numbers will show the immensity of the first power in comparison to the second. . . . I say that the power of population is indefinitely greater than the power in the earth to produce subsistence for man.

Malthus rejected the idea of artificial birth control methods on theological grounds. He approved of delaying marriage and moral restraint, although he said, "Passion between the sexes is necessary." He did not want to limit family size severely because large families were needed for economic reasons. At that time, children went to work in the mines and factories at a very early age, and they were needed to help on farms as well. The more children working in a family, the higher the family income, which in turn helped the overall economy.

Malthus was giving a warning: Unchecked population growth would eventually outstrip the food supply on earth. Then warfare, famine, and disease would kill enough people to correct the situation. To avoid much human misery, overpopulation should be avoided. As the world's population grew, however, new and better food-producing techniques were invented, and most of the modern world has avoided the problem that Malthus predicted. His ideas are important, however, for explaining some of the problems, especially food shortages, in the underdeveloped areas of the world.

Examples of densely populated regions, the landscapes of large urban centers. (*Top left*) a portion of Tokyo, Japan, in the East Asia cluster. Photograph by Kazuyo Nakao. (*Top right*) Paris, France, viewed from the top of the Eiffel Tower. France is in the European population cluster. (*Bottom right*) New York city, viewed from the top of the Empire State Building. New York has the largest population of any city in the eastern North American population cluster.

by about 775 million people. The most heavily populated countries in this region are Great Britain, France, Germany, Italy, and the Soviet Union. The smallest population cluster is in eastern North America, where nearly 200 million people live. This region extends between the St. Lawrence River on the north to North Carolina on the south and runs inland to a line between Milwaukee and St. Louis.

Heavy concentrations of people exist other than those mentioned above, but they occur in isolated pockets. For example, the Mexico City region is heavily populated, as are the Rio de Janeiro and Buenos Aires regions. The entire island of Java in Indonesia has a population density of 1,000 people per square mile. None of these outlying concentrations, however, matches the size and continuous population densities of the four major regions. Besides the clusters and the isolated pockets, the remainder of the world is largely empty by comparison. In many areas, extreme climatic conditions discourage human habitation. Topography, especially mountains, also limits the number of people that can live in some areas. In general, most of the world's people live in moist, lowland areas and not too far from an ocean.

Another important aspect of the geography of world population is the uneven distribution of growth rates. The **rate of natural increase (RNI)** of a country's population is the birth rate minus the death rate, without regard to migration. As noted above, the natural increase rate for the world is 1.7 percent annually. For the most well developed and richest countries, the rate is about 0.6 percent, but for the less developed and poorest countries the rate is 2.4 percent. Most of the rapidly growing but poor countries are located in the tropics, while the slower-growing, rich countries are found at higher latitudes in the northern hemisphere.

The distribution of population growth rates tells us that the countries least able to support a bulging population are the countries that are growing most rapidly. One of the highest growth rates in the world, for example, is Kenya's 4.2 percent, while the 1983 per capita gross national product for Kenya was the equivalent of about U.S. $340. On the other hand, the growth rate for Japan is 0.7 percent, while that country's per capita gross national product is over U.S. $10,000. Rapid rates of population growth in the poor countries of the world cause many problems. Poor countries have difficulty feeding their people, so chronic malnutrition prevails. Crowded conditions and lack of proper health care allow diseases and early death to decimate the population. Unemployment adds to the poor economic conditions, creating even more poverty. Many countries of the world are hopelessly poor and overcrowded.

■ Demographic Terms

Three basic measures are especially important for understanding statistical summaries of population. These are the crude birth rate, the crude death rate, and the natural growth rate.

Crude birth rate
$$(CBR) = \frac{\text{Number of births per year}}{\text{Total population}} \times 1000$$

Crude deathrate
$$(CDR) = \frac{\text{Number of deaths per year}}{\text{Total population}} \times 1000$$

$$\text{Natural growth rate} = \frac{CBR - CDR}{10}$$

rate of natural increase (RNI) A country's annual rate of population growth without regard for net migration. Calculated as the birth rate minus the death rate.

crude birth rate The annual number of births in a country for every 1,000 people.

crude death rate The annual number of deaths in a country for every 1,000 people.

The **crude birth rate** is calculated by dividing the total number of births in one year by the total population for that year and then multiplying the result by 1,000. For 1986 the crude birth rate for the world was 27. This means that for every 1,000 people on earth there were 27 babies born. The **crude death rate** is calculated by dividing the number of deaths by the total population and multiplying the result by 1,000. The crude death rate for the world in 1986 was 11. The natural growth rate is determined by subtracting the number of deaths from the number of births and

Some factors that lead to high mortality rates are evident in this scene from an Afghan refugee camp in western Pakistan. Cleanliness, sewage disposal, clean drinking water, and general high standards of living are all lacking in this South Asia village. Photograph by Jim Curtis.

mortality rate The annual number of deaths in a country for every 1,000 people.

demographic transition A four-stage change process over time in birth and death rates that typically occurs when less developed nations become more developed.

dividing the result by 10. The rates for all the countries of the world are shown on the map in Figure 3.1.

When the birth rate exceeds the death rate, the population is growing. If the birth and death rates are equal, "zero population growth" is attained. If the death rate exceeds the birth rate, the population is on the decline. For most of history, **mortality** (death) stayed even with fertility (childbearing), and the total population grew little or not at all. Both the birth rate and the death rate hovered around 50 per 1,000. The recent population explosion is the result of a drastic decline in the death rate. Once again we can note the current world rates of 27 per 1,000 for births and 11 per 1,000 for deaths. In many of the poor, rapidly growing nations, the birth rates are still above 50 per 1,000, with death rates around 15 or 16 per 1,000.

■ *Demographic Transition*

Demographers have discovered that, as nations modernize, national population structures often change in similar ways. The change is called the **demographic transition.** The first stage of the transition is when the birth rate and the death rate are both high. In the next stage, modernization

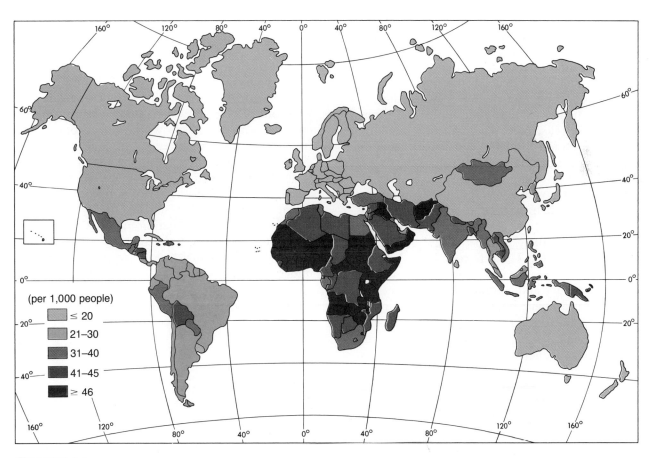

FIGURE 3.1
World Birth Rates, 1986. The crude birth rate is the number of live births for every 1,000 people in each country.

The Human Dimension

brings a drop in mortality, causing a period of rapid population growth. A third stage sees a decline in fertility, and in the final stage the birth rate and death rate become similar again, but at much lower levels than before.

Three major factors contribute to the second stage of the demographic transition, the decline in mortality: improvements in nutrition, improvements in health through better hygiene, and improvements in health stemming from the control of diseases. A general world demographic transition began in Northern Europe when nutritional improvements were made through the introduction of new crops, especially corn and potatoes from the United States. Then new tillage methods and new equipment were invented, resulting in better production of crops. Increased trade helped to reduce the shock of local crop failures, and migration allowed people to escape the poor areas of Europe. All these factors helped to reduce the number of deaths related to malnutrition. In underdeveloped countries, the demographic transition begins in a similar manner.

The second improvement that helped reduce the death rate was a general improvement in the standard of living. Better houses were built, and better clothes were made, because of the general economic development. The industrial revolution and urbanization brought numerous economic changes and generally more wealth to the people. Education and communication improved, and medical experts were able to spread the information that cleanliness helped to reduce communicable diseases. Widespread use of soap resulted, and clean drinking water, along with better sewage disposal, became important. These factors led to a steady decline in the mortality rates beginning in the seventeenth century and continuing until today.

The third major improvement that helped to reduce death rates was the remarkable advancement in medical science and technology. Discoveries such as Jenner's smallpox vaccine in 1796, Pasteur's antirabies vaccine in 1885, Nicolle's typhus vaccine in 1909, and Fleming's discovery of penicillin in 1928 are but a few of the many improvements that helped save lives. Reduction in mortality from disease is the most recent of the three improvements and perhaps the most important. Medical science continues to find new ways

population pyramid A graph depicting the age structure of a country's population. The percentage of males and females in categories of five-year increments are shown. Although the graphs now do not usually resemble pyramids, the traditional name has been retained.

age structure The relationship between the percentage of people in various age categories in a country. For example, in countries with high fertility rates, childen make up large proportions of the population, giving a pyramid shape to the age-structure graph. In low-fertility nations each generation is a smaller size, giving the population pyramid a boxier shape.

to save lives, and that in turn continues to reduce death rates. Figures 3.1 and 3.2 show world birth rates and death rates by country, respectively.

One interesting aspect of the demographic transition is that birth rates begin to decline at stage three in the transition. In the Western world, this decline began after the industrial revolution, as a result of changing values and attitudes. Industrialization changed not only the economic structure but also ideas about family size. Large families went out of vogue. Farmers needed large families to help with the farm work, but people who worked in factories needed less such help, so children became an expense, not a labor asset. In addition, as women achieved more independence and began to enter the work force, the role of women changed. The result was that fewer babies were born. A transformation in social values had brought about a lower birth rate. The poor and overpopulated countries of the world have not yet undergone such societal changes. Dramatic declines in mortality were forced upon them from the outside, by disease control and other factors, but fertility rates in these countries remain high.

▪ Population Pyramids

A **population pyramid,** is a graphic representation of a country's **age structure** that shows the recent demographic history of a country or region and the social and economic conditions there. Such graphs tell us about a population's past mortality,

dependency ratio The ratio of people younger than 15 and older than 65 to those who are between 15 and 65 years of age.

migration, and especially fertility, and this information makes it possible to predict future demographic patterns.

Population pyramids are visual representations of the percentage of the population in age-group categories called *cohorts,* which are divided between males and females (see Figure 3.3). The pyramids can convey much important information. For example, a country's **dependency ratio,** an important measure of a population's age structure, can be determined by looking at a population pyramid, but the ratio is calculated by dividing the number of people younger than 15 and older than 65 by the number of people between 15 and 65 years of age. The young and the old are the people "dependent" on the remainder of the population for their welfare, so the ratio is related to economic conditions.

In the less developed countries, where fertility is high, the bottom of the population pyramid is wide, compared with the remainder of the graph. On a worldwide basis, the dependency ratio for these countries is 0.94, which means that there is nearly one dependent for each productive worker. In the more developed countries, where fertility rates are lower, the bottom of the population pyramid is narrow, or about the same width as the remainder of the graph. The dependency ratio for these more developed countries is about 0.53, or

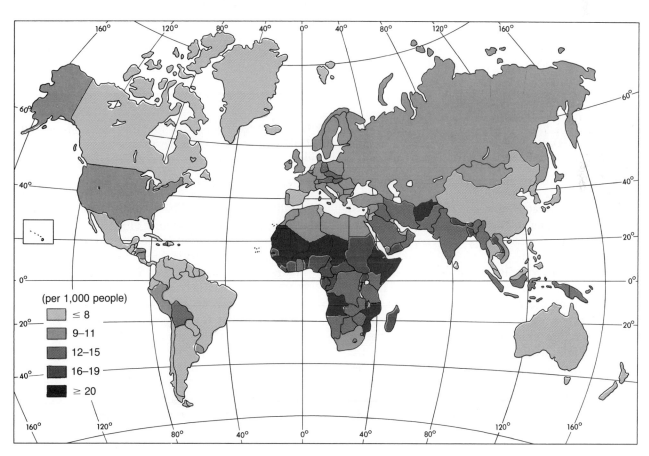

FIGURE 3.2
World Death Rates, 1986. The death rate is the number of deaths for every 1,000 people in each country.

The Human Dimension 33

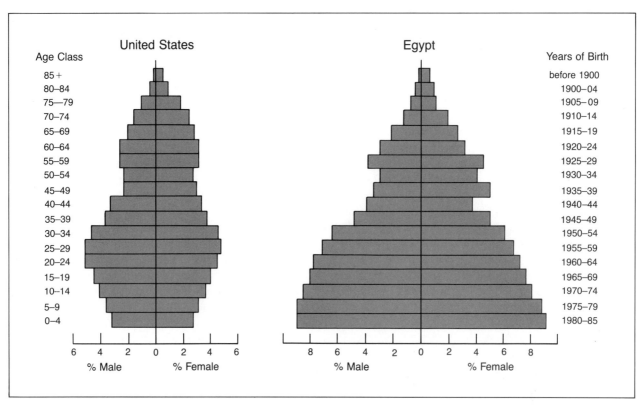

FIGURE 3.3
Population Age Structures for the United States and Egypt. Since the "baby boom" of 1955–59, the rate of population increase in the United States has declined steadily, so that the U.S. is now a slow-growth nation. By comparison, Egypt, whose population continues to increase, is a rapid-growth nation.

about half as great as that for the less developed countries. This means there are about two productive workers for each dependent in the more developed countries.

Population pyramids for individual countries can provide detailed information about the historical aspects of the countries' population age/sex structures. See, for example, the population pyramids for Egypt and the United States in Figure 3.3. In a high fertility country, such as Egypt, the younger age cohorts on the pyramid are wide, indicating that rapid growth is occurring, and in fact the fertility rate is 5.3 and the doubling time is about 27 years. By comparison, the graph for the United States indicates low fertility. The actual rate is 1.8, and the doubling time is 99 years. Closer inspection of the graph for the United States, however, indicates a narrow portion in the middle-age cohorts. This is a result of low fertility during the Great Depression of the 1930s, followed by only slightly higher rates during World War II. The bulge in the lower age cohorts represents the "baby boom" that followed World War II. The narrowing again at the base of the graph indicates the recent emphasis in the United States on having smaller families. The dependency ratio for Egypt is about 0.75, and that for the United States is about 0.52. A more significant difference, however, is in the age of the dependents. In Egypt, about 39 percent of the population is less than 15 years old, and only about 4 percent is 65 or older. In the United States, those age 15 and under comprise about 22 percent of the population, but those 65 and over make up about 12 percent. The difference in the ages of the dependents puts different strains on the countries. The need for schools and future employment

> **mutation** A sudden change in an inheritable characteristic of a plant or animal that reappears in all following generations.
>
> **natural selection** The survival of certain forms of plants and animals that have adjusted best to the conditions under which they live.

is much stronger in Egypt than in the United States, while in the United States the need for certain kinds of health services and housing are stronger than in Egypt. Overall, then, the age/sex structure of a population can not only reveal current problems but also indicate future needs.

Racial Types

In the simplest classification of the human races there are three major divisions, based on skin color: Black, White, and Yellow. The Yellow includes not only the Mongoloids of Asia but also the Indians of North and South America and is sometimes called the "Yellow-Red races." Although the three divisions are usually called "races," they probably should be called "stocks," because there are further subdivisions. Anthropologists consider races to be divisions of people that have enough inherited traits to set them apart as distinct groups, and the traits must be physical, not cultural. There are about 30 subdivisions of the Black, White, and Yellow basic groups. Variability within each group is great, and no "pure race" exists.

A number of factors are thought to have contributed to the evolution of racial types. These include variation, selection, heredity, and geographic isolation. *Variation* is what causes every individual in a family to differ from his or her brothers and sisters. Sometimes sudden variations occur, and these are called **mutations**, but because mutations are rare they create only gradual changes in large groups of people. Each person has about 44,000 pairs of *genes* (heredity carriers), and only about one mutation per person can be expected. Furthermore, only about 1 in 30 mutations will affect a characteristic that is visible.

Selection can take place in a number of ways. **Natural selection** occurs when environmental conditions cause anatomical changes. Extreme climatic conditions may produce different types of physiques. For example, Eskimos have very narrow noses, and some people believe that characteristic is an adaptation to breathing cold air. Others argue that the bulky body and narrow eyes of the Mongolic people of Siberia also are adaptations to the cold. Other selection processes include *physiological incompatibility*. For example, if a mother who is Rh-negative (that is, does not carry what is called the Rh factor in her blood) carries the baby of a man who is Rh-positive, serious medical problems can arise; without close medical attention, the baby usually dies or is weakened. Thus, Rh-negative people tend to disappear from the population. Finally, *social selection* can occur— that is, if certain physiological characteristics are in vogue within a certain group, people without the characteristics are unlikely to marry and reproduce. Some groups perform human mutilations in order to conform to what is thought to be attractive to the opposite sex. The bound feet of Oriental women, the artificially enlarged lips of some African tribal women, and silicon breast implants or breast reductions of modern American women are examples of people trying to overcome the problem of social selection. The characteristics are not inherited, but if people married only people with the desired features, over time the less desirable features would disappear.

Heredity perpetuates the basic racial patterns; variation and selection tend to change them. It is believed that both mental and physical traits may be inherited, and we do know indisputably that some physical characteristics are. For example, skin color, hair texture, and eye color are all known to be inherited traits. The color of the skin has been a basic criteria in racial classification for a long time, probably because it is easy to observe and easy to measure. The percentage of the colors known as yellow, black, and white in a person's skin can be obtained by comparing skin color with some standard guide. Human hair texture can be straight, wavy, curly, frizzly, and wooly. The straightness or curvature of the hair root sac can be determined, as well as whether hair strands are circular or oval. Finally, the presence or absence of

In the following four photographs examples of racial types in children from around the world are shown. (*Above*) Children of the United States. Note the racial influences from Northern Europe. These children are tall and slender, with light hair and skin colors.

Children of Japan. Note the black hair, dark skin, and short, solid body trunks. Photograph by Kazuyo Nakao.

an *epicanthic fold* in the eyelid is an easily observable measure of inheritance. When there is no fold in the eyelid, it is called an "oriental eye."

Other observable physical traits important in racial classification include the shape of the nose, the form of the teeth, the shape of the lips, and stature. A trait that can be observed but that is better identified through measurement is the shape of the head. The width of the head divided by the length provides a ratio called the **cephalic index.** Anthropologists consider the index to be a valuable standard for distinguishing between similar people. It is clearly an inherited trait, and it does not vary with age, sex, or other factors.

Many inherited traits are not observable. For example, color blindness occurs only in males but is transmitted by females who have normal color vision. Certain characteristics of blood also are inherited; the blood types O, A, B, and AB are inherited, for example, but all racial groups have all four blood types. Some racial tendencies do exist, however. The people in a group of polar Eskimos living in northern Greenland all have type-O blood, but they are the only pure blood types on earth. The "sickling" trait of red corpuscles is an inherited trait, and why the red blood cells of some people are sickle-shaped rather than the normal disk-shaped is not clear. The trait was first discovered among people living in Tropical Africa, but it is also common in Jews, Greeks, and Turks living along the Mediterranean Sea. People with the sickle-shaped cells enjoy a special resistance to malaria, but inbreeding can lead to "sickle-cell anemia."

A number of diseases occur much more frequently among people of some races than others.

cephalic index   A number obtained by dividing the width of the cranium by its length and multiplying by 100. The number 80 is usually used as a base measurement to distinguish between long, narrow heads (<80) and heads that are more round (>80).

Children of Costa Rica. These Latin American children are mixtures of Spanish and Indian ancestors, and the black hair and dark skin are results.

Children of South Asia. These Muslim boys of Pakistan share the fine features, dark hair, and light skin color developed over the centuries since the Persian (Aryan) and Arab (Semitic) invasions of their homeland. Photograph by Jim Curtis.

Homo sapiens The various races of humans regarded as a single species. *Homo* is from the Greek word *homos,* meaning "the same," *sapiens* is from the Latin word *sapientia,* meaning "wisdom" or "knowledge," a main factor that separates humans from other animals.

For example, cancer occurs more frequently in white races. The Navaho Indians of Arizona have the lowest cancer rate of any group on earth. Most Eskimos had very low incidences of cancer until European racial mixing began; now their cancer rate is similar to that of whites. And diabetes is more common among Jews than among members of the other white races. But many diseases are related more to location than to race. Studies on the ecology of diseases show links between the environment and many diseases. Thus, it is difficult to sort out whether a particular disease is inherited or comes from the environment.

Finally, because the processes of variation, selection, and heredity continue undisturbed among isolated groups of people, *geographic isolation* is important in racial classification. For the most part, inbreeding is harmful, because it perpetuates weaknesses, so small, isolated groups can die out without "new blood" entering in the gene pool. On the other hand, some isolated groups have adapted to their environments through the natural selection process and become weakened by "new blood." The distribution of some of the major racial groups can be seen on the map in Figure 3.4.

Overall, the number of physical differences among the various races of the world is insignificant compared with the similarities. Between 95 percent and 99 percent of the 44,000 genes in each person are shared by all people. Thus, humans are basically the same all over the world. **Homo sapiens** is one species. This means that a fertile person from any race can mate with a fertile person from any other race and the union may produce fertile offspring.

The Human Dimension 37

■ Cultural Traits

The word *culture* is used to designate the learned traits of people. The ideas and concepts that make up such traits are not inherited but are passed down from one generation to the next through various types of teaching. Variations in culture occur through *inventions*. A group of people may incorporate new ideas into their way of life and thus make them part of their culture. This process is called **acculturation.** Culture includes almost everything about a people that is not a physical trait and is not instinctive. All beliefs, attitudes, and customs are cultural factors. People's techniques for making tools, houses, fences, monuments, and other material items also are part of their culture, as are art and artifacts as well as religions and artifices.

acculturation The transfer of certain cultural elements from one social group to another by contact between the groups.

Cultures evolve, and some change and grow while others decline. For instance, it appears that early tribes had to pass through three successive stages to leave behind their animal-like existence. In the earliest stage of early cultural development, food was obtained through hunting and gathering. In the second stage, the herding culture, the prolonging of lactation through continued milking of the female animal was an important concept to be learned. The final stage was that of learning the farming culture. The concepts having to do with

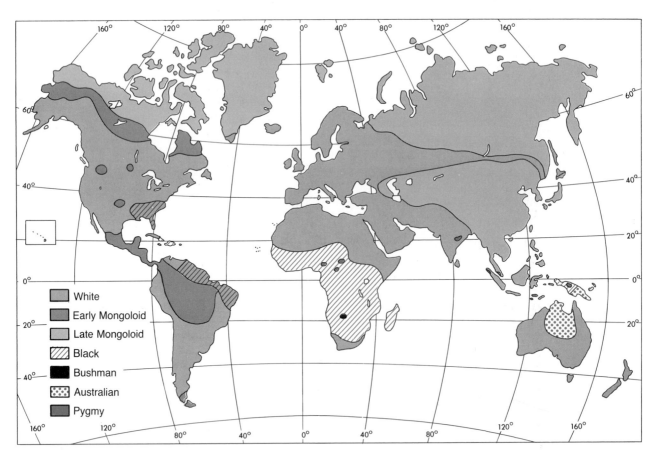

FIGURE 3.4

World Racial Group Regions. The regions where the seven major racial groups predominate are generalized. The boundaries are indistinct, and areas of overlap exist.

World Levels of ECONOMIC DEVELOPMENT

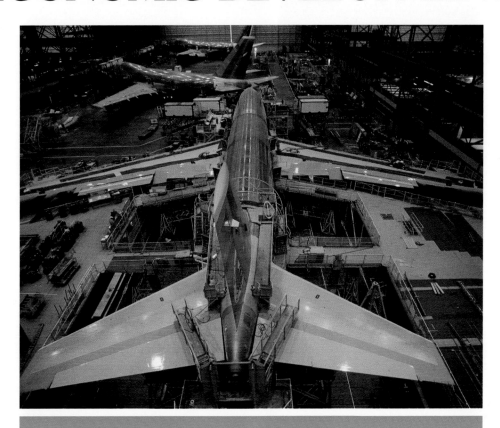

ALTHOUGH PEOPLE EVERYWHERE are quite alike in many ways, they have not all achieved similar levels of economic development. Anglo-Americans and Europeans, for example, live in highly developed culture regions in which they make use of many earth materials and have many complicated tools and machines. Other people, such as those in the tropical areas of Africa, have not achieved the same levels of economic development, nor do they enjoy the same material well-being as the people who live in Anglo-America and Europe.

When various levels of economic development are mapped on a world basis, large regions are created. These regions appear because numerous contiguous countries have similar levels of economic development. Usually three or four regions can be determined, depending on the criteria used for mapping. Four levels of economic development are indicated in the map on the following page. Names for the four regions could be the First, Second, Third, and Fourth Worlds, but in terms of economic development they are:

1. Highly Developed
2. Moderately Developed
3. Developing
4. Underdeveloped

World Levels of Economic Development

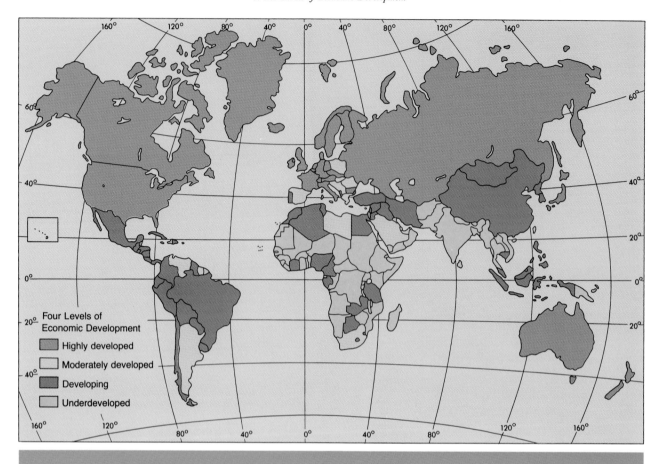

The different levels of economic development shown on the above map were determined by three important criteria:

1. Per capita income. Per capita income is computed by dividing the aggregate money income of all persons in a country by the total population of the country.

2. Percent agriculture. Percent agriculture refers to the proportion of the labor force employed in farming.

3. Per capita energy. Per capita energy is computed by dividing the total kilowatt hours of electricity produced in the country by the total population of the country.

World Levels of Economic Development

(Above) The **highly developed world** is depicted in this photograph of the Ginza in Tokyo, Japan. Property fronting on the Ginza is the most expensive in the world, costing about twice what a choice location in Manhattan would cost. It has been established that Tokyo has more neon lights than any other city in the world, and some of the most congested and noisiest automobile traffic.

(Right) Scenes indicating extreme contrasts in economic development are common in the **moderately developed world.** This squatter's shack overlooking modern apartment buildings is in Sao Bernardo do Campo, a suburb of Sao Paulo, Brazil. Large Latin American cities are social centers for the rich and elite, but they also are magnets for the poor, who leave the countryside to live in poverty on the edges of the cities.

World Levels of Economic Development

- *(Above)* Many people of the **developing world** live in crowded, low-cost housing areas. The street scene shown here is in Shanghai, China. Notice how close the houses are to each other, the mud walls, and the open sewer running down the middle of the street. Also, note the modern apartments in the near background.

- *(Right)* People of the **underdeveloped world** live under the worst conditions found anywhere in the world. The people in this picture are famine victims living in the Korem refugee camp in Ethiopia. The famine was caused by a series of droughts that started in 1972. Hundreds of thousands of people have died, mostly from starvation. A worldwide relief effort began in 1984 that created tent cities such as the one shown here.

Examples of written languages from various parts of the world. (*Above*) Examples of written French as seen from the window of a pastry shop in Lille, France. (*Top right*) Examples of Japanese script as seen on the labels of containers lined up at a Yokohama dock awaiting shipment. Photograph by Kazuyo Nakao. (*Bottom right*) Examples of written Spanish as seen on this anti–United States poster on a wall at the University of Costa Rica in Heredia.

preparing the soil, planting seeds, and tending the crops are fairly sophisticated, and farming takes much planning and considerable work.

Culture is a complex set of human conditions established over long periods of time, adjusted to fit new discoveries, and clinging persistently to certain regions and peoples. Some basic characteristics distinguish one culture from another, and there is much variation from culture to culture throughout the world.

■ Language

Language is the ability to communicate with sounds and written symbols. Humans first learned to communicate by using signs, then they used their voices to make sound symbols, and later they learned to make written symbols. The ability to communicate with sounds and symbols is uniquely human and is the foundation of culture.

Language makes it possible for ideas and experiences to be transmitted from one generation to the next, and in this way it serves to perpetuate individual cultures. The worldwide variability in languages corresponds closely to the variability in cultures.

Currently, there are 2,796 active languages in the world, but only 164 of them are spoken by a million or more people. Some order can be brought to this confusion by classifying the languages into families. Eight major language groups have been identified, and the languages within each group are related, probably coming from a common source. The eight major language groups are the Indo-European, Hamito-Semitic, Ural-Altaic, Japanese and Korean, Malayo-Polynesian, Bantu, Dravidian, and Sino-Tibetan. Geographic conditions account for differences between languages within a language group. Separation of people by migration or by political boundaries encourages the development of different word usages, tones,

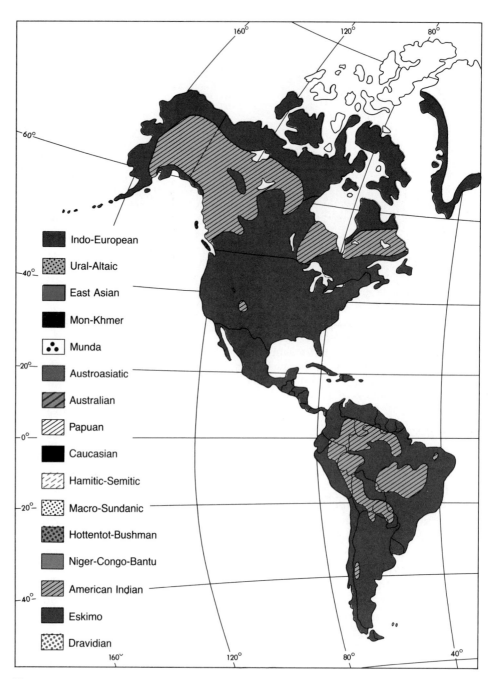

FIGURE 3.5

World Language Regions. World language-use regions are generalized areas where most of the people use one of the major languages of the world.

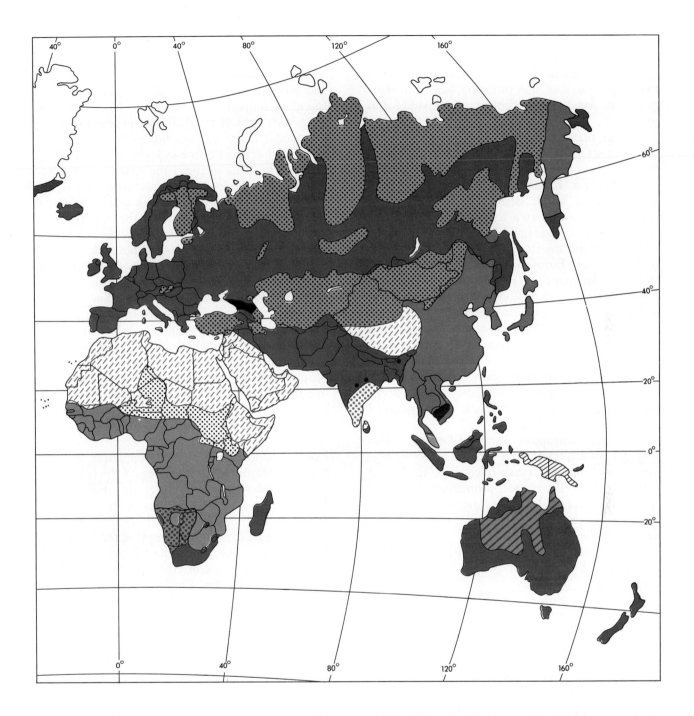

and inflections. Languages that were once very similar gradually become incomprehensible to people outside the local group.

Each language family contains a number of branches, and each branch contains a number of individual languages. The *Indo-European* family of languages is the largest and most widespread. The Germanic branch of the Indo-European family includes English, Dutch, Danish, Swedish, Norwegian, Flemish, and German. These languages are found throughout Anglo-America, northwestern Europe, Australia, and New Zealand. People that understand any one of the Germanic languages will recognize certain words and phrases from all the others. But normal conversation between people that use different Germanic languages is usually impossible without study and practice.

Other branches of the Indo-European family of languages are the Romance languages spoken in the countries of southwestern Europe. The Portuguese, Spanish, Italians, and French have a common language bond. Besides these well-known countries, people of two lesser-known areas also speak Romance languages. These are the Walloons of Belgium and the Romanians. It may seem odd that the Eastern European nation of Romania has a Romance language as its national tongue, but historians tell us that Roman troops were once stationed in the region of present-day Romania. The Spanish and Portuguese also spread their languages during their days of conquest, most notably to Latin America and Africa. The people in most of the countries between Mexico and Argentina speak Spanish, with the notable exception of Brazil, where Portuguese is the national tongue.

The Indo-European family of languages also includes Slavic, Iranic, and Indic. The Slavic languages are found in Eastern Europe and the Soviet Union; Russian is the most prominent Slavic language. The Iranic languages are spoken in the region between eastern Turkey and western India, and the Indic group of languages are found primarily in northern India. Iranian (Farsi) and Hindi are spoken by the most people in each group respectively.

Languages within the *Hamito-Semitic* family are unrelated to those in the Indo-European family. The Hamito-Semitic languages are found in the desert regions of Northern Africa and the Arabian Peninsula. More people speak Arabic than any other type of Hamito-Semitic language. Both Arabic and Hebrew are forms of Semitic languages, but Yiddish is a variant of German, so it is Indo-European in origin but written with Hebrew characters.

South of the Sahara in Africa, in what is referred to as Black (or Tropical) Africa, the *Bantu* language family predominates. About 85 languages used by separate tribes make up this group. Congolese is the most common language in western Africa, Swahili in eastern Africa, and Zulu in southern Africa. The *Dravidian* language family is found in southern India and the *Ural-Altaic* languages are spoken by groups of people located in the harsh climatic regions north and south of the region where most of the Indo-European speaking people live. This includes the sparsely populated areas of the deserts in the southern part of the Soviet Union and the tundras near the Arctic Circle in the northern part of the country. The Ural-Altaic family also includes the languages found in Turkey, Hungary, and Finland.

The oriental language families of East Asia include the *Sino-Tibetan* and the *Japanese-Korean* families. Mandarin Chinese, of the Sino-Tibetan family, is spoken by more people than any other single language of the world (see Table 3.1). It was originally spoken primarily in northeastern China, but since its adoption as the official language of the People's Republic of China it has spread throughout the country. The Japanese and Korean languages are not related to Chinese in origin and in the spoken form; they are from different language families. However, both the Japanese and the Koreans adopted the written characters for their languages from the Chinese.

The outstanding feature of the *Malayo-Polynesian* family of languages is its geographic distribution. People who speak these related languages live on the tropical islands of the Indian and Pacific oceans. The region extends from Madagascar, off the east coast of Africa, to Hawaii, in the middle of the Pacific. This longitudinal extent is more than halfway around the world and constitutes the largest single region for any world language family. The most common language in the group is Indonesian, which is spoken primarily on the heavily populated island of Java.

■ Religion

The second major component of culture is religion. Religion is the organized system of worship of one god, groups of gods, or the supernatural in general. Although from time to time religion gains

TABLE 3.1

Principal Languages of the World

LANGUAGE	SPEAKERS (in millions)	BRANCH	FAMILY
Mandarin	755	Chinese	Sino-Tibetan
English	409	Germanic	Indo-European
Russian	280	Slavic	Indo-European
Hindi	275	Indic	Indo-European
Spanish	275	Romance	Indo-European
Arabic	166	Semitic	Hamito-Semitic
Bengali	160	Indic	Indo-European
Portuguese	157	Romance	Indo-European
Malay-Indonesian	122	Austro	Malayo-Polynesian
Japanese	121	Japanese-Korean	Sino-Tibetan
German	118	Germanic	Indo-European
French	110	Romance	Indo-European

SOURCE: World Almanac, 1988, p. 245.

and loses its emotional impact, it remains a very strong force in determining human behavior. Religion has been and continues to be a major cause of both internal national strife and international wars. According to Van Valkenburg, a political geographer, religion has caused more wars than any other single factor except hunger. In some cultures, religion is the primary trait, the one around which all aspects of life revolve; in other cultures, religion is looked upon merely as a superstition. Religious differences, however, are at the root of the problems in many of the world's trouble spots today.

As with language, the major world religions can be categorized into families that contain groups of religions that came from similar origins. Although many more could be used, we will use the six families of religion shown on the map in Figure 3.6.

The *Christian religion* is by far the largest, both in terms of membership and in terms of world distribution. About one billion people combine to form the Roman Catholic, Protestant, and Eastern Orthodox churches. Most of the people in Europe, the Americas, Australia, New Zealand, and the Philippines belong to the various Christian sects and denominations that make up the Christian faith. The largest Christian religion, again in terms of both membership and distribution, is the Roman Catholic. Since the Protestant breakaway from the Catholic church in the early 1500s, Catholics have remained strongly unified. The Protestants, on the other hand, have continued to split into a myriad of sects and denominations. The largest of these groups is the Baptist church in the United States, which has about 30 million members.

The second largest religion in world membership is *Islam*. About 750 million followers of the Prophet Muhammad live primarily in a string of countries ranging from the west coast of North Africa, through the Middle East, and into South Asia and Southeast Asia. Most Muslims are of the Sunni sect, but about 10 percent are Shiites. The Shiites are found mostly in Iran, although some are in Iraq.

Christianity, Islam, and *Judaism* all originated in the Middle East. They are all monotheistic religions—that is, their adherents believe in one god. Judaism is the oldest of the three, dating back to more than 4,000 years ago. It has about 15 million members and is found in all parts of the world, but the Jewish people consider Israel their homeland. Jerusalem is a holy city for Christians

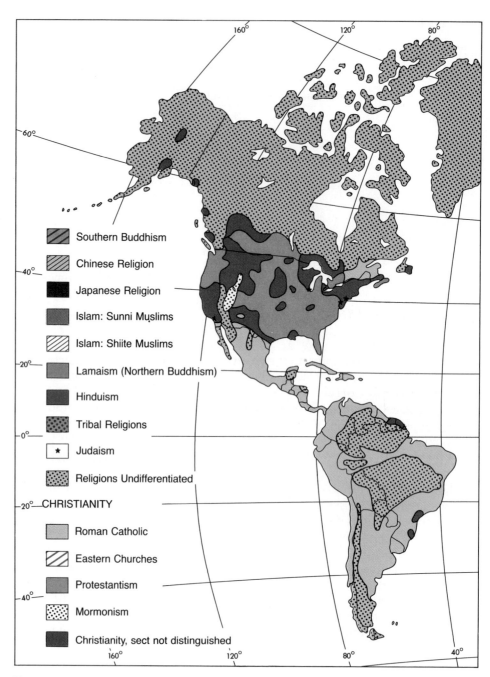

FIGURE 3.6
World Religions. As with race and language, world religion regions are generalized, but most of the people in the areas adhere to one of the major world religions.

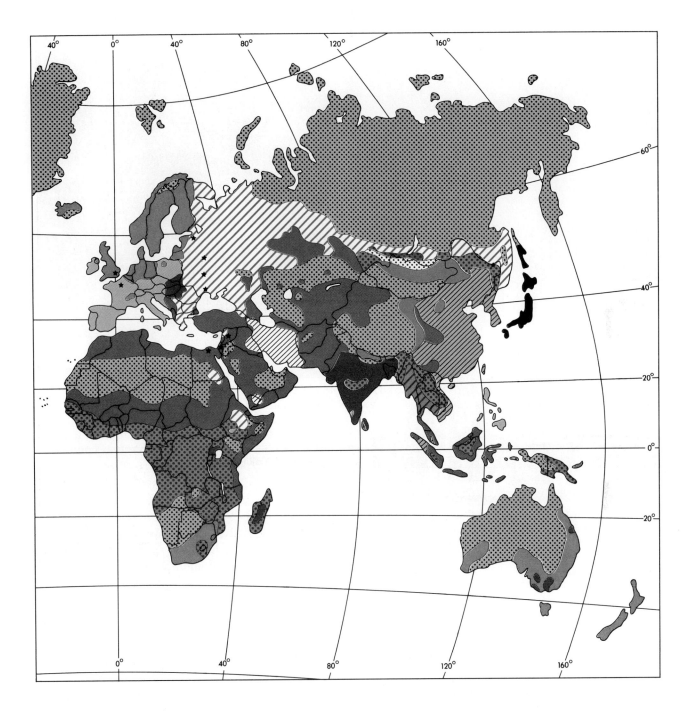

and followers of Islam, as well as for Judaism, although Mecca (in Saudi Arabia) is the most important religious center for Islam. About half the world's Jews live in the United States, and New York City has the largest concentration.

As the Middle East was the spawning ground for Christianity, Islam, and Judaism, the Punjab and Ganges plains of northern India were the source areas for two other great religions of the world. The Punjab is the core for *Hinduism,* a faith

The Human Dimension 45

This cathedral in Salisbury, England, is a symbol of the historical influence of the Protestant religion in that country. The cathedral was completed in 1259, but it is still used today on a regular basis for Anglican church services.

Religious shrines in Japan are located throughout the islands, but the largest concentration is in and around Kyoto. They are revered by the devout and highly popular among sightseers. The small shrine in this picture is a Shinto holy place near Kyoto. Photograph by Kazuyo Nakao.

that began about 4,000 years ago and spread from the Punjab to dominate the Indian subcontinent. It is the major religion of India today, with a membership of about 600 million. *Buddhism* began as a reform movement from Hinduism in about 500 B.C. The core area for Buddhism, the foothills of the Himalayas, lies just to the east of the Punjab Plain. Unlike Hinduism, however, Buddhism spread well beyond India. Buddhism is now found not only in Southeast Asia but also in China, Japan, and Korea, where it has blended with such native religions as Confucianism, Taoism, and Shintoism. It is also the primary religion of Tibet and Mongolia and has even won converts in the United States. Total membership is difficult to determine, because Buddhism blends in so well with other religions. Estimates range from 200 million to 500 million.

The last important world faith is not really a religion, at least in the ordinary sense. *Animism* is the belief that certain inanimate objects, such as rocks, trees, mountains, rivers, the wind, the moon, and the sun, possess spiritual powers. Individual groups, especially in the tropical areas of Brazil, Central Africa, and on the island of New Guinea, have specific objects in which they believe that spirits dwell. Usually a tribal leader acts as an intermediary between the believers in animism and the spiritual objects.

This Roman Catholic church in Cartago, Costa Rica, is the center of social activities as well as a place of worship. A Sunday afternoon soccer game is in progress in the foreground, and a carnival with a Ferris wheel also draws people to the location.

The Islamic mosque, like many Christian church buildings, is a place for people to gather for activities unrelated to their religion. Some of the Arabs in this picture are praying on the steps of a mosque in Sousse, Tunisia, others are merely sitting and watching the traffic on a busy street.

The Human Dimension

Economic Geography

Agriculture

Aside from the primitive hunters and gatherers, the farmers of the world live closer to the physical environment than anyone else. Farmers must constantly adapt to the whims of nature in order to survive. In response to the physical forces, farmers produce commodities that vary greatly from one part of the world to another. They grow crops and animals that are best suited to their local physical environment. Farming is also influenced by traditions (cultural factors), markets (economic factors), and trade agreements (political factors). Thus, the varying patterns of agricultural production found over the surface of the earth can be attributed to a number of influences. See Figure 3.7 for a map showing the agricultural areas throughout the world.

Agriculture is a type of *primary* economic production, which means that it deals with the first step in the production process. The categories of agriculture are based on what the farmers do and how they do it. "Subsistence producers" use nearly everything they produce to satisfy their basic needs for food, clothing, and shelter. "Commercial producers" generate a surplus they can trade or sell to obtain their needs. Agricultural production can be divided into two other general categories. One agricultural group deals primarily with animals, the other group spends most of its energy raising crops. These divisions of activities give us four categories of agriculture: (1) subsistence herding, (2) subsistence farming, (3) commercial herding, and (4) commercial farming. Al-

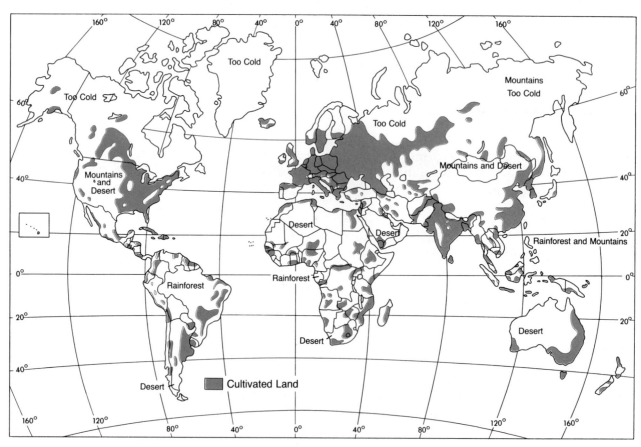

FIGURE 3.7

World Areas of Cultivated Land. Note the large areas of the world that are not suitable for agriculture of any kind.

48   Chapter 3

though there are some overlaps among the categories (e.g., mixed farming) and some gaps (e.g., hunting and gathering), these four categories provide a framework for studying most types of agriculture found in the world.

Subsistence Herding. The largest economic regions on earth are made up of areas where subsistence herding is the predominant economic activity. These places are all arid regions, where the lack of precipitation supports only sparse vegetation. The lack of vegetation limits the number of animals that can be accommodated, which in turn limits the number of people. Although subsistence herding occurs over large portions of the world, relatively few people are involved in the activity.

Subsistence Farming. Like subsistence herding, subsistence farming is a primitive economic activity found in large areas of the world. Unlike subsistence herding, however, many people are involved in this type of farming. The occupation of farming ranks number one in the world, based on the number of people involved, and many of these people live in the heavily populated areas of southern and eastern Asia. The regions where subsistence farming predominates are located primarily in the rainy, tropical areas of the western hemisphere, Africa, and Southeast Asia. The warm, moist conditions (Af and Am Köppen-type climates) encourage plant growth, but they cause meat and animal products, such as milk, to spoil quickly. The environmental conditions are more favorable for plants than for animals as the major source of food, clothing, and shelter.

Subsistence farmers today live much the same way their ancestors did thousands of years ago. Their food preferences remain stable over time, and they cultivate plants that were domesticated from wild plants found locally. Maize (corn) is the leading grain in the western hemisphere, rice is the staple of South and East Asia, and millet (a cereal grass) is most common in Africa. All three of these plants originated in the areas where they are now the leading subsistence crops. Maize was first domesticated in Mexico, rice in Thailand, and millet in Uganda.

Commercial Herding. The two major commercial herding activities are livestock ranching and dairy farming. As with the subsistence herders, animal husbandry is the major concern of the commercial herders. The animals used are generally sheep or cattle, and the major products are wool, hides, meat, and milk. The flocks and herds are painstakingly protected against theft, parasites, predators, and disease. Natural food and water supplies for the animals usually are augmented by feed purchased or grown especially for the animals. Owners also provide water from wells rather than letting their animals search for it.

Water is critical for the commercial herders, especially the ranchers, because their herding activity usually takes place in the dry regions ("BS" Köppen climate) of the world. The lack of moisture tends to exclude crop agriculture, and the carrying capacity for animals is low—several acres are needed to support each animal. Consequently, the human population in such areas is low also. Every continent (except Antarctica) has extensive areas of arid and semiarid regions where livestock grazing predominates.

World milk production does not occur in climate areas used mostly for livestock ranching. Dairying is associated with cool, humid regions ("Cf" and "Df" climates). The three major areas are northwestern Europe, the north-central United States, and southeastern Australia. Numerous other pockets of high production exist throughout the world, many around large urban centers. Milk was a subsistence product through most of history, but for the last 150 years the majority of the world's supply has come from commercial dairy farms. Commercial herders who specialize in the production of milk sell it for use as fluid market milk, to be consumed directly, and manufactured milk products, such as butter, cheese, cottage cheese, yogurt, and ice cream.

Commercial Farming. Commercial agriculture is characterized by specialization. Usually only one or a limited number of commodity types is produced. Specialization does not mean exclusivity, however, because numerous auxiliary animal and crop products usually are grown along with the primary crop. Wheat is the world's most important commercial crop. More rice is produced than wheat, but rice is used more as a subsistence crop than is wheat. Numerous other commercial grains are grown, including corn, barley, soybeans, and oats.

Commercial farming is found in many diverse areas of the world. For example, bananas grow best in the rainy, tropical areas, whereas wheat is a mid-latitude product because it prefers much cooler and drier conditions. Coffee and tea grow best in the upland areas of the tropics, while sugar cane and pineapple do best in the lowlands. World patterns of crop production are influenced to a great extent by the physical environment, but human behavior also dictates where things are produced. Sometimes it is a matter of preference in taste, sometimes religion may dictate what not to eat, sometimes politically motivated boycotts of certain crops occur—but for whatever reason, agriculture is controlled by human behavior.

Mineral Production

Mineral production—the extraction of valuable fossil fuels and metals from the earth—is a primary economic activity and the foundation for manufacturing. *Manufacturing* is changing one product into another product in order to increase its value. It is a secondary economic activity because it does not deal directly with the utilization of earth materials and because it relies on mineral production.

The energy materials usually are referred to as "fossil fuels" because they are the remains of plants and animals that existed during a former geologic time. The fossilized plants and animals changed form over time, but retained the energy they had received from the sun. They are valuable as a source of energy because the energy can be released by burning the materials. The most common fossil fuels are coal and petroleum. Coal is found worldwide, but some deposits are much thicker and richer than others. Petroleum also is common in the upper strata of the earth's crust, but large local pools have created a myriad of economic and political problems for the world's people.

Iron is the most important metal. The discovery of uses for iron changed civilization completely. The Iron Age began about 3,500 years ago, when the pure metal was obtained from meteorites. After methods for extraction from ores were discovered, iron became increasingly available for arms, tools, and then machinery. Throughout the centuries, iron has increased in importance, especially after the discovery that steel could be made by combining alloys with the pure iron. Steel is the basic, indispensable metal of modern civilization, and iron ore production—along with its companion, coal production—is a prime indicator of the material or economic progress of any nation. See Figure 3.8 for the world's coal and iron ore areas.

Industry

The turning point in the human endeavor to produce uniform, quality goods in quantity occurred during the industrial revolution. The period from 1750 to 1850 saw a rapid change from individually handcrafted goods to uniform, machine-made products. The first real machines were invented in England for use in the textile industry. At first, waterwheels were used for power, until the steam engine was invented and patented in 1769 by James Watt. Then the search began for sources of fuel to power the engines and for metals from which to make them and other machines.

Today, industrial activity usually is divided into light industry and heavy industry, and sometimes a third type, known as high-technology, is included. Industrialization in England began with the light industries, and each country today usually begins with light industry on its way to further industrial development. The light industries, which are now widespread throughout the world, have certain advantages in common that make them a good first step into industrial activity. First, they supply products that have universal demand, such as textiles for clothing, slaughtered animals for meat, and processed crops for food. In addition, the light industries use local raw materials, simple manufacturing processes, and lightweight automatic machinery or abundant cheap labor, or both. Many Third World countries have not progressed beyond the stage of light industry.

Heavy industry includes and depends on the manufacturing of steel and steel products, so heavy industries are not found throughout the world. Steel production occurs where abundant supplies of iron ore are located near large coal deposits, or

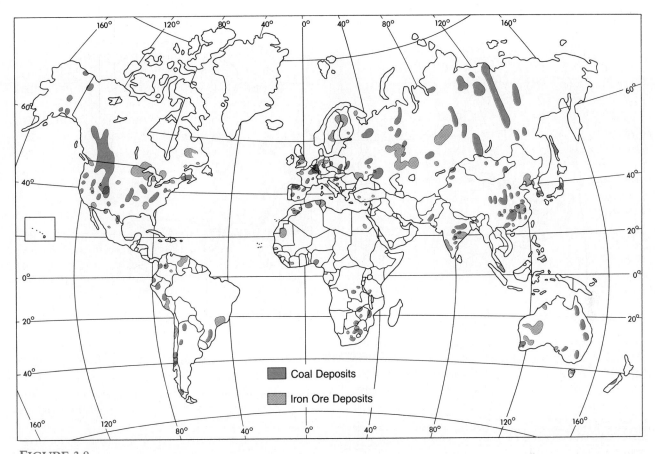

FIGURE 3.8
Coal and Iron Ore Deposits of the World.
SOURCE: U.S. Bureau of Mines, *Mineral Yearbook* (Washington, D.C.: Government Printing Office, 1985.)

at least in places where the two commodities can be brought together cheaply. Coal is used for the energy to heat the blast furnaces for changing iron ore into steel. A well-developed iron and steel industry is the initial and most crucial phase for establishing other heavy industries, such as automobile, train, and airplane manufacturing. The world's manufacturing areas today are shown in Figure 3.9.

Trade, Transportation, and Communication

Geographers are interested in trade because it reflects differences from area to area and sets up patterns of movement over the earth. The most obvious place-to-place differences have to do with (1) the physical environment, particularly the climatic and geological aspects, (2) the socioeconomic environment, or level of development, (3) the differential distribution of people, both in total numbers and in density, and (4) the degree of accessibility permitted by natural features and transportation routes. These factors help explain trade patterns. Other factors include international relations, tariffs and embargoes, foreign investments, differences in technology, and even such things as war and peace.

Trade keeps the world economy moving because only a very few people are completely self-sufficient. No nation has a range of resources wide enough to bring its standard of living to the top level. The resources a nation needs and does not have within its own boundaries must be taken by force, acquired by trade, or done without. The

The Human Dimension

FIGURE 3.9

Manufacturing Areas of the World. Compare the manufacturing areas with the locations of coal and iron ore deposits in Figure 3.8.

more complex an economy becomes, the wider the range of resources required. Great industrial nations must scour the earth for the fuels, minerals, forest products, and foodstuffs they need to maintain their accustomed levels of living. Once the materials are found, trade must be negotiated and carried out.

Transportation includes not only the worldwide movement of commodities, but also the local movement of people and goods. These movements vary considerably from one country to another, not only in the volume of people and commodities moved, but also in the ways they are moved. In some areas, rivers and canals are important carriers, but in other areas trucks and trains are more important. The industrial nations also have the most highly developed transportation systems, and they use the rivers and canals as well as trucks and trains. The most highly developed countries rely heavily on their airlines, especially for the movement of people and light goods, such as mail. The developed countries of the world are tied together through the major shipping routes of the oceans (see Figure 3.10). Thus, transportation is a fundamental concern of geographers, both for the place-to-place differences in modes of transportation and for the volume of movement.

Progress anywhere depends largely on effective communication, which is the only way people can get new ideas and new products. The movement of ideas also is important for economic and political transactions and for a myriad of other purposes. Communication varies from area to area. There are simple forms, such as drumbeats in

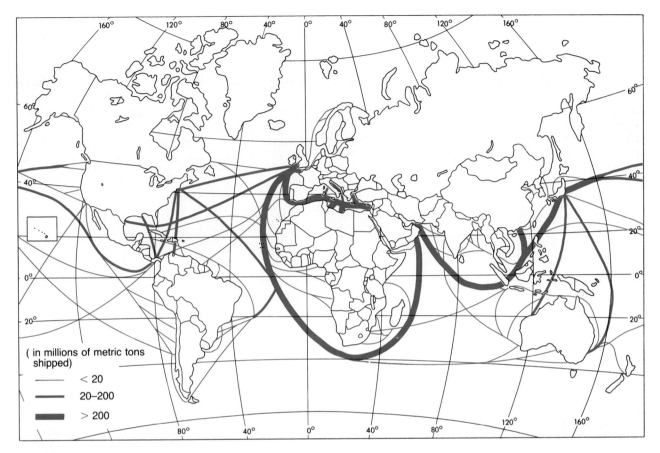

FIGURE 3.10
World Shipping Routes. Notice the heavy traffic lanes between the developed countries of the world.

the tropical rainforest, and complex forms, such as radio beams from satellites. A country's economic development can be judged by such statistics as the number of telephones per person and/or the number of radio and television stations available. Communication is a vital part of human geography.

Summary

This chapter is an introduction to the regional chapters that follow. Because some of the information presented here will not be repeated in the regional chapters, the reader will want to refer to this chapter from time to time. Most of the above topics relating to human geography are discussed in the regional chapters, but topics that pertain to particular regions are also included in the following chapters. In addition, each regional chapter concludes with a look at the political geography of the region.

Key Words

acculturation
age structure
cephalic index
crude birth rate
crude death rate
demographic
   transition
dependency ratio
Homo sapiens
mortality rate

The Human Dimension

mutation
natural selection
population pyramids
rate of natural increase (RNI)

References and Readings

Clark, Audrey N. (ed.). *Longman Dictionary of Geography: Human and Physical*. White Plains, N.Y.: Longman, 1985.

Goudie, Andrew S. *The Human Impact on the Natural Environment*. Cambridge, Mass.: MIT Press, 1986.

Hall, Peter. *The World Cities*. New York: St. Martin's, 1984.

Jackson, John B. *Discovering the Vernacular Landscape*. New Haven: Yale University Press, 1984.

Johnston, Ron J., et al. (eds.). *The Dictionary of Human Geography*. New York: Free Press, 1981.

Newman, James L., and Matzke, Gordon E. *Population: Patterns, Dynamics, and Prospects*. Englewood Cliffs, N.J.: Prentice-Hall, 1984.

Chapter 4

CONCEPTS AND TOOLS

■ Spatial Distributions

One of the main concerns of geographers is the location of things on the surface of the earth. Where things are located is important, but why they are located where they are is just as important and makes geography a scientific discipline. It is fairly easy to memorize where things are from a map, but it is usually more difficult to figure out why they are located there. One concept that geographers use in determining why things are where they are is the idea of **spatial distributions.** Nearly everything geographers study is distributed in space in some manner; if there is no distribution involved in the study, then it is of little concern to geographers. Maps are important to geographers as tools used to portray spatial distributions.

The elements common to all spatial distributions are pattern, density, and dispersion (see Figure 4.1). These three elements describe the comparative locations of the things being studied, but they are independent of each other. Let us consider rural settlement—specifically the homes of individual farm families—as an example. We can call the farmsteads "geographic facts" because they are all the same type of object (homes) and because

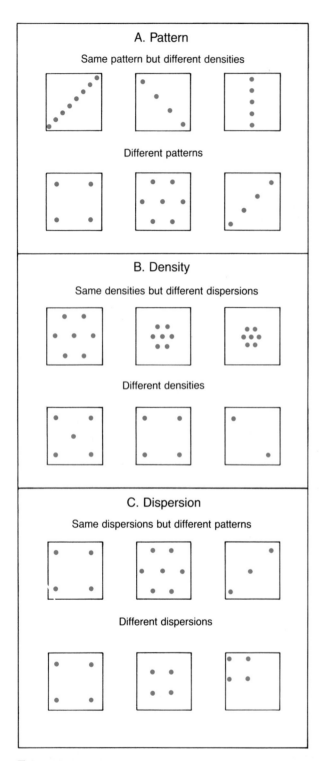

FIGURE 4.1
Pattern, Density, and Dispersion. The three elements of all spatial distributions can be compared in this sequence of diagrams.

spatial distributions The distribution of things in space. Can be three-dimensional, but usually refers to the geometric arrangement of geographic phenomena on the surface of the earth.

agglomerated Refers to a spatial distribution where the items of concern are gathered together in a mass or separated only by short distances.

they all have specific locations. The geometric arrangement, or *pattern,* of the farmsteads can be seen either from an air photograph or from a map. The pattern may be linear, centralized, uniform, or random.

The number of farmsteads per unit area (for instance, a square mile) indicates the *density* of the distribution. Because it is numerically based, density can be given in more accurate terms than pattern, but both are important for comparing one spatial distribution with another. One location may contain four farmsteads per square mile, while another may contain forty farmsteads per square mile. The difference in density is obvious and can lead to many other important comparisons. The area with forty farmsteads, for example, would have a much greater population density than the one with only four farmsteads. The people in the more densely settled area would require more roads and other facilities than the one with only four farmsteads. In addition, the individual farm plots would be smaller and the agriculture more intensive than in the less heavily settled area. Thus, much information can be gained by merely comparing one distribution with another.

The third element of spatial distributions, *dispersion,* measures the degree to which objects are separated. Some unit area must be considered when discussing dispersion too. For example, the farmsteads within a square mile can be either agglomerated or dispersed. If they are **agglomerated,** the space between them would be at a minimum; if dispersed, they are separated. Various degrees of dispersal are possible, but maximum dispersion occurs when the farmsteads are all separated from each other to the maximum distance possible for the area being considered. It should be noted that, because maximum disper-

spatial interaction Movement (interaction) through space between sets of spatial distributions.

sion creates a hexagonal pattern, pattern and dispersion are not totally unrelated.

■ Spatial Interaction

Movement on the surface of the earth can be shown on maps in numerous ways, but direction arrows are the most common (see Figure 4.2). The movement of people—migration—from one place to another can be shown by arrows, and the sizes of the arrows can indicate the magnitudes of the movements. Similar types of arrows can show traffic flows, world trade, and many other types of movement. Movements are usually not random, and because they occur for some reason an order can be found. **Spatial interaction** is the phrase geographers use to describe movement; it also implies that the places involved "complement" one another—one place has a surplus of something, the other place has a need for it. The concept

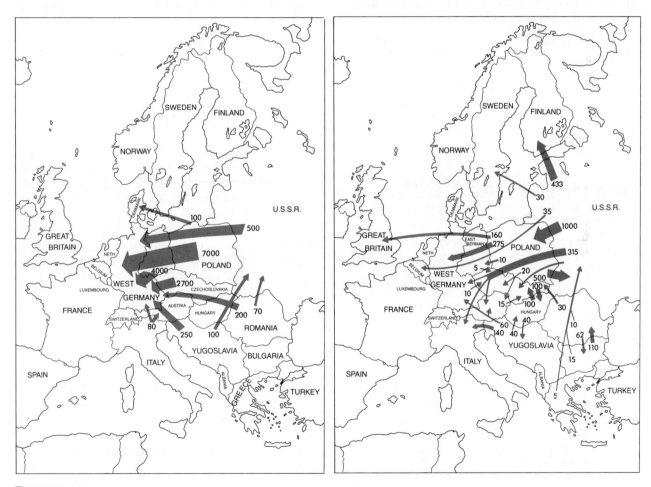

FIGURE 4.2

Redistribution of Population. Direction arrows indicate the volume of migrants produced by World War II. The figures are in thousands of people and do not include movements within countries. A: The transfer, evacuation, and flight of ethnic Germans. B: The movement of non-Germans.

SOURCE: Robert E. Norris and Lloyd Haring, *Political Geography* (Columbus, Ohio: Merrill, 1980), p. 145. Used by permission.

of spatial interaction is especially important for understanding trade, but it also applies to many other types of movement, such as migration, commuter trips, school bus routes, deliveries of wholesale goods and mail, and other daily movements.

▪ *Spatial Diffusion*

Another way to indicate movement with maps is to show a sequence of maps from different time periods. A single map portrays a spatial distribution at one point in time, producing a static representation of something that is changing, but a sequence of maps can indicate movement of the spatial distribution. An important concept in geography is that distributions usually are not static, but are either increasing or decreasing in their areal extent. This idea is important because it is involved with the processes of dispersal and/or disappearance of distributions. By discovering the process, we can get a better idea of the cause. Some spatial distributions have existed for a long time, and it is difficult to determine how they originated; others are just beginning to appear on the surface of the earth. A spatial distribution that is spreading is said to be "diffusing," and we call the process "spatial diffusion." When a distribution is disappearing, the process is called "reverse diffusion."

Spatial diffusion can be thought of as a special type of spatial interaction. The idea that one place is tied to another through something that is moving between them is the basis of the concept. But diffusion is different from interaction because it implies that a "spreading" is occurring. Many examples of diffusion have been studied by geographers and others. Epidemiologists study the diffusion of diseases in people, and plant pathologists study the diffusion of diseases in plants. Physical geographers and geomorphologists study the spreading of alluvial fans and rock glaciers. The spread of insects, especially "killer bees" and fire ants, has been watched closely. All these forms of diffusion leave some physical evidence behind in the wake of their spread, but ideas or information usually do not leave any physical evidence.

The *diffusion of an innovation* is the spreading of a new idea through a population. It is definitely a

---

spatial diffusion   The dispersal of items by moving them through space, creating a new spatial distribution.

---

movement process, but it is one that can be difficult to trace because it may not leave evidence of where it has been. Gossip is a good example of the diffusion of an idea through a group of people, and one that leaves no observable pattern. On the other hand, such innovations as AM radio broadcasting and, later, television broadcasting, and, later still, FM radio broadcasting leave physical evidence—broadcasting towers and broadcasts that can be seen or heard. (See Figure 4.3 for the example of diffusion of FM radio stations.)

Most of the many things that are spreading or have spread across the surface of the earth can be classified as having moved in one of three ways: through expansion, through relocation, or in a hierarchical manner (see Figure 4.4).

Expansion diffusion is the process of the movement of something such as a disease or an innovation through an existing population. Initially, one person at one place gets the disease or innovation from a source outside the region under consideration. That person is a "sender" of the disease or innovation. The initial "receivers" are the people located nearest to the sender, so distance is always a factor in expansion diffusion. The innovation or disease moves outward from the origin in a wave-like manner, going from senders to receivers until the area becomes saturated with whatever is being diffused.

The best example of *relocation diffusion* is probably the migration of people, especially the movement of settlers into previously unoccupied lands. Not only are the people diffusing, but everything they carry with them is diffusing too. An epidemic, for instance, usually starts when a disease is transferred from one region to another by a "carrier," or infected person or animal. Innovations also move through space with carriers of the idea by the relocation diffusion process.

The *hierarchical diffusion* process is similar to expansion diffusion, except the movement is not between the nearest senders and receivers. Move-

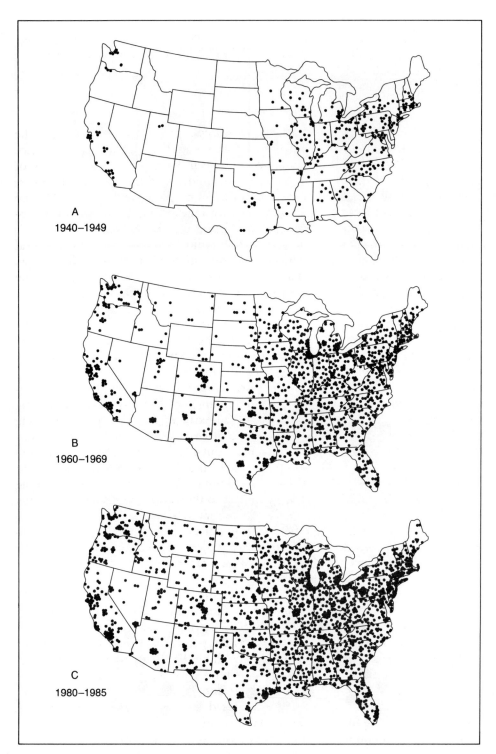

FIGURE 4.3

Diffusion of FM Radio Stations in the United States. The points on the maps indicate the locations of FM radio stations for 1940–49 (Map A), 1960–69 (Map B), and 1980–85 (Map C).

SOURCE: John Colburn, "The Diffusion of FM Radio Stations in the United States" (Master's thesis, Oklahoma State University, 1987). Used by permission.

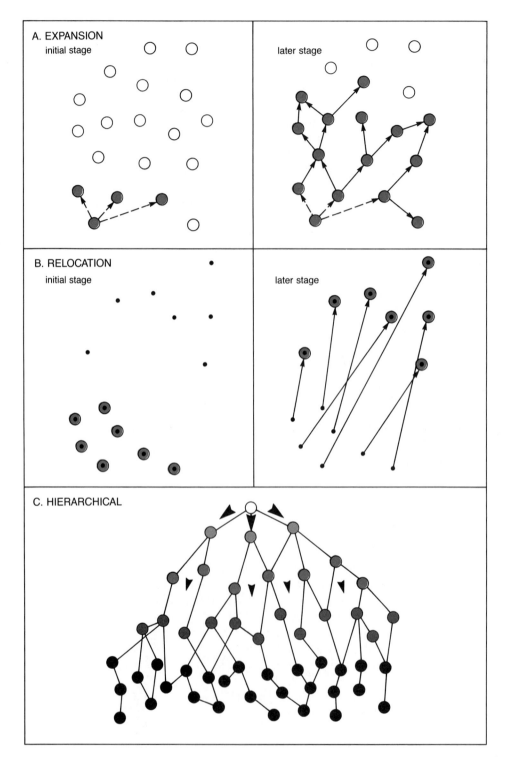

FIGURE 4.4

Types of Spatial Diffusion. (A) Expansion, (B) Relocation, and (C) Hierarchical.

SOURCE: Adapted with permission from Peter Gould, *Spatial Diffusion* (Association of American Geographers, Resource Papers for College Geography, no. 4, 1969), pp. 4–6.

logistic curve A curve that rises slowly at first from left to right, then more quickly, and finally slows down again and flattens out at the upper bounds so that the curve has an S-shape. Sometimes called an S-curve, Lorenz curve, or sigmoid curve.

spatial models Method used for understanding why things are located in particular places. Models can be verbal (descriptive), visual (maps and graphs), or symbolic (mathematical expressions).

ment occurs between the largest places on a hierarchy first, then between the next largest places, and so forth, until all potential receivers are contacted. An example of this might be designer clothes. A new design appears in New York City first, then in Chicago, and next in Los Angeles. Then the idea filters back to Phoenix from Los Angeles, and to Denver or Dallas from Chicago. Finally, the idea spreads throughout the United States, but only after appearing in the larger places first. The process is described as a "trickle down" from larger places to smaller places.

Understanding the diffusion processes is important because sequential geographic patterns can help predict future patterns. In addition, the speed of the diffusion can help predict when a disease, insect, or innovation may arrive at a certain point. An underlying regularity in the diffusion of innovations makes it possible to determine how long it will take for new ideas to catch on in various regions. The regularity can be shown by what is known as the "S-curve," also called the **logistic curve** or Lorenz curve, which is produced when the cumulative percentage of adopters of an innovation is plotted against time (see Figure 4.5). By comparing the curve to a normal (bell-shaped) curve, the percentage of adopters can be divided into categories. The first people to adopt the new idea are called "innovators," then the "early majority" adopt, followed by the "late majority." Finally, the "laggards" are the last to adopt. The people in these categories usually have certain personality traits in common and conform to certain social, economic, and educational conditions. The innovators, for example, usually are the wealthier and better educated people in the region, while the laggards are usually older, less educated, have lower incomes, and are more withdrawn than others in the region.

The diffusion processes are spatial in nature, but many underlying social, cultural, and economic factors are tied to the explanation of the processes. Diffusions of innovations will occur at different rates in regions of unlike cultures and in regions with different economic conditions. In spite of the differences, the processes are similar; only the temporal (time) dimensions vary. That is, the logistic curve can be used to help explain the spread of the use of a new tool in a primitive society as well as the spread of the adoption of FM radio in a modern society (Figure 4.3).

■ Spatial Models

Spatial models are used to indicate the relationships between the factors influencing the location of human activities. They help explain the location

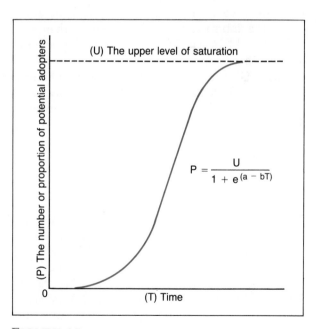

FIGURE 4.5

The Logistic Curve. An S-shaped curve characterizes many diffusion processes, especially the diffusion of such innovations as FM radios.

of things on the landscape. Humans are generally quite orderly in their activities, but sometimes the order is not apparent until it is studied. Spatial models can help in discovering this order.

One of the first spatial models was developed to explain the location of economic activities. It was designed during the 1820s by Johann Heinrich von Thünen (1783–1850), a German economist, landowner, and farmer. Von Thünen lived on the flat, agricultural lowlands of northern Germany near the present town of Rostock. He was interested in the production of crops and had noticed that crops were grown in definite spatial patterns around towns and small settlements. These land-use patterns consisted of concentric zones, with different crops predominating in each zone outward from the settlement.

Rather than merely describing what he saw, von Thünen designed a spatial model to *explain* the recurring circular patterns. He did this by constructing an imaginary landscape. Von Thünen called his series of books *Der Isolierte Staat* (The Isolated State), because he wanted to isolate his imaginary landscape from outside influences, such as regional or national economies. The isolated state consisted of one **market city** and the agricultural region surrounding it. To ensure that crop yields would be similar throughout the region, von Thünen assumed a **uniform plain,** which eliminated river valleys, hills, and mountains, where crop yields might vary. Furthermore, von Thünen assumed that the farmers would transport their own products and would use only horses and wagons. The roads of the imaginary region all led directly from the farms to the market city in a radial pattern. Thus, transportation costs could be related directly to distances from the market (see Figure 4.6).

By controlling the myriad of factors involved in making a profit from farming, von Thünen was able to concentrate on the relationship between three important aspects: (1) the *prices* farmers receive for their products, (2) the *distances* the farms are located from the market city, and (3) the *profit* farmers make from a unit of land. Over time, the prices paid to the farmers for various crops tended to stabilize; thus, the distance at which a particular crop can be grown from the city, for a profit, also becomes fixed. Furthermore, products that perish

market city In an imaginary landscape, used for modeling of land-use patterns, a city that serves as the single market for all goods produced.

uniform plain An imaginary plain that has no variation in the physical or human environments, used for modeling spatial distributions.

most quickly and those that yield high returns would be grown nearest to the market—for example, vegetables and dairy products. Products with low yields and products that do not spoil quickly would be grown at greater distances from the market city. By combining the economic factors with the distance factors, von Thünen was able to explain the concentric zonation of crop production.

Von Thünen used his spatial model to describe five zones surrounding the market city. The belt nearest the city (zone one) was an area of intensive agriculture, where the products were mostly perishable; these included fresh vegetables and milk. The second zone outward from the city was not farmed, but left forested. Wood was an essential product at the time, both for construction and for fuel. The wood products were heavy and not easily

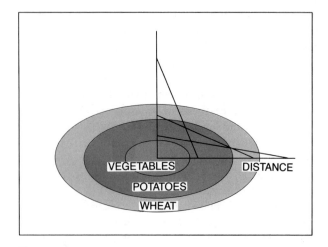

FIGURE 4.6

Von Thünen's Landscape. Von Thünen hypothesized that concentric zones of land use would exist around a market town in his idealized landscape.

central place model A model designed in the 1930s by Walter Christaller to describe the location of market cities using marketing principles and distance.

unbounded plain An imaginary plain not constrained by boundaries, used to model the location of market cities.

uniform distribution A spatial distribution where all the items are equally spaced or have equal distances between them. A completely uniform distribution forms hexagonal patterns among the items.

---

transported, so trees were grown as close to the market as possible. The next belt outward after the forest was a region of extensive agriculture and less perishable vegetables. The products from this zone three, then, included such things as corn, wheat, and potatoes. The fourth zone was an area of grassland, where the major products were grazing animals for meat, or less perishable dairy products, such as cheese. The outermost region, zone five, was a area of wasteland, where no crops were grown.

Although the "Thünen rings" are theoretical, they have been found to exist in many areas of the world, including the pre–industrial revolution types of economies as well as the more developed nations. But modern devices, such as refrigeration and improved modes of transportation, obviously upset the concentric zone theory. Von Thünen also realized that the perfectly concentric zones did not represent reality. This led von Thünen to design a second model (see Figure 4.7). He introduced other modes of transportation, such as a railroad and a canal, then he added another market city and some diversity in the uniform plain. The second model recreates an actual landscape much better than the first one.

About 100 years after von Thünen, another spatial model was developed to help explain patterns found on the landscape. Walter Christaller's 1933 **central place model** was designed along the lines of von Thünen's. Christaller too began with a series of assumptions, one of which was that of the **unbounded plain.** The unbounded plain contained a **uniform distribution** of population and a *uniform transportation network*. Christaller was concerned with the consumption of goods and services, and he tried to explain where settlements (central places) would occur based on consumption trends and distances.

Christaller assumed that people would purchase both goods and services from the closest source, seeking to minimize the distance they must travel. The uniform distribution of population and uniform transportation network assumptions assured a uniform demand and uniform cost of

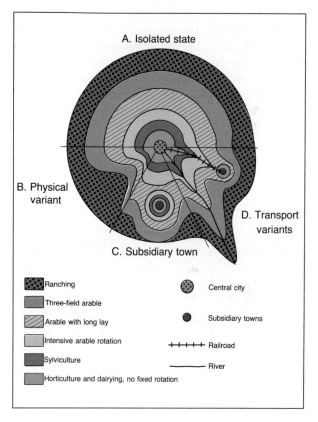

FIGURE 4.7

Distorted von Thünen Landscape. The distortions to the concentric zonation of the idealized landscape make it appear more similar to a real landscape. The isolated state (A) would produce concentric zones, but a physical variant (B) such as soil or climate differences would cause distortions. Also, if a subsidiary town (C) were added to the model, it would create its own concentric zones and distort those of the first market town. Moving goods on a river or canal costs about one-tenth what it would to move the same amount of goods on land, so adding waterways (D) also would distort the original hypothesized zones.

Concepts and Tools

traveling. The central places would draw upon their surrounding areas (hinterlands) for customers. The landscape that would result from Christaller's assumptions would contain a uniform distribution of central places that would be small and spaced equidistantly from one another. These small central places are called "first order" towns and would be arranged in a hexagonal pattern (see Figure 4.8).

According to Christaller, a minimum level of demand for a particular good or service was necessary for that good or service to exist in all the orders of towns. The minimum level is called the **threshold** and can be expressed in terms of population—that is, it takes a certain number of people to create a sufficient demand in order for a shop to carry certain items. For instance, all people need food daily, but a hoe handle would be needed only occasionally. The size of a central place, in terms of the number of goods or services offered, would depend on the threshold population the place was to serve. The **range** of a particular good or service is the distance people are willing to

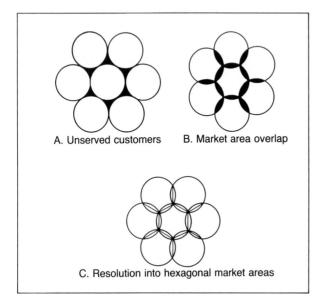

FIGURE 4.8
Christaller's Market Area Development. Areas of market voids and market overlaps are resolved by using hexagonal market areas.

---

threshold The minimum price and quality of goods needed to bring a market into existence and keep it functioning. Usually measured in terms of potential buyers (people) in the area surrounding the market location.

range The average distance that people are willing to travel in order to purchase a particular type of good or service.

---

travel in order to purchase it. The threshold (number of people) and range (distance) are the two critical factors that determine the size, spacing, and number of central places.

Christaller's hypothetical landscape had a uniform distribution of first-order central places offering a limited number of goods and services (see Figure 4.9). Then a hierarchy of fewer but larger central places would develop, where each larger place carried a greater variety of goods and services than the smaller places. The second-order central places would carry all the goods and services available in the first-order places, plus other items purchased less frequently. Consumers would be willing to travel farther to purchase such items, and so the range would be extended and as a result the threshold would be larger for items purchased less frequently. Christaller identified a total of seven orders of settlement in his study area of southern Germany.

Although Christaller's model may be too rigid and too narrow to be of help today, such phenomena as retail outlets in a modern urban center can be studied using Christaller's ideas. Small "mom and pop" or convenience stores exist in much larger numbers and are more closely spaced than the large supermarkets. The small markets carry fewer goods, and specialize in the more frequently purchased items, such as bread, milk, and gasoline. These first-order goods can be bought at the larger stores, and usually at a lower price than at the small markets, but the small markets rely on the short "range" of the good to stay in business. Modern shopping centers are designed to offer a large number of goods through many specialty shops or a few large department stores. They count

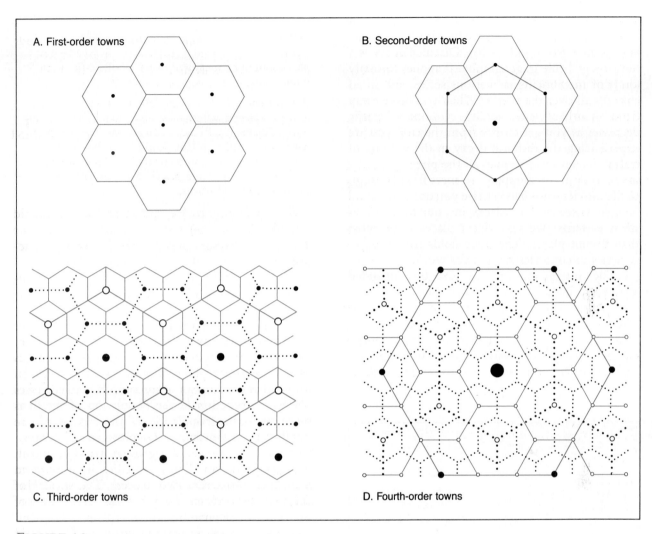

FIGURE 4.9
Christaller's Hexagonal Market Areas. The trade areas for first-order towns (A), second-order towns (B), third-order towns (C), and fourth-order towns (D) are progressively larger.

on both a sizable threshold and a considerable range, as well as multiple-purpose shopping trips.

■ Spatial Interaction Model

The concept of spatial interaction discussed above now can be considered in terms of an explanatory model. Instead of the graphic models shown for von Thünen's and Christaller's ideas, a mathematical model will be used. The first expression to consider is:

$$I = f(1/D)$$

The above expression states that interaction (I) is equal to some function (f) of the inverse of distance (D)—or that as distance increases, the amount of movement (interaction) decreases. This is what geographers call **distance decay,** meaning that interaction "decays" (is reduced) with increasing distance (see Figure 4.10). Distance decay is a fundamental rule in geography. It applies to many kinds of movements, at various scales, and in all cultures all over the world. When you drive away from an urban center and the volume of traffic decreases as you get farther from the city, you are experiencing the distance decay in the volume of traffic. If you were to map all the places you visit in one month for shopping, to meet friends, to go to the movies, or whatever else you do, you would be able to see distance decay in your own movement patterns. We visit closer places more often than distant places. The same holds true in other societies in countries around the world.

The interaction model can be expanded slightly to include other factors, such as population. More traffic is generated, more telephone calls are made, more people migrate, and more of most all other types of interactions occur between the two large cities, such as New York and Chicago, than between two smaller cities, such as Toledo and Tulsa. The idea can be expressed as follows:

$$I = \frac{k(P_1 \times P_2)}{D^\alpha}$$

The formula is similar to Isaac Newton's expression for calculating the force of gravity between two bodies in space. In this case, population (P) is used as an expression of mass, so the formula means that the amount of interaction between two places is a function of the population of the first place times the population of the second place, divided by the distance (D) between the places. Geographers call this the **gravity model,** and it has been used many times to predict the amount of movement between two places. The k in the formula is a constant used to scale the prediction up or down, depending on the type of movement being considered. The exponent on the D permits the distance factor to be manipulated—for example, if the actual mileage between two cities is 100, and we wanted to predict the volume of traffic between the two places but the major highway bridge is washed out by a flood, the *friction of distance* would be greater than 100. Alpha could be set at 2, and the resulting distance factor would be 10,000 (100 squared). This would decrease the predicted interaction between the two places, because the larger the number in the denominator, the smaller the result when the calculation is finalized. This can be seen as follows:

Population of place one (P_1) = 10,000

Population of place two (P_2) = 5,000

---

distance decay The normal decrease in human activities with increased distance away from a central place. Could be measured in terms of traffic, telephone calls, migrants, and so forth.

gravity model Social models based on Newton's gravity concept, where mass and distance are the important elements. Bodies of large mass (large cities) have more attraction (for people) than bodies of less mass (small towns), so there are more people who visit and move to large places than people who visit or move to small places.

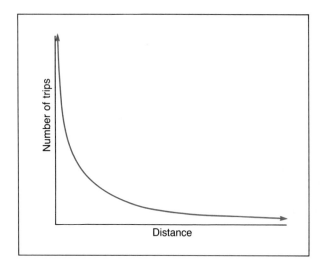

FIGURE 4.10

Distance-Decay Curve. In comparing the amount of spatial interaction (number of trips to a city) with distance from the city, many more trips are made over short distances than over long distances—that is, the volume of traffic decreases (decays) with increasing distance.

Distance between place one (P_1) and place two (P_2) = 100 miles

Predicted Interaction = $\frac{10{,}000 \times 5{,}000}{100}$ = 500,000

Interaction with bridge out = $\frac{10{,}000 \times 5{,}000}{10{,}000}$ = 5,000

As with the distance decay concept, the gravity model can be used at various scales and for various activities and different cultures all over the world. People everywhere usually behave in predictable ways. The gravity model concept may be one of the most important principles in geography because of this universal applicability. The basic idea that nearly all movement decreases with distance underlies both the distance decay model and the gravity model concepts.

■ Maps and Mapping

This book uses many maps that portray various large and small regions around the world. It would be interesting if we could fly over the places and look at them firsthand, or could look at aerial photos, but we must use the next best thing—maps. The purpose of maps is to show the location of things on the surface of the earth at a scale we can see and understand. Numerous types of maps have been designed, but the two most important types for our purposes here are area maps and thematic maps.

Area maps are used to portray the earth and certain parts of the earth (see Figure 4.11). They can be quite complex because the information shown on them can be extremely detailed. Usually, the more information an area map has, the better, but there is a point when too much information renders the map illegible. In a road atlas of the United States, the maps contain not only roads but also towns, rivers, lakes, and mountains, as well as other things. Each of these maps is a composite of numerous spatial distributions—the roads make up one distribution, the towns another, and so forth. An area map, then, indicates the areal relationship among numerous sets of spatial distributions. Locations on area maps usually are designated with longitude and latitude (see Focus Box).

The second type of map is the *thematic map*. Thematic maps tend to be less confusing than area maps because only one spatial distribution is shown on each. The four basic types of thematic maps are the statistical map, the dot map, the choropleth map, and the isoline map. Although the maps look much different from each other, they all portray the same information. We will use a rural settlement in a "congressional township" as an example. Congressional townships were established in the United States in 1789 to help simplify and clarify land ownership among the homesteaders. A township is a square unit measuring 6 miles on each side and consisting of 36-square-mile sections of land. Each square-mile section of land within the township consists of 640 acres. When rural settlement occurred each homesteader was usually allotted 160 acres, so each square mile usually contained four farmsteads.

The first type of thematic map is called a *statistical map* because it contains data or "statistics" (Figure 4.12). The number of farmsteads in each

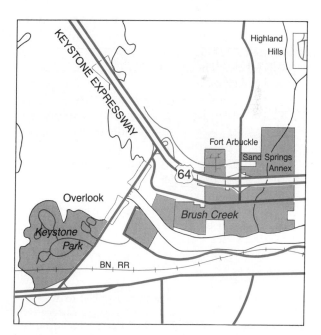

FIGURE 4.11

An Area Map. A section of a map depicting an area near Tulsa, Oklahoma, illustrates some features that might be found on an area map.

Focus Box

Latitude and Longitude

Latitude lines and *longitude lines* are imaginary lines found on all globes and most maps. The lines were designed as an aid to finding locations on the surface of the earth. Latitude lines, or "parallels," run east and west on the globe and are used to measure distances north and south of the equator. Thus, the equator is 0 degrees latitude. The North Pole is located at 90 degrees north latitude, and the South Pole is located at 90 degrees south latitude. The equator, then, divides the globe into a northern hemisphere and a southern hemisphere. Latitude lines are all parallel to each other, so the circles they make on the globe decrease in size with distance away from equator. Although the poles are located at 90 degrees latitude, only a point, not a line, indicates the location.

Longitude lines, or "meridians," run north and south, from pole to pole and are used to measure distances east and west of the "prime meridian." The prime meridian was selected arbitrarily as a starting line. It is located just east of London, England, at the Royal Naval Observatory in Greenwich. The line divides the globe into eastern and western hemispheres. Longitude lines are not parallel; they get closer together toward the poles. Longitude lines also differ from latitude lines in that their maximum value is 180 degrees instead of 90. The "international date line" is located at 180 degrees (east or west).

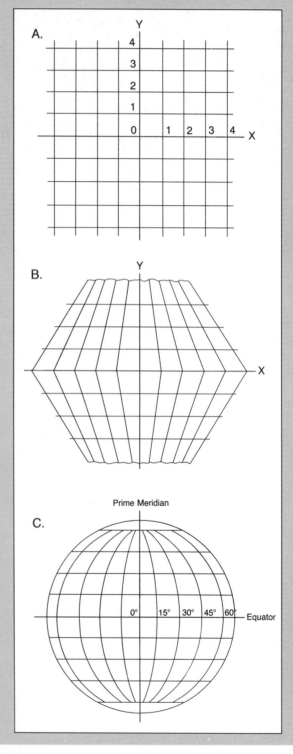

Figure 4.A

Cartesian Coordinate System Applied to a Sphere. A flat surface (A) with a grid pattern on it can be wrapped around a sphere, and by pulling the ends together at the top and bottom (B) the two-dimensional surface can be applied to the three-dimensional surface (C) of the globe. The latitude and longitude system is an adaption of the Cartesian system.

Latitude and longitude are given in degrees because the measurement is of angles. The angles are those formed from the center of the earth (see Figure 4.A). When the line from the center is rotated, it inscribes a line on the surface of the spheroid (the earth). Any number of these lines can be used, but those marked for reference on a globe are usually about 10 to 15 degrees apart. A particular location on the globe can be designated as the place where latitude and longitude lines intersect. A location, then, must contain both the latitude and longitude measurements and the hemisphere involved. By convention, latitude is given first. For example, Tulsa, Oklahoma, is located at 36 degrees 8 minutes north latitude, and 95 degrees 58 minutes west longitude. Usually, such a designation is given as "36-08 N, 95-58 W" because the degrees and minutes are understood. For more precise locational information, the minutes are sometimes divided into seconds, but such precision is not necessary on world maps.

3	5	7	6	5	2
4	8	8	8	6	5
5	10	9	6	4	2
4	6	7	5	5	2
4	5	5	4	3	1
3	3	4	4	1	1

FIGURE 4.12

Schematic of a Statistical Map. The numerical values are the amounts for the items being mapped, and the small unit areas could be countries or some arbitrary grid.

SOURCE: Harold H. McCarty and James B. Lindberg, *A Preface to Economic Geography* (Englewood Cliffs, N.J.: Prentice-Hall, 1966), p. 31. Adapted by permission of Prentice-Hall, Inc.

square mile of the township is placed in the middle of the appropriate section on the map. The result is a spatial distribution of the frequency with which farmsteads occur. This distribution can be compared with that in other townships, or it can be compared with other distributions in the same township. Population density would be one possible comparison, but we would have to have another statistical map showing that density in order to make the comparison.

Another type of thematic map is the *dot map* (Figure 4.13). Dot maps are fairly easy to make, and the visual impression is better than that of a statistical map. The dots can be placed on the map at the exact locations of the farmsteads, or if the exact locations are not known the dots can be placed randomly within each square mile. In the latter case, the exact location information is lost, but the overall distribution shown on the map is retained. Dot maps are effective in showing the density of a distribution. Often each dot is used to indicate more than one item—for example, each dot in a distribution might represent 1,000 bushels of wheat harvested or 500 head of cattle. It should be remembered, however, that every dot on any one dot map must represent the same type of object. In other words, wheat dots cannot appear

Concepts and Tools

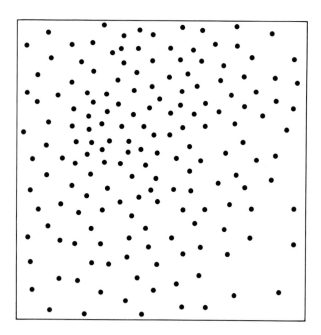

FIGURE 4.13

Schematic of a Dot Map. The dots represent each occurrence of the items being mapped, or they could stand for a group of items—for example, each dot might represent 1,000 head of cattle, or 100,000 bushels of grain.

SOURCE: Harold H. McCarty and James B. Lindberg, *A Preface to Economic Geography* (Englewood Cliffs, N.J.: Prentice-Hall, 1966), p. 33. Adapted by permission of Prentice-Hall, Inc.

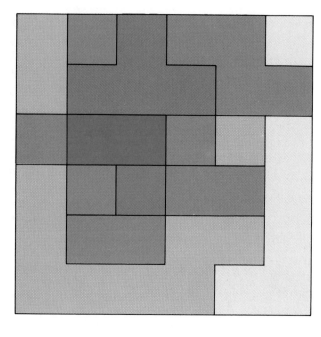

▢ 1–2 ▨ 3–4 ▨ 5–6 ▨ 7–8 ▣ 9–10

FIGURE 4.14

Schematic of a Choropleth Map. Shades or colors are used on choropleth maps to represent categories (intervals) of the numerical values.

SOURCE: Harold H. McCarty and James B. Lindberg, *A Preface to Economic Geography* (Englewood Cliffs, N.J.: Prentice-Hall, 1966), p. 34. Adapted by permission of Prentice-Hall, Inc.

on the same map with cattle dots, unless the difference is indicated separately by color, shape, or some other method.

A special type of thematic map, designed to increase the visual impact of a distribution, is the *choropleth map* (Figure 4.14). The word "choropleth" comes from the two Greek words *choros* and *pleth,* meaning "place" and "fullness" respectively, and each area (place) on a choropleth map is shaded completely ("with fullness"). The original information on the number of farmsteads per square mile can be grouped into categories, and the categories (called intervals) are assigned shades, meaning that each shade on the resulting map indicates a range of quantities, rather than exact amounts. For example, all the sections of land that contain one or two farmsteads might be one color, those with three or four another color, and so

forth. By convention, geographers usually use light shades for the smaller amounts or values, and dark shades for the larger amounts or values, so increasing densities are portrayed with increasingly darker colors.

In choropleth and statistical mapping, the internal boundaries of the mapped area are retained, but they are removed for dot maps and *isoline maps*. Isoline maps (Figure 4.15) are constructed by connecting points of equal value. The term "isos" is a Greek word that means "equal." In drawing isolines, interpolations usually are made in order to place the lines between known values. The resulting lines are contours that indicate the shape of the distribution. Probably the most common contour maps are those that show elevations above sea level. These are called *topographic maps*. Isolines, however, are not restricted to topo-

map scale The relationship between distances on a map and the actual distances they represent on the surface of the earth.

projections Methods of changing the curved surface of the earth so it can be represented on the flat surface of a map.

graphic maps. They can be used to portray numerous things, and our farmstead data are but one example.

Interpretation of the various types of maps is important in geography. For example, when the lines are close together on an isoline map, it means that a dramatic change in the magnitude of the observations occurs over a short distance. When the lines are farther apart, the change over distance is less severe. Isoline maps also may contain shaded zones that can be thought of as regions because they contain an element that is somewhat homogeneously distributed throughout the zone. In our farmstead example, the zones contain a similar number of farmsteads per square mile.

■ Map Scales and Projections

The relationship between the size of a map and the size of the actual area of the earth that the map portrays is known as **map scale** (Figure 4.16). This relationship should be given on every map, either in the legend or in some conspicuous place, using either a bar scale with actual distances, or a ratio. The ratio is a *representative fraction* that indicates the relationship between distances on the map and distances on the earth. For example, the fraction 1/24,000 means that each unit of measurement (inches or centimeters) on the map, represents 24,000 inches or centimeters on the earth. The fraction is usually written as 1 : 24,000. Maps with this ratio would be called *large-scale maps* because the fraction is large; such maps are used to portray *small* areas. *Small-scale* maps, on the other hand, have small representative fractions, such as 1 : 75,000,000; these maps are used to portray *large* areas. One inch on a large-scale map may represent 2,000 feet, but one inch on a small-scale map could represent 2,000 miles.

The appearance of any map of a region or a mapped spatial distribution will vary according to how closely the maps are examined. Thus, understanding the use of map scale is crucial to understanding what the maps portray and to comparing one map with another. Therefore the first thing to look for on a map is its scale, but the second thing to look for is the map's orientation. By convention, most maps are oriented so that North is at the top of the page, but this is not always true. Orientation can be determined either by the *direction arrow* or by latitude and longitude lines.

Another aspect important in the understanding of maps is **projection.** Besides globes and some other models of the earth, most maps are flat—that is, they are two-dimensional. The earth, however, is spheroid, or three-dimensional. When an area from the surface of a three-dimensional object is placed on a two-dimensional surface, distortion

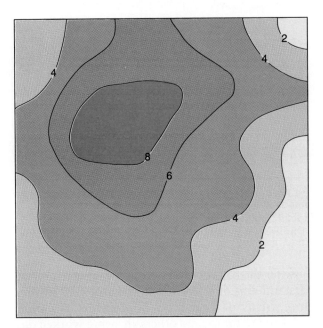

FIGURE 4.15
Schematic of an Isoline Map. The lines are drawn on the map by connecting points that have equal numerical values. The lines become boundaries between areas that are similar.
SOURCE: Harold H. McCarty and James B. Lindberg, *A Preface to Economic Geography* (Englewood Cliffs, N.J.: Prentice-Hall, 1966), p. 34. Adapted by permission of Prentice-Hall, Inc.

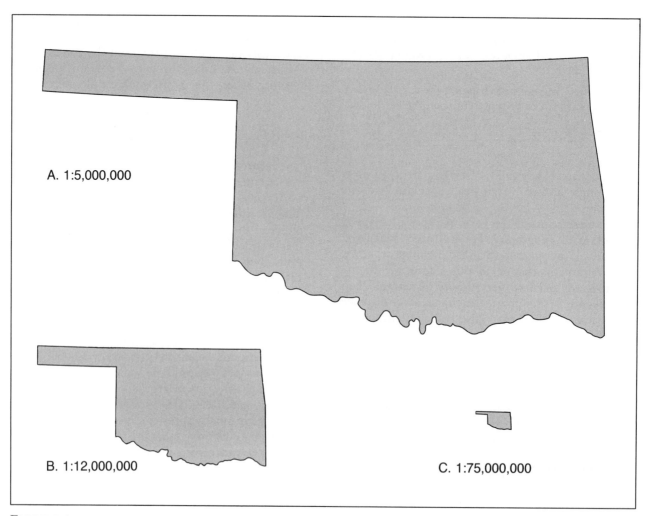

FIGURE 4.16

Map Scale. The size of the area mapped increases with increasing scale values, so that large-scale maps portray smaller areas of the earth than small-scale maps. Here, Oklahoma is shown at scales of (A) 1 : 75,000,000, or one inch on the map represents about 1,200 miles on the earth; (B) 1 : 12,000,000, or one inch represents about 190 miles; and (C) 1 : 4,000,000, or one inch is the equivalent of 64 miles.

occurs. The problem is similar to trying to flatten a basketball on a tabletop so that all areas of the ball are touching the table. Cartographers have developed many methods for minimizing the distortion created by the projection problem. The term "projection" comes from the idea of using light to project the earth's surface onto a flat map surface. If the basketball were a clear plastic ball with lines drawn on it, we could shine a light through it and project the lines onto the table. But such problems as how far the light should be from the ball, and how far the ball should be from the table to get the least amount of distortion in the projected image, arise. Many of these types of problems have been worked out for us, and others are being worked on.

Many types of projections could be made, but we will consider only three (see Figure 4.17). A *cylindrical projection* is one where the surface of the sphere (the earth) is projected onto a cylinder. Because the light source is considered to be at the center of the earth, it projects outward in all directions. The problem with the cylindrical pro-

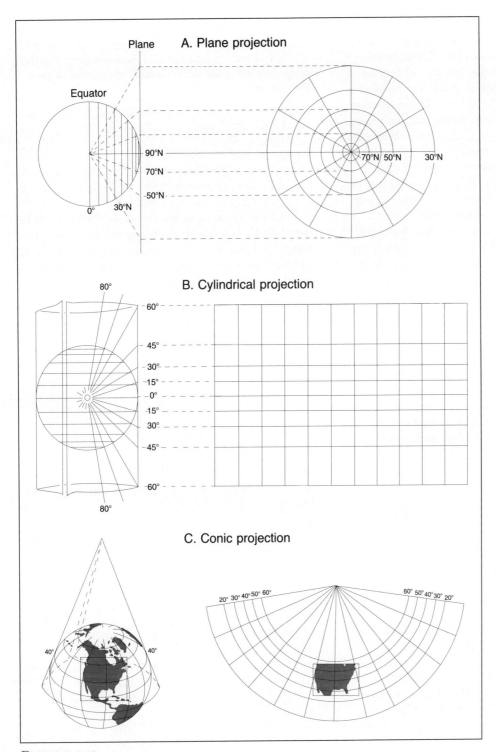

FIGURE 4.17

Map Projections. The three major types of map projections are (A) the plane projection, derived by placing a plane tangent to the globe; (B) the cylindrical projection, derived by wrapping the globe in a cylinder; and (C) the conic projection, made by fitting a cone over the surface of the globe.

jection is that distortion increases away from where the cylinder touches the sphere. Many world maps are made with a cylindrical projection, which means that the polar regions are greatly distorted. A *plane projection* is made by placing a flat surface or plane onto the sphere. The distortion associated with the plane projection also increases with distance from the point of contact between the two surfaces. Polar regions are sometimes shown with a plane projection, using either the North Pole or the South Pole as the point of tangency. The third and most common type of projection is the *conic*, from the word "cone". A cone is used to make the transition from the curved surface to a flat one. Usually numerous cones are used, making a *polyconic projection*. The word "poly" means many, so "many cones" are placed on the spheroid, and projections are made onto each, which holds distortion to a minimum. Most maps today are made by using polyconic projections.

Maps can convey a lot of information, but in order to use the information accurately the scale, orientation, and projection of each map should be considered. Remember also that a map contains either one spatial distribution (a thematic map) or numerous spatial distributions (an area map).

■ Computer Maps

One major advantage of computers is their ability to manipulate information (data) at extremely high speeds, permitting rapid summary and analysis. The idea of programming a computer to arrange data according to spatial coordinates came early in the era of computer usage. Today, computer maps can be produced with impressive speed and accuracy. Maps that could take weeks to make by hand can now be ready for use in a few minutes.

The process is not quite as simple as it may sound. The information to be mapped must be entered into the computer, either through tapes, disks, punch cards, or directly from a terminal. With large data sets, direct terminal input can take a long time. Most geography departments now have a collection of computer programs for producing various map types and covering numerous regions of the world. Once the data are in the

remote sensing   Gathering information by means of remotely located instruments. Usually refers to gathering data from the earth's surface by means of satellite photography.

computer and the appropriate map is selected, the computer cartographer must select the best orientation, scale, and projection to be used, just as when making a map by hand.

All the types of maps mentioned above can be made with computer cartography, but the computer is used to its best advantage in making thematic maps. Recently, printers have been developed that can produce maps in color. This adds another dimension to map-making, but care must be taken in selecting the colors because they are bright and usually do not blend easily. Although computer cartography has not eliminated the need for hand cartography, computers have greatly reduced the time needed to produce certain types of maps.

■ Remote Sensing

Computer technology, combined with advances in photography, has given geographers another tool for studying the earth. In **remote sensing,** information is obtained through a device that is not in direct contact with the object being studied. An example of remote sensing is when a photograph is taken or an image is produced by means of an electronic device. Interpreting information from photographs taken from the air is a science developed by military intelligence. The value of observations on troop deployments and movements made from balloons was greatly enhanced when the camera was invented. It meant that the information could be taken back to military headquarters and studied closely.

Balloons were used as military observational platforms as early as the eighteenth century, but practical use through photography was not possible until Talbot's camera improvements in 1835. The first air photograph was taken from a balloon in France in 1858, and the military use of air photography was greatly expanded during the

Civil War in the United States (1861–65). Many improvements in cameras were made during the later part of the nineteenth century, so when airplanes were invented aerial photography became possible. During both World War I (1914–19) and World War II (1941–45), air photo interpretation was improved dramatically. By the 1950s, photographs taken from high-flying aircraft, such as the U2, were capable of showing objects on the ground as small as golf balls.

Air photographs have provided vast amounts of information about the earth. The discipline of *photogrammetry* is the science of obtaining accurate measurements from air photographs. When the

---

FOCUS BOX

The Metric System

The Metric Conversion Act became law in the United States in 1975. The purpose of the law was to convert all units of measurement to the metric system from the American modification of the British Imperial system. Most countries of the world, including Great Britain, have accepted the use of metric measurement. The units of pounds, bushels, and inches are difficult to use because they are not based on the unit ten, but they have become traditional in the United States, a part of our culture, and we cling to them. Along with Burma, the United States is the only country in the world to maintain attachment to the British Imperial system. In spite of numerous attempts to use the metric system, such as on automobiles and in the teaching of metrics in some schools, Americans have mostly ignored the Metric Conversion Act.

Converting from inches to feet requires dividing by 12, and converting from ounces to pounds requires dividing by 16, but in the metric system all conversions are made by dividing by 10, so conversions can be made by merely moving the decimal point. For example, how many feet are there in 57 inches, or how many pounds in 39 ounces? By comparison, 500 centimeters is 5 meters, or .005 kilometers. All the other measurements in the metric system, such as those for volume, area, and velocity, are similarly based on the decimal system.

As early as 1790, Thomas Jefferson proposed that the United States convert to the metric system, but at that time Great Britain was America's major trade partner. Jefferson's proposal was rejected on the grounds that severe economic problems would occur if we abandoned the British system. John Quincy Adams's proposal in 1821 was argued down for similar reasons. The U.S. Constitution charges Congress with establishing the nation's weights and measures, so in 1866 Congress declared the metric system valid for "contracts, dealings, or court proceedings." Then, in 1893, Congress officially defined the foot, pound, and inch in metric equivalents. By the 1960s, most of the nations of the world had converted to metric. Austrians gave up the *becher* and the *pfiff* as units of volume; the Soviets no longer use the *zolotnik* and the *funt* for weight.

After the 1975 Metric Conversion Act, Americans were told to "Think metric." But it will take more than that to get most Americans to convert to the system. Conversion is similar to learning a new language, except it affects everything from the size of a window, to how much milk is in a carton, to how far it is from here to there. These things affect our daily lives, and to force Americans to change their thinking on such daily matters is in fact to force them to change their culture. It will not be done quickly, and it may take decades or even centuries to change the United States to the metric system. Both the British Imperial system and the metric system are used in this book.

focal length of the camera and the height of the camera above the ground are known, accurate scale measurements can be made, so that accurate maps can be made from the air photographs. When two pictures are taken of the same area but from different angles, a three-dimensional effect can be produced by viewing the two photographs through a stereoscope (an optical instrument). This allows the photogrammetrist to map vertical distances as well as horizontal distances, and so contour maps can be made.

Since the development of satellites, "remote sensing" has taken on a new meaning. Today, the phrase almost always refers to the rapidly developing science of analyzing information transmitted back to earth from these orbiting platforms. The Landsat series of satellites, launched between 1972 and 1983, have provided huge amounts of information. These satellites sweep over the earth at an altitude of about 559 miles (900 km), picking up light reflected from the various surfaces on the earth. One image covers an area of about 13,182 square miles (34,140 km^2). The image is divided into 7,581,600 **pixels**, which are small units about 62 × 86 yards (57 × 79 m). Each pixel image is recorded as an average light-reflecting value (number) in the visible and near infrared range. The numbers are transmitted to a receiving station on earth, where they are converted back to color. It is easy to understand why computers are important to the analysis of Landsat data—because each image contains millions of pixels, and thousands of images have been made.

pixels Originally, the dots that make up a picture on a television screen. Used in remote sensing to describe the small unit areas (62 × 85 yards) that make up the image recorded by satellite cameras.

Earthsat, a newer version of Landsat, provides pixel images that are about four times as sharp as those from Landsat. Remote-sensing technology continues to improve, and its possibilities seem to be limited only by the imagination. One can even count people by using satellite images, and the diffusion of diseases through fields of crops or through forests of trees can be traced. Infrared images show things that give off heat that are usually invisible to the eye, so buildings in need of insulation, or hidden army tanks, or rocket ignitions can be identified. Satellite pictures can monitor a building storm or identify fault lines. Both human and physical geographers have found many uses for satellite images.

The modern geographer uses other concepts, theories, and tools, but those described above are some of the basics. These ideas will be used throughout the remainder of the book, and this chapter should be referred to from time to time. A major theme of the ideas presented here is that geography is essentially a search for order, but that sometimes the order is obvious and at other times it must be discovered.

■ Key Words

agglomeration
central place model
distance decay
gravity model
logistic curve
map scale
market city
pixels
projections
range
remote sensing
spatial diffusion
spatial distributions
spatial interaction
spatial models
threshold
unbounded plain
uniform distribution
uniform plain

■ References and Readings

Abler, Ronald, et al. *Spatial Organization: The Geographer's View of the World.* Englewood Cliffs, N.J.: Prentice-Hall, 1971.

Muehrcke, Philip C. *Map Use: Reading—Analysis—Interpretation.* Madison, Wisc.: JP Publications, 1986.

Norris, Robert E., et al. *Geography: An Introductory Perspective.* Columbus, Ohio: Merrill, 1982.

Robinson, Arthur H., et al. *Elements of Cartography.* New York: Wiley, 1984.

Chapter 5

THE UNITED STATES AND CANADA

Introduction

The United States and Canada are two of the largest and wealthiest countries in the world. They are located adjacent to each other on the continent of North America (see Figure 5.1). Their combined area is slightly larger than 8 million square miles (21 million km^2), occupied unevenly by about 267 million people. The cultural background of these countries is varied, but because of considerable British influence the region sometimes is referred to as "Anglo-America." The wealth of both countries is tied to their physical environments, as both have vast amounts of natural resources.

Physical Geography of the United States and Canada

Landforms

The topography of the United States and Canada is noted for its diversity. Geomorphologists have divided Anglo-America into numerous regions, called **physiographic provinces** (see Figure 5.2). These regions contain internal similarities in the appearance of the landscape and the underlying rock structures. A complete inventory of physiographic provinces might include as many as 30 or 40 regions, but only 9 will be discussed here.

The Appalachian Highlands. The mountainous part of the eastern United States and Canada known as The Appalachian Highlands begins near Birmingham, Alabama, and runs northeastward through Newfoundland. The Appalachians are old mountains that have been eroded and are now only remnants of what were once high mountain chains. This physiographic province includes a number of subregions. The *Appalachian Plateau* is an uplifted area that was once flat but has been dissected so that it now appears as a mountain area. The *valley and ridge section* contains linear folded

---

physiographic province A region of the earth's surface that is somewhat uniform in terms of the underlying rock and surface configuration.

---

mountains and corresponding valleys, creating a "washboard" landscape. The *Blue Ridge Mountains* are a range of mountains that extends from Georgia to Pennsylvania. They contain the tallest peak in the east, Mount Mitchell (6,684 feet, or 2,037 meters) in North Carolina. The *Piedmont section* slopes south eastward from the foot of the Blue Ridge Mountains to the coastal plain and parallels the mountains along their entire length. The Fall Line is a noted boundary between the Piedmont

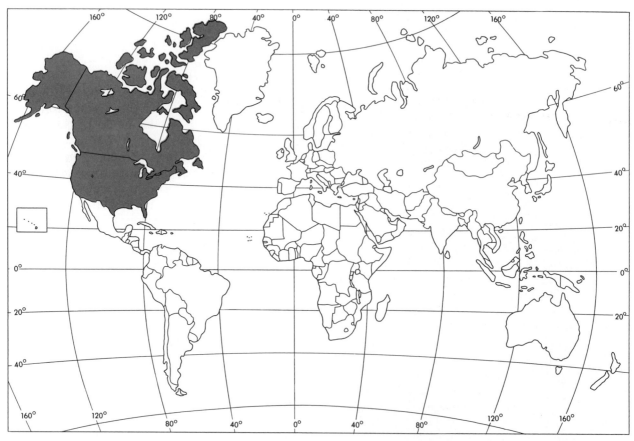

FIGURE 5.1
World Location of the United States and Canada.

> **permafrost** Subsoil that is frozen throughout the year.

and the coastal plain to the east. The *Adirondack Mountains* subregion is a nearly circular, dissected dome about 100 miles in diameter located in northern New York. And the *New England-Canadian* section contains small mountain ranges, such as the Green Mountains of western Vermont, the Hoosac Mountains and the Berkshire Hills of western Massachusetts, the White Mountains in east-central New Hampshire, and the Notre Dame and Chic Choc ranges of Canada. The Aroostook Plain of Maine and New Brunswick is one of the numerous low-lying areas of the region. Isolated mountains in the province are called *"monadnocks,"* after Mount Monadnock in New Hampshire.

The northern part of the Appalachian Highlands province was scoured by the continental glaciers of the Pleistocene period. The resulting landscape reflects the historical geology of the region, as many lakes and other remnants of glacial features exist today. Bunker Hill, noted as a Revolutionary War battlefield, is a pile of glacial debris. The southern limit of the glaciers' advance is marked by a terminal *moraine,* which is now a hummocky ridge running east-west and composed of sand, gravel, boulders, and finely ground rock flour. And the glaciers also are responsible for the very thin and rocky soils of this region.

The Gulf and Atlantic Coastal Plains. The Gulf and Atlantic Coastal Plains in the United States usually are referred to simply as the "Coastal Plains." These plains run from the southern margins of New England southward along the coast of the Atlantic Ocean, and then westward along the coast of the Gulf of Mexico. The Coastal Plains are relatively narrow from New Jersey to Georgia, but they widen out along the Gulf Coast. The plains include the entire lower reaches of the Mississippi Valley, where they are at their widest. They extend outward under the ocean surface, where they become the *Continental Shelf,* which in some places is as wide as 250 miles (400 km). The plains are generally very flat, with only a few *cuestas* breaking the flatness. It is possible to drive from Brownsville, Texas, to Miami, Florida, without rising more than a few feet above sea level. More than half of the entire plain is less than 100 feet (31 m) in elevation.

The shore along the Coastal Plain is slightly more than 3,000 miles (4,827 km) long. Many of the sandy beaches have been set aside as national seashores. Inland from the beaches are numerous swamps, such as the Dismal Swamp in North Carolina and Virginia, and the Everglades region of Florida. Numerous offshore islands that run parallel to the coast have been built up by sediment deposits and wave action. Padre Island, off the Texas coast, is the longest and probably the most famous.

The Arctic Coastal Plains. The Arctic Coastal Plains include the North Slope of Alaska, all the large islands of northern Canada, and the southern shore of Hudson Bay. Underlying this entire region is a vast extent of permanently frozen ground called **permafrost.** On top of the ground is a treeless, tundra landscape with some shrubs and grasses. The land is not as flat as the Gulf Coastal Plain because the extremely cold climate causes a freeze-thaw condition that creates low mounds and "patterned" ground. Very few people inhabit the plains, and much of the region remains unexplored. The Alaskan oil discovery, however, has indicated that resources may be available in other parts of the region.

The Canadian Shield. The Canadian Shield is an uplifted area located in a semicircular zone running from Labrador on the east to the Northwest Territory on the north. Landforms of the shield have been ground, molded, and shaped by the great continental glaciers. The glaciers originated near Hudson Bay and extended outward in all directions to cover the entire shield before moving farther south. These massive ice sheets covered as much as 5 million square miles (13 million km^2) at their greatest extent. As they moved they caused major landscape changes through large-scale erosion and huge glacial deposits. They plucked large boulders out of the ground and moved them many miles. These massive rocks,

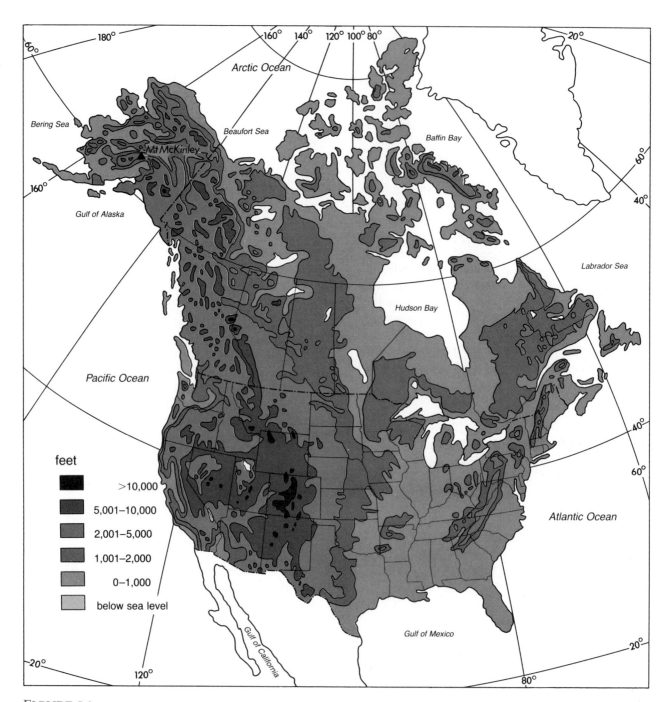

FIGURE 5.2

Land Elevations (A) and Physiographic Regions (B) of the United States and Canada.

called "erratics," lie strewn across the landscape. The glaciers changed the drainage patterns of the rivers and created thousands of lakes, including the Great Lakes.

Today the soils of the Canadian Shield are generally shallow, but in most places they are deep enough to support vast forests that are regularly interrupted by a myriad of lakes. Northern Minnesota, "the land of ten thousand lakes," lies on the southern fringe of the shield. Most of the region is not suited for agriculture. It is difficult even to travel through because of the lakes.

The Interior Plains. The Interior Plains region is the largest of the physiographic provinces. It runs from the Great Lakes to the Rocky Mountains, through the interior part of the United States and on northward through western Canada. The region is often called the Interior Lowlands because of the low average altitude and lack of local relief. The Interior Plains are hillier than the Coastal Plains, but there is very little rough terrain in the entire region. The great midland rivers, such as the Missouri, the Ohio, and the Mississippi, have dissected the region. Most of the northern part of the Interior Plains was glaciated, but the glaciers tended to dump their loads here, leaving thick soil layers and topographic variation.

One notable highland interruption to the Interior Plains is the *Ozark Plateau*. This fairly rugged, dissected upland is located in southern Missouri, eastern Oklahoma, and northwestern Arkansas. The plateau covers about 100,000 square miles (259,000 km^2) and includes the Ouachita, Boston, and Saint Francois mountain ranges.

The Interior Plains region is known as the "Heartland of America." It includes the agricultural cores of both the United States and Canada. The rich economies of both countries are influenced by the resources from this region. Agriculture is balanced by mineral production, manufacturing, and transportation. The lack of topographical features makes east-west travel relatively easy, and numerous highways and railroads crisscross the region. North-south transportation in the United States was historically, and still is, enhanced by the Mississippi River and its tributaries.

The Great Plains. The Great Plains are located along the eastern edge of the Rocky Mountains and were built of sediments eroded from that great mountain system. The entire region slopes eastward from about 5,500 feet (1,676 m) elevation at the base of the mountains to about 2,000 feet (610 m) along the boundary with the Interior Plains. The boundary between the Interior Plains and the Great Plains runs roughly parallel to the 100th meridian. This boundary not only separates the physiographic provinces, it is also the 20" (51 cm) annual precipitation line. This gives the Great Plains a steppe climate that is drier than the Interior Plains. The climate change creates another line that separates the tall prairie grass region on the east

This view of farm land in Iowa shows the lack of local relief and gives an idea of how flat parts of the Interior Plains region are.

Palo Duro Canyon was cut through the Llano Estacado of the Great Plains region of the Texas panhandle by the Prairie Dog Town fork of the Red River.

from the short steppe grass region of the Great Plains. Corresponding to the climate and vegetation changes, the soil types also change near the 100th line of longitude. Because of all the changes that occur in the natural environment, numerous changes also can be seen in the cultural landscape. Some people claim that the American West begins at the 100th meridian.

Parts of the Great Plains, also called the High Plains, contain extensive areas of sand dunes. Some of the dunes are covered with grass, as in western Nebraska, while others form small seas of exposed sand. The Black Hills interrupt the otherwise featureless landscape of western South Dakota, and the Edwards Plateau in central Texas does the same for that area. The Llano Estacado, or Staked Plain, of eastern New Mexico and the Texas panhandle is one of the most featureless regions of the United States. When exploration began by the Spanish, there were so few landmarks on the plains that they were forced to drive stakes in the ground to mark their trails across the prairie. The name of the region came from the Spanish explorers.

The Rocky Mountains. The Rocky Mountain system runs in a belt from northern New Mexico to the Bering Strait off the northwest coast of Alaska. It is a part of the great mountain system that extends from Alaska to the southern tip of South America. The region contains numerous individual mountain ranges but can be divided generally into the Southern, Middle, and Northern Rockies. Many individual mountain peaks exceed 14,000 feet (4,267 m), and the local relief is extreme, varying from bases to summits by as much as one-half of the total elevations. This rugged region contains mostly shallow soils and large areas of bare rock.

The Rocky Mountain region has been noted for its mineral wealth and scenic beauty. Most of the mountains in both the United States and Canada are in national forests, parks, or wilderness areas. The mountains have been a barrier to trans-

This view near Durango is of the Rocky Mountain foothills of southwestern Colorado.

portation. All the passes through the Southern Rockies, for example, are above 9,000 feet (2,743 m). Most of the cities of the region are located along the foot of the mountains, where transportation is easier, soils are better, and water can be obtained from the higher elevations.

The Basin and Range. The Basin and Range province is a broad plain on the south from the Mexican border to eastern Oregon; then it narrows, but continues northward through western Canada, then broadens again to include most of Alaska. The plain is continually interrupted by many short linear mountain ranges, all of which are aligned north-south. The mountain ranges average about 50 miles (80 km) in length and about half that in width. They rise to heights of 2,000 to 3,000 feet (600 to 900 m) above the surrounding plains. Much of the Basin and Range country has *interior drainage*, where the small rivers and intermittent streams drain into lakes and dry sinks, rather than into the oceans. East-to-west transportation routes follow a zigzag pattern in order to go around the isolated mountain ranges.

Two large plateaus rise above the surrounding Basin country. The Colorado Plateau is centered on the "four corners," where Colorado, New Mexico, Arizona, and Utah meet. The famous Grand Canyon has been cut through the Colorado Plateau by the Colorado River. The Columbia Plateau is centered in eastern Washington and Oregon, but it also runs eastward through southern Idaho. The Columbia Plateau also has been dissected by numerous tributaries of the Columbia River, and the most notable dissection is Hells Canyon, which runs along the Oregon-Idaho border. The two plateaus look somewhat similar on the surface, but they are geologically quite different. The Colorado Plateau is an uplifted area, while the Columbia Plateau was built from layer upon layer of lava beds, reaching thicknesses of as much as 3,000 feet (920 m).

The Pacific Mountains and Valleys. Along the Pacific coast, from San Diego to the Aleutian Islands, is a series of mountain ranges interspersed by broad, flat valleys. The mountain ranges along the coast are generally low, rising on the average

The Basin and Range country of southern Arizona is shown in this scene near Prescott.

no more than 3,300 feet (1,000 m). These linear mountains follow the major fault lines, and the region is noted for its frequent earthquake activity. The great valleys also are linear, and they lie between the coast ranges and the higher mountain ranges to the east. These interior lowlands include the Central Valley of California, the Willamette Valley of Oregon, and Puget Sound of Washington and British Columbia. There are no large valleys along the Canadian coast north of Puget Sound, but large valleys reappear on the west coast of Alaska.

The mountains along the eastern border of the province contain the tallest peaks in Anglo-America. Most of the mountain ranges were created by uplifting and tilting of the earth's crust, but individual volcanic peaks also are common. The Sierra Nevada Mountains of California, the Klamath Mountains of Oregon, and the Cascade Mountains of Oregon and Washington are similar in structure. They are known as *fault block mountains* because they were created by the tilting of large blocks of the earth's crust. Because the tilt is toward the west, the highest parts of the mountain ranges are on the east. The eastern front of the mountains is steep, while the western slope is less severe. The eastern summit of the Sierra is about 11,000 to 14,000 feet high (3,350 to 4,270 m).

No interior fault block mountains comparable to those in the United States exist in Canada. The eastern foothills of the Canadian coast ranges butt against the foothills of the Northern Rockies. On the west, the Canadian coast ranges run down to the waters of the Pacific Ocean. This Canadian coast, as well as the southern panhandle of Alaska, is one of the most fractured coastlines in the world—there are thousands of inlets, bays, and islands. The coastal mountain ranges continue from Canada into Alaska, and the Alaska Range contains the tallest peak in Anglo-America. The top of Mount McKinley is 20,320 feet (6,194 m) above sea level.

Climate

Climatic patterns in the United States and Canada are the result of interaction of numerous geographical factors, including topography. The climatic regions of Anglo-America are therefore related to the nine physiographic regions. The most important locational control on climate is latitude, since the amount of annual heat energy from the sun that a place receives is directly related to its latitudinal location. Because Anglo-America stretches across 50 degrees of latitude from Key West, Florida, to Point Barrow, Alaska, this variation in latitude produces variation in the climates. Places in the lower latitudes have long hot summers and short mild winters; in the higher latitudes the summers are short and cool and the winters are long and severe. Places located between the two extremes (mid-latitude) experience both severely cold winters and hot summers.

Another temperature control related to location is the relationship between land and water. Land, being a solid surface, will heat up and cool down faster than water. As a liquid, water tends to absorb the sun's rays and diffuse them internally, so the heat is retained. This means that water bodies tend to stabilize the temperature of the air that lies over it, and in turn moderate the climate of nearby land. The farther away from the oceans, then, the more severe the climate, in both the cold season and the warm season. This is called the *continental influence* on climate. In addition, the larger the land mass, the greater the influence becomes. North America is a very large land mass, so the interior of the continent experiences annual temperature differences much greater than coastal locations at the same latitude. Large bodies of water also influence precipitation levels. Coastal locations almost always receive more rain than places inland (see Figure 5.3). Thus, temperature and precipitation are related to continentality.

Topography also influences climate, causing higher elevations to be cooler than the surrounding lowlands. Heat decreases with elevation at an average rate of 3.6° F for every 1,000 feet (6.5° C per km). This is called the **normal lapse rate.** Besides the influence on temperature, topography may also affect the flow of air, which in turn affects

normal lapse rate The constant decrease in temperature with increased elevation above the surface of the earth. Usually given as 3.6° F for every 1,000 feet.

The Central Valley of California is broad and flat, with a mountain backdrop on both the east and the west. This view of cropland in the valley (note the rice paddies in the background near the mountains) is near Sacramento.

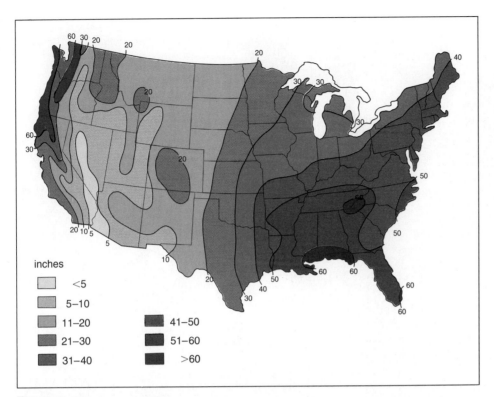

FIGURE 5.3

Patterns of Annual Precipitation in the United States.

the distribution of precipitation. If a mountain chain lies perpendicular to the flow of air, the wind will be forced to rise over it. This rising causes cooling, and in turn may cause condensation of the water vapor that the wind carries. The result is that the windward side of the mountain chain may receive moisture, while the leeward side is drier. Because air movement across North America is from west to east, the western deserts of North America lie on the eastern (or leeward) side of the mountain chains.

Vegetation

The type of natural vegetation found in a region is closely associated with its climate, because plants adapt to long-term conditions of temperature and precipitation. Other factors, such as the type of soil, the amount of drainage, and the seasonality of precipitation, also are important to plant growth, but the primary concerns for any plant type are temperature and moisture. The size of the plants that grow in a region depends on the amount of moisture received. For example, large trees require large amounts of precipitation, smaller trees require less, tall grass less yet, and so on. In very dry areas only small, highly specialized plants can survive. Geographers usually regionalize natural vegetation, then, according to four broad categories of plants: forests, grasslands, scrublands, and tundra.

Much of the natural vegetation in Anglo-America has been removed and replaced with crops or other types of vegetation introduced from other areas. Forests once covered much more land area than they do now, but generally the forest regions correspond to the high precipitation regions. Large needleleaf trees were native to three distinct areas: (1) a narrow band along the west coast, (2) a broad belt running through the southern United States, and (3) another wide region across Canada. Both needleleaf and broadleaf trees

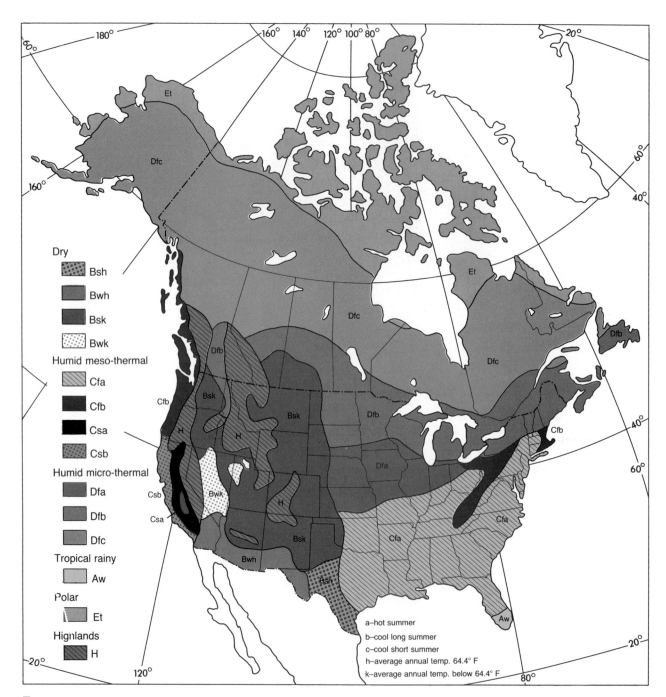

FIGURE 5.4

Köppen Climate Regions of the United States and Canada.

The foothills of the Sierra receive enough precipitation to support a covering of pine trees. Note the volcanic peak in the background.

The desert valley in Arizona contains sparse vegetation that is all xerophytic, but the valley is irrigated so that citrus trees can grow.

grow in the eastern United States, but farther west the forests become mostly broadleaf trees. The broadleaf trees take over in the region of the Ohio and Mississippi river valleys (see Figure 5.5).

Most of the *grasslands* of Anglo-America are found in a broad belt east of the Rocky Mountains. The belt runs from the Texas coast on the south to the middle of the Canadian provinces of Alberta and Saskatchewan on the north. The region corresponds to the location of the Great Plains and Interior Plains physiographic provinces. It is a semiarid region, where precipitation amounts are too low to support forests. The Central Valley of California was once a grassland area, but the grass has been replaced by crops. Another grassland exists along the eastern slopes of the Cascade Mountains in Washington and Oregon.

Where precipitation amounts are insufficient for forests or grasslands there are *scrublands*. The true desert scrub includes the various types of cacti found in the dry, southwestern part of the United States. Other types of scrubland vegetation include the mesquite of the brushland of Texas and the chapparal found along the slopes in southern California.

Tundra vegetation includes bunch grass, moss, and lichen found in the extremely cold areas of northern Canada and Alaska. It is also the predominant vegetation found above tree line in most mountain areas. Tundra vegetation is hardy but delicate—that is, it survives the cold temperatures but is easily disturbed. Because of the very cold and dry conditions under which it grows, it takes a long time to replace itself after it is destroyed.

Soils

Soil is a mixture of finely ground (weathered) rock material and organic matter (decayed vegetation). It develops out of the interaction of such things as the parent rock, the climate, the vegetation, and the topography, and the different soil types are related to those factors. The five major categories of soil types are based on the amount of heat and moisture available for soil development (see Table 2.2). *Lateritic soils* form in regions that receive a surplus of precipitation and that are usually warm all year. Thus, they are found in the southeastern part of the United States. Where winters are long and cold, *podzol soils* predominate, so they are

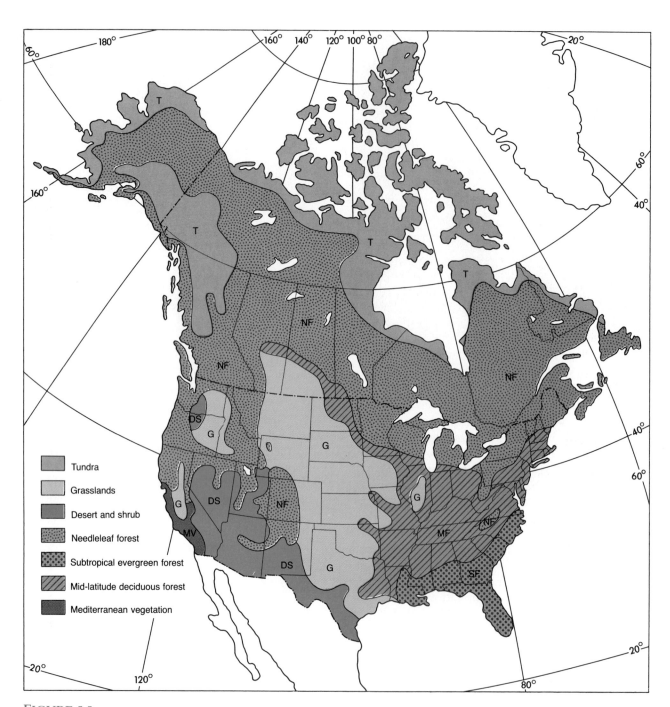

FIGURE 5.5

Natural Vegetation Regions of the United States and Canada.

found in the northern sections of the United States and in most of Canada. Where drainage is poor and the soil remains saturated, the *gley-type soils* form; they are therefore found in low-lying areas anywhere, but are common where precipitation totals are high, as in the southeastern United States. *Calcification* and *salinization* are the main soil-forming processes in the dry regions of the world and in the grasslands. For this reason, soils formed from calcification predominate in the grassland areas of North America. See the map in Figure 5.6 for an overview of the soils found in the United States and Canada.

Human Geography of the United States and Canada

Except for the approximately 800,000 native American Indians, the current inhabitants of the United States and Canada came from all regions of the world. They came to America seeking a life they could not find elsewhere. Today the descendants of the original settlers are a small minority of the diverse populations of both countries. Only about 14 percent of the combined populations are of true Anglo origin. And still new immigrants arrive daily.

Approximately 400,000 people legally enter the two countries each year, although part of that figure includes Canadians who immigrate to the United States. The annual number of illegal entries is difficult to estimate, but some guesses run into the hundreds of thousands. Most of the illegal immigrants are Mexicans coming into the United States, although Mexico does provide a large number of legal entries. After Mexico and Canada, Asia is the major source for legal immigrants to the United States, with India, Hong Kong, Korea, and the Philippines all among the top ten nations that provide immigrants.

With such a large group of diverse people spreading across the land and seeking jobs, climates, social groups, or lifestyles of their preference, clear generalizations concerning the human geography of Anglo-America are difficult. Only the most distinct patterns can be described.

Population

According to the latest estimates, the United States has 241 million people and Canada has about 25.6 million. The birth rates, death rates, and annual rates of natural increase are all similar for the two countries. Of the countries of the world, the United States ranks fourth in population, but its population density of about 68 people per square mile is not especially high by world standards. Canada is even more sparsely populated, with about 6.3 people per square mile.

The age structures of the populations of the two countries also are similar. In both places, about 22 percent of the people are less than 15 years old, and about 11 percent are over the age of 65. Overall, both the United States and Canada should be considered relatively young countries. A life expectancy of 75 years prevails in both countries, but the infant mortality rate in Canada is 8.1, compared with 10.5 for the United States.

A general description of the population distribution in the United States is that it is most dense in the Northeast (see Figure 5.7). It decreases toward the south and the west and is relatively sparse west of the 100th meridian, except for pockets of high density along the west coast. State population densities averaging greater than 800 per square mile in the east decrease to 20 per square mile in the south and midwest, and fewer than 5 in the Great Basin. The most densely populated areas of Canada are along the boundary with the United States and extending only a few miles north of the border. Beyond that narrow strip, Canada is nearly devoid of habitation except for a few miners, trappers, and subsistence herders and fishermen.

In the United States, 74 percent of the people live in urban areas. New York City is one of the world's largest cities, with a metropolitan population of more than 15 million. Chicago has about half that many, followed closely by Los Angeles. According to the 1980 census, there were 40 cities in the United States that had metropolitan areas containing more than one million people, and another 40 cities had populations between 500,000 and one million. In Canada, Toronto is the largest city, followed closely by Montreal. Each has more than 3 million people, and the total population of

(text continues on page 100)

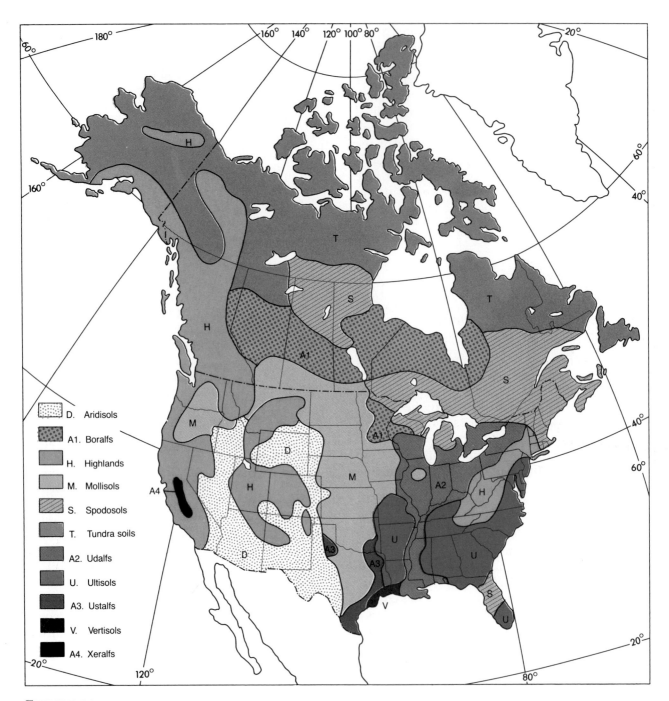

FIGURE 5.6

Soil Regions of the United States and Canada. Numerous soil classifications have been made. The one shown here is based on the Seventh Approximation, or the most recent survey made by the U.S. Department of Agriculture.

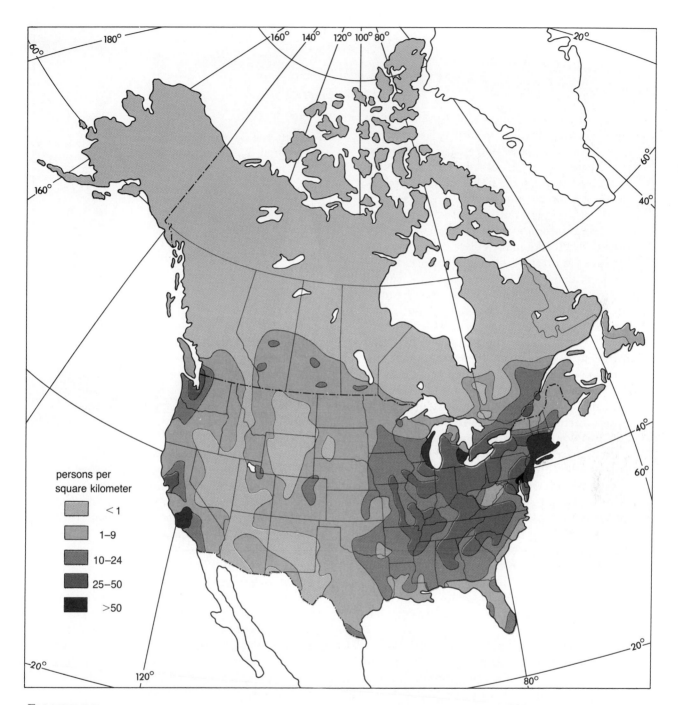

FIGURE 5.7
Population Distribution of the United States and Canada.

FOCUS BOX

Sports in America

Sports are an integral part of American culture. Most children learn to play at least one sport at an early age, and many continue to participate in sports in adulthood. A few adults even earn their living playing professional sports, but most sport enthusiasts are the kind that play touch football in the park on Sunday afternoon, or take part in skins-shirts basketball games, or are colorfully clad for hacking on the golf course. Sports activities provide fun for leisure time and harmless competition in a competitive society. In addition to all the people engaged in sports that require equipment (such as a ball or something to strike a ball), thousands of others take to the streets and roads to walk, jog, or run for pleasure and conditioning. Most Americans are either actively engaged in a sport or are fans or spectators of at least one sport. Sports often provide a common ground for conversation between strangers.

Sports are not only a part of the American culture, they are intrinsically tied to the economies of both the United States and Canada. When one considers the gate receipts for professional sporting events, team salaries, equipment costs, travel costs, betting, and all the other money exchanges related to professional sports, the total for one year in the United States and Canada is many billions of dollars. The numbers of major professional sports teams in the two countries are shown in Table 5.A.

Professional athletes also participate in individual sports—for example, bowling, rodeos, golf, automobile racing, boxing, tennis, and horseracing. Such professional sports serve as entertainment for the public and provide relatively large earnings for the participants. For example, each year the National Association of Stock Car Auto Racing sanctioned 26 automobile races of 400–500 miles

TABLE 5.A

Professional Sports Teams in the United States and Canada

GAME	U.S.	CANADA
Baseball	22	2
Basketball	23	0
Football	28	9
Hockey	16	6
Soccer	14	0
Total	103	17

in length, and the leading drivers earn well over a million dollars each. Professional golfers participate in 63 tournaments each year (36 for men, and 27 for women). The average prize money offered for each tournament is about $100,000 for the men's tournaments and about $50,000 for the women's, and the leading money winners average about $500,000 for the men and about half that for the women. The leading money winners in tennis pocket about $1.5 million each year for both the men's and women's matches. Professional sports salaries and earnings increased dramatically during the 1980s, and an annual income of $1 million is common.

Nonprofessional athletes participate in the same types of sports as the professionals, but the number of participants increases downward in pyramid fashion. In football, for example, the 37 professional teams play an average of 16 games a season for a total of 592 games. In college football there are 200 Division I teams, 500 small college teams, and 176 junior (two-year) college teams that play an average of 10 games a season, or a total of 8,760 games. In *A Geography of American Sport,* John Rooney reports that 15,274 high schools have football teams in the United

A full football stadium means income for the university. This one is Lewis Field at Stillwater, Oklahoma, where the local Oklahoma State University Cowboys are playing the University of Oklahoma Sooners. Photograph by Sheri Helt.

States and Canada. Each team plays about nine games a season, or about 137,466 games a year. Approximately one million high school boys participate in football each year, or about one of every nine in school.

In the United States, more high school students participate in football than in any other sport, but the figures for the other sports also are impressive. About 782,300 high schoolers play basketball each year, and about the same number go out for track and field (these figures include the girls that participate too). Another 400,000 boys play on high school baseball teams, but some of these are the same boys who play basketball and/or football. Besides the major sports, high school athletes participate in badminton, bowling, cross-country, curling, decathlon, fencing, field hockey, golf, gymnastics, ice hockey, lacrosse, riflery, rugby, skiing, soccer, softball, swimming, tennis, volleyball, water polo, and wrestling. Along with the large variety of team and individual sports, many schools also have physical education classes that provide exercise for students and teach students how to play certain sports.

Untold millions of fans attend the sports events in the United States. Fan fever can be just as strong with viewers seated on the wooden bleachers at a six-man football game in western Oklahoma as it is with the people in the 106,721 seats of the Rose Bowl in Pasadena, California. In fact, the word "fan" comes from "fanatic," which means "unreasonably enthusiastic." Fans follow their favorite teams through winning seasons or losing seasons, always hoping the team will provide some vacarious thrills, but settling for enjoyment and something to talk about. Sporting events in America are major topics of conversation, and every newspaper contains a sports section. In addition, there are many magazines devoted to sports, and many that are dedicated to only one sport, so fans can follow sports events even if they cannot attend in person.

The huge volume of materials written on sports includes the mass of records and statistics compiled for each participant and team. Americans seem to savor sports statistics, and records are kept for nearly every event. Major league baseball especially is known for the records kept on all games. It is possible to find out how many times the most obscure player struck out in any particular season (or even game) back to before 1900, or how many pitches were thrown in yesterday's Cubs-Braves game. Batting averages are reported daily during the season and memorized by thousands of fans. Every baseball fan knows that 1941 was an important year, not necessarily because of the bombing of Pearl Harbor, but because that summer was the season Ted Williams hit .401 and Joe DiMaggio hit in 56 straight games.

Sports reporting also includes the thousands of television and radio broadcasts of live events. In America today, there is one major television channel devoted entirely to sports coverage, and two or three additional channels have been proposed. The variety of sporting events covered and the amount of coverage on television and by radio are incredible—and difficult for people from other countries to understand. It is possible to watch or listen to almost any major sports event that takes place in the United States, and coverage of important world sporting events, such as the Olympic Games, is also quite complete.

Sports are such an integral part of the American culture and economy that an America without sports is difficult to imagine. Without sports, both the United States and Canada would be entirely different. Sports are important in other countries of the world, but the participation levels, media coverage, and money spent are much less, in comparison with the two North American countries. For this reason, sports should be considered in any study of the people of the two countries.

the two cities constitutes about 27 percent of the total population of the country.

Cultural Influences

The cultural influences in most countries can be divided between the traditional and the modern. In the United States and Canada this dichotomy corresponds to the rural and urban differences, and somewhat to the east and the remainder of the country. Traditional societies are somewhat leery and may be afraid of modern influences. In both the United States and Canada, American rural culture is wary of urban influences. And regionally, the people in the southern and western parts of the United States maintain cultural attributes different from those found in the east. Regional differences do exist, in spite of the tremendous homogenizing effect of national television broadcasts, chain stores, and franchised outlets.

Religion

The primary religion of Anglo-America is the Christian religion, which is divided among the various churches that make up Christianity (see

This view of the New York City skyline indicates the crowded conditions on Manhattan Island. New York's financial district, with the World Trade Center, is visible in the background.

Figure 5.8). The combined churches of the Protestant denominations and sects are number one, but outside Protestantism the Roman Catholic church is the single largest organized group, with approximately 50 million members in the United States and another 5 million to 6 million in Canada. The largest church in Canada, however, is the Anglican church, inherited from England. The majority of people who are members of a church in all the New England states, New York, New Jersey, Louisiana, and the southwest are Roman Catholics. In Canada, Catholics are concentrated in the province of Quebec, where the French influence is strongest.

The second largest church in the United States is the Baptist church. Its 25 million members are the dominant group in a broad belt across the southeast from the Atlantic coast through Texas. The Methodists predominate in another belt, between the Baptists and just south of the Great Lakes, reaching from the Atlantic Ocean to the Rocky Mountains. The northern tier of states contains mostly Lutherans, although there are pockets of Roman Catholic majorities throughout. From the Rocky Mountains west, migration from other parts of the United States has resulted in a mixing of religious groups, so that none predominates. The only exceptions are the concentrations of members of the Church of Jesus Christ of Latter-Day Saints (the Mormons) in Utah and a fairly clear-cut majority of Catholics in California.

Language

The official language of the United States is English, although there is a distinct "American" accent in the spoken version. Minority language islands do exist, however, especially in some large cities. The greatest variation in languages may be heard in New York City, where every major language group is represented. But the largest non-English-speaking groups reside on the west coast, where Chinese in San Francisco and Spanish in the Los Angeles area are common. About 98 percent of the people in the United States are literate.

Some regions of the United States contain identifiable areas where non-English-speaking people are common. French-speaking areas exist along the northeastern boundary with Canada and

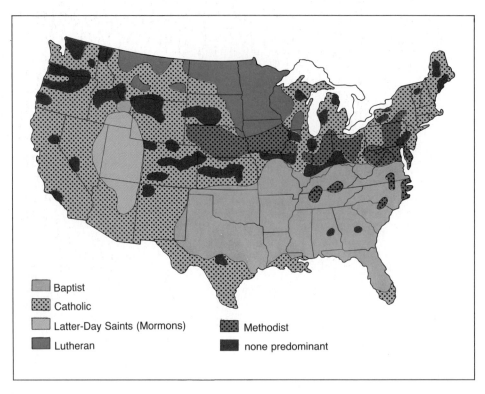

FIGURE 5.8

Religious Affiliations in the United States. The regions shown are generalized, but they indicate the predominant religion within some areas of the United States.

SOURCE: Adapted from the 1974 survey by the National Council of Churches.

in large parts of Louisiana. Besides the Los Angeles area, Spanish is common all along the Mexican boundary from southern California to Texas. Cuban immigration brought numerous Spanish-speaking people to live in the Miami area. Some native American Indians, especially on reservations in the west, still speak Indian dialects. Many commercial radio stations cater to the people living in these linguistic islands, and some businesses hire only bilingual workers.

Unlike the United States, Canada has two official languages: French and English. About 80 percent of the people of Quebec are of French descent and have retained their native language. The remainder of the country is predominantly English-speaking. All official and government materials must be printed in both languages, and politicians running for national office are advised to speak both French and English. The linguistic division in Canada has caused recurring problems. Proposals to divide the country into two nations have been considered, but so far they have been voted down.

Ethnicity

Most Anglo-Americans are white and have little detectable trace of ethnic origin in their speech or in how they look. Much mixing has occurred, both racially and ethnically, especially since World War II. Racially mixed marriages are not the stigma they once were. Great Britain is usually considered the mother country for both the United States and Canada, but since the mid-1800s immigrants from other European countries have outnumbered British immigrants. From 1820 to 1984,

core area The place of origin and usually where the heaviest concentration of certain types of people live. Could be either a region or small area within a city.

cultural diffusion The geographic dispersion of a culture or cultural trait.

---

for example, 9.7 percent of all immigrants to the United States came from Great Britain. During the same period, 13.5 percent came from Germany, 10.2 percent from Italy, 9.1 percent from Ireland, and 8.3 percent from Austria-Hungary. These white Europeans, from only five countries, make up over half of all the immigrants and are the basic stock of America. They are the people most instrumental in settling the country and making it prosper.

The term "ethnic group" refers to a group whose members can be classed according to common traits, such as race, language, customs, or social views. Most members of the minority ethnic groups in Anglo-America can be traced to **core areas,** or centers of population where the densities of ethnic types are greatest. From these "home bases," members of the ethnic minorities have spread to other locations. Through the process of **cultural diffusion,** they have spread their culture beyond the limits of the original core areas.

One of the most common cultural diffusions occurred with food specialties. In most U.S. cities there is a selection of restaurants specializing in the various ethnic food types—for instance, Chinese, Japanese, Mexican, or Italian, and some that specialize in "soul food." Even Vietnamese restaurants are common in some places. Other cultural artifacts that have diffused include architectural styles, place names, art forms, vocabularies, and religions. In Canada, the French Canadians are the largest ethnic minority group. Unlike the movements of minority groups in the United States, French Canadians have tended to remain in the Quebec area, and therefore their cultural traits have not diffused throughout the country.

The most distinct ethnic minorities in the United States are the blacks, Mexicans, native American Indians, Filipinos, Vietnamese, Chinese, Japanese, and Puerto Ricans. Pockets and sometimes regions exist, however, where other groups predominate. These areas include people from nearly every European country, as well as other countries of the world.

The black ethnic core area is in the south, where the blacks were brought originally to the United States as slaves. Since World War II, however, they have moved out of the core into the cities of the North and West. Blacks are now the majority in Washington, D.C., Newark, Baltimore, Detroit, Richmond, and Oakland, as well as in Atlanta, Birmingham, and New Orleans in the south. They make up nearly half the populations of Chicago, Cleveland, Memphis, Philadelphia, and St. Louis. In spite of the emigration from the South, that region is still home to about 45 percent of the blacks in the United States. As of mid-1986, the black population of the United States was estimated at just over 28 million, or about 11.6 percent of the total population.

After the blacks, the second largest ethnic minority in the United States is the Mexicans. The official count is about 16 million, accounting for about 6.6 percent of the total U.S. population. Unofficial estimates run much higher, however, because illegal entry from Mexico goes on daily. The Mexican core area is the southwest, especially southern Texas, central New Mexico, and southern California. Most Mexicans are Roman Catholic and speak Spanish, so the ethnic core area corresponds to the Catholic and Spanish regions of the southwest. Today, the only city with a Mexican majority is San Antonio, but the largest concentration of Mexicans is in Los Angeles, where about one million reside. There are Mexicans, or more appropriately "Mexican Americans," living in every state, but about 87 percent still live in the core area of the American southwest.

The number of American Indians in the United States at the time the first white people arrived has been estimated at about 850,000, but wars, diseases, malnutrition, and massive disruptions of lifestyles caused the native American population to decline to about 240,000 by 1900. Now, however, the Indian population is almost back to where it was in the fifteenth century. Indians once lived in most all regions of the United States and Canada and are found today in every state and province. The highest population densities are in the southwest—especially Oklahoma, Arizona, California, and New Mexico. Outlying core areas include North Carolina, South Dakota, and Washington. Indian cultural items are quite common in the United States, but the most visible Indian cultural influence must be with place names. The list is very long, but some examples include Iowa, Omaha, Wichita, Mississippi, Tucson, Yosemite, Dakota, Wapsipinicon, and Arapaho.

The first Chinese and Japanese immigrants to the United States came to California and Hawaii. Most were imported to work in the sugar cane fields of Hawaii and in the mines and railroad construction in California. Although the histories of the two ethnic groups are different, they have essentially the same core areas. In the ten-year period after gold was discovered in California, the Chinese population increased from about 900 in 1850 to more than 60,000 in 1860. This rapid increase brought hostility from the whites, and in 1882 the Chinese Exclusion Act curtailed their entry into the United States. Their numbers held steady until the Immigration and Naturalization Act of 1952, when a rapid increase in immigration from China began again. During the lull in immigration, the Chinese spread throughout the United States to create secondary core areas in New York, Chicago, Oregon, and the state of Washington. The Chinese-Americans now number about 500,000.

The first Japanese to arrive in the United States were 180 contract laborers brought to Hawaii in 1868. Other groups followed, and some filtered on to the mainland, especially to the San Francisco Bay area. By 1900 there were about 60,000 Japanese in Hawaii and about 10,000 in California. During World War II, the Japanese Americans were forced to live in camps, and all their lands and properties were confiscated for fear that they would side with the enemy, the government of Japan. The Japanese now live in secondary core areas similar to those of the Chinese. About one-third of the population of Hawaii is now Japanese American, and the total for the United States is about 250,000.

The core area for Puerto Ricans is New York City. About 65 percent of the 1.2 million Puerto Ricans reside in the metropolitan New York area. About 60 percent were born in Puerto Rico, and of those only about 25 percent ever return to their homeland either permanently or for a visit. Only about 10 percent of the Puerto Ricans that migrate to the United States return to their home island to stay, but it must be remembered that Puerto Rico is part of the territory of the United States. The secondary core areas for Puerto Ricans are Chicago, Los Angeles, and Miami.

Besides the six major ethnic groups, pockets of many other minorities exist in the United States, and to a lesser degree in Canada. Most are found in the urban areas. Many of the people are refugees from political oppression, such as the Vietnamese, Cubans, and Haitians. Many—for example, the Mexicans, Filipinos, Chinese from Hong Kong, and Koreans—come for economic reasons. Each group contributes to the rich fabric of American culture.

■ Economic Geography of the United States and Canada

Agriculture

Anglo-American agriculture is best characterized by specialization, mechanization, and high productivity. The number of people engaged in agriculture is less than 5 percent of the total population. The acreage planted in crops decreases each year; in 1983 it was down from 327,554,000 in 1950 to 283,408,000. Still, American farmers supply food for the world's most diverse market, export millions of tons of food, and usually have a surplus to be stored. In a world where millions of people face starvation daily, many Americans are overweight. In a world of shortages, American farmers frequently find that their incomes are lower because they produce more food than can be consumed.

The immense productivity of the American farmers can be attributed to a number of factors. One of these is the geographical location of the farming region (see Figure 5.9). The major portion of American cropland is located between the Appalachian Mountains and the 100th meridian, and between the latitudes of 30 and 55 degrees north. This is the Interior Plains province, the region of

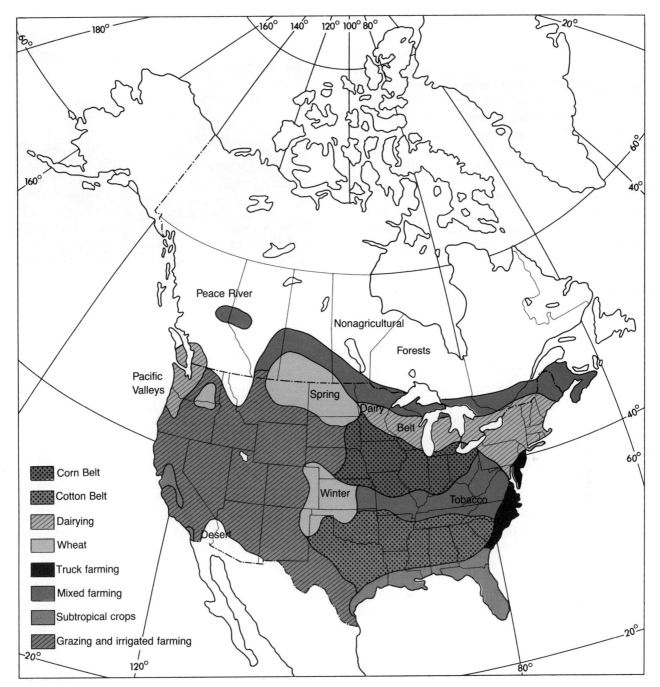

FIGURE 5.9

Agricultural Regions of the United States and Canada.

natural grasslands. The surface is flat or gently rolling, so machinery can be used to cultivate the rich soils. It is a zone with hot summers, but located so that air masses bring summer rain. A dense network of hard-surface, farm-to-market roads connects each farm with a local market town. And the entire region is connected to the eastern and world markets by highways, waterways, and railroads.

The histories and economies of the United States and Canada have also affected farm productivity in those countries. In both countries, the inclination was to distribute land to individuals, and the owner-operated farm became a tradition. Many of the small farms have been consolidated, but more than half of all farms are still owned by the operators. The free market economies of both countries have also contributed to farm success. It should also be mentioned that Americans are noted for working very hard, and American farmers carry on with this strong work ethic.

Agriculture in the United States is a specialized activity that has resulted in subregions, or belts, of production. Each subregion specializes in the product that yields the greatest return at that location, not necessarily the crop that might grow best. For example, wheat in the Corn Belt will produce larger yields per acre than where wheat is actually grown, and records for corn production have been set in the Cotton Belt. The specialization is the result of the transportation network that totally integrates the country and of an economic system that encourages farmers to produce the crop most prized by the consumer. The boundaries of the agricultural belts fluctuate, but the core areas remain quite stable.

The Cotton Belt is a broad zone along the Gulf Coastal Plain from western Texas eastward to the Atlantic coast. It does not extend all the way to the coast because near the water the fall months are too humid to allow the cotton to ripen properly. The belt is restricted on the north by the line marking the end of the 210-day growing season. Cotton is grown outside this region under irrigated conditions, especially in Arizona and southern California. Although much crop diversification has occurred within the region, cotton is still a major crop, in spite of the numerous cotton substitutes, such as polyester. The United States produces

Besides cotton, the Gulf Coastal Plain is noted for producing cattle and petroleum, both of which are part of this scene near College Station, Texas.

about 10 million bales of cotton each year—about 12 percent of the world's total—and about one-third of that comes from Texas. Other agricultural revenues generated in the Cotton Belt come from cattle, fruit, nuts, tobacco, vegetables, rice, and chickens.

The Corn Belt is north of the 210-day growing season line. The most typical Corn Belt states are Iowa, Illinois, Indiana, and Ohio, but the region extends into Nebraska, Missouri, Minnesota, and South Dakota. The Corn Belt is the zone where grains and meat animals are produced. Corn is the major crop, but it is challenged for the number-one position by soybeans. A major source of income throughout the region comes from the sale of animals raised for their meat, especially cattle and hogs. Much of the crops produced, particularly corn, is fed to these meat animals. Iowa usually leads the nation (followed closely by Illinois) in the production of corn, soybeans, and hogs. Annual production averages close to 8 billion bushels for corn and 2 billion bushels for wheat, and each year about 55 million hogs are found on American farms. The Corn Belt is frequently cited as the best example of agricultural efficiency. Almost all labor is performed with machinery, and modern equipment characterizes the entire operation. Each farmer's home is sur-

> **winter wheat** Wheat that is planted in the fall, starts to grow but lies dormant over the winter, then continues to grow the following spring, giving it a head start on the growing season. Used at lower latitudes, where the ground does not freeze.
>
> **spring wheat** Wheat planted in the spring in the higher latitudes, where the ground freezes during the winter and prevents fall planting (winter wheat).

rounded by a cluster of silos, corn bins, cattle and hog barns, toolsheds, and gardens and may remind the traveler of a medieval village.

Where average summer temperatures drop below 70° F (21° C), corn does not mature properly and is usually cut for ensilage: instead of using just the ear of corn, the entire stalk, leaves, and ears are chopped to make ensilage. The temperature line (isotherm) corresponds with the southern boundary of the Dairy Belt. The core of the Dairy Belt is Wisconsin, but it includes most of Michigan, all of the northeastern part of the United States, and southeastern Canada. Most of this belt is also highly industrialized, so much of the dairy production is for the urban workers of the area. The Scandinavian and German immigrants to Wisconsin found a cold climate and thin soils, so they turned from farming to dairying for their livelihoods. Wisconsin leads every state in almost all categories of dairy products—fresh milk, canned milk, butter, and cheese. Other leading dairy states are Michigan, Pennsylvania, New York, and New Jersey. In Canada, the St. Lawrence Valley is an important dairy production area, and much of the output is exported to cities in the northeastern United States. Secondary core areas for dairy products are the Willamette valley–Puget Sound regions of the northwestern United States, coastal British Columbia, and California.

Along the 100th meridian in mid-continent, wheat is the main farm crop. Two distinct wheat belts appear. **Winter wheat** is grown in the southern part of the region—namely, Kansas, Oklahoma, and Texas. It is planted in the fall, grows some during the winter, then makes its rapid growth during the spring and is harvested as early as June. **Spring wheat** is produced in the northern part of the belt in a region that extends from South Dakota northward into the prairie provinces of Canada (Manitoba, Saskatchewan, and Alberta) and westward into Montana and Washington. Spring wheat is planted in the spring and harvested in August and September. Kansas is by far the leading wheat-producing state. In 1984 the United States produced 66.1 million metric tons of wheat, of which 58.7 percent was exported. In the same year, Canada produced 26.9 million metric tons, exporting 80 percent of that total. The two countries provide the world with more than half of all the wheat exported.

In the Rocky Mountains and the Basin and Range country, livestock grazing is the major agricultural activity. The mountainsides are frequently forest clad, but even in the mountains there are open grasslands used by ranchers. Most of the irrigated valleys are devoted to alfalfa and other crops for livestock feed. The typical ranch is quite large, and a few exceed 250,000 acres. The United States has about 100 million beef cattle each year and about 10 million sheep. Between 1975 and 1985, the steady drop in market prices caused the number of beef cattle to decline by 20 percent, and the number of sheep declined by 80 percent from 1945 to 1985 because of the increased use of synthetic fibers.

The west coast of the United States has no distinctive types of agriculture that dominate, except for the dairy production regions already noted. But California requires special mention because it is the leading agricultural state. The varieties of soils and climates in the state allow a diversity of crops unequaled anywhere in the world. About 40 percent of the state's agricultural income is from fruits and vegetables. California ranks first in the production of oranges, peaches, pears, prunes, strawberries, grapes, carrots, lettuce, cantaloupes, tomatoes, eggs, sugar beets, rice, and turkeys, and second nationally in the production of beans, onions, hay, sheep, barley, and cotton. The total value of California's farm products sold each year amounts to billions of dollars, surpassing the total gross national income of most countries of the world. Although most of the agricultural land is in the Central Valley, every coastal valley lowland specializes in some crops. In addition to the large local market,

California supplies the remainder of the United States.

Mineral Production

The United States and Canada are both blessed with a large variety of high-quality minerals, and the utilization of these minerals has been a major factor in the development of both countries. The basis of industrial strength is the energy minerals: coal, oil, and natural gas, along with an abundance of iron ore for making steel. Both countries have plenty of all these minerals.

Coal is widely distributed throughout the United States and Canada, but most of it comes from only a few major coalfields (see Figure 5.10). The Appalachian field, especially West Virginia, Kentucky, and Pennsylvania, and the interior field in Illinois, are the leading producers. Canadian coal is located mainly in the western prairie provinces and is too soft for **coking** and not convenient for marketing. Coal was the major source of energy during the formative years of America, but its use declined during the 1950s and 1960s in favor of petroleum. As political and economic factors changed and oil prices increased, coal once again became the major source of energy (see Table 5.1).

coking coal A type of bituminous coal that has been heated to remove the gases and is used for industrial fuel.

The largest petroleum-producing region in the United States and Canada is the mid-continent field of Kansas, Oklahoma, and Texas (see Figure 5.11 on page 111). The crude oil is either refined locally or piped to ports along the Gulf Coast and sent by tanker to the markets in the northeast. The second largest field is the Gulf Coast itself, with Louisiana and coastal Texas being the major producers. As a result, many of the Gulf Coast cities, such as Houston, Galveston, Port Arthur, and Beaumont in Texas, and Lake Charles and New Orleans in Louisiana, are processors and shippers of oil and petroleum products.

A large pool of oil extends under the Central Valley of California southward to Los Angeles, making California another major producer. Large refining facilities are located in Long Beach and in

Farm land in the Central Valley of California. One of the specialty crops of the valley—rice—can be seen in this view.

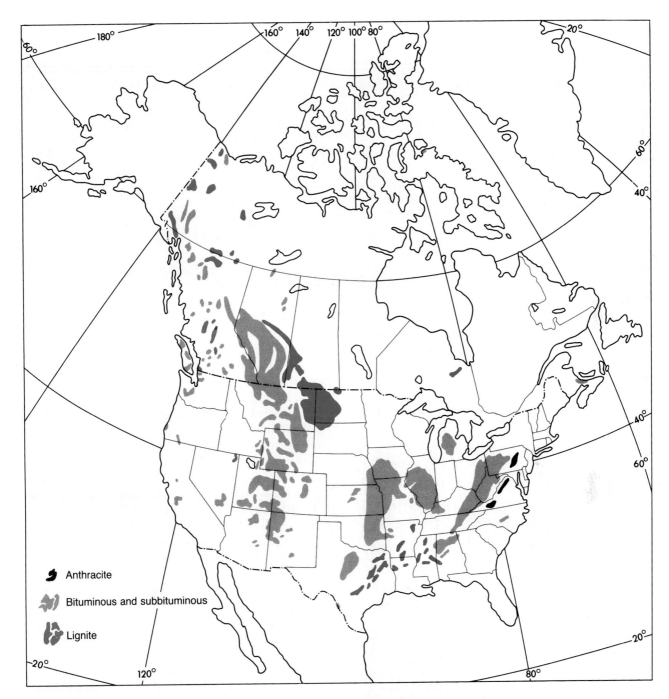

FIGURE 5.10
Major Coalfields of the United States and Canada.

TABLE 5.1

Total Energy Production for the United States, 1950–1985 (in quadrillion Btu)

YEAR	COAL		OIL		GAS		HYDRO		NUCLEAR		TOTAL
1950	14.1	(42)	11.5	(34)	7.1	(21)	1.4	(3)	0	(0)	34.0
1960	10.8	(26)	14.9	(36)	13.1	(32)	1.6	(4)	0.1	(0.2)	41.5
1970	14.6	(24)	20.4	(33)	24.1	(39)	2.6	(4)	0.3	(0.4)	62.1
1980	18.6	(29)	18.3	(28)	22.1	(34)	2.9	(4)	2.8	(4.0)	64.8
1985	19.8	(30)	18.6	(28)	20.1	(31)	3.4	(5)	3.6	(5.0)	65.5

Note: Percentages are given in parentheses. Totals may not equal the sum of parts due to rounding.
SOURCE: Energy Information Administration, U.S. Department of Energy (Adapted from: *The World Almanac,* Newspaper Enterprise Assoc., 1986, p. 126.)

the San Francisco Bay area. The eastern field extends through southern Illinois eastward through Kentucky to Pennsylvania, but Illinois is the largest producer in the region. The single largest oil-producing state, however, is Alaska, where the north slope production surpasses all other fields. In Canada, Alberta is the leading producer, followed by Saskatchewan.

Natural gas is made from organic material in a manner similar to coal and petroleum and is found in the same general locations. In both liquid and gas forms, natural gas supplies about 31 percent of America's energy needs, so it is an important product. Thousands of miles of natural gas pipelines cross the United States, moving the product from the mid-continent and Gulf Coast fields to the north and east. California and Canada are both tied in to the network of gas pipelines.

The mineral iron, the essential ingredient in steel, is found everywhere on the earth's surface, but it is economically useful to mine it only when ore concentrations are 25 percent or more. Iron ore is of adequate quality and concentration in several locations in North America (see Figure 5.12). For years the Mesabi Range of the West Lake Superior District supplied the smelters of the Lower Great Lakes with ore that was more than 50 percent pure iron. It probably produced more iron ore than any other single range in the world, but today it is nearing depletion. Minnesota, however, is still a major source of iron ore. Scattered deposits exist in the Appalachian Highlands, the Adirondack Mountains, and in Alabama, especially in the Birmingham area. In the west, large deposits have been found at Iron Springs, Utah, and at Fontana, California. In Canada the major fields are located in scattered deposits in northeastern Quebec and Labrador, around the northern edge of the Great Lakes, and near the Vancouver Island region in the west. The Quebec-Labrador deposit is the largest, and the ore is transported by railroad to the St. Lawrence River, where it is loaded onto barges for delivery to the smelters along the lower Great Lakes.

This oil refinery near Beaumont, Texas, produces various grades of gasoline and oil.

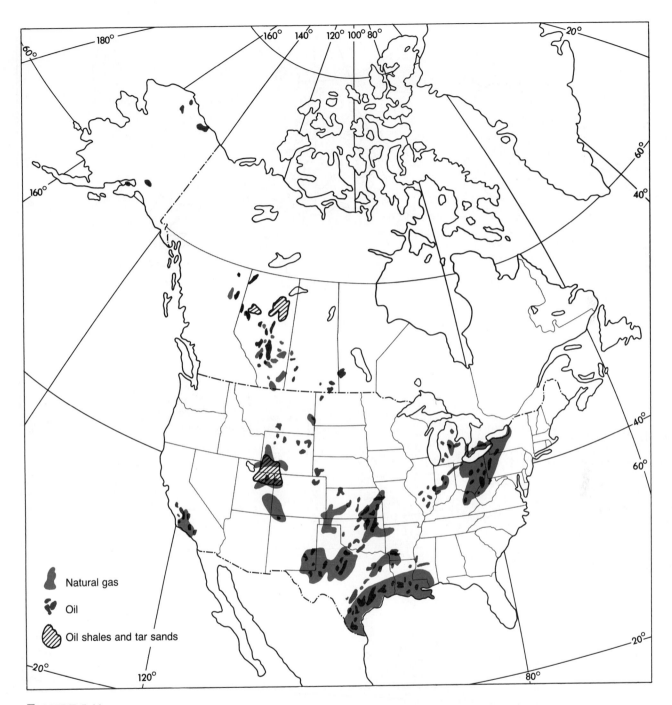

FIGURE 5.11
Petroleum-Producing Areas of the United States and Canada.

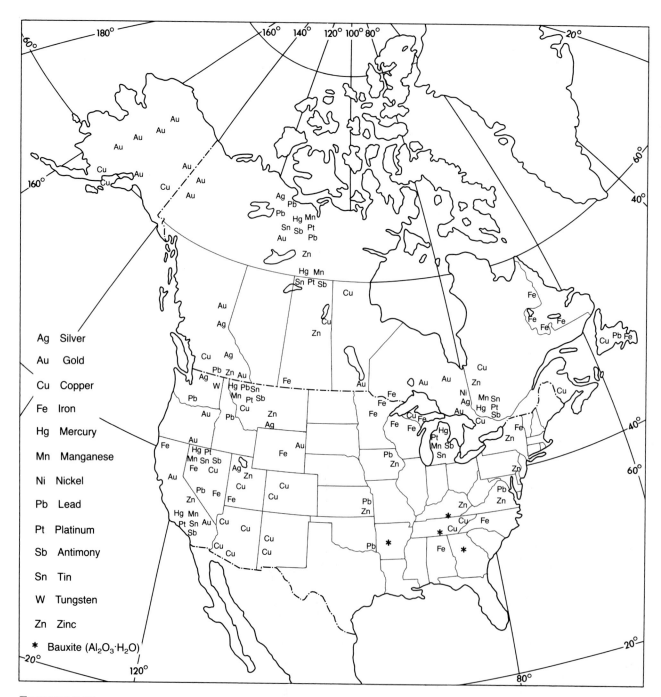

FIGURE 5.12

Major Nonfuel Mineral Deposits of the United States and Canada.

Other minerals are distributed widely across the continent. The deposits may appear to be randomly scattered, but closer examination shows that many of them are in the western mountains, the Appalachian Highlands, and along the Canadian Shield. The United States and Canada lead the world in the production of some nonfuel minerals, and each country ranks highly for a number of them. The most important nonfuel minerals in terms of value are ranked as follows: (1) clay, (2) crushed stone, (3) sand and gravel, (4) copper, (5) phosphate, (6) magnesium, (7) gold, (8) lime, (9) sulfur, and (10) salt. The leading producing states, with their percent of the U.S. total, are California (8.7 percent), Texas (7.4 percent), Minnesota (7.3 percent), Florida (6.5 percent), and Arizona (6.4 percent). Some of the leading states are not located in the three regions mentioned above because they are rapid-growth states and require large amounts of building material.

The United States is deficient in a number of essential minerals. Chromium, for example, must be imported from South Africa or Zimbabwe. Cobalt, used in jet engine parts, is found in the United States, but because it is low grade it must be imported from Zaire or Zambia. The United States uses about one-third of the total world's consumption of cobalt. Columbium (niobium), used in making steel for pipelines, is not mined in the United States but is found in Canada. Manganese and nickel, both essential to iron and steel production, also are not found in quantities that are worth mining in the United States. Platinum and tantalum are not found in sufficient quantities in the United States to support the manufacture of medical and dental supplies, automobiles, and airplanes. Thus, many thousands of tons of essential minerals used to make common items for daily use must be imported because they are not found in the United States, and in some cases not in Canada either.

Industry

Not only is the United States the world's largest industrial power, it is also the most efficient producer in terms of work-hours per unit of commodity produced. Combined with the Canadian output, the Anglo-American region is a massive

An open-pit mine located near Superior, Arizona, produces high-grade copper ore.

producer. The success is related to a combination of factors, but the fact remains that about 5.5 percent of the world's population produces more than one-third of the world's manufactured goods. It was not very long ago that these Anglo-Americans produced more than half the world's manufacturing output. For an overview of the industrial regions of the United States and Canada see Figure 5.13.

The leading industrial zone is in the northeastern United States and southeastern Canada. The zone could be enclosed by drawing a line from Boston to Montreal, then to Milwaukee and St. Louis, and back to Boston through Baltimore. The resulting polygon would enclose the source of most of America's iron and steel, automobiles, clothing, light metal goods, and machinery of various types. In addition, most of the railroad cars and equipment, machine-shop products, textile machinery, agricultural machinery, and electric machinery of the United States are produced in the region.

Some areas are noted for certain types of goods. New England, for example, specializes in woolen textiles and shoes. The Boston area especially is noted for these products. South of Boston, through Rhode Island and up the Connecticut River valley, is a zone where light metal goods are produced. Products include everything from pins to watches and guns, with the most prominent items being a full range of electrical machines and appliances. The Lower Great Lakes cities are known for heavy metal items. Most of the nation's iron and steel originates in a triangular region from Buffalo to the upper Ohio River and back to Chicago. Iron, coal, and limestone are combined here to pass through the smelters, blast furnaces, and finally the rolling mills, often without cooling in the process. The manufacturers of railroad tracks, locomotives, and equipment, and agricultural machinery, and other heavy users of fabricated metals and machinery, try to locate as close to the rolling mills as possible. Thus, one industry feeds off another.

The nation's largest manufacturing employer, the automobile industry, is centered in the Detroit area. About half the automobile workers live in southern Michigan, mostly in Detroit and its

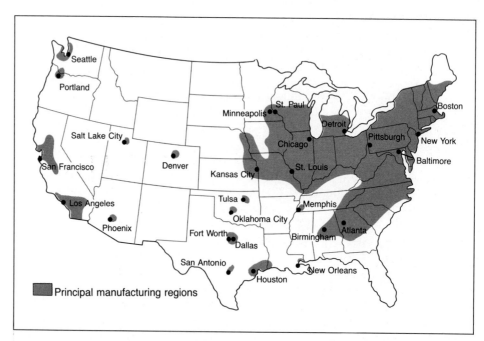

Figure 5.13

Industrial Regions of the United States and Canada.

suburbs. Most of the remainder live in a zone from Janesville, Wisconsin, to Cincinnati, to Buffalo, although there are automobile assembly plants outside that region. These individual plants are located in the outlying market areas along the east coast and in California.

New York City is the largest single industrial producer in the United States and Canada. Every class of product is made there, but the most characteristic items are clothing and printed materials. The city has a huge local market, a very large supply of labor, a magnificent port facility, and transport connections with the interior of the country. It is unlikely that New York will ever be surpassed as America's largest city.

Not all industry has its center of production inside the northeastern core region. For example, cotton textile producers have migrated from New England to the Piedmont, partly in search of cheaper labor but also to be located closer to the cotton-producing region. Since shortly after World War II, the primary zone for the cotton textile industry has extended from Virginia to Georgia. Another major industry not concentrated in the northeast is aircraft manufacturing. The raw materials for an airplane are relatively light and inexpensive, and the finished product can be delivered in hours to any market in the world. Airplane companies can therefore afford to locate outside the major manufacturing region. The primary locations for airplane manufacturers are southern California, Seattle, Washington, and Wichita, Kansas. About one-third of all aircraft workers live in the Los Angeles area.

As with the cotton textile industry, the chemical industry also has moved during the last 40 years. It was once centered in the Lower Great Lakes region, where the by-products of the iron smelters were utilized. Most new chemical plants, however, have located along the Gulf Coast, where petroleum, salt, and sulfur deposits supply the raw materials.

It is impossible to mention all industries and industrial sites in this brief description, but it should be noted that every part of the United States has some significant industrial product. Generally, the amount of industry corresponds closely to the population of the area. The United States has about 20 million people working in manufacturing, and about one-third of these workers live in New York, Pennsylvania, Ohio, Illinois, and Michigan. California, however, has a wide variety of manufacturing plants and ranks second nationally in the number of industrial workers.

A recent trend in American manufacturing is for companies to send parts to other areas of the world for assembly, with the intention of bringing the finished products back to the United States for sale. The purpose of this trend is to take advantage of the cheaper labor in foreign countries. For example, the El Paso–Juarez area contains over 500 *maquiladora* plants that are mergers of American capital and technical knowledge with relatively inexpensive Mexican labor. Other similar operations are located in South Korea, Hong Kong, and Denmark.

Transportation

The general image of Anglo-America is one of mobility. People are always on the move, both in terms of migration and just in daily commuting. The transportation systems of the United States and Canada are modern and well integrated. Every type of modern conveyance is available—highways, railroads, waterways, airways, and pipelines. The transportation system in the United States is the most extensive of any country in the world.

Canada is faced with special problems that make it difficult to construct an overall transportation system. Its great east-west extent of 4,000 miles (6,400 km) is interrupted by mountains and lakes, but the major problem is the sparse population. As a result, most of Canada's roads, railroads, and pipelines are limited to the country's southern edge. But Canada's communication lines do cross the international boundary to connect with those in the United States, and international travel between the two countries is easy.

The United States has about one-third of the world's railroad mileage. The total track mileage has decreased by about 500 miles (805 km) a year since the 1930s. Most of the current 200,000 miles (321,800 km) of track are in the eastern half of the country, although Texas leads the nation with about 13,000 miles. Chicago is the major terminal, with about 30 lines focusing on the city, and Illinois is the second leading state in terms of miles of track. Other cities considered railroad centers are St. Louis, Kansas City, Minneapolis, Memphis, Atlanta, and Dallas. Six railroad lines stretch across the west, and some wind through the Rocky Mountains to connect the west coast with the eastern system. Canada has about 43,000 miles (70,000 km) of railroad track, including two transcontinental lines and numerous spurs reaching northward to the interior. See Figure 5.14 for the locations of the railroads of the United States and Canada.

The American road system is an intensive network supported by governments at various levels, including federal, state, county, township, and individual cities. The federal government in the United States supports two systems of roads. The Federal Aid Highway Act, signed into law on June 29, 1956, inaugurated the Interstate Highway System. This system makes it possible to drive between and through nearly every major city in the country without encountering a stop sign, stop light, or toll booth. The east-west routes of the interstate system are identified with even numbers, beginning with I-10 from Los Angeles to Jacksonville, Florida; the most northerly route is I-90, between Seattle and Boston. The north-south lines have odd numbers: Interstate 5 connects San Diego with Seattle, and I-95 runs from Miami to northern Maine. These limited-access, four-lane, hard-surface roads make up about 260,000 miles (416,000 km) of the network, but there are also another 600,000 miles (960,000 km) of connecting and feeder lines, secondary roads that make up another system supported by the federal government. In addition to these federal systems, many thousands of miles of state and local roads have been built.

In spite of the intensity of the road system, it seems at times that there are not enough roads. Traffic around and in most urban centers can become quite congested. The United States has about 170 million registered automobiles, buses, and trucks and about 160 million licensed drivers. California leads the nation in both categories. About 10 million new cars, buses, and trucks are manufactured each year, with Ford, Chrysler, and General Motors contributing about 85 percent of

The Golden Gate Bridge is a vital traffic link between San Francisco and the communities to the north in Marin County.

the total. Besides these local manufacturers, Honda and Nissan, also have factories in the United States. Added to this are the thousands of motor vehicles imported to the United States each year.

Waterways were vital to the early development of America and still play a large part in the country's transportation system (see Figure 5.15). About 25,000 miles (40,225 km) of navigable waterways are used in the United States today. Most traffic is on the Mississippi-Missouri-Ohio river system between the gulf port of New Orleans and numerous upriver ports, such as St. Louis, Minneapolis, Kansas City, Omaha, Chicago, and Pittsburgh. The old Erie Canal system from New York City to Buffalo is still followed now in part by the New York State Barge and Canal Line. This 340-mile (547 km) route, along with the paralleling rail lines and highways, is the most traveled transportation corridor in the United States. Oceangoing ships travel inland as far as Albany.

The Intercoastal Waterway is another major U.S. shipping lane. It is made up of the lagoons

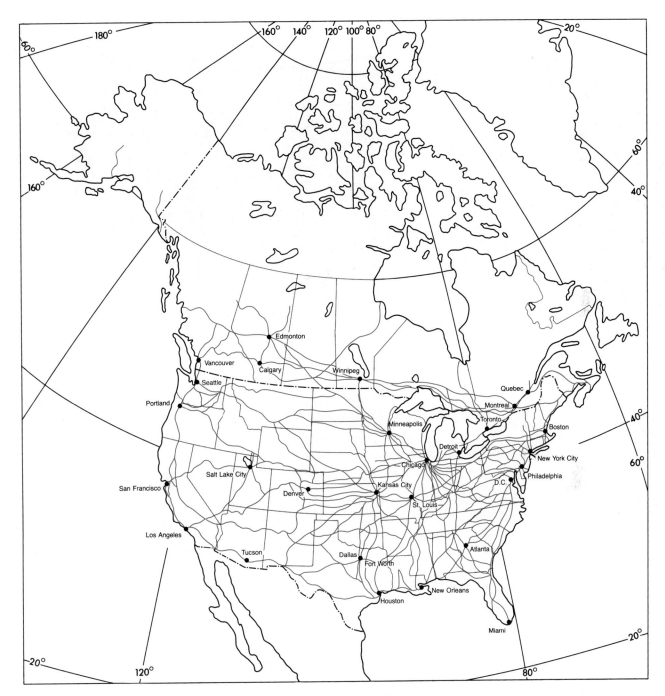

FIGURE 5.14
Major Railroads of the United States and Canada.

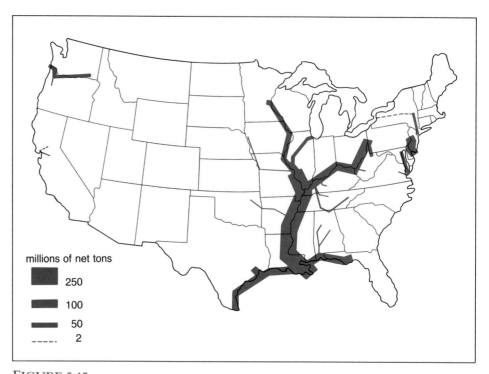

FIGURE 5.15

Inland Waterway Traffic in the United States. The volume of traffic is shown by the width of the lines that indicate the flow areas.

and canals between the land and offshore sandbars that extend from New York to Miami, then continue along the Gulf coast to ports in Texas. But the most heavily used inland waterway is the St. Lawrence River and Seaway in Canada. The route opens the Great Lakes to oceangoing vessels and has a tremendous volume of traffic. Iron ore, coal, timber, grain, and petroleum products make up the most tonnage. More ships pass through the Sault Sainte Marie Canals between Lake Superior and Lake Huron than through any other canal in the world.

America's great distances, rugged terrain, integrated business interests, and natural inclination to hurry have stimulated North Americans to prefer airways to other public carriers. Routes connect every major city, and most smaller places are connected to the network with "commuter airlines." About 85 percent of the airline business is in passenger service, but freight is a growing source of revenue. As with railroads, Chicago is the hub of air traffic. O'Hare International Airport records about 750,000 takeoffs and landings each year, but Atlanta International is a close second, with about 700,000 a year. Other city airports that record more than 500,000 takeoffs yearly include Van Nuys (California), Los Angeles International, Dallas–Fort Worth Regional, and Denver Stapleton International. About 350 million passengers enplane each year in the United States, and about 8.2 billion ton-miles of cargo is transported by air.

America's transportation system is greatly enhanced by a mode that is mostly underground—the pipeline system. Pipelines are the cheapest form of land transport, roughly one-seventh the cost of rail freight. The United States is laced with about one million miles (1,609,000 km) of pipelines (see Figure 5.16). Because most of these lines are buried in the ground, they are the least visible of the transport modes. The major lines carry natural gas and other petroleum products between the mid-continent and Gulf Coast oilfields and the markets of the north and east. A number of pipelines also serve the west coast, and a few cross the U.S.-Canadian boundary. One of the most important recently constructed pipelines is the

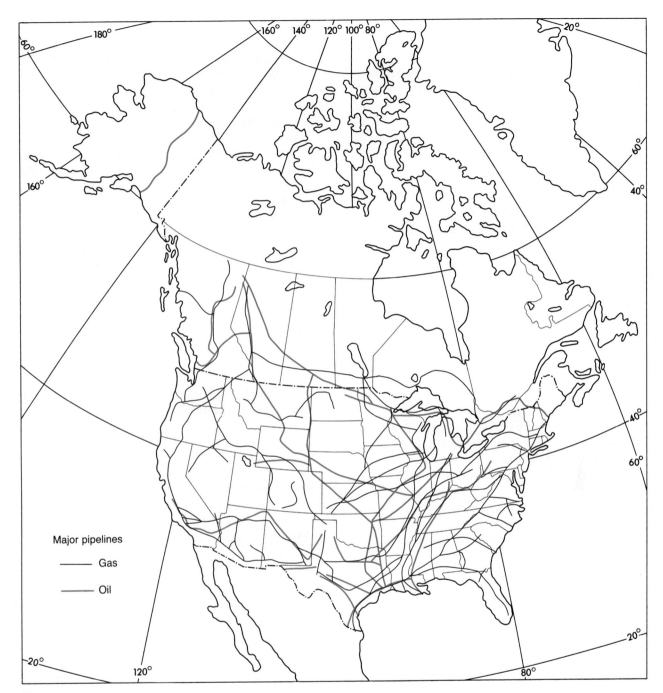

FIGURE 5.16

Major Pipelines of the United States and Canada.

Trans-Alaskan, bringing crude oil from the north slope to the southern port of Valdez. Besides petroleum products, some pipelines carry acids and hydrocarbon-based chemicals. Some pipelines carry solid commodities, such as coal (in solution, called "slurry"), fish, coffee beans, wood pulp, wheat, and even sealed containers. The use of these inexpensive carriers probably will increase in the future.

Political Geography of the United States and Canada

Canada officially remains part of the British Empire, but Canadians vote for their own elected officials and have nearly complete sovereignty over their own affairs. The United States is the undisputed world's leader of democratic ideals. The U.S. Constitution is the oldest document of its type in the world. It outlines the democratic form of government as well as the rights and privileges of American citizens. The United States has accepted immigrants from every other country of the world, many of whom were searching not only for a better way of life but also for political, religious, or economic freedom. There is a line of people waiting to become U.S. citizens.

The most complex problems facing both the United States and Canada are associated with foreign affairs. Each country has been involved in numerous wars, but they have always been on the same side and have never fought each other. The very delicate balance in world affairs can be easily upset, and in a world with so many varied cultures, ideals, economies, and political factions it is difficult to maintain a balance. If the United States supports one country, that action could displease the governments of other countries—that is, merely protecting what is called the "national interest" of the United States could antagonize many other countries of the world.

The complexity of the world problems that affect the United States means that the political friends and foes of the United States can change dramatically over time. France, which had been a good political ally of the United States since the Revolutionary War, recently would not permit U.S. combat planes to fly over France. Under the Shah, Iran was a close friend of the United States, but Iran's present leaders claim that the United States is Iran's political enemy. Yet, the two countries carry on trade, and Iran sends students to the United States to be trained. Japan and Germany fought a long, bitter war against the United States, but each is now a political friend and good trade partner with the United States. The United States has a critical need for many minerals produced in South Africa, but still it leads a worldwide attempt to boycott that country because of the government's apartheid policy. The United States backed the establishment of the country of Israel, but it also tries to maintain close political ties with Israel's Arab neighbors, who hate Israel's existence.

The United States and Canada are wedded in location, politics, and economics to such an extent that the well-being of each affects the other. The international boundary between the countries has never been protected by army troops, and no foreign power has violated either country for more than 150 years. Except for slight adjustments along the Rio Grande resulting from changes in the course of the river, no territory has changed hands in Anglo-America since 1867. The people of the United States and Canada are by nature isolationists but will undoubtedly play a significant role in world affairs for some time to come.

Key Words

coking coal
core areas
cultural diffusion
normal lapse rate
permafrost
physiographic province
spring wheat
winter wheat

References and Readings

Allen, James P., and Turner, Eugene J. *We the People: An Atlas of America's Ethnic Diversity.* New York: Macmillan, 1987.

Atlas of North America: Space-Age Portrait of a Continent. Washington, D.C.: National Geographic Society, 1985.

Atlas of the United States: A Thematic and Comparative Approach. New York: Macmillan, 1987.

Bennett, C. F., Jr. *Conservation and Management of Natural Resources in the United States.* New York: Wiley, 1983.

Birdsall, Stephen S., and Florin, John W. *Regional Landscapes of the United States and Canada.* New York: Wiley, 1985.

Blackbourn, Anthony, and Putnam, Robert G. *The Industrial Geography of Canada.* New York: St. Martin's, 1984.

Borchert, John R. *America's Northern Heartland.* Minneapolis: University of Minnesota Press, 1987.

Brunn, Stanley D., and Wheeler, James O. (eds.). *The American Metropolitan System: Present and Future.* New York: Halsted, 1980.

Chudacoff, Howard P. *The Evolution of American Urban Society.* Englewood Cliffs, N.J.: Prentice-Hall, 1981.

Clark, David. *Post-Industrial America: Geographical Perspectives.* New York: Methuen, 1985.

Dicken, P., and Lloyd, P. E. *Modern Western Society: A Geographical Perspective on Work, Home, and Well-Being.* New York: Harper & Row, 1981.

Garreau, Joel. *The Nine Nations of North America.* Boston: Houghton Mifflin, 1981.

Guinness, Paul, and Bradshaw, Michael. *North America: A Human Geography.* Totowa, N.J.: Barnes & Noble, 1985.

Jakle, John A. *The American Small Town: Twentieth-Century Place Images.* Hamden, Conn.: Shoe String Press, 1982.

Malcolm, Andrew H. *The Canadians.* New York: Times Books, 1985.

McCann, Lawrence D. (ed.). *Heartland and Hinterland: A Geography of Canada.* Scarborough, Canada: Prentice-Hall, 1982.

McKee, Jesse O. (ed.). *Ethnicity in Contemporary America: A Geographical Appraisal.* Dubuque, Iowa: Kendall/Hunt, 1985.

Mitchell, Robert D., and Groves, Paul A. (eds.). *North America: The Historical Geography of a Changing Continent.* Totowa, N.J.: Rowman & Littlefield, 1987.

Palm, Risa. *The Geography of American Cities.* New York: Oxford University Press, 1981.

Paterson, John H. *North America: A Geography of Canada and the United States.* 7th ed. New York: Oxford University Press, 1984.

Pred, Allen. *Urban Growth and City Systems in the United States.* Cambridge, Mass.: Harvard University Press, 1980.

Raitz, Karl B., et al. *Appalachia: A Regional Geography.* Boulder, Colo.: Westview, 1984.

Robinson, J. Lewis. *Concepts and Themes in the Regional Geography of Canada.* Vancouver, Canada: Talon, 1983.

Rooney, John F., Jr., et al. (eds.). *This Remarkable Continent: An Atlas of United States and Canadian Society and Culture.* College Station: Texas A&M University Press, 1982.

Thernstrom, Stephen (ed.). *Harvard Encyclopedia of American Ethnic Groups.* Cambridge, Mass.: Harvard University Press, 1980.

Vale, T. R. *Plants and People: Vegetation Change in North America.* Washington, D.C.: Association of American Geographers, 1982.

Walker, D. F. *Canada's Industrial Space-Economy.* Toronto: Wiley, 1980.

Weinstein, Bernard L., et al. *Regional Growth and Decline in the United States.* 2d ed. Westport, Conn.: Praeger, 1985.

White, C. Langdon, et al. *Regional Geography of Anglo-America.* 6th ed. Englewood Cliffs, N.J.: Prentice-Hall, 1985.

Yeates, Maurice H. *North American Urban Patterns.* New York: Halsted, 1980.

Chapter 6

EUROPE

◼ Introduction

Europe has been called a continent for so long that the term is generally accepted. In reality, however, Europe is a peninsula jutting off the Asian land mass (see Figure 6.1). This relatively small world region contains less than 4 percent of the earth's land area but about 10 percent of the earth's people. The global impact that Europeans have had is immense. For 500 years it was European scholars who wrote most of the world's books, surveyed and mapped the world, and charted routes across the oceans. Commerce and politics followed the ships, until finally European nations held political control over 90 percent of the earth's surface. The European nation-state became the model for all modern nations. In fact, the phrase "Western civilization" has come to mean "European culture."

Europe is fractured both physically and politically. The European peninsula is fragmented into numerous bodies of water, islands, and smaller peninsulas, and sovereignty over parts of the area has changed hands many times throughout history. Two world wars and hundreds of lesser wars have been caused by ownership quarrels. A failure to join together has resulted in many small states

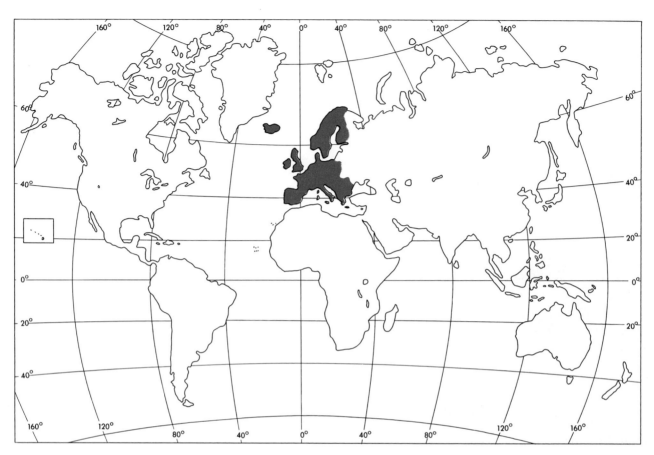

FIGURE 6.1
World Location of Europe.

TABLE 6.1

The Nations of Europe, by Region

NORTHERN EUROPE	SOUTHERN EUROPE	WESTERN EUROPE	EASTERN EUROPE
Denmark	Albania	Austria	Bulgaria
Finland	Greece	Belgium	Czechoslovakia
Iceland	Italy	France	East Germany
Ireland	Portugal	West Germany	Hungary
Norway	Spain	Luxembourg	Poland
Sweden	Yugoslavia	Netherlands	Romania
United Kingdom		Switzerland	

whose areas and populations are not large enough for efficient economies. An imaginary average European nation would be about the size of Nebraska, with a population similar to that of New York State. The major nations of the area are grouped according to regions in Table 6.1. (See also Figure 6.9.)

The regional groupings of national units shown in Table 6.1 and Figure 6.2 are based both on location and on politics. For example, Poland might fit best regionally with the Northern European group, but because of the country's political alignment with the Soviet Union it is placed with the other Soviet bloc countries of Eastern Europe. The mini-states of Europe—Andorra, Liechtenstein, Malta, Monaco, San Marino, and Vatican City in Rome—are not included in the groupings.

FIGURE 6.2

Divisions of Europe. The divisions on this map correspond to the discussion in the text. The only political significance of the divisions is the break between Eastern Europe (the Communist countries) and the remainder of the continent.

Europe 125

Physical Geography of Europe

Landforms

Although Europe's topography is complex, a basic pattern becomes clear when one recognizes only four general physiographic provinces. Each province is composed of many small units, but there is enough internal uniformity for each to be called a region. The basic component of uniformity is the geologic structure. Most of Europe lies in an unstable zone of the earth's crust, which means that there has been much tectonic activity in the past. Much of the area also was glaciated, so the mountain-building was counteracted by the smoothing action of the glaciers. The four physiographic provinces are the Northwestern Uplands, the Central Lowlands, the Central Plateaus, and the Alpine System (Figure 6.3).

The Northwestern Uplands. The Northwestern Uplands include the mountainous terrain of Scandinavia and Finland, northwestern Great Britain, Ireland, and western France. Although these regions are separated by bodies of water, their underlying rock structures are similar, and in many places the landscapes are similar too. The regions all face the Atlantic Ocean on the west and are washed by the warming waters of the Gulf Stream and its continuation, the North Atlantic Drift. They also face the eastward-moving storms loaded with moisture off the Atlantic. Heavy and constant precipitation which creates windswept, rain-drenched, and often swampy conditions, is common. None of these areas has been heavily utilized, and as a result the population densities are low.

The surface configuration of the Northwestern Uplands is more plateaulike than mountainous, but dissection from both rivers and glaciers has created steep slopes and rugged terrain, giving the area a mountainous appearance. Flat land occurs only along the larger rivers. The highest part of the region is about 8,000 feet (2,440 m) above sea level, in southern Norway. Most of the coastline is extremely irregular, partly the result of rising ocean water after the ice sheets melted and partly the result of the land sinking from the weight of the glacial ice. *Fiords*—former river valleys, once deepened by glaciers and now invaded by sea

A natural bay at Plymouth, England. Much of the coastline of southwestern England is idented with such bays and inlets. Plymouth was the port from which the *Mayflower* sailed across the Atlantic to America in 1620.

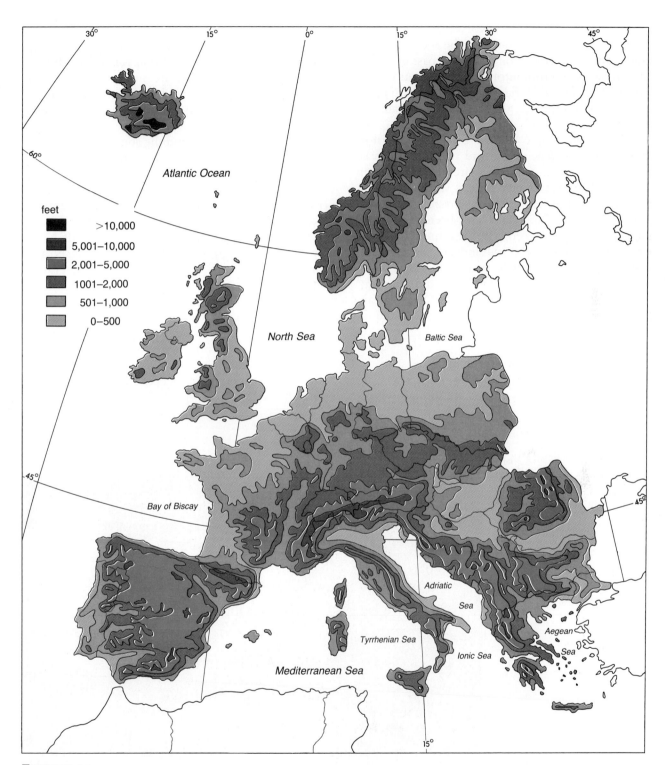

FIGURE 6.3
Land Elevations and Water Bodies of Europe.

water—exist all along the Norwegian coast and reach far inland. In Brittany on the west coast of France and in Cornwall on the west coast of Great Britain, rocky promontories shelter numerous bays and inlets. In Scotland, narrow lochs (lakes) extend inland much like the fiords (glacier valleys) of Norway.

The native people of these areas are noted for their political independence and for the preservation of ancient customs. The Bretons of western Brittany still speak a Celtic dialect, which is related to the Gaelic speech of the Irish, Welsh, and Scotch uplanders. These languages have been largely replaced by English, but the Welsh sometimes challenge England for political independence based primarily on language differences. Most road signs in Wales are printed in both Welsh and English, and most of the middle-aged people of Wales speak both languages. The Irish language was limited to isolated pockets in the mountains of the west but has recently been revived.

The economic basis of life in these sparsely populated Northwestern Uplands is the sea. The world's greatest fishing regions are located where warm water from tropical areas meets cold water from the polar regions. This mixing of the waters allows plankton and other food for fish to thrive. The oceanic circulations are such that the warm Gulf Stream meets the cold Arctic waters near the Northwestern Uplands, creating these conditions. The relatively shallow depths of the English Channel and the North Sea, and the additional fish food brought by the rivers, also contribute to the fishing potential of the area. The result is the numerous fishing villages that dot the bays and inlets nearly everywhere along the coast of the upland region. Fishing boats roam the entire area for all types of fish—cod from the cold waters off Norway, herring from the southern parts of the North Sea, sardines from the coastal waters off France and Portugal.

Fishing dominates life along the coast, but on the inland uplands sheepraising is the most common activity. The dark-green steep slopes of Brittany, Wales, and Scotland are covered with grazing sheep. The sheep support the mills that produce the famous woolen products of Brittany, Scotland, and Wales.

The city of Paris, located in the Paris Basin, is bisected by the Seine River. The island at this point in the river is known as the Allée des Cygnes.

The Central Lowlands. The Central Lowlands (the North European Plain) consist of low-lying terrain where elevations rarely exceed 500 feet (150 m) and the local relief is never greater than 100 feet (30 m). The lowland extends from the Pyrenees Mountains northward through western France, the low countries (Belgium, Luxembourg, and the Netherlands), and Germany and Poland, from where it broadens into the plains of the Soviet Union. The ancient ice sheets left the rocks bare in the north, but most of the region has soils sufficiently deep to sustain agriculture. In small but important regions, coal, iron ore, and other minerals have been found. The level landscape has made extensive transportation systems possible. The broad, slow-moving rivers and thousands of miles of interconnecting canals add to the dense transportation network of railroads and roads.

Most of Europe's high population densities are in the Central Lowlands. Manufacturing, great harbors, local resources, highly developed transportation facilities, good agricultural production, and tolerable climates are all conducive to high densities of population. Although the land is level, these lowlands are not uniform and monotonous. The different climates create variations in native vegetation and agriculture, which adds variety to the landscape. Local customs and the availability of local building materials affect the types of architecture and to a lesser extent the transportation facilities, and all this adds to the variety in the visual landscape. Thus, there is great variation in both the cultural landscapes and the physical landscapes.

Some of the Central Lowland areas include large basins and river valleys (see Figure 6.4). The *Aquitaine Basin* (or Aquitanian Lowland) of western France is one of the foremost wine-producing regions of the world. Grapevines are especially well adapted to the warm climate and fertile soils of the basin. Grapes grow nearly everywhere, and in many places they dominate the landscape, especially in the *Garonne Basin*. The golden brandies and dark-red Bordeaux wines bear the names of small villages and medieval castles and enter the wine markets of the world. The Garonne Basin connects with the *Paris Basin*, where a gently rolling plain is dominated by the city of Paris. The basin itself is circular, and a series of concentric escarpments slope gently toward the city but have steep sides pointing outward. Both the Loire and the Seine rivers flow through the basin and find their way independently to the Atlantic coast.

The *Lowland of Brabant*, which includes the Flanders Plain, is located in Belgium, but it resembles the physical landscape of the Paris Basin. The loess soils of the region are fertile, and large fields of grain are broken by plots of fruit and vegetable gardens. Belgian farmers live on isolated plots instead of crowding together in villages, as farmers do in France, so the cultural landscape is less cluttered than in France. The Latin and Germanic cultures, vying for political supremacy, have clashed more than once in this lowland region. Flanders is a noted World War I battlefield, and the Battle of the Bulge of World War II was fought nearby in eastern Belgium.

The lowland of France and Belgium reappears in England. The shallow English Channel and North Sea are drowned portions of the same physical structure. Except for being a similar type of lowland, rural England looks much different from either France or Belgium. Again, the variation in culture is evident. Almost all the fields in England are enclosed with hedges, and clumps of

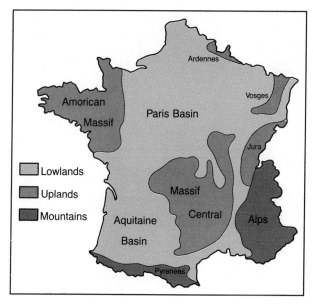

FIGURE 6.4
Physiographic Regions of France.

large old trees are common. The farmers' cottages differ architecturally from those on the continent, and clusters of trees indicate the existence of villages, castles, or cathedrals.

Back on the mainland, the Central Lowlands also include the coastal lowlands along the North Sea, and the glacial moraine areas of southern Scandinavia. The **polders** of the Netherlands coast are old lake beds next to the sea that have been reclaimed by building dikes and pumping out the sea water. The *Baltic Lake Plain* extends eastward through Germany and Poland. The name indicates that the plain contains numerous lakes and is located along the Baltic Sea. The lakes are indicators of past glaciation in the region. The coastal lowlands are cut by some of the great rivers of North Europe, including the Rhine, the Elbe, the Oder, and the Vistula. The glacial moraines created more local relief than that found on the coastal lowlands. The east-west trending hills are similar to those found in the moraine areas of North America. The Central Lowlands broaden on the east and extend from the shores of the Arctic Ocean to the Black Sea and the Caucasus Mountains. Most of this part of the region is in the Soviet Union.

The Central Plateaus. The central upland region, or the Central Plateaus, lies between the Central Lowlands on the north and the Alpine System on the south. It is an almost continuous zone of basins, plateaus, and low mountain ranges, but the sections are separated from each other. The region extends from the Iberian Peninsula on the southwest to Poland on the northeast (see Figure 6.5). Although it is an upland region, it has never been a barrier to transportation because it is dissected in numerous places by great river valleys.

The Spanish portion of the Central Plateau region is called the *Spanish Meseta* and is made up of the Plateau of Castile, the Sierra de Guadarrama, and the Cantabrian Mountains. The plateau quality of the Meseta is shown by the steep escarpment that surrounds the region on the north, east, and south. In France, the Central Plateau region is called the *Central Massif* and is outlined by a steep escarpment on the south. This southern wall, called the **Cevennes,** runs parallel to the

---

polder A region of low land reclaimed from the sea by using dikes or dams to hold back the water.

Cevennes Pronounced ca-*ven*. The southern edge of the Central Massif, located between Montpellier and Valance in southern France.

Mediterranean coast and about 50 miles (80 km) inland from the coast. On the north and west this large limestone plateau slopes gently away from the southern crest. The Central Massif is dissected in numerous places by deeply eroded, steep-sided canyons. The Tarn River, for example, runs through one such canyon. In spite of the steepness of the canyon walls, many of them are terraced, and grapes as well as other fruits and vegetables are grown.

The Central Plateau region runs northeastward from the Central Massif to include the Jura Mountains, the Central German Upland, the Bavarian Plateau, and the Bohemian Massif. The *Jura Mountains* extend about 200 miles (322 km) along

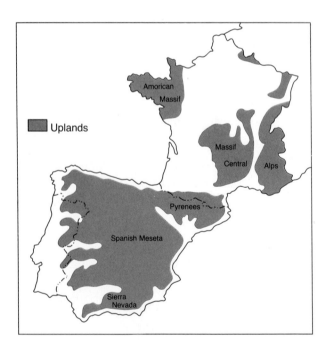

FIGURE 6.5

The Upland Regions of Spain and France.

> **Basques** A distinct race of people whose original home was in the region of the western Pyrenees Mountains near the Bay of Biscay in Spain and France.

the French-Swiss border. The highest peak in the range is Mount Reculet, at 5,462 feet (1,665 m). The *Central German Upland* is a region of glaciated granite where clear glacial lakes are surrounded by stretches of conifers. Old volcanoes dot the landscape and can be recognized under their covers of pine trees by their conical shape. The Central German Upland includes the Ardennes region of Belgium. The upland is also associated with some of the most important rivers of Central Europe. The Muese borders it on the west, and the Oder on the east. The Weser and Elbe are largely products of the upland, but the most important river is the Rhine. It flows through a *graben,* or rift valley created by diastrophism rather than by erosion. This gorge is one of Germany's most popular tourist attractions, and the river itself handles a continual procession of steamers, tugs, and barges carrying products up and down the river.

The *Bavarian Plateau* is a large, elevated basin between the Jura Mountains and the Alps. Drainage out of the basin is toward the east because it contains the headwaters of the Danube River. The city of Munich is located in the middle of the basin. The *Bohemian Massif* includes the Bohemian Basin and the low mountains surrounding it. The crests of these mountains form the boundaries between Czechoslovakia and Austria, Germany, and Poland. The city of Prague is located in the basin between the mountains.

The Alpine System. The Alpine System includes most of southern Europe and contains great complexities of relief and cultural landscapes. Mountains are the dominant feature, but plateaus, basins, wide fertile valleys, and coastal plains also are numerous throughout the region. Geologically, the Alpine System is a region, but isolated mountain chains break the region into numerous parts. The mountains give a complexity to the distribution of people. The highly populated coastal lowlands and intermontane basins contrast sharply with the emptiness of the high mountain ranges.

The western edge of the mountain ranges in the Alpine System begins with the Sierra Nevada of southern Spain. These mountains were the last stronghold of the Moors, and the land still reflects the influence of the Arab invaders from Africa. Orchards are common in the valleys, and grove after grove of dates and olives can be seen. This is the citrus region of Spain, but sugar cane also is produced. The Iberian Peninsula is separated from the remainder of Europe by the wall of the Pyrenees Mountains. Here the French and Spanish cultures tend to merge, but they are kept separated by the mountains. The tiny country of Andorra is located in one of the mountain valleys. In the western part of the mountains, the **Basques,** a racial remnant of ancient times, continue to exist as a thorn in the side of the Spanish government as they push from time to time for independence. Neither Andorra nor the Basques would have survived as separate entities if the mountains had not been there.

The Alps are Europe's greatest mountain range. They lift upward out of the Mediterranean, with the south-facing slopes along the Riviera coast, and extend into higher elevations to the north and east. The northeastern extension of the mountains stops where the Danube Valley separates the Alps from the Carpathians. The Alps carry a blanket of snow all summer, and many peaks are mantled with a coat of glacial ice. The Vienna Basin marks the transition of the Alps to the Carpathians, and the depression allows the Danube to flow eastward toward the Black Sea. The Carpathians form a large loop that encloses the Hungarian and Transylvanian basins, and meets the Balkan range at the *Iron Gate.* The gate is another gorge of the Danube River and has been used as a transportation route for centuries. At the eastern end of the Alps proper, the mountains extend southeastward, making up the Dinaric Alps, which run through Yugoslavia and the Pindus Mountains of Greece. From the Riviera coast eastward, a narrow coastal mountain range broadens into the Apennine range, which forms the backbone of Italy.

FIGURE 6.6

Mountains and Rivers of Europe. The names of the physical features of Europe should be compared with the land elevations as shown in Figure 6.3.

The British countryside is sectioned off with rows of hedges, and the location of trees indicates settlements. Part of the village of Launceton is shown here.

Climate and Vegetation

The various climates of Europe are controlled by factors similar to those mentioned for Anglo-America. Latitude, elevation and location with respect to bodies of water and to moving air currents all play a part in determining the amounts of both precipitation and heat energy received from the sun. Located between 35 degrees and 70 degrees north latitude, and positioned on the west side of a continental land mass, most of Europe lies in the zone of prevailing westerly winds. The winds bring the marine influences far inland and tend to soften the severity of the winters and diminish the summer heat. The marine influences are controlled by offshore warm waters brought by the Gulf Stream from tropical areas. The zone of the westerly winds also is where cyclonic storms develop. These whorls of air are born from the contact between the cold arctic air and warm subtropical air and are effective producers of precipitation.

Europe may be divided into three basic climatic regions (Figure 6.7). One has a marine-type climate (Köppen's Cfb), and is located along the Atlantic shore from Spain to northern Norway. The region includes the British Isles and southern Iceland and extends inland as far as central Poland. Most of the people of this region experience mild winters where long freezing periods are rare, and the summers are cool. The precipitation is regular and distributed throughout the year. Winds are usually strong, often reaching gale force during the winter months. The second climate region of Europe lies east and inland from the marine climate and is removed from the influence of the sea. This continental climate (Köppen's Dfa) is noted for its short warm summers and long cold winters. Cyclonic storms provide year-round moisture, but there is a summer maximum, and in the winter the precipitation almost always comes as snow. The northern part of the region, in Scandinavia, is covered with *taiga,* a coniferous forest found only in cold regions. The southern part of the climate zone is a vast agricultural area.

The third climate region of Europe is decidedly different from the others. It is called a Mediterranean climate (Köppen's Csa) because it is

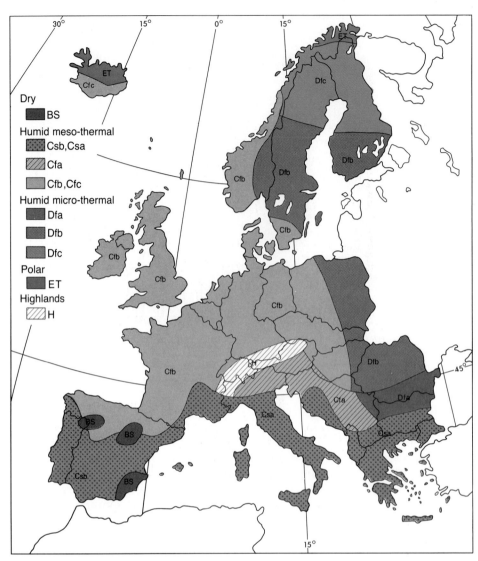

FIGURE 6.7

Köppen Climate Regions of Europe.

found all along the shores of the Mediterranean Sea. The region is protected from the cold of Northern Europe by the mountains, but it is too far south to benefit from the summer cyclonic storms. These two factors give the climate its uniqueness. The winters are delightful, with mild temperatures and sunny weather interrupted only occasionally by cloudy days and rain. The summers also are mild, but very dry. The natural vegetation in the region tends to turn brown and lie dormant during the summer drought, then comes back lush and green during the fall and winter rainy period. The growing season is year-round, with the help of irrigation in the summer. The Mediterranean region is the classic area for growing citrus fruits, grapes, olives, and dates, but

World Levels of AGRICULTURE

IN THE ADVANCED NATIONS, the proportion of farm workers in the total labor force is low. In nations which have low levels of economic development, the proportion for farm workers in the total labor force is high. Farmers in the countries of the **highly developed world** can afford to purchase mechanized equipment (as well as commercial fertilizers, insecticides, hybrid seeds, and numerous other commodities that help grow crops), and as farm work becomes more mechanized, fewer and fewer farmers are needed. In the United States, for example, the numbers of farms and farmers have steadily decreased over the last 40–50 years, but the total farm production has steadily increased. In most of the **highly developed world**, the number of farmers in the labor force is less than 10 percent. In the **underdeveloped world**, on the other hand, many of the people live directly from farming (subsistence agriculture), and many others work on farms to help feed the people of the country. Farmers in these countries constitute over 90% of the total work force. In fact, farming is the number one occupation in the world today.

World Levels of Agriculture

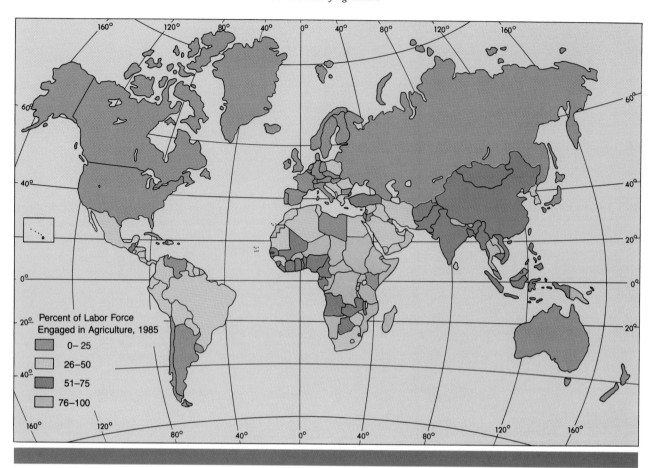

The above map contains one of the factors used to judge the levels of economic development in the world: the percentage of farmers per country. The categories for the map are based on the percentage of the labor force engaged in agriculture and are: (1) 0 to 25%, (2) 26 to 50%, (3) 51 to 75%, and (4) 76 to 100%. The range varies from 1% in Luxembourg to 95% in Bhutan.

World Levels of Agriculture

(Left) Most agriculture in the **highly developed world** is commercial, or products are produced for sale rather than for home consumption. This family on the northern Japanese island of Hokkaido is harvesting sugar beets, which they will sell to a local refinery. The sugar will eventually reach the tables of the urban people on the heavily populated southern islands. The small structures in the background are ensilage that has been cut and piled into stacks, and is curing before it is fed to the farmer's livestock.

(Below) Countries of the **moderately developed world** produce some commercial crops, but much of their agriculture is of the subsistence type. Here plantation workers in Brazil remove cacao beans from the pods in which they grow. Once the beans have dried, they will be exported to the large cocoa and chocolate markets of the northern hemisphere.

World Levels of Agriculture

(Left) Some farmers of the **developing world** enjoy a food surplus, so they can produce commercial crops that help supplement their meager incomes. This Malaysian woman, with the classic Oriental straw hat, is planting rice in a paddy field. Notice the lush, tropical growth in the background, where many useful food items are found.

(Right) Much of the farming in the **underdeveloped world** can be conducted only after clearings in the forest have been made. This "slash and burn" form of agriculture is very destructive to the tropical rainforests. It is a concern to everyone, because the rain forests are major producers of the world's oxygen. Here, fire is used to remove the cut trees and undergrowth in the rainforest near the Amazon River in Brazil. Notice how the larger trees, those most difficult to cut down, are left standing.

European Economic Community (EEC) A group of 12 Western European countries that have abolished customs, duties, and other trade restrictions in order to promote the exchange of goods among themselves. The organization started in 1957, when the Benelux countries (Belgium, the Netherlands, and Luxembourg) joined with France, Germany, and Italy in signing treaties expanding the 1951 agreement on the free movement of coal and steel. Denmark, Ireland, and the United Kingdom joined the organization in 1973, Greece became the tenth member in 1981, and Spain and Portugal joined in 1986. The EEC created a large market that competes with the United States and Japan, but the EEC members have been plagued in recent years by unemployment. Workers also are free to move among the countries, so some areas have too many people looking for jobs.

Benelux A forerunner of the EEC, Benelux is an organization based on treaties with Belgium, the Netherlands, and Luxembourg that allowed the lowering of trade restrictions among the three countries so that local resources (especially iron ore and coal) could be better utilized.

Council for Mutual Economic Assistance (CMEA) The organization of the Communist countries of Eastern Europe based on the idea of the EEC and designed to lower trade restrictions among the member countries.

found in the lower elevations. Along the shores of the Mediterranean Sea from Portugal to Greece, the natural vegetation used to be scrub-tree or brush-type evergreen forests, but most have been removed for agricultural use of the land. In California, similar cover is called "chaparral," but in Europe it is called "maquis," "makis," or "macchia," depending on the region.

Some natural grasslands exist in Europe, but such vegetation is not extensive. In Spain, steppe grass forms a natural pasture for sheep and goats on the Meseta. In Eastern Europe a similar vegetation extends from Romania along the north edge of the Black Sea into the Soviet Union. The best natural grass was found in the Hungarian Basin, but only small parts of the steppe still remain for wandering herds and flocks. The excellent prairie grass once distracted the Huns who were invading the Roman Empire, and many of them settled along that section near the Danube, rather than continue toward Rome. Today the "puzta" is cultivated into immense fields of corn and wheat, with vineyards along the sandy hillsides.

■ Human Geography of Europe

After centuries of bickering, invasions, and wars, some countries of Europe seem to be moving toward a united continent, although a major split still exists between the east and the west. Within the countries of each region, workers, raw materials, and finished products are moving across borders with increasing ease. Both the socialist states and the states of Western Europe have set up trade associations that benefit their members. The European Common Market **(European Economic Community, or EEC)** began as a small-scale trade cooperative between the **Benelux countries** (Belgium, the Netherlands, and Luxembourg). It has grown to include most of the major countries of Western Europe. The **Council for Mutual Economic Assistance (CMEA)** is a Moscow-based "common market" for the Soviet satellite countries of Eastern Europe. Trade and defense organizations help make Europe a more integrated region than it ever was in the past.

many other crops are grown too. Wheat fields are found throughout the region.

Precipitation is adequate for crops over most of Europe. The rainfall pattern suggests that the entire region should be covered with some form of forest. At one time it was, but today most of the trees have been cut to clear the land for agriculture and other human activities. The only remaining forests are found in the rough hill and mountain country, across Finland and in the northern Scandinavian taiga. Most of the trees of these forests are needleleaf evergreens, which grow in pure stands. Similar trees occupy the mountain slopes in southern Europe, but broadleaf deciduous trees are

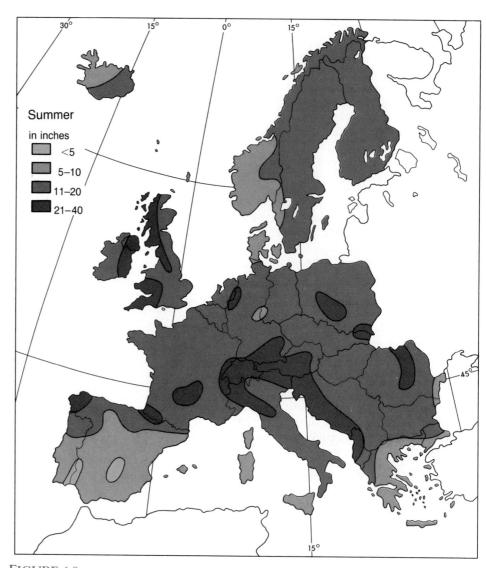

FIGURE 6.8

Summer (A) and Winter (B) Precipitation Patterns of Europe.

Population

The heavily populated part of Europe can be described as a broad belt running southeast from Manchester and Liverpool in England, across the English Channel into northern France, and into the Benelux countries, where it divides. One branch then runs east along the North European Plain into Poland, and the other extends south to Florence, Italy. Outside this population belt, heavy concentrations of people are found only in a few isolated urban areas. The core population area includes Belgium, the Netherlands, and the Rhine River valley to Stuttgart, Germany. This small core averages about 1,000 people per square mile and contains some of the great cities of Europe.

The most densely populated countries of Europe are the tiny states of Malta and San Marino.

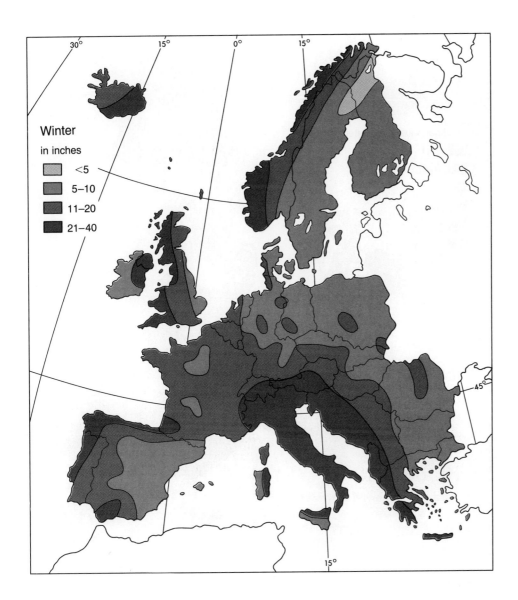

Each square mile of the islands of Malta averages 2,673 people, and the density in San Marino is about 1,000 people per square mile. The real core of Europe's population, however, is the Netherlands, where the population density is about 1,100 per square mile. Outward from there, the densities decrease with increasing distance. The average population density per square mile for the Scandinavian countries is about 50, with the low in Iceland of 6. In Eastern Europe the average density is about 300 people per square mile, with all countries near that average. In Southern Europe, excluding Malta, the population densities average about 275 people per square mile, with the lowest in Greece (194) and the highest in Italy (488).

About 73 of every 100 people in Europe live in urban areas—approximately the same urban percentage as in Anglo-America—but there is considerable variation from one country to another. Generally, the Eastern European countries are the least urbanized, and those of the population core are the most urbanized. For example, the lowest

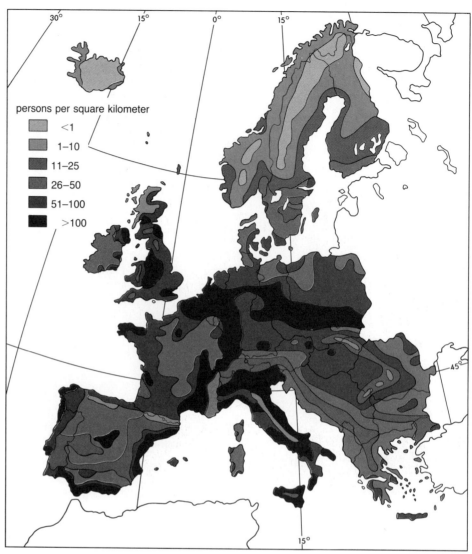

FIGURE 6.9

Population Density of Europe. The map was constructed by using the isopleth technique. The high population densities are shown in the darker shades, while the low densities are indicated by the lighter shades.

urban percentage is for Albania (34 percent), followed by Yugoslavia (47 percent), while the most urbanized countries are Belgium (95 percent), the United Kingdom (90 percent), and the Netherlands (88 percent). As with other developed regions, the percentage of the population engaged in agriculture is very low. Most Europeans live and work in the urban areas.

Most of the cities of Europe are very old, and many have familiar names (see Figure 6.10 and Table 6.2). The largest European cities are London, England; Madrid, Spain; Berlin in East and West Germany; Rome, Italy; Paris, France; and Athens, Greece. Except for West Germany, these great cities are all capitals that dominate their respective countries not only politically, but also

This view of Marseille on the south coast of France indicates the density of population in this large port city.

Picadilly Circus is a street intersection in the central business district of London, England. Note the double-decker bus and the advertisement for an American product.

Europe

FIGURE 6.10

Capitals and Other Major Cities of Europe.

economically and culturally. London and Paris are located on the banks of large navigable rivers (Thames and Seine), giving them access to world trade. Rome and Athens lie near seacoasts, but goods transported by ocean ships to Berlin must be changed from ocean going vessels at Hamburg to river (Elbe) and canal boats that connect with Berlin. The cities, then, are noted for their influence locally and for their connections with world markets. Table 6.2 contains a list of the largest cities of Europe.

Europe has the lowest annual rate of population increase of any region in the world (see Figure 6.11). The average rate of natural increase as of the middle of 1986 was 0.3, compared with the world rate of 1.7. The populations are declining in Denmark, Hungary, and West Germany and have stabilized (that is, the birth and death rates are the same) in Austria, East Germany, Luxembourg, and Sweden. This slow growth, combined with the highest life expectancy rates in the world, means that the populations of most European

TABLE 6.2
Largest Cities of Europe with 1986 Populations

COUNTRY	CITIES AND POPULATIONS IN THOUSANDS			
Northern Europe				
Denmark	Copenhagen	642	Arhus	250
Finland	Helsinki	485	Tampere	170
Iceland	Reykjavik	85		
Ireland	Dublin	525	Cork	140
Norway	Oslo	450	Bergen	210
Sweden	Stockholm	650	Goteborg	430
United Kingdom	London	6,700	Birmingham	1,100
Western Europe				
Austria	Vienna	1,500	Linz	125
Belgium	Brussels	1,000	Antwerp	500
France	Paris	2,300	Marseille	1,000
Luxembourg	Luxembourg	80		
Switzerland	Bern	100	Zurich	400
The Netherlands	Amsterdam	700	Rotterdam	600
West Germany	Bonn	400	West Berlin	1,800
	Hamburg	1,660	Munich	1,200
Southern Europe				
Albania	Tirana	275	Durres	130
Greece	Athens	3,020	Thessaloniki	800
Italy	Rome	2,900	Milan	1,600
	Naples	1,200	Turin	1,100
Portugal	Lisbon	815	Porto	150
Spain	Madrid	3,200	Barcelona	1,800
Ygoslavia	Belgrade	1,300	Zagreb	700
Eastern Europe				
Bulgaria	Sofia	1,100	Plovdiv	310
Czechoslovakia	Prague	1,100	Brno	385
East Germany	East Berlin	1,170	Liepzig	560
Hungary	Budapest	2,100	Mislsolc	220
Poland	Warsaw	1,600	Lodz	845
Romania	Bucharest	1,980	Brasov	350

Note: The capital of each country is listed first, regardless of its size, and Italy and West Germany each have four cities with over one million population, so they are all listed in the table.

SOURCES: (1) *Demographic Yearbook, 1987*, New York: United Nations Publications, 1988, (2) *Demographic Statistics, 1981*, Washington, D.C.: European Community Information Services, 1983, and (3) *The World Figures*, Detroit, Mich.: The Economist, 1981.

> ## Focus Box
>
> ### Working Mothers and Child Care
>
> The increase in the number of working mothers throughout the world has created a need for child-care centers. More than 100 countries now have national policies concerning such centers. The United States is not among these countries, and it lags behind most industrial nations in offering an adequate child-care system. Many European countries, however, have extensive networks of child-care centers that are subsidized and regulated by the government.
>
> In France, working mothers are legally entitled to at least 16 weeks of maternity leave. They receive 84 percent of their salaries while on leave. When they start back to work, they can choose from 1,494 state-run child-care centers that are open 11 hours a day at an average cost of about $10 a day. The centers are called "crèches," but numerous *haltes garderies* also exist, where children up to 5 years old can be left to play for a few hours at a time. Nonworking mothers who care for other children in their homes are called *nourrices*. In addition to the crèches, the *haltes garderies* and *nourrices* also are subsidized by the French government. If the local crèches are oversubscribed, the government pays as much as $340 a month to parents who hire help to stay with their children in their own home.
>
> In Sweden, new parents are guaranteed a one-year leave of absence after childbirth, and Social Security pays 90 percent of the mother's salary during that time. Denmark offers working mothers only 24 weeks of maternity leave, but the country has a much larger network of child-care centers than Sweden. About half the Danish children younger than age 3 are enrolled in public day-care facilities. The percentage increases to about 70 percent for children between the ages of 3 and 4. The monthly fee for day care in Denmark is only about $115, and the demand is greater than the supply, so the Danish equivalent of the French *nourrices* is becoming popular.

countries are getting older. The percentage of the European population that is 65 years old and older is 13 percent, compared with 6 percent for the entire world. The country with the oldest population in Europe is Sweden (17 percent), followed closely by Denmark, Norway, the United Kingdom, and West Germany, all with 15 percent of their population at the age of 65 or older.

Religion

Three major religious orientations exist in Europe, all of which are Christian. These are Protestantism, Roman Catholicism, and Eastern Orthodoxy. In the southeastern countries of Yugoslavia, Romania, Bulgaria, and Greece, most of the people are of the Eastern Orthodox faith. The Eastern Orthodox church is closely related to the Roman Catholic church, but it became a separate branch of Christianity when the ancient Roman Empire was divided. The western branch of the Christian church followed the lead of the church in Rome, under the pope, while the eastern branch developed without a central authority. Southeastern Europe today has many Roman Catholics as well as a few Muslims. In fact, Muslims outnumber all others in Albania, and many are found in Yugoslavia.

The second major split from the Roman Catholic church in Europe can be traced to October 31, 1517. On that day a German Catholic monk and priest, named Martin Luther, nailed his strong criticisms of the Roman church to the door of the church in Wittenberg. This small but historic town is located in what is now East Germany, about halfway between Berlin and Leipzig. Luther's pro-

Protestant Reformation The religious movement in Western Europe, beginning in the early sixteenth century, which led to the formation of the various Protestant churches. Martin Luther led the movement by attacking the Roman Catholic church's concepts of paying for indulgences for church favors, the veneration of the Virgin Mary and the saints, and the practice of clerical celibacy.

tests and his defense of them produced a cleavage between Protestant Northern Europe and Catholic Southern Europe that has lasted to the present day. This religious upheaval, which came to be known as the **Protestant Reformation,** has touched the lives of millions of worshipers throughout the world. The boundary between the Roman Catholic faith and the Protestant faith is roughly the same as the battle line across Europe at the end of the Thirty Years' War in 1648. This series of wars

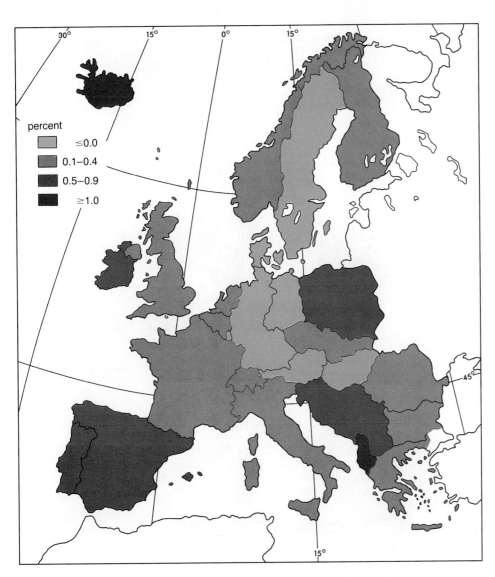

FIGURE 6.11

Population Increase in Europe for 1986. The rate of natural increase for each country is the surplus of births over the number of deaths, without regard for migration and standardized to a rate by using the total per 1,000 people.

was fought originally between German Catholics and German Protestants, but later it involved Sweden, France, and Spain as well.

Today, the countries of Europe that are predominantly Roman Catholic are Ireland, the Benelux nations, most of southern Europe, and a belt of countries running from northern Italy through Lithuania in the Soviet Union, including most of Switzerland, Austria, Czechoslovakia, southern Germany, Poland, and western Hungary. Overall, there is a slight Protestant majority in both Germany and Switzerland. Protestants make up the majority in the Scandinavian countries, Great Britain, northern Germany, northern Poland, and Finland. The Protestant denominations include Lutherans in all the Scandinavian countries, Finland, and northern Germany; the Anglican church in England; and the Presbyterian church in Scotland (see Figure 6.12).

FIGURE 6.12
Religions of Europe.

Romanic (or Romance) languages The Indo-European languages originating from the countries that succeeded the Roman Empire, especially Italy, Spain, Portugal, France, and Romania.

Teutonic languages The languages of the Northern European people, including the Germans, the Scandinavians, the Dutch, and the English. Sometimes called Germanic languages.

Slavic languages A group of related languages spoken by the Slavic people, with an alphabet based on Greek, as opposed to the alphabet of the Romance languages, which is based on Latin. The Slavic people include the Russians, Poles, Bohemians, Moravians, Bulgarians, Serbians, Croatians, Wends, Slovaks, and other small groups.

Language

Most of the languages of Europe can be divided into three main branches of the Indo-European family according to linguistic differences. Although the picture is complicated by the existence of other miscellaneous languages, in general Europeans speak a Romanic language, a Teutonic language, or a Slavic language.

People who speak a **Romance language** are found in southwestern Europe, primarily France, Italy, Spain, and Portugal. The speech of Romania, however, also is a Romance language. All five of the Romance languages are historically related to Latin, the language of the old Romans. The region of Romance-speaking people today is a geographical reflection of the ancient Roman Empire. Still, there are within the region numerous enclaves where small groups of people speak a language other than the official language. One of the most noted is in the Basque country in the border region between France and Spain. The Basque language is not only not a Romance language, it is also unrelated to any other language in the world. The Basques have been somewhat of an anthropological mystery because they differ from all other people of Europe in facial features and domestic customs, as well as in language. The Basques seem to represent a relic of some former population that has maintained its uniqueness by being geographically isolated within the rugged mountains where they live.

Other linguistic enclaves within the Romanic region of Europe include the people that speak Breton (Celtic) and live along the Brittany coast of France. Also, in southern Italy and Sicily there are small areas where the people speak Albanian, a language of the Illyrian family spoken mainly in Albania. In Romania, both German and Hungarian are spoken in small regional pockets. Besides these enclaves within the Romance-speaking region, there are also exclaves (small areas outside the main region) where one of the five main Romance languages is spoken. In southern Belgium and western Germany, for example, the French Walloons speak a Romance language, and in southern Switzerland both French and Italian are spoken (see Figure 6.13).

The region where the Teutonic languages are spoken is primarily northwestern Europe. The **Teutonic languages** include Icelandic, Swedish, Norwegian, Danish, English, German, Dutch, and Flemish. The name of each of these languages corresponds to that of the country where it is spoken with the exception of Flemish, which is found in northern Belgium and the southern part of the Netherlands. Anomalies in the northwest are the Celtic languages of the British Isles (Irish, Gaelic, and Welsh) and the small groups of people in northern Scandinavia who speak Lapponian. The Lapp language is of the Uralic group and is unrelated to the Teutonic languages.

The third major language group of Europe includes the **Slavic languages** of Eastern Europe. Branches of the Slavic family are spoken in Poland, Czechoslovakia, Yugoslavia, and Bulgaria. The Slavic-speaking people of the Soviet Union feel a close relationship to the "Little Slavs" of eastern Europe. It might be noted that "Yugo" means "south,"—thus "Yugoslavia" means "land of the south Slavs." Czechoslovakia was created by combining the western half of the country, where Czech was spoken, with the eastern half, where Slovak was spoken. The countries of Eastern Europe where the Slavic languages are not spoken are Albania, Finland, Greece, Hungary, and Romania. As noted above, Albanese is an Illyrian tongue, and Romanian is a Romance language. Finnish and Hungarian are Uralic languages orig-

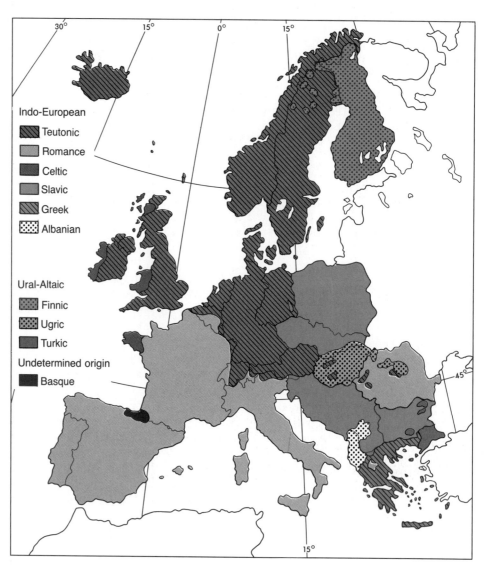

FIGURE 6.13
Languages of Europe

inating in Central Asia. They also are historically related to the Lapp language. Greek is a Hellenic branch of the Indo-European family but otherwise unrelated to any other language in Europe.

The hodgepodge of languages in Europe may appear to be a "Tower of Babel," but there are some relationships between historical factors and geography. The Celtic, Illyrian, Slavic, Romance, Hellenic, and Teutonic languages are all in the Indo-European family of languages; Hungarian, Lapp, and Finnish are in the Uralic family; and the Basque language is in a group of its own. People that speak one of the languages within a group usually have little trouble learning other languages of the group. A few large regions of Europe are somewhat internally homogeneous because of the similarity of the languages within them. While anomalies do occur, a basic correspondence exists between languages and national boundaries (see Figure 6.13).

Ethnicity

The core areas of ethnicity in Europe are the individual countries. Most of the historical friction areas, or zones where cultural clashes have occurred, are along the political boundaries. Europe has many political boundaries, so the potential for trouble has been great. Also, the lowering of restrictions for the movement of workers has created new ethnic troubles. This movement has been an economic boon, but it has created problems within countries where the ethnic minorities have congregated. The breaking up of colonial empires and the resulting free movement of people from India and Africa into Europe also has created local ethnic problems. For example, the Indians, Pakistanis, and blacks from the West Indies have not been integrated into English society. France has been experiencing recurring problems with the blacks from Northern Africa, especially those from Algeria.

The European ethnic situation is complicated by the fact that people living in a transition zone often feel different from the people of the two adjoining countries. The Basques, between France and Spain, are the classic example of such a group, but the Catalonian people of northeastern Spain also seek independence based primarily on the difference between their language and that of the

---

FOCUS BOX

The Balkans

A complex of geography, races, languages, religions, economics, politics, and history exists in the southern part of Eastern Europe. The region known as the Balkans consists of Yugoslavia, Albania, Romania, and Bulgaria. "Balkan" is a Turkish word for mountain, and the Balkan mountain range runs east-west through the middle of Bulgaria. Traditionally, the Balkan Mountains separated the maritime civilization on the south from the barbaric tribes to the north, but they were not effective as a political divide. The name also has given rise to the term "Balkanization," which means to separate a region into smaller units.

The Balkan countries were created after World War I, and in fact were "Balkanized" from the Austro-Hungarian Empire. Today the four countries stand on the ruins of ancient and medieval empires and contain a diversity of cultures and ethnic minorities. The region has been influenced by the Romans, the Goths, the Huns, the Avars, the Slavs, the Magyars, the Byzantines, the Tatars, the Turks, the Venetians, the Austrians, the Russians, and the Germans. Unfortunately, however, and perhaps because of the ethnic diversity, the Balkan countries do not have a tradition of good government. The four Balkan countries are now under strict Communist rule, but revolts and dictatorships have punctuated their history.

Private property and private enterprise are controlled by the Communist Party in each country. Farm workers are allowed to own small plots of land, some artisans can own small shops, and professional people usually own their own homes, but rigid controls of these types of activities are maintained. Control is also held over all money transactions, such as rents, prices, taxes, and credit. In spite of the controls, considerable industrial development has occurred recently as the younger, better educated, and more Westernized leaders gained political control. The changes are most evident in the cities. Material improvements in the rural areas come slowly, and each of the four countries remains a predominantly agricultural society, so the changes are not affecting many people.

remainder of Spain. In Belgium, there has been friction between the French-speaking Walloon minority and the Flemish Dutch-speaking majority. The Alsace-Lorraine region of eastern France is another transitional area. The people still speak a German dialect, and the villages and towns show German characteristics. The same is true in parts of northern Italy, where German ethnics have been living since the region belonged to Germany decades ago. In Yugoslavia, the Serbs, Croats, and Slovenes each have moved at various times toward political confrontation with their government.

Areas that are historically or ethnically related to one country but governed by another are called **irredentas.** European ethnic minorities often have sought to sever their portion of territory from their country in order to unite with the country of their national brethren, and their own governments often abetted their plans. Minority groups have claimed persecution, and national governments often have been forced to coerce their minority citizens in order to maintain social order. Irredentism, the desire of a country to incorporate irredentas, has encouraged political turmoil.

■ *Economic Geography of Europe*
Agriculture

The history of agriculture in Europe shows two great tendencies. One is the spread of intensive agriculture from the borders of the Mediterranean northwestward to the North Sea region. The other is the change from subsistence farming to commercial farming, where the farmer raises very little for his own use. Until the end of the Middle Ages, most farming in Europe was subsistent. The land was owned by the nobility or the church, and the peasants were concerned merely with survival. Little trade existed, and techniques for preserving food were not developed. The crops grown were mostly grains, such as barley and wheat. No fertilizers were used, and little attention was paid to the land itself. The French Revolution (1789), however, brought enormous changes to agriculture in Western Europe. It marked the emergence of individual land ownership, which in turn introduced farming as an occupation rather than just a means of survival.

> **irredenta** Literally, unredeemed. A region containing people (irridentists) who advocate policies that seek to reincorporate the region into the country of their national origin. The term was applied originally to Italians living within Austrian territory. *Italia irredenta* means "unredeemed Italy."

About the same time as the French Revolution, the industrial revolution came to England. The farm workers began to migrate to the cities for jobs, and an ever-growing demand for food began. Farmers started to see the advantages of producing a surplus, and with the help of newly invented tools and scientific techniques, they began to produce more. The idea of crop specialization came from the Mediterranean region, where intensive commercial raising of grapes, olives, and a few vegetables had existed since Roman times. Crop specialization outside the Mediterranean area occurred first in the Flanders region of Belgium. That was the core area for the spread of modern agriculture in Europe, and all agricultural land there remains heavily utilized today.

This farmer on the Spanish Meseta is using primitive methods for plowing his field, much like the peasants of the Middle Ages. Although most farming in Europe is mechanized, occasionally a scene like this is found. Photograph by Louis Seig.

Agriculture in Europe varies considerably between the crops grown in the Mediterranean coastal areas and the remainder of the continent. Along the Mediterranean, agriculture is intensive and commercial, where relatively small, highly utilized plots of land are important. Grapes, olives, and vegetables remain the most important crops, but the region also produces citrus fruits, wheat, and barley. Mediterranean diets are heavy with pastas, wine, and vegetable salads. Agricultural exports include wine from France, Italy, Spain, and Portugal. Olives and their products, such as olive oil, come from Spain, Italy, and Greece. Spain and Portugal export cork, which comes from the cork oak trees that are native to the region. Spain and Italy produce some rice. Livestock consists mostly of sheep and goats, but cattle are grazed in all countries that have any grasslands at all. Hogs are common, especially in France, where they sometimes are used to hunt truffles and to scavenge for acorns in the oak forests.

Cash-crop farming and commercial livestock raising are common in a broad belt north of the Mediterranean region from northern Spain across

Outdoor markets like this one in Gerona, Spain, are common in Europe and provide a means for farmers to display the variety of crops they produce.

the continent to the Soviet Union. The crops of this region include wheat, rye, oats, barley, and corn (maize). Although the summers are too cool for wheat and corn to mature properly, government subsidies encourage their production. France is the leading producer of wheat and corn, and Italy is second (see Figure 6.14).

Besides the grains, root crops play an important role in the agriculture of the middle agricultural belt of Europe. Sugar beets, turnips, rutabagas, and potatoes are grown extensively throughout the region for both human and animal consumption. Potatoes, native to South America, are so commonly identified with Ireland that they are often called Irish potatoes. In 1845 a blight struck the Irish potato crop and lasted through three seasons, creating a famine. This disaster illustrated the importance of potatoes to the European diet. Potatoes were the main source of food for the rural people, and almost the only food for the very poor. As a result of the famine, about 1.5 million Irish starved to death, and another million were forced to emigrate. The population of Ireland today is about 3 million less than

This farm building in the Alsace region of northeastern France near Cirey-sur-Vezouze provides shelter for both humans and animals. The barn is on the left, living quarters for humans are on the right.

Europe 149

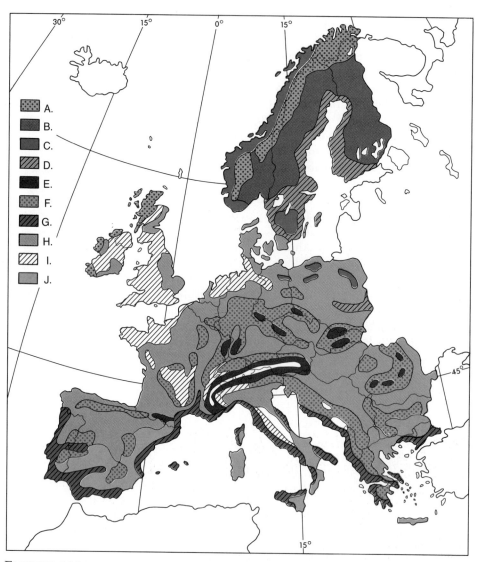

FIGURE 6.14

Agricultural Regions of Europe. Because Europe has been settled for so many centuries, much of the natural vegetation has been removed for farming or fuel. The agricultural regions, however, are fairly distinct because the farmers have adapted to the environmental conditions under which they live. The land-use regions are (A) tundra, with little vegetation and no farming, (B) northern coniferous forests with little farming, (C) mixed coniferous and deciduous forests with farming on small clearings only, (D) zone where the forests have been cleared and farming is fairly extensive in the cleared areas, (E) mixed forests and grasslands of the alpine regions with transhumance livestock grazing and some hay grown, (F) mixed woodlands and clearings along with mixed farming, (G) zone of intensive cultivation where grains predominate, (H) zone of intensive cultivation where the cattle industry predominates, (I) zone of Mediterranean agriculture that is nonproductive because of altitude or cultural attitudes, and (J) zone of Mediterranean agriculture devoted to grapes, orchards, and truck farming. The white areas in the alpine regions on the map are places that are unsuited to vegetation growth.

transhumance The seasonal movement of livestock to and from mountain regions to utilize the mountain grasses for summer grazing and to protect the animals at lower elevations during the winter.

it was 140 years ago. Other crops of middle Europe include fruit (especially apples), tobacco, sunflower seeds, and rape, a plant in the mustard family whose seeds are used for cooking oil.

The most common livestock produced in the middle agricultural belt of Europe are cattle, hogs, and sheep, but chickens and domestic ducks, geese, and rabbits also are found on most farms. The livestock are produced primarily for local consumption, but many are sold to the large urban markets. Sheep are especially numerous in Scotland, Wales, and the Spanish Meseta. Goats are found primarily in the Mediterranean areas.

The third great agricultural belt of Europe is the dairy region, located in northwestern Europe along a narrow strip that includes the coastal sections of the British Isles, northern France, Belgium, the Netherlands, Germany, Poland, and the southern parts of the Scandinavian countries. In this region, beef cattle take the place of dairy cows. The region produces all dairy products, and some of the cheeses enter the world market. Well-known dairy breeds, such as the Jersey, Guernsey, Ayrshire, Holstein, and Brown Swiss, were developed in the region and are now used in all the other dairy regions of the world.

Denmark is the leading exporter of dairy products in Europe. There the individual farms are quite small by Anglo-American standards. Many are smaller than 2 acres (5 hectares), but the cows are only housed and milked on these small plots. Most feed for the cows must be purchased. Other areas of Europe with intensive dairying are the extreme northern section of Italy, southern Austria, and Switzerland. In these mountainous regions, **transhumance** is a distinctive activity. In the summer the dairy herds are moved into the

In this agricultural valley in Spain, the slopes are covered with rows of grapevines while wheat is grown on the flat valley floor. The bare patches in the middle of the picture are threshing floors, where the wheat is separated from the straw. Photograph by Louis Seig.

high mountain pastures; in the fall they are brought down to the valleys, where hay and root crops are used for feed through the winter.

Mineral Production

The minerals of Europe are the foundation for the manufacturing that is the hallmark of the continent. The industrial revolution began in Great Britain during the seventeenth century and spread east and south across Europe. The most highly industrialized section today is the northwestern part of the continent, including Great Britain, France, West Germany, and the Benelux countries. The industrial belt extends into southern Norway and Sweden, Polish Silesia, Czechoslovakia, East Germany, and northern Italy. The manufacturing regions correspond primarily to the areas noted for their high densities of population. Nearly every type of industrial product is manufactured in the region.

The basis of European industry is coal, and the most significant products are iron and steel. Coal is a cheap source of fuel for the world as a whole, and the countries with an ample supply have an advantage in producing steel (Figure 6.15). The steel is used in further manufacturing to make thousands of products, such as machinery of all types, moving equipment, military devices, agricultural equipment, and a host of other important goods. Even where electricity is produced extensively, coal is usually more important than water for turning generators, but the mountainous countries of Norway, Sweden, Switzerland, Austria, and to some extent Italy depend heavily on hydroelectric energy. France is one of the world's leading users of nuclear power, and more than half the country's electrical needs are met with nuclear energy.

Iron and steel production in Europe is dominated by West Germany, but recent trends show a decline in output. The United Kingdom was once a leading producer of iron and steel, but economic problems have produced an even greater decrease in output than in West Germany. Italy has surpassed both France and England to become Europe's second leading steel producer. In spite of sharp declines in output, Western Europe still leads the continent in crude-steel production (see Table 6.3 for trends). The combined European production of pig iron and steel is approximately 35 percent of the world's total.

Although France is not the leading steel producer in Europe, it does lead all European countries in deposits of iron ore. The Lorraine district in the northeastern corner of the country is the most well known, and the deposits extend into Belgium and Luxembourg. The ore from this field is easily mined and has an iron content of about 30 percent. The district supplies the smelters in France, Germany, Luxembourg, and Belgium. France usually exports more than 40 percent of its iron ore, mostly to its three neighbors. Noted European iron ore deposits are also located in West Germany, Sweden, Spain, and the United Kingdom (see Figure 6.16).

Coal is the necessary energy source in the smelting of iron, and Europe is a coal surplus area. In Germany, the Ruhr region ranks supreme as a coal producer. Thick seams of high-quality coking coal are shaft-mined at about one-half mile below the surface. The region takes its name from the Ruhr River, a tributary of the Rhine. The entire region between the Ruhr and the Lippe rivers eastward from the Rhine for about 50 miles (80 km) is one large industrial complex. All these industries are based on the iron and steel industry, which in turn relies on the coal. The industrial cities of Duisburg, Essen, and Dortmund lie within the Ruhr region. Another famous coal-producing area of West Germany is the Saar, but coal from the Saar region must be mixed with that from the Ruhr region for proper quality for coking. The Saar too takes its name from a river and is located west of the Rhine near Luxembourg. The Ruhr and the Saar are two of the world's most important and best known industrial regions. Other centers of importance in Europe are the Paris Basin, especially from Paris northward to the Belgian border, the Thames Valley in England, and the Milan-Turin area of the Po Valley in Italy.

Eastern Europe has good-quality coal deposits in Upper Silesia. These thick, easily mined seams are shared by East Germany and Poland. Czechoslovakia leads the European socialist nations in the production of crude steel (see Table 16.2), and that production relies on local supplies of coal. However, much of the coal in Eastern Europe is lignite,

North Sea oil Petroleum produced by using drilling-platform stations offshore in the North Sea region.

which has a lower carbon content and is less useful than the bituminous coals of western Europe.

The world's first oil discovery was in Romania, and the fields around Ploesti produced nearly half the European output for many years. Commercial production of the **North Sea oil** began in 1975, and shortly after that the United Kingdom and Norway both surpassed Romania as Europe's leading producers of crude oil. All the countries bordering the North Sea now have proven reserves of petroleum. The United Kingdom and Norway remain the leaders in north sea oil production, followed by West Germany and Denmark. Other than the North Sea and Romanian fields, Europe has very few known oil deposits, so it is deficient

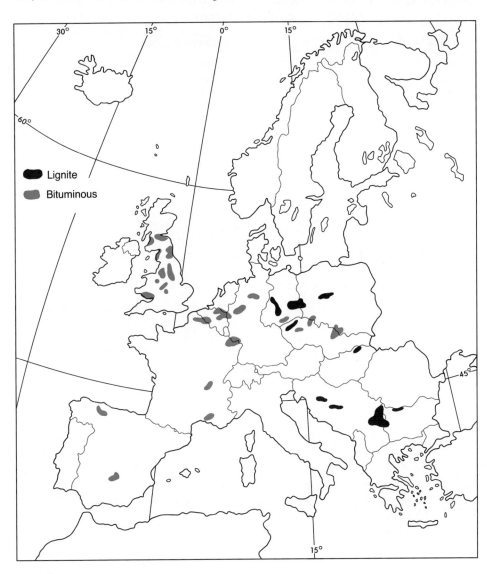

FIGURE 6.15
Major Coalfields of Europe.

Europe 153

TABLE 6.3

Crude-Steel Production in Europe (in millions of metric tons)

COUNTRY	1978	1983	COUNTRY	1978	1983
Northern Europe			**Eastern Europe**		
Denmark	0.86	0.50	Bulgaria	2.60	2.50
Finland	2.30	2.40	Czechoslovakia	15.40	14.90
Ireland	0.07	0.03	East Germany	6.90	7.10
Norway	0.79	0.70	Hungary	3.90	3.70
Sweden	4.30	8.20	Poland	19.50	16.50
United Kingdom	20.30	13.60	Romania	11.80	13.60
Total	28.60	25.40	Total	60.10	58.30
Western Europe			**Southern Europe**		
Austria	4.30	4.20	Greece	1.00	0.90
Belgium	12.60	9.90	Italy	24.30	23.90
France	22.80	18.40	Portugal	0.63	0.40
Luxembourg	4.80	3.50	Spain	11.30	12.70
Netherlands	5.60	4.30	Yugoslavia	3.50	2.80
Switzerland	0.70	0.90			
West Germany	41.30	36.30			
Total	92.10	77.50	Total	40.70	40.70

Note: Data not available for Scotland and Albania.
SOURCE: *International Iron and Steel Institute Yearbook, 1980, 1986.*

in petroleum (see Figure 6.17). Large amounts must be imported from Africa and the Middle East.

Other minerals important to European industry include bauxite ore (aluminum), found primarily in France. Smaller deposits also have been found in Italy, Yugoslavia, Romania, and Greece. Zinc is produced in Ireland, Sweden, and Finland, as well as in all the Eastern European countries. Italy is the leading producer, with large zinc deposits in the Po Valley and on the island of Sardinia. Chromium ore, used to make special hard-steel products (chrome), is mined in Southeastern Europe, in Albania, Greece, and Yugoslavia. Other steel alloys are mined in various regions of Europe,—tungsten in Portugal, manganese in Romania, and nickel in Greece.

Trade and Transportation

Europe is noted for the vast amount of movement of both goods and people that occurs on a regular basis. Trade and transportation are two activities that Europeans have accomplished very effectively for a long time. Trade and transportation depend on each other. Most of the trade in Europe goes on among the countries of the region, so good internal transportation networks are important.

Some trends in European trade can be seen in Table 6.4. For example, the major trade partner in terms of exports for all the Eastern European countries is the Soviet Union. This is not surprising, because these countries are Soviet satellites, but the volume of trade is interesting. More than one-third of every Eastern European country's exports, except Romania's, goes to the Soviet Union, and nearly half of Bulgaria's exports. The leading European trader by far is West Germany. Besides Finland, Iceland, and the Soviet bloc countries, West Germany is a major trade partner of every country in Europe and is the primary partner for eight countries. The next leading traders, after West Germany, are France and the United Kingdom.

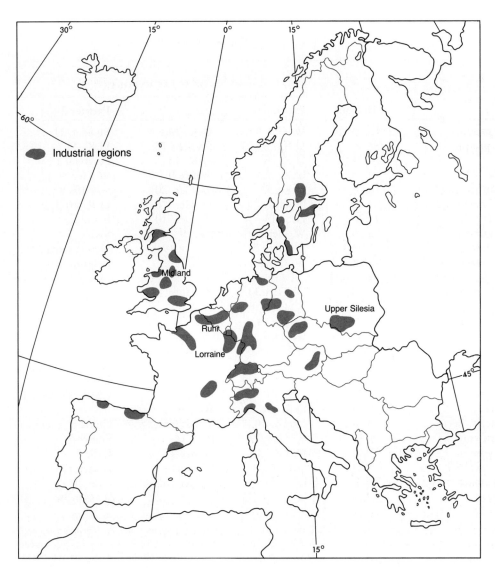

FIGURE 6.16
Industrial Regions of Europe.

The United States is a major trade partner only with Iceland, Italy, Spain, and the United Kingdom (see Table 6.4). One reason for the lack of trade with the United States is the fact that many European exports are manufactured items that are not needed in the United States. Another reason is that steel is a major export item in seven countries, and the United States is a leading steel-producing nation.

Transportation facilities vary considerably throughout Europe, but the overall movement of goods and people is excellent (Figure 6.18). Northwestern Europe has a superb systems of railroads, roads, waterways, and air routes, but in Scandinavia, Spain, Portugal, southern Italy, and some places in Eastern Europe these systems are less well developed. Most of the freight in Eastern Europe moves by railroad (see Table 6.5, especially Poland), but the networks are thin and overloaded with traffic. The Balkan countries and part of the Iberian Peninsula still have the older wide-gauge track, so freight must be unloaded and reloaded

TABLE 6.4

European Exports: Commodities, Amounts, and Major Partners for 1984

			% OF TOTAL EXPORTS		
Country	Exports	$ Billions	Partner 1	Partner 2	Partner 3
Denmark	Dairy products	15.90	W. Ger. 17	U.K. 14	Sweden 11
Finland	Forest products	13.40	U.S.S.R. 27	Sweden 12	U.K. 11
Iceland	Fish	0.94	U.S. 28	U.K. 13	Portugal 12
Ireland	Potatoes	9.60	U.K. 39	France 9	W. Ger. 9
Norway	Oil, timber products	18.90	U.K. 37	W. Ger. 20	Sweden 9
Sweden	Steel, forest products	29.30	Norway 11	W. Ger. 10	U.K. 10
United Kingdom	Vehicles	93.70	U.S. 13	W. Ger. 10	France 8
Austria	Steel, grain	15.70	W. Ger. 29	Italy 9	Switz. 7
Belgium	Steel, meat	51.90	W. Ger. 20	France 19	Netherlands 14
France	Steel, meat	97.50	W. Ger. 14	Italy 11	Belgium 8
Luxembourg	(Trade included with Belgium)				
Netherlands	Metals, grain	65.70	W. Ger. 30	Belgium 14	France 10
Switzerland	Watches	25.50	W. Ger. 18	France 9	Italy 9
West Germany	Ships, automobiles	171.70	France 14	Netherlands 8	Belgium 7
Bulgaria	Chemicals	11.20	U.S.S.R. 48	E. Ger. 6	
Czechoslovakia	Wheat	15.70	U.S.S.R. 38	E. Ger. 10	Poland 7
East Germany	Steel, grain	24.00	U.S.S.R. 35	Poland 10	Hungary 5
Hungary	Steel, meat	8.50	U.S.S.R. 34	W. Ger. 7	E. Ger. 6
Poland	Ships, grain	4.90	U.S.S.R. 30	E. Ger. 6	Czechoslovakia 6
Romania	Steel, corn	11.50	U.S.S.R. 18	W. Ger. 7	
Albania	Chemicals, beets	0.20	Czech. 10	Yugoslavia, 9	China 8
Greece	Textiles, corn	4.40	W. Ger. 19	Italy 14	France 7
Italy	Steel, wine	73.30	W. Ger. 16	France 15	U.S. 7
Portugal	Cork, textiles	5.20	U.K. 15	W. Ger. 13	France 13
Spain	Olives, machinery	23.50	U.K. 21	W Ger. 18	U.S. 10
Yugoslavia	Wood products	10.20	U.S.S.R. 33	Italy 9	W. Ger. 8

SOURCE: *Foreign Trade by Commodities, Series C,* Washington, D.C.: OECD Publications, 1984; *Handbook of International Trade and Development Statistics,* New York: United Nations Publications, 1986; *International Trade, 1982/83,* Ann Arbor, Michigan: UNIPUB, General Agreement on Tariffs and Trade, 1984; *Monthly External Trade Bulletin,* Washington, D.C.: European Community Information Service, (March 1984); *Monthly Statistics of Foreign Trade, Series A,* Washington, D.C.: OECD Publications and Information Center, (June 1984).

before going to other parts of Europe. The passenger rail traffic also focuses on the industrial cities of the northwest, and express trains move thousands of people between the major cities. The express train between Paris and Lyon is the fastest in the world and careens along the route at more than 200 miles per hour. France also leads the continent in the number of passenger-kilometers per year. Precise railroad schedules generally are maintained in most of Europe, and the systems run with efficiency and precision. The famous Orient Express that runs from Calais to Istanbul still provides a delightful tourist excursion across Europe.

The highway systems in Europe are similar to the railroads in that they are well developed in the northwest but deteriorate somewhat toward the south and east. In the northwest, a network of fine concrete roads carries large volumes of fast automobile and truck traffic. The German roads, especially the autobahns, may be the best in Europe,

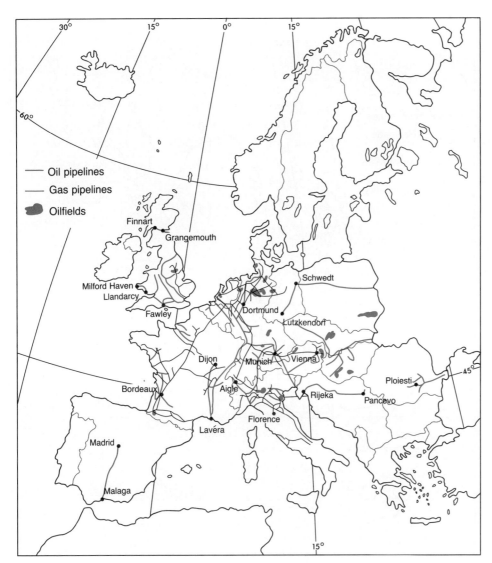

FIGURE 6.17

Petroleum-Producing Areas and Pipelines of Europe.

although the toll roads in France and the "motorways" in England are nearly as good. The major difference American drivers find in Europe is the lack of high-speed roads within the urban areas. The highways between European cities are very good, but they end when one comes to an urban area, and no city loop roads have been built. European cities are old, and most have narrow, winding streets that were not intended for fast vehicular traffic. Road travel in Europe, then, is always slowed considerably at nearly every village, town, and city.

Inland water transportation in Europe is aided by excellent port facilities, many deep rivers, a vast canal network, and historical momentum. Large amounts of freight are moved by barges and river boats, especially in France, West Germany, Belgium, and the Netherlands The Rhine River, Europe's busiest inland waterway, has been used for centuries. The long Rhine barges, pushed by tug-

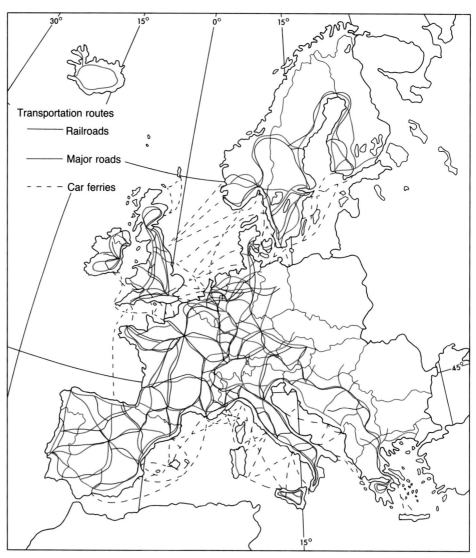

FIGURE 6.18
Transportation Links of Europe.

boats, fly flags from many nations, showing the international character of the river. Besides the barge traffic, a boat trip on the Rhine is a favorite tourist attraction in Europe. In the east, the Danube is a major transport artery, and East Germany and the Bohemian section of Czechoslovakia use inland waterways, especially the Elbe River. The canal system in England has been nearly abandoned for hauling freight, but the canals are still open. They are used now mostly for long, narrow houseboats leased to tourists.

Ocean traffic in Europe is served not only by outstanding port facilities, but also by the convenient location of the ports. Here too, however, Eastern Europe is at a disadvantage. The ports along the inland seas, such as the Black Sea and the Baltic Sea, are not as accessible for world trade as those along the North Sea and the English Chan-

TABLE 6.5
Transportation in Europe: Modes and Amounts

Country	RAILROAD (IN MILLIONS OF PERSON- AND TON-KILOMETERS PER YEAR)		VEHICLES (IN MILLIONS)		AIRLINES (IN MILLIONS OF PERSON- OR TON-KILOMETERS PER YEAR)	
	Passenger	Freight	Cars	Trucks	Passenger	Freight
Denmark	4.200	1.600	1.300	.400	2.900	.100
Finland	3.300	8.000	1.300	.200	2.500	.700
Iceland			.090	.009	1.400	.300
Ireland	.800	.600	.700	.070	2.300	.800
Norway	2.200	2.500	1.300	.200	4.100	.200
Sweden	6.400	14.900	2.900	.200	5.500	.200
U.K.	30.100	17.100	16.000	1.800	43.900	1.200
Total	47.000	44.700	23.600	2.900	62.600	3.500
Austria	7.300	10.100	2.300	.200	1.200	.200
Belgium	6.900	6.700	3.200	.300	5.200	.500
France	56.800	61.100	20.600	3.200	37.800	2.200
Luxembourg	0.300	.600	.200	.010	.600	.020
Netherlands	9.300	2.800	4.600	.400	16.200	1.000
Switzerland	9.000	6.400	2.500	.200	12.200	.500
West Germany	39.000	55.000	24.000	1.500	22.800	4.200
Total	128.600	142.700	57.400	5.800	93.000	8.600
Bulgaria	8.000	18.000	.800	.200		
Czechoslovakia	19.000	66.100	2.400	.400	1.500	.200
East Germany	22.600	54.800	2.900	.300	2.000	.700
Hungary	11.100	23.200	1.200	.200	1.100	.300
Poland	48.900	112.600	2.800	.600	.800	.006
Romania	25.700	73.300	.300	.100	2.300	.040
Total	135.300	348.000	10.400	1.800	7.700	1.300
Albania			.003	.001		
Greece	1.500	.600	1.000	.600	5.500	.700
Italy	37.300	15.800	19.600	1.600	12.600	.600
Portugal	5.400	1.000	1.200	.200	4.100	.100
Spain	15.000	10.500	8.700	1.600	15.500	.400
Yugoslavia	10.800	25.800	2.700	.300	2.800	.060
Total	70.000	53.700	33.200	4.300	40.500	1.900

SOURCES: *World Motor Vehicle Data, 1983–84,* Detroit, Michigan: Motor Vehicle Manufacturers Association of the United States, 1985; *World Road Statistics, 1978–82,* Washington, D.C.: International Road Federation, 1983; *World Statistics in Brief: U.N. Statistical Pocketbook,* New York: United Nations Publications, 1983; *World Tables: The Third Edition,* Baltimore, Maryland: Johns Hopkins University Press, 1984.

The Train à Grande Vitesse (TGV), the world's fastest train, speeds through the French countryside on its way from Paris to Lyon. Photograph courtesy of the French National Railroads.

The Gutenfels Castle overlooks the Rhine River in Germany. Notice the steady stream of barges and boats carrying goods and people up and down the river. Photograph courtesy of the German Information Center, New York City, and is used by permission.

nel. A large part (27 percent) of the world's merchant fleet belongs to Europe, and Greece is the leader in most categories (freighters, bulk carriers, and tankers). With only a few exceptions, the great cities of Europe serve as harbors.

Air transportation connects all major cities in Europe, flights are frequent, and facilities are good. Most countries in both Eastern and Western Europe subsidize their airlines quite heavily, so modern equipment is common. Service and terminal facilities in the major cities compare favorably with those in the United States, but it is not possible to fly to the smaller towns and to go nearly anywhere at any time, as it is in the United States.

■ Political Geography of Europe

Much of the bitterness has been removed from the bickering among the European countries. Many political problems still exist, but they are now being resolved through negotiation rather than warfare. The older people of Europe remember the ravages caused by the wars of the past, and

This canal in southwestern England near Bude is one of the thousands of small waterways that have been dug in Europe.

The harbor at Barcelona, Spain, is typical of many outstanding facilities for shipping found in the port cities of Europe.

they lead in the attempts to maintain the peace. All countries remain armed, however, and the East-West split creates a definite barrier to the free movement of goods and people.

Europeans are sometimes dismayed at the confrontation between the United States and the Soviet Union because they feel that their own countries, located between the two superpowers, are being treated as pawns. Many Europeans, as well as the Soviets, would like to have American missiles removed from Europe.

Political strength in Europe was at one time tied to the number of overseas possessions. The colonies provided many European countries with raw materials and resources not found in Europe, and these materials were used to boost the economies, which in turn provided for the cost of arms. As the great empires began to dissolve, European nations had to look elsewhere for economic strength. Thus, the "common markets" were born from a need to fill the void left by the crumbling empires. Today, because of the economic unification, Europe is united politically more than it ever has been. When goods and people can move freely between countries, the countries are more likely to negotiate political problems than to declare war. The individual countries in Europe may not be any stronger in terms of world politics than they were 20 years ago, but the region as a whole is considerably stronger politically and economically.

Key Words

Basques
Benelux countries
Cevennes
Council for Mutual Economic Assistance (CMEA)
European Economic Community (EEC)
irredentas
North Sea oil
polders
Protestant Reformation
Romance languages
Slavic languages
Teutonic languages
transhumance

References and Readings

Bamford, C. G. *Geography of the EEC*. Plymouth, Eng.: MacDonald & Evans, 1983.

Berg, V. den L., et al. *Urban Europe: A Study in Growth and Decline*. Oxford: Pergamon, 1982.

Bogardus, James Furnas. *Europe: A Geographical Survey*. New York: Harper, 1934.

Burtenshaw, David, et al. *The City in West Europe*. New York: Wiley, 1981.

Clout, Hugh D. *Regional Variations in the European Community*. Cambridge: Cambridge University Press, 1986.

———. (ed.). *Regional Development in Western Europe*. New York: Wiley, 1981.

Clout, Hugh D., et al. *Western Europe: Geographical Perspectives*. White Plains, N.Y.: Longman, 1985.

deBlij, Harm J. *Wine: A Geographic Appreciation*. Totowa, N.J.: Rowman & Allanheld, 1983.

Demko, George J. (ed.). *Regional Development Problems and Policies in Eastern and Western Europe*. New York: St. Martin's, 1984.

Diem, Aubrey. *Western Europe: A Geographical Analysis*. New York: Wiley, 1979.

Embleton, Clifford (ed.). *Geomorphology of Europe*. New York: Wiley, 1984.

Faringdon, Hugh. *Confrontation: The Strategic Geography of NATO and the Warsaw Pact*. Boston: Routledge & Kegan Paul, 1986.

Fodor's Guide to Europe. New York: McKay, 1959.

Foster, C. R. (ed.). *Nations Without a State: Ethnic Minorities in Western Europe*. New York: Praeger, 1980.

Freeman, Edward A. *The Historical Geography of Europe*. New York: Longman, Green, 1920.

Gottman, Jean. *A Geography of Europe*. New York: Holt, Rinehart, & Winston, 1969.

Hall, Ray, and Ogden, Philip. *Europe's Population in the 1970s and 1980s*. Cambridge: Cambridge University Press, 1985.

Hoffman, George (ed.). *A Geography of Europe: Problems and Prospects*. 5th ed. New York: Wiley, 1983.

———. *Eastern Europe: Essays in Geographical Problems*. New York: Praeger, 1977.

Hudson, Ray, and Lewis, Jim (eds.). *Uneven Development in Southern Europe*. New York: Methuen, 1985.

Ilbery, Brian W. *Western Europe: A Systematic Human Geography*. New York: Oxford University Press, 1986.

John, Brian S. *Scandinavia: A New Geography.* New York: Longman, 1984.

Jones, E. L. *The European Miracle: Environments, Economies, and Geopolitics in the History of Europe and East Asia.* Cambridge: Cambridge University Press, 1981.

Jordon, Terry. *The European Culture Area.* New York: Harper & Row, 1973.

Knox, Paul L. *The Geography of Western Europe.* Totowa, N.J.: Barnes & Noble, 1984.

Lyde, Lionel William. *The Continent of Europe.* London: Macmillan, 1930.

Minshull, G. N. *The New Europe: An Economic Geography of the EEC.* 2d ed. London: Hodder & Stoughton, 1980.

Morris, Jan. *The Matter of Wales: Epic Views of a Small Country.* New York: Oxford University Press, 1985.

Musto, Stefan A., and Pinkle, Carl F. (eds.). *Europe at the Crossroads: Agendas of the Crisis.* Westport, Conn.: Praeger, 1985.

Parker, Geoffrey. *The Logic of Unity: A Geography of the European Economic Community.* 3d ed. New York: Longman, 1981.

Pinchemel, Phillipe, et al. *France: A Geographical, Social, and Economic Survey.* Cambridge: Cambridge University Press, 1987.

Pounds, Norman J. G. *Eastern Europe.* Chicago: Aldine, 1969.

———. *Europe and the Soviet Union.* New York: McGraw-Hill, 1966.

Rogers, Rosemarie (ed.). *Guests Come to Stay: The Effects of European Labor Migration on Sending and Receiving Countries.* Boulder, Colo.: Westview, 1985.

Rotkin, Charles E. *Europe: An Aerial Close-Up.* Philadelphia: Lippincott, 1962.

Rugg, Dean S. *Eastern Europe.* White Plains, N.Y.: Longman, 1986.

Sears, Dudley, and Ostrom, Kjell (eds.). *The Crises of the European Regions.* New York: St. Martin's, 1983.

Shackleton, Margaret R. *Europe: A Personal Geography.* New York: Longmans, Green, 1936.

Short, John R., and Kirby, Andrew M. *The Human Geography of Contemporary Britain.* New York: St. Martin's, 1984.

White, Paul. *The West European City: A Social Geography.* New York: Longman, 1984.

Wild, M. Trevor. *West Germany: A Geography of Its People.* New York: Barnes & Noble, 1980.

Woodell, Stanley R. J. (ed.). *The English Landscape: Past, Present, and Future.* New York: Oxford University Press, 1985.

Chapter 7

THE SOVIET UNION

■ Introduction

The Soviet Union was the first major Communist nation, and it still leads the international Communist political community. Leadership is provided through the Communist Party, the only legal party in the country. The internal political structure of the Soviet Union consists of 15 "Union Republics." These political units are subdivided into "Autonomous Soviet Socialist Republics," "Autonomous Regions," and "National Districts." The Russian republic, which is by far the largest and most populous of the union republics, provided the country with its historical name: Russia. The official name of the country is the Union of Soviet Socialist Republics or, in Russian, Soyuz Sovetskikh Sotsialisticheskikh Respublik, but it is usually referred to merely as "the Soviet Union or the U.S.S.R."

The Soviet Union is the largest political unit in the world—in fact, it is more than twice as large as any other country. The official area is 8,649,490 square miles, but some comparisons will help make this giant, sprawling size more meaningful. For example, it is farther from Moscow to Vladivostok than it is from Moscow to New York. The east-west extension of the country covers 160 degrees of longitude—nearly halfway around the world. This means that in traveling across the entire country you would pass through 11 time zones, compared with 4 for the continental 48 states of the United States. If the Soviet Union were divided into states the size of Texas, it would contain 32.5 of them. The Soviet Union is larger than all of Latin America and much bigger than the combined areas of China and India. The European section, the area west of the Ural Mountains and north of the Caucasus Mountains, is larger than the remainder of Europe.

The relative location of the Soviet Union may also be somewhat surprising (see Figure 7.1). It is a country of the northland. For example, while population density in North America declines dramatically north of 50 degrees latitude, most of the people in the Soviet Union live north of that line. The "sunny" Black Sea resort of Odessa is at about 46 degrees latitude, in the southern part of the country, but that latitude is north of Minneapolis. The latitude of Moscow is similar to that of Sitka, Alaska. This means, of course, that the Soviet Union does not have a "sunbelt," or tropical region. In political space, the Dezhneva Cape (or East Cape) lies only about 50 miles (80 km) across the Bering Strait from the Seward Peninsula of Alaska. The Soviet Union shares international boundaries with 12 other countries and consequently has more immediate neighbors than any other country.

Some information about the Soviet Union is difficult to obtain, and many of the economic and production figures can only be estimates. Ameri-

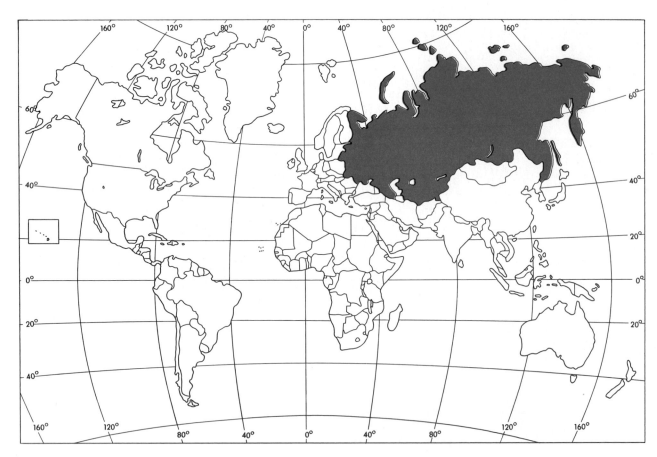

FIGURE 7.1
World Location of the Soviet Union.

166 Chapter 7

can citizens are not allowed to travel to many parts of the Soviet Union, and the Soviets refuse to release most information that may be related to military strength. Much information is available, however, especially on the physical geography of the country. With the future of the world at stake, it is important that Americans be as informed as possible about their major adversary.

Physical Geography of the Soviet Union

Landforms

Although there is great variation in the landscape of the Soviet Union, the topography may be described in relatively simple terms. For the most part, the variation is found along the southern margins of the country. Within the interior, there are vast areas with similar topography.

The general landforms of the Soviet Union can be considered in four sections: (1) the East European Plain, located west of the Ural Mountains and bordered on the south by the Carpathian, Crimean, and Caucasus mountains; (2) the West Siberian Lowlands, between the Urals and the Yenisey River, and its southern extension—the Turan Lowland—which is bordered by the high Central Asian mountains; (3) the East Siberian Highlands, and (4) the Middle Asian Mountains.

East European Plain. The East European Plain, or Russian Plain, is a rolling lowland in which the elevations never exceed 1,500 feet (457 m). This vast lowland undulates from the western boundary of the country to the Ural Mountains. Glacial ice radiating from Scandinavia once covered the northwestern and middle sections of the plain, as well as the adjoining Polish and German lowlands. The advancing and retreating ice scoured, polished, and denuded the underlying rocks and deposited huge amounts of clay, sand, gravel, and even boulders. Parallel series of terminal moraines lie across the northwestern portion of the plain. Thick deposits of loess—the wind-blown, fine-grained, dustlike clay—cover large areas of the plain. Because the loess is easily eroded, many ravines and gulleys have been created in the otherwise rolling surface. After the Ice Age, the large amounts of meltwater created an inland sea that covered most of the plain and connected the Arctic Ocean with the Caspian Sea.

The Ural Mountains form a narrow, linear, north-south trending range that bounds the East European Plain on the east. The mountains are not tall—the tallest peak is Naradnaya, at 6,214 feet (1,894 m)—but they stand out from the lowlands on either side. The linear formation has been used as the dividing line between the continents of Europe and Asia. The Ural Mountains region is noted for the large variety and amount of minerals found there.

West Siberian Lowlands. South of the Urals, the East European Plain joins with another vast level area known as the Turan Lowland. That plain continues northward along the eastern side of the Urals and merges with the West Siberian Lowlands. This vast lowland is the largest nearly flat plain on earth. The Ob' River, and its major tributary the Irtysh, drain most of the West Siberian Lowlands. The huge, flat plain would seem to be ideal for agriculture, but it is largely unproductive because of the severe dry and cold climate. The ground is frozen for many months, and then turns swampy when it does warm up during the brief summer.

The Caucasus Mountains are located on the Crimea peninsula between the Black Sea and the Caspian Sea. The range is about 700 miles (1,126 km) long and is mostly of volcanic origin. Many peaks exceed 15,000 feet (4,572 m), and the tallest, Mount Elbrus, is 18,481 feet (5,633 m). The range is crossed by high mountain passes, and the two best known are the Daryal Pass and the Mamison Pass.

East Siberian Highlands. The Yenisey River forms a natural boundary between the West Siberian Lowlands and the East Siberian Highlands. Eastward from the Yenisey, the ground elevations begin to rise and the land becomes low mountain and hill country. The region is interlaced with broad, fluvial lowlands along the Siberian rivers. In the far eastern section of Siberia, the mountains become taller and more rugged. The tallest peak in the eastern part of the country is Gora Chen, at

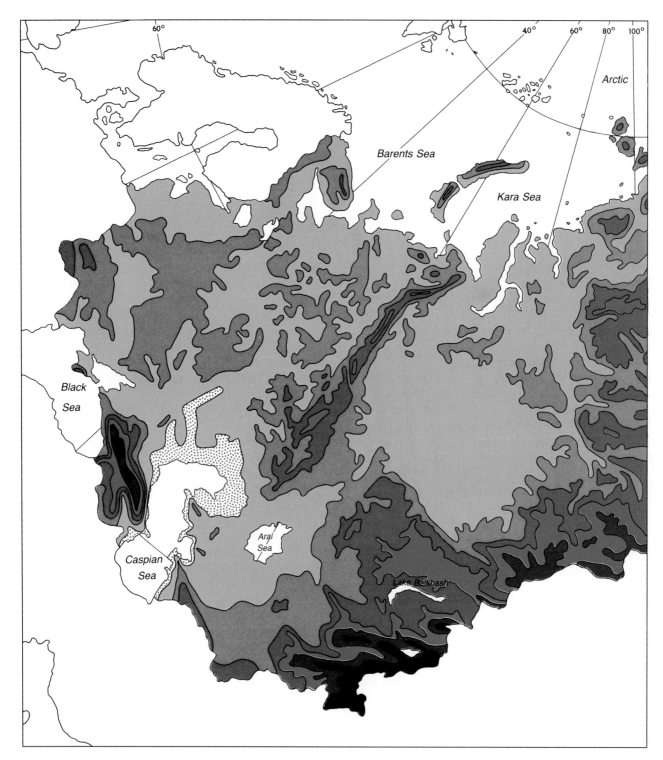

FIGURE 7.2
Land Elevations and Water Bodies of the Soviet Union.

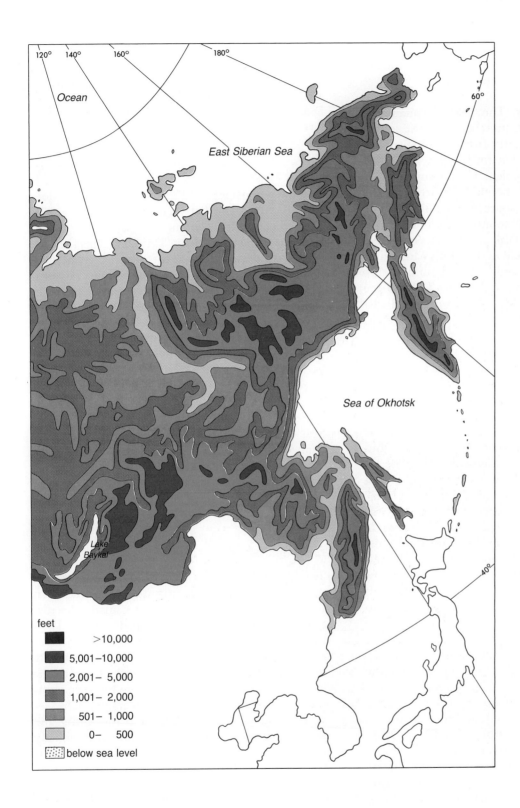

10,171 feet (3,100 m), located just south of the Arctic Circle at about 142 degrees east longitude in the Khrebet Cherskogo range.

Middle Asian Mountains. The Asian Mountain section of the Soviet Union stretches along the southern boundary of the country from the Caspian Sea to the Sea of Okhotsk. The topography along this section is extremely complex. The mountain crests form the international boundaries between the Soviet Union and its southern neighbors. The tallest mountain in the country is Communism Peak, once called Stalin Peak, which rises to 24,590 feet (7,497 m). It is located in the Pamir Mountains near the Afghanistan border. The boundary between the Soviet Union and China follows the crest of the Tien Shan Mountains north and east from the Pamirs. Then, farther north, the Soviet, Chinese, and Mongolian boundaries converge near Belukha Peak, the highest point in the Altai Mountains (14,783 feet, or 4,507 m). Eastward from Belukha, most of the border area between the Soviet Union and Mongolia is rugged mountain country. Lake Baikal is nestled in one of the mountain valleys and is notable because it is the deepest inland body of water in the world and contains more fresh water than any other lake.

Many of the important rivers of the Soviet Union flow northward because the high mountains are on the southern border and the flat plains on the north gradually slope toward the north. The Ob', the Yenisey, and the Lena are among the longest rivers in the world, and together with their tributaries they drain vast regions of the eastern part of the country. This means that more than half the territory of the Soviet Union is drained into seas that are frozen during the greater part of the year. Another one-quarter of the country is drained by rivers that flow into inland basins, such as the Caspian Sea and the Aral Sea. Despite the thousands of miles of coastline, the Soviet Union is essentially a landlocked country as far as outlets to world ports are concerned, and this has been a problem throughout its history. (See Figure 7.3).

Some other Soviet rivers should be mentioned, because of their importance to the country. Historically, the rivers have served as navigation routes for opening the country, and most major cities are built along the banks of rivers. The

---

continental climate The climate designated by Köppen as "D," located where extreme continental influences create hot summers and very cold winters. No D climate exists in the southern hemisphere because there are no large landmasses to retard the stabilizing influence of the oceans.

Danube is considered a European river, but it does form the boundary between the Ukrainian S.S.R. and Romania. It flows into the Black Sea, as do both the Dnieper and the Dniester. The Dnieper and its tributaries drain most of the Ukraine, and the great city of Kiev is located along its banks. The Caspian Sea, the world's largest saltwater lake, receives the waters of the Volga River and the Ural River. The Volga is the most important river in the Soviet Union. It drains the heartland, and it is used as a transportation artery, for generating hydroelectricity, and for irrigation. The capital and largest city, Moscow, is located along the banks of the Moskva (or Moscow) River, which is connected by means of canals to the Volga, the Baltic Sea, and the Arctic Ocean. The Volga also is connected to the Black Sea by canal route to the Don River. In the far east, the Amur River forms much of the boundary between the Soviet Union and China and is one of the principal rivers of the world.

Climate and Vegetation

The Soviet Union is a high-latitude country, and most of its territory lies north of the latitude that separates the United States and Canada. Its distance from the equator, combined with its large landmass, creates a **continental climate** over most of the country (see Figure 7.4). The climate is characterized by the lack of moderating oceanic influences, so there is a wide range between summer and winter temperature averages. The severely cold winters are long, and the spring and fall seasons are short. Precipitation comes mostly in the summer, with low total amounts for the year. All these features of the climate result from the continental location.

Along the southern border of the Soviet Union, stretching from the Black Sea to Mongo-

permafrost Subsoil that is frozen throughout the year.

lia, the climate is arid (Köppen's BW). Approximately 750,000 square miles (1,950,000 km^2) of this region is true desert and receives less than 10 inches of annual rainfall. These deserts of the Turkistan surround the Aral Sea. The steppes (with 10 to 20 inches of annual precipitation) surround the desert on all sides. Agricultural has been carried on in the steppes for centuries, but always a precarious balance exists between drought and barely sufficient rainfall for crop survival. Water from the southern mountains is important for irrigation in the region.

The northern fringe of the Soviet Union has a polar climate (Köppen's ET). This tundra stretches from the White Sea on the west to the Bering Strait on the east and includes most of the land north of the Arctic Circle. Winter monthly temperatures in the tundra average below $-20°$ F ($-29°$ C), and some stations report January averages of below $-60°$ F ($-51°$ C). Much of this land contains **permafrost,** or permanently frozen ground. In the interior, permafrost is as deep as 1,500 feet (457 m), but along the coast it remains shallow because of the moderating effect of the water. The top few inches of the permafrost thaw during the short arctic summer, and the delicate tundra flowers complete their life cycles during this time. The ground is frozen rock-hard during the winter but is soft, spongy, and swampy during the summer. Generally, trees do not grow in the tundra regions.

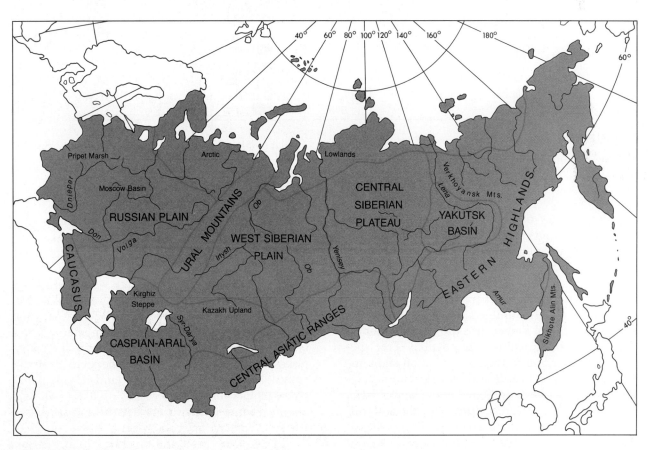

FIGURE 7.3
Locations and Names of Landforms of the Soviet Union.

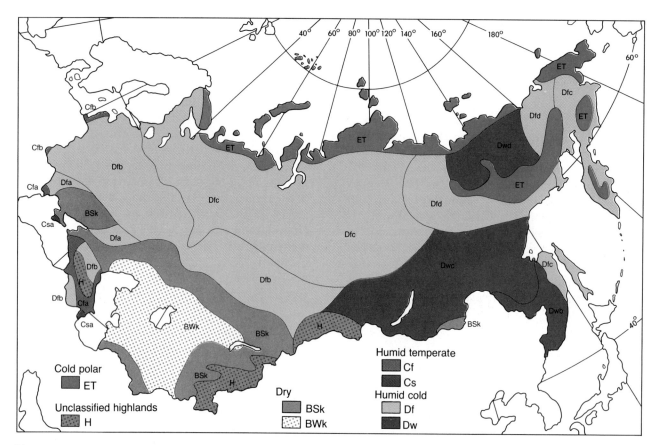

FIGURE 7.4

Climate Regions of the Soviet Union.

Vast reaches of broadleaf continental forests and taiga-type evergreen forests lie between the arid southern regions and the tundra on the north (see Figure 7.5). The taiga is located in a subarctic climatic region, which has extremely cold winters as well as all the other characteristics of a continental climate. The summers, however, are somewhat warmer and longer than those in the tundra. Some types of agriculture are possible. Tree species vary according to local precipitation and temperatures. Mixed broadleaf deciduous and needleleaf evergreen forests are found in the wetter west, only needleleaf evergreen forests in the interior, and needleleaf deciduous trees in the northeast. The taiga contains some of the world's largest stands of single species of trees. More than one-quarter of all the timberland in the world is found in the Soviet Union. The forests are one of the country's rich natural resources.

In addition to the north-to-south increase in average temperatures in the Soviet Union, there is also a gradual increase in average winter temperatures from east to west. Cities along the Baltic, for example, have January averages of 32° F (0° C), but the eastern interior cities have temperatures far below zero. The coldest permanently inhabited places in the world are the northeastern Siberian communities of Verkhoyansk and Oymyakon, where temperatures have reached −108° F (−78° C). Summer temperatures in northeastern Siberia, however, often average above 60° F (16° C).

Most of the precipitation throughout the Soviet Union occurs during the six summer months. During the winter, extreme high pressure, referred

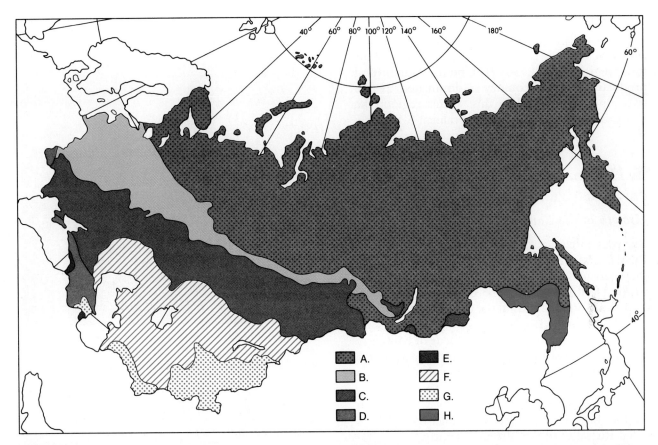

FIGURE 7.5

Natural Vegetation and Agricultural Zones of the Soviet Union. The eight zones are: (A) The northern forests and tundra that have podzol and tundra soils and very little agriculture. (B) The mixed forest zone with mostly podzol soils and mixed farming. Crops in this zone include the hardy types, such as rye, potatoes, barley, and flax as well as hay and livestock. (C) Primarily steppe, but also stands of trees with chernozem soils. Agriculture in this zone is predominantly wheat-growing, but corn, sugar beets, and sunflowers are common too, as is livestock. (D) Mountain forests with thin, poor soil development and very little agriculture. (E) Humid subtropical climate region with chernozem soils and subtropical crops, such as citrus fruits, grapes, tobacco, tea, vegetables, as well as corn, livestock, and some rice. (F) Desert climate region with desert soils, used mostly for grazing but also for some irrigated agriculture. (G) Mountains and mountain foothills with semiarid climates and serozem soils. Some dryland wheat farming, but mostly irrigated cotton and hay. The livestock industry uses the transhumance method of grazing. (H) Low mountains and basins of the Far East with cool monsoon climate, mixed forests, and steppe soils. Mixed agriculture including wheat, soybeans, sugar beets, rice, and livestock.

SOURCE: Map adapted from Paul E. Lydolph, *Geography of the U.S.S.R.* (New York: Wiley, 1964), p. 18. Used by permission.

to as the "Siberian High," builds up over the Asian landmass. This dense, stable air does not allow cyclonic storms to penetrate the interior. Both the summer rain and the winter snow decrease from west to east across the country. It can snow anywhere in the Soviet Union and, in large areas of the country, snow is possible during any month. However, the heavy snow covers are generally found west of the Urals because the moisture supply from the Atlantic Ocean is closer to that region.

The continental influence on the Soviet climate is weakest in the East European Plain. The region receives the most direct effect of the moderating cyclones from the Atlantic Ocean. Because precipitation usually averages about 30 inches (76 cm) annually, the area is the most dependable agricultural region in the Soviet Union. Agricultural production for the country is dominated by the two republics located in this wetter region: the Byelorussian S.S.R. and the Ukrainian S.S.R.

Human Geography of the Soviet Union

Growth of the Soviet Union

The present city of Kiev was the capital of the first Russian state. It was established in A.D. 862, and the rulers in Kiev dominated Russia from the ninth to the twelfth century. The principality of Muscovy was founded during the twelfth century, and out of it grew the Russia prior to World War I, and the Soviet Union of today. The focus of Muscovy was the small walled settlement of Moscow, which grew with the empire to become the country's major metropolis. During the thirteenth century, however, the new state of Muscovy was overrun by the Tartars, and these Mongolian conquerors ruled Russia until the fifteenth century.

The Russians finally drove off the Mongols in 1480, and the first czar, Ivan the Great, assumed control of the country. He expanded the empire by taking over neighboring territories, and his grandson, known as Ivan the Terrible, continued the expansion when he came to power. After the Ivans, the empire weakened, and in 1610 the Polish army marched into Moscow. The Poles soon were defeated, however, and Russian retaliation added the Polish-held region of the Ukraine to the Russian Empire. The later agricultural and industrial developments in the Ukraine have proven that the acquisition was vital to the Soviet Union.

The next conquerors of Moscow and its loosely structured empire were the Swedes, but in 1709 Peter the Great (1672–1725) led a peasant army of Russians that was successful in defeating and evicting the Swedes. Peter is credited with changing Russia from a very primitive oriental country into a more modern European empire. A later czarina, Catherine the Great (1729–1796), continued the Western orientation and expansion

Red Square in Moscow faces east toward the Kremlin, which is a triangular fortress that for several centuries was the residence of ruling families and is now, under the Soviets, the headquarters of government officials and a meeting place for Soviet administrative councils. Photograph by Colin Hagarty.

of the empire. She is noted for abolishing capital punishment and introducing exile to Siberia in its stead, a practice that continued in the Soviet Union until recently. Napoleon (1769–1821), the French military and political leader, ended Russian expansion temporarily when his troops marched to Moscow in 1812. (See Figure 7.6).

The twentieth century brought revolution to Russia. A revolt failed in 1905, but the second one began in October 1917 and was successful. The leaders of the revolt called themselves "Communists." The last czar, Nicholas II, was murdered, and the revolt leader, Vladimir Lenin, became the ruler of the country. Civil war raged for three years, but by 1920 the Communists had firm control of most of the country. When Lenin died in 1924, the power was passed to the second-in-command, Leon Trotsky, but a shrewd, ruthless son of a Georgian shoemaker, Joseph Stalin, won control of the government. Trotsky was forced to flee Russia and he eventually was murdered in

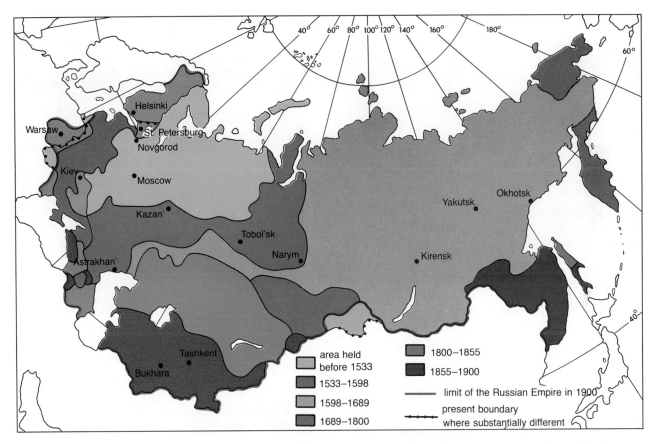

FIGURE 7.6

Historical Expansion of the Territory of the Soviet Union.

Mexico in 1940. Stalin ruled the Soviet Union with an iron fist from 1927 until his death in 1953.

Since Stalin's death, some of his most ruthless acts have been denounced by the country's new leaders. A recent succession of Soviet leaders has also relaxed some of the repressive policies started by Stalin, although the country is still far from a democracy. In fact, the country's international position—as a vowed enemy of democracy and with expansionist tendencies—seems quite clear.

Population

The population of the Soviet Union in 1986 was estimated by the Population Reference Bureau as being 280 million. The natural rate of increase was 0.9 percent, or just slightly higher than the rate for the United States and Canada. From 1950 to 1980, the population of the Soviet Union increased by about 2.1 million people a year. A spatial variation exists in the population increase, because some Soviet republics have been growing much faster than others. The most rapidly increasing region is the southern part of the country, adjacent to the Caspian Sea. The Armenian republic west of the Caspian, and the Kazakh, Kirgiz, Tadzhik, Turkmen, and Uzbek republics east of the Caspian all doubled their populations between 1950 and 1980 (see Table 7.1). This mountain and desert area traditionally has had a low population density, and the population increases indicate an effort by the Soviets to populate the region and to encourage the people to make it more productive. (See Figure 7.7.)

The region in the Soviet Union with the slowest population growth rate is located in the northwestern part of the country, near the Baltic

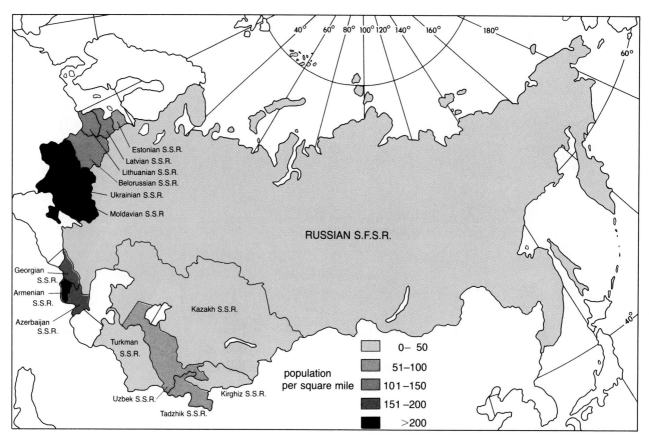

FIGURE 7.7

Population Density of the Soviet Union.

Sea. From 1950 to 1980, the Byelorussian republic only grew from 9.3 million to 9.6 million. The Latvian, Lithuanian, and Estonian republics are the other slow-growth states. The Russian republic is by far the largest in the Soviet Union in terms of both population and area. It contains about 52 percent of the Soviet people and about 77 percent of the land area. Traditionally, the Ukraine has been the most densely populated of the 15 republics. In terms of natural resources, climate, and agriculture, it is the most favored region of the Soviet Union.

The overall density of population in the Soviet Union is about 30 people per square mile. The most densely populated parts of the country are the west and the southwest, especially the Ukraine, Moldavia, and Armenia. In these republics, the density exceeds 200 people per square mile. The sparsely populated areas are in the east—especially in Siberia, where the density figure is about 1 or 2 people per square mile. The few permanent settlements in the region are almost all found along the rivers.

Urbanization

Urbanization is a striking feature of recent Soviet history. Not only have the older cities grown rapidly, but completely new industrial cities have been built. The overall urbanization factor is about 65 percent for the entire country, but exceeds that in some republics. (See Figure 7.8).

Moscow is by far the largest city in the Soviet Union, with a population estimated at 8.3 million

TABLE 7.1

Population Figures for the Soviet Union in Millions, 1950–1980

REPUBLIC	1950	1960	1970	1980	TOTAL INCREASE	PERCENT INCREASE
Armenia	1,500	2,253	2,493	3,031	1,531	102
Azerbaijan	3,300	4,802	5,111	6,028	2,728	83
Byelorussia	9,300	8,744	9,003	9,559	259	3
Estonia	1,200	1,294	1,357	1,466	266	22
Georgia	3,600	4,611	4,688	5,016	1,416	39
Kazakh	6,600	12,413	12,850	14,685	8,085	123
Kirgiz	1,500	2,704	2,933	3,529	2,029	135
Latvia	2,100	2,285	2,365	2,521	421	20
Lithuania	3,000	3,026	3,129	3,399	399	13
Moldavia	2,700	3,425	3,572	3,948	1,248	46
Russia	116,400	127,312	130,090	137,552	21,152	18
Tadzhik	1,500	2,654	2,900	3,801	2,301	153
Turkmen	1,200	1,971	2,158	2,759	1,559	130
Ukrainia	40,800	45,966	47,136	49,757	8,957	22
Uzbek	6,000	10,896	11,963	15,391	9,391	157
Total	199,350	234,356	241,748	262,442	63,092	32

SOURCE: *United Nations Demographic Yearbook, 1952, 1962, 1976, 1983.*

in 1983. Located near the center of the Russian heartland, it has been the capital for centuries. Major roads and railroads emanate from the city like spokes on a wheel. In 1930, Moscow had less than 2 million people, but when Stalin began an industrial drive, workers flocked to new jobs in the city, and the population doubled in less than 10 years. The population of Moscow declined during World War II but has increased steadily ever since. Today, Moscow is a modern city with museums, parks, restaurants, stadiums, and theaters. The Kremlin, the seat of the Communist government, is located in the center of the city.

Leningrad is the second city of the Soviet Union, located in the northwest, near the Baltic Sea. The city dates from 1703, when Peter the Great started its construction on the low, swampy land at the eastern terminus of the Gulf of Finland. Peter called the city "Saint Petersburg" and used it as his capital and "window on Europe." Today, Leningrad is an economic center of gravity, a major transportation center as well as an industrial giant. The population in 1983 was estimated at 4.7 million.

At least a dozen other cities in the Soviet Union exceed 1 million in population, and most of these have fewer than 2 million people. Gorky and Kuibyshev are located east of Moscow, on the banks of the Volga River. The major cities of the Ukraine are Kharkov and the ancient settlement of Kiev. Sverdlovsk and Chelyabinsk are relatively new mining and industrial centers on the eastern slopes of the Urals. Tashkent, another old settlement that has seen recent growth, is located near the western end of the Tien Shan mountain range in Central Asia (Uzbek S.S.R.). Novosibirsk grew up along the Trans-Siberian Railroad, where the railroad crosses the Ob River. The city, also a relatively new industrial center, is located in the Kuznetsk Basin, which is noted for its natural resources. Minsk is the capital of the Byelorussian republic and is located halfway between Moscow and Warsaw, Poland. Tbilisi is the capital of Georgia, and Odessa is a Black Sea resort city in the southern Ukraine. Nearly every city mentioned is located on the banks of a river, but with the exception of Leningrad none has direct access to the open sea. Another main port in the Soviet

FIGURE 7.8

Major Cities of the Soviet Union. Note the greater density of large cities in the western half of the country.

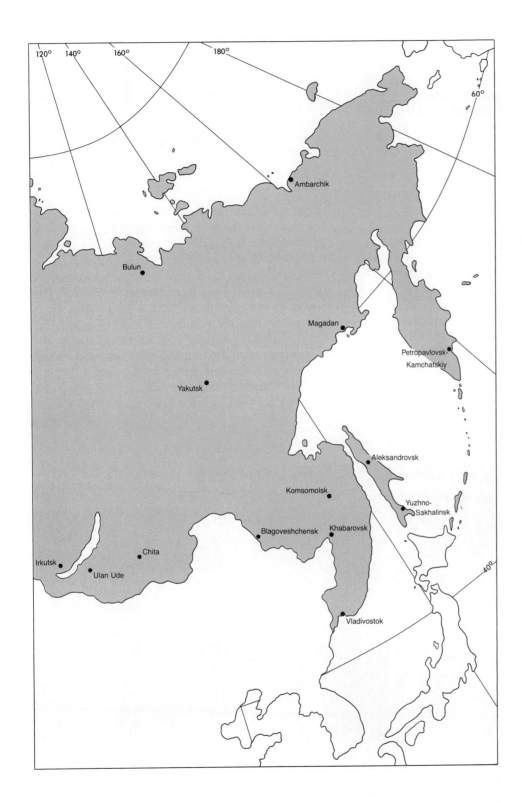

Union is Vladivostok, located at the eastern terminus of the Trans-Siberian Railroad near the Sea of Japan. It is, however, far from the industrial and population centers of the country. One other port city is Murmansk, but it too is located far from the heartland, on the Barents Sea. Even though it is north of the Arctic Circle, Murmansk is an ice-free port.

Religion

Atheism is the official state doctrine in the Soviet Union. When the Communists gained control in 1917, they saturated the general public with antireligious propaganda, closed theological schools, and did not allow churches to publish religious books or church magazines. Communism would stand no rival for the loyalty of the people. This was in keeping with the opinion of Karl Marx, who wrote: "Religion is the sign of the oppressed creature, the sentiment of a heartless world, and the soul of soulless conditions. It is the opiate of

Russian Orthodox church   A branch of the dominant Christian body of Eastern Europe, the Orthodox church. Church members are in union with the "ecumenical patriarch" (the Orthodox patriarch of Turkey) and do not recognize the pope in Rome as their leader. Otherwise, the rituals are similar to those of Roman Catholics.

the people." During World War II, however, Stalin began to ease some of the religious restrictions. Although he was a devout atheist, he knew that the Soviet Union had to present a united front to the Germans, and the strength of religious traditions remained a serious internal problem. The Soviet churches took advantage of the war situation to reunite their members. The Russian Orthodox church won the most favors from the Communists, and it remains the strongest today, even though it is controlled by the state.

The **Russian Orthodox church** (which is one of the churches in the branch of Christianity

A modern streetcar moves through a somewhat deserted street in Moscow. Photograph by Lawrence Gibbs.

An inside view of the famous G.U.M. department store in Moscow. Shopping is difficult in Moscow because consumer items and shopping facilities are scarce. Photograph by Colin Hagarty.

declining, or at least not making any gains. Nearly every settlement has a church building, and some continue as working churches, but many are closed, some are being restored as architectural monuments or museums, others are used for meeting halls and clubhouses, and some are merely crumbling buildings. Not much is known about anti-Semitic attitudes in the country, except that most Jews have either asked to leave or have already left. Until 1970 they were not allowed to emigrate, and estimates put the total number of Jews who have left since then at more then 130,000. Not much is known about the everyday religious attitudes of the common people of the Soviet Union either, but clearly most are nonbelievers.

known as Eastern Orthodox) is similar to the Roman Catholic church, except that it denies the authority of the pope in Rome. Although the Russian Orthodox church has the largest following in the Soviet Union, nine branches of Christianity are represented in the country. In addition to the Christian groups, the second largest religious group consists of the followers of Islam. Muslims are concentrated along the southern border of the Soviet Union, in the mountain and desert regions. Numerous Buddhists live in the same areas as the Muslims, and a few groups of Jews are scattered in the urban centers.

Despite the religious revival that began during World War II, there have been no outstanding gains in converting younger people to religious teachings. Members of the Russian Orthodox church enjoy a certain amount of freedom to worship and are not persecuted, but other religions may be

St. Basil's Cathedral near Red Square in Moscow has the onion-shaped domes characteristic of Orthodox churches. The building is used as a museum today. Photograph by Lawrence Gibbs.

The Soviet Union

> # FOCUS BOX
>
> ## Christianity in the Soviet Union
>
> Christianity in the Soviet Union dates from the summer of A.D. 988, when mass Christian baptisms were conducted in the Dnieper River at Kiev. Prior to these conversions, the Russians were pagans who used human sacrifices in their godless rituals. Prince Vladimir, of the small Slavic state of Kievan Rus', was looking for a religion to unite the Slavic tribes under his rule. The prince's emissaries traveled to Constantinople, the center of Eastern Orthodoxy, where they were impressed by the trappings of Byzantine Christianity. Vladimir converted, primarily so he could marry the sister of a Byzantine emperor. He then ordered all his subjects to convert to Christianity, and they obeyed his orders.
>
> When Constantinople fell under the rule of the Muslim Turks in 1453, Moscow became the center of the Eastern Orthodox branch of the Christian faith. Christianity flourished in Russia until 1721, when Czar Peter the Great removed the church leaders and gave control of the church to a political friend. Religion then became an arm of the state and never again attained its previous power. By the time of the Communist revolution in 1917, the Russian Orthodox church owned a good deal of land but was weak and somewhat corrupt. Thereafter, the Communist leaders, especially Lenin and Stalin, led vicious attacks against Christians. An estimated 50,000 priests, monks, and nuns were killed outright, and others were sent to die in camps. Untold thousands of lay Christians also were killed. After World War II, Nikita Khrushchev continued the persecution by imprisonment, rather than murder, and closed many of the remaining churches. Before the Communist takeover in Russia, about 50,000 Eastern Orthodox churches were functioning, but fewer than 7,000 exist today.
>
> The Eastern Orthodox church split from the Roman Catholic church during the eleventh century. The two branches of Christianity agreed on ceremonial matters and on doctrine, but they disagreed on the authority of the pope. Today, the Russian Orthodox church has about 50 million members in the Soviet Union, while Roman Catholics number about 10 million. Protestants, who make up the third branch of Christianity, total about 4 million in the Soviet Union, with the largest groups being the Baptists and the Pentecostalists. All told, about one-fourth of the population of the Soviet Union are Christians, but many are forced to attend underground churches. Religious persecution is beginning to ease, however, especially with the policy of *glasnost* (openness) promoted by General Secretary Mikhail Gorbachev. Many religious prisoners have been freed, and some underground churches have been allowed to emerge. Thus, during its one-thousandth year (1988) on Russian soil and after 70 years of extremely oppressive state atheism, Christianity tenaciously endures in the Soviet Union.

Language and Ethnicity

The language patterns of the Soviet Union follow the ethnic settlement patterns. About three-fourths of the Soviets are Slavs, and of that total most are Great Russians. The remainder of the Slavic group consists of Ukrainians and White Russians. Most Slavic people speak Russian, one of the **Slavic languages** in the Indo-European family. Others speak Ukrainian (the second most common language), Byelorussian, or Polish. These people are located primarily in the European part of the country, west of the Urals. They also have made most of the settlements in the new industrial cities

Slavic languages A group of related languages spoken by the Slavic people, with an alphabet based on Greek, as opposed to the alphabet of the Romance languages, which is based on Latin. The Slavic people include the Russians, Poles, Bohemians, Moravians, Bulgarians, Serbians, Croatians, Wends, Slovaks, and other small groups.

Turkic languages Subfamilies of the Ural-Altaic languages spoken by people who live in the vast region ranging from the Adriatic Sea to the Okhotsk. Although the people are of similar Mongol ancestry, they are racially mixed, but the numerous languages are remarkably similar and are noted for agglutination and vowel harmony.

east of the Urals and along the Trans-Siberian Railroad. In fact, most of the people in Vladivostok, in the far eastern part of the country, are of Slavic stock.

The remaining quarter of the population consists of about 175 ethnic groups and about 150 different languages. The most widespread, besides Russian, is the **Turkic language,** which is spoken by most of the people of Turkistan and throughout East Siberia. In the far east, the Mongolian and Manchu languages are common. In the north, the nomadic people have their own languages. The nomads in the far northeast speak a language similar to that of the American Eskimo, while those in the northwest speak Finnish.

Because about 25 percent of the Soviet population are ethnic and linguistic minorities, these people are an important part of the country. Russian-language schools are common in the ethnic regions as the Soviets try to integrate the minorities into the national body. As noted, most of the minority regions are located on the fringes of the country, either in the arid south, the frozen north, or the far east (see Figure 7.9). Ethnic uprisings do occur occasionally, but they are quickly suppressed.

Education and Health

The government of the Soviet Union provides free education, and it is compulsory for all children between the ages of 7 and 16. As a result, illiteracy has

The Mongolian ethnic background of these Soviet women is evident. Note also the difference in attire between the older women and the younger one in the middle of the group. Photograph by Lawrence Gibbs.

The Soviet Union

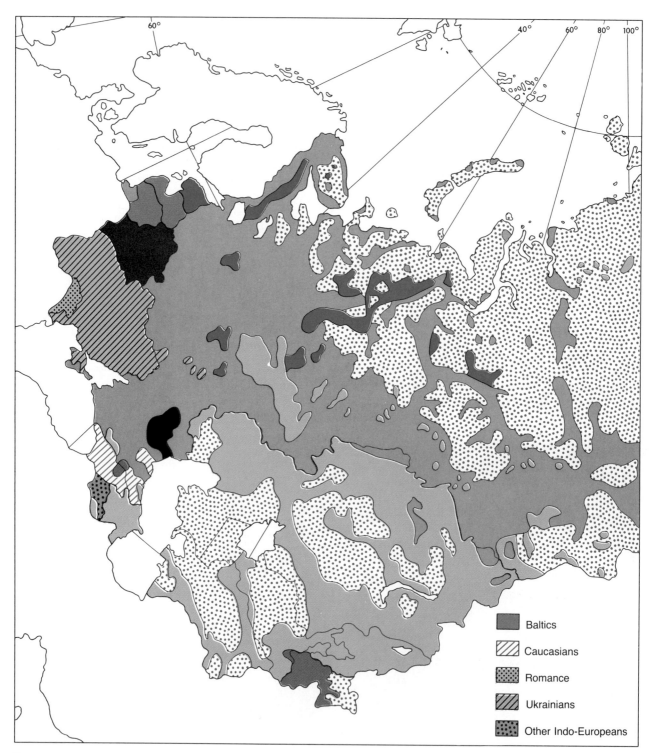

FIGURE 7.9
Ethnic Regions of the Soviet Union.

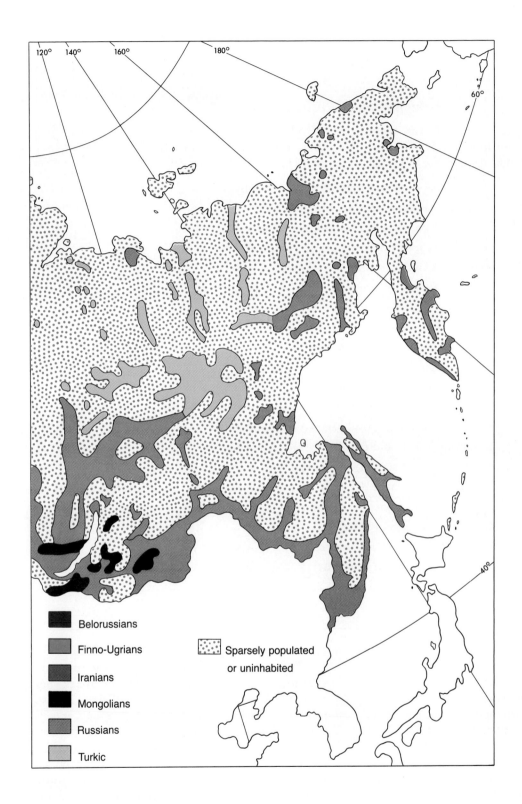

Focus Box

Pronunciation of Russian Words

On first encounter, Russian words may seem to be very difficult to pronounce. The following list gives the pronunciation for some of the place names in the Soviet Union. With these as a guide, the pronunciation of other words may be determined.

Chelyabinsk (city and region) = Chi-lya'-byinsk
Donbas (manufacturing region) = Dun-bas'
Donets (basin, river, and region) = Du-nyets'
Gorky or Gorki (city and region) = Gar'-ki
Irkutsk (city and region) = Ir-kootsk'
Karaganda (city and region) = Ka'-ra-gan-da'
Kazakh or Kazakhstan (republic) = Ku-zak'
Kharkov (city and region) = Kar'-kof
Kiev (city and region) = Ke'-yef
Kirgiz or Qirghiz (republic) = Kir-geez'
Kuibyshev (city, once Samara) = Kwe'-bi-shef
Kuzbas (manufacturing region) = Kooz-bas'
Kuznetsk (city, basin, and region) = Kooz-netsk'
Novosibirsk (city and region) = No'-vo-se-birsk'
Rostov (city and region) = Ru-stof'
Sverdlovsk (city, once Ekaterinburg) = Svurd-lofsk'
Tadzhik or Tadzhikistan (republic) = Ta-jeek'
Tashkent (city) = Tash-kent'
Tbilisi (city, means "warm springs") = T-pi'-li-si
Uzbek or Uzbekistan (republic) = Ooz'-bek

SOURCE: *Geographical Dictionary* (Springfield, Mass.: G.&C. Merriam, 1960).

---

been reduced to less than 2 percent of the population. There are many colleges and universities throughout the country, but there is also a strong emphasis on technical training. Both men and women attend college, and about 60 percent of the medical students are female. Every citizen is entitled to free public health service. The life expectancy rate is lower than the rates for Europe and Anglo-America and has been attributed in part to problems related to alcoholism (see Focus Box on page 187).

The state provides Soviet workers with a paid vacation, and free transportation to Black Sea resorts for some workers and officials is also provided. In the Soviet Union, no one travels beyond his or her home region without first obtaining a permit from the government. Besides the public health service, workers are provided with sickness insurance, and all workers may retire with a pension at age 60 for men and age 55 for women. These provisions apply to farm workers

communism A doctrine and program based on socialism that calls for regulation of all social, economic, and cultural activities through a single authoritarian party.

state farms Large single-unit farms run as cooperatives by farmers who share the workload. The state owns all the land and equipment.

---

as well as to workers in the urban factories. About 17 percent of the total population receives some type of pension from the state.

Economic Geography of the Soviet Union

Agriculture

The Soviet Union is larger in land area than any other country, and much of it is vast stretches of flat land, but insufficient food production has been a chronic problem throughout the country's history. Part of the problem stems from the inability of the political leaders, both the czarists and the Communists, to organize a satisfactory agricultural system, but another important factor is the natural environment.

Prior to the Communist takeover in 1917, the Russian farmers were essentially peasants, producing very little surplus. Much like their forefathers of the Middle Ages, they scraped an existence from the land. Under **communism,** however, extreme changes were made in the Soviet economy. The objective was to force industrialization on the country, but to do it through socialist reforms. Laborers in industrial cities need to be fed, so part of the process was to collectivize the farms in order to force the farmers to produce a surplus. This meant that all the small holdings were to become large state-owned farms. The plan met with considerable resistance because the farmers did not want to give up their meager holdings. A series of *Five-Year Plans* were prescribed, and by the late 1930s about 90 percent of the peasants' holdings were consolidated into collective farms. These huge **state farms,** called *sovkhozes,* were found to be unwieldy and are now being replaced with smaller collectives called *kolkhozes.*

The environmental factors that have handicapped Soviet agricultural production include the dry and cold climates. In the southern part of the country, where the growing season is long, the

---

FOCUS BOX

Potatoes and Vodka

Much of the enormous Soviet potato crop is used in making vodka, but the colorless and unaged liquor is also distilled from the mash of rye or wheat. Vodka has been a popular drink in Russia since the time of the czars. Russian vodka and caviar are considered delicacies and enjoy popularity in other countries of the world. In the Soviet Union, however, vodka is considered the working-man's drink.

One of the most prevalent diseases among the Soviet working class is alcoholism. People apparently seek relief from the cold and dreary weather and from boredom by excessive drinking. This takes its toll on the state by lowering production due to absenteeism, and it causes much human suffering from liver problems and early death. Life expectancy in the Soviet Union is 69 years, below that of the other industrialized nations of the world and in the class of some Third World countries. Soviet authorities attribute the low life-expectancy rate to the drinking problem.

In order to try to control alcoholism, the Soviet leaders have raised the price of vodka. In 1987 a half-liter bottle sold for five times the price in 1970—the equivalent of about $15. But the price increase apparently has not decreased sales. Soviet citizens still buy vodka on a regular basis regardless of the price, and Soviet production of potatoes remains at an all-time high.

The Soviet Union

rainfall generally is scant and unpredictable. In the north, where moisture is more certain, the growing season is extremely short. Also, most of the soils of the Soviet Union are not known for their high fertility. The gray podzols predominate over most of the eastern part of the country. Because of their high acid content, these are not good soils for farming. Only the Ukraine and the dry steppes have the darker, more productive chernozem soils.

The cultivated land in the Soviet Union exceeds 500 million acres (200 million hectares). The land is divided into about 40,000 *kolkhozes* and about 13,000 *sovkhozes*. A *kolkhoze* usually consists of a village or group of villages, with about 10,000 acres (4,000 hectares) under cultivation. On the collective farms, farmers do not work for wages. They divide the seasonal proceeds according to the amount of work each person does. The division comes only after the state has taken its share and after another percentage has been taken for improving the farm and buying equipment. As everything belongs to the state, all equipment is shared. The government does allow individual farmers to hold small plots of land near their homes for their own use. These private plots are much like gardens, but some animals can be raised along with the fruits and vegetables. These usually are the only fenced areas in the farming district.

The leading agricultural region of the Soviet Union is a belt running south of Moscow from the eastern slope of the Carpathian Mountains to the Irtysh River valley east of the Urals (see Figure 7.10). This region contains most of the large collective farms, and the work is highly mechanized. The Ukrainian, Azerbaijan, and Russian republics are the leading wheat producers. Wheat products are an important part of the Soviet diet, and much time and effort is devoted to its production. However, the agricultural belt produces many types of products besides wheat. Potatoes, for example, are grown everywhere possible. Potatoes are not only important for food for animals and humans, they also are used to make vodka. The annual potato crop is about five times that of the United States.

Other crops grown in the main agricultural belt of the Soviet Union include sugar beets, flax, a variety of grains, and sunflowers. Soybeans were introduced in the late 1960s as a source of protein for the human diet and as a feed grain for animals, but acreage is still limited. Some corn is grown, but it is used mostly to feed hogs. In the southern

Some farm products are sold in open markets in Soviet cities, but little variety exists. In spite of the meager number of choices, this merchant is obviously proud of his offering. Photograph by Lawrence Gibbs.

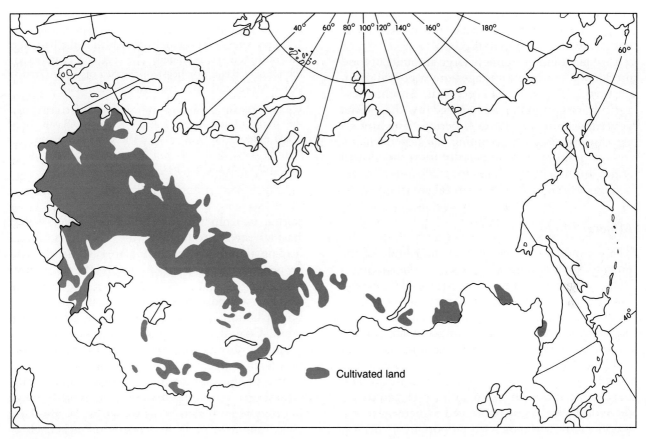

FIGURE 7.10

Agricultural Areas of the Soviet Union. Compare this map with the one in Figure 7.5.

part of the country, where the climate is warmer, tea, rice, grapes, and other fruits are produced. The republics of Georgia, Armenia, Turkmen, and Uzbek are also the main producers of some non-food items, such as cotton and tobacco. Most of these crops are irrigated. The Uzbek republic also produces silk and lambskins for clothing, and nearly all the country's hemp. Through the middle part of the country, the farming is local—that is, production is limited to areas near settlements and is used mostly for local consumption. The short growing season allows the production only of such hardy crops as potatoes, some grains and vegetables, and dairy products. In the far north and east of the Soviet Union, only the most rapidly maturing vegetables can be grown in the very short summer season. The diet in the region is supplemented by hunting and long-distance trucking from the agricultural belt.

The Soviet Union ranks among the world leaders in the livestock industries. The amount of milk and meat produced is similar to the amount produced in the United States. The most recent (1980) production figures indicate that the Soviets produce about 7 million metric tons of beef, 5 million metric tons of pork, and 1 million metric tons of lamb each year. They also produce three times as much wool as the United States, but that is partly because they do not utilize synthetic fibers as much as U.S. textile industries do. The Soviets are second only to Australia in the number of sheep raised, and second behind China in hog production.

Another important economic activity in the Soviet Union that helps supplement the diet is fishing. Only China and Japan exceed the Soviets in the tons of fish caught. Soviet fishing boats roam the world's oceans, and the catch from

The Soviet Union

interior waters is also important. Sturgeon eggs for caviar are significant mainly for domestic use, but partly also for exportation.

In spite of the seemingly large amounts of food production, serious shortages are reported from the Soviet Union nearly every year. Another new agricultural program was started by the Soviet government for the 1982–90 period. The Soviets are concerned about becoming too dependent on foreign grain imports, especially from the United States, but the best planning possible cannot overcome the limits set by the natural environment.

Mineral Production

The continued struggle to feed the people of the Soviet Union is somewhat offset by the country's vast amounts of mineral resources. The Soviet Union is perhaps the richest country in the world, in terms of minerals. It has an abundance of the ingredients for steel production and is the world's leading steel producer. Outputs for some years run as high as 25 percent of all the steel made in the world. The steel industry relies on iron ore and coal, and the Soviets claim to have 41 percent of the world's known iron ore and 57 percent of the world's coal. Manganese, an important element in the manufacture of alloys for iron, aluminum, and copper, is also abundant. The Soviets have about 90 percent of the world's known supply of Manganese.

One problem with the mineral deposits is that they are scattered (see Figures 7.11 and 7.12). It requires much time, effort, and expense to overcome the distributional problem. As a result, industry tends to grow at the point where transport modes meet or where one commodity can be found, while other commodities are shipped to the location. Usually the iron ore is shipped to the coalfields for steel production. The southeastern Ukraine region, just north of the Sea of Azov, has been a notable mineral-producing area for many years. Large deposits of lignite, bituminous, and anthracite coal are located in the Donets Basin. The industrial region is called the **Donbas,** and the mining activity supports a string of cities along the Northern Donets River. Kharkov, Donetsk, Rostov, and Lugansk are all important industrial cities of the region. Iron ore is brought to the Donbas from deposits near Kiev, and manganese is trans-

---

Donbas region The manufacturing region of the Ukraine where high-grade coal, which is the basis for the local industry, is mined. The region is centered in the Donets Basin northeast of the Sea of Azov (Azovskoye More) and east of the city of Donetsk.

Kuzbas region An area of heavy manufacturing in the Soviet Union, located in the Kuznetsk Basin about 1,200 miles (2,000 km) east of the Ural Mountains in Central Asia. Both coal and iron ore have been found and utilized in the Kuzbas.

---

ported in from the lower Dnieper River valley. Besides steel production, the region is also a leading aluminum-processing area. Hydroelectricity generated at Volgograd supports the aluminum-processing.

The second major coal-producing area of the Soviet Union is the region along the Ural Mountains. Coal deposits have been found in many locations throughout the southern part of the mountain range. The city of Sverdlovsk, at the center of this mining region, boasts one of the largest geology museums in the world. Huge deposits of both iron ore and manganese also have been found, as well as deposits of nickel, chromium, copper, and zinc. Even bauxite (aluminum ore) and some uranium are mined in the Ural Mountains region. The mineral wealth of the Urals has been recognized since the eighteenth century—the original idea for exile to Siberia was influenced by a need for mine workers.

Three other major coal-producing areas of the Soviet Union are (1) the Kuznetsk Basin near the city of Novosibirsk, (2) the Karaganda fields of the central part of the Kazakh republic, and (3) the region west of Lake Baikal, near the city of Irkutsk. All three regions have very large deposits of coal, but the Lake Baikal area also has iron ore, manganese, tungsten, mica, uranium, and petroleum. The industrial area of the Kuznetsk Basin is known as the **Kuzbas,** and the high-grade coal mined there is the basic source of energy for the local steel mills. The iron ore comes from near the city of Abakan, located about 200 miles (322 km) to the south. Most of the coal mined near the Kazakh city of Karaganda is shipped elsewhere for use, so the local area is not being promoted as a place to settle. The city of Irkutsk is growing

> **Ural-Volga region** Two linear regions of manufacturing that run along the eastern and western slopes of the Ural Mountains. The regions are noted for their oil production, manufacturing, and transportation connections with the Moscow area.
>
> **Povolzhye region** The region in the Soviet Union that extends along the Volga River. A major contribution is the transportation of foodstuffs and raw materials on the river, but the region is also the country's major source of petroleum and natural gas. Water transport extends from the Volga to the Don River and Black Sea through a series of canals, and pipelines for moving petroleum products extend outward to Moscow and the Ukraine.

rapidly, in spite of its remote location. Growth there is attributed to the exploitation of local minerals, especially the coal, and the resulting industrialization.

In addition to the five major coalfields mentioned above, numerous other scattered deposits exist in the Soviet Union. One deposit in the upper basin of the Lena River in East Siberia has estimated reserves that contain as much as half the already proven reserves for the entire country. Besides the vast amounts of coal, the Soviet Union has more than half the world's potassium salts, about one-third of the world's phosphates, and 13 percent of the world's gold.

The Soviet Union also is self-sufficient in petroleum, with an estimated reserve of 70 billion barrels of crude oil. One major producing area lies west of the Caspian Sea and directly north of the oilfields of the Middle East. The major producing area is near the city of Baku, but production occurs at scattered locations on all sides of the Caspian and along the northern slope of the Caucasus Mountains. Another major oil field, known as the **Ural-Volga oil region,** is located in the southern part of the Ural Mountains. Major fields also exist in scattered locations along the Urals northward from the Ural-Volga. The third large oil-producing area of the Soviet Union is located along the Ob' River near the center of the West Siberian Lowlands. The three major production areas are connected by pipeline to the urbanized and industrialized regions of the country. Besides the major oil-producing areas, other important oilfields are scattered throughout the Soviet Union (see Figure 7.12).

Industry

The mineral resources of the Soviet Union give the country a solid base for industry. Industrial output doubled during the decade of the 1970s, and lessened somewhat during the first half of the 1980s, but it continues to expand at impressive rates. Production is geared primarily to heavy industry rather than to smaller consumer products, such as cars, stoves, and refrigerators. Prices for consumer items remain high because of their scarcity and because they are considered luxuries. Even clothes and shoes are expensive by Western standards, and people usually have to stand in long lines to purchase goods.

The most famous industrial regions of the Soviet Union are the Donbas and the Kuzbas, the oldest and the traditional areas of heavy industry, but manufacturing occurs in numerous other locations (see Figure 7.13 on page 196). The major manufacturing region of the country is called the *Moscow Region*—an area including the capital city and its hinterland. The region extends from the city of Gorky east of Moscow, to Tula on the south, Rzhev on the west, and Yaroslavl on the north. Gorky is the major automobile-producing city of the Soviet Union; most of the rifles and other small arms for the Soviet army are made at Tula; and Yaroslavl is a major rubber-products manufacturing center. Other manufacturing cities in the Moscow Region include Ivanovo, noted for textiles; Kolomna, which has railroad shops and munitions factories; Serpukhov, a textile and cloth-dyeing center; and Kalinin, which produces machinery, textiles, and leather goods.

The Volga River region, known as the **Povolzhye,** is another important manufacturing area of the Soviet Union. Manufacturing towns are found all along the Volga, from Volgograd on the south to Berezniki on the north. The region was pushed into industrial production during World War II as the Soviets moved some of their industries eastward, away from the advancing German armies. Later, petroleum was discovered there, along with other minerals, so the area was encouraged to grow. Government projects named "the Great Volga" supplied money to build dams for

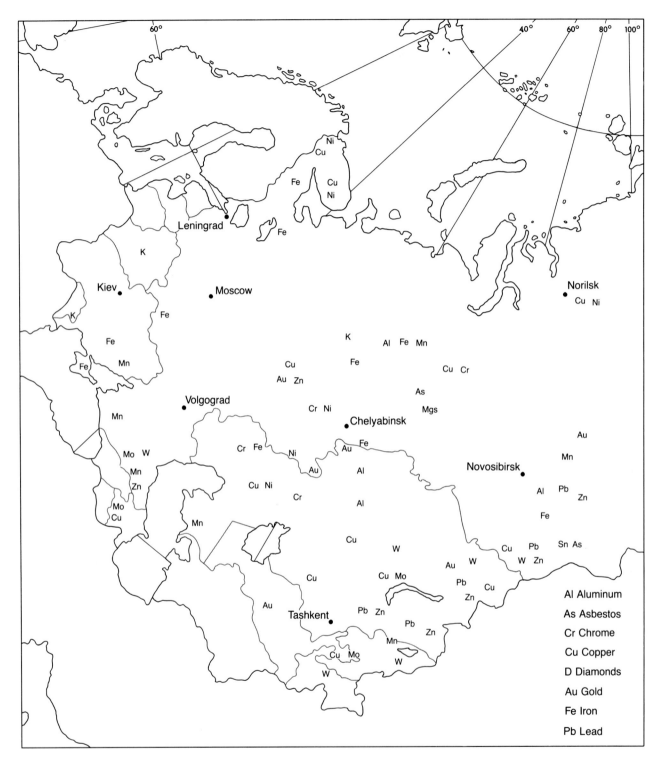

FIGURE 7.11
Mineral Deposits of the Soviet Union.

192   Chapter 7

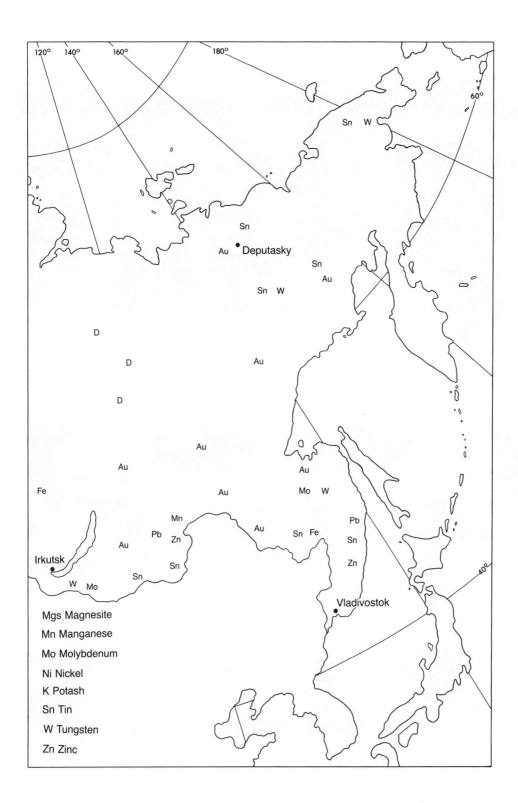

The Soviet Union

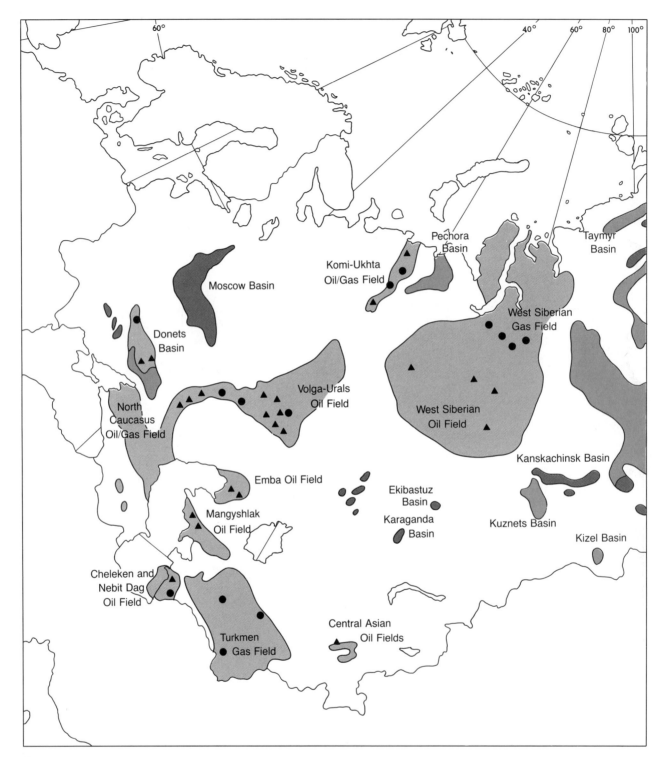

FIGURE 7.12
Coal and Petroleum Deposits of the Soviet Union.

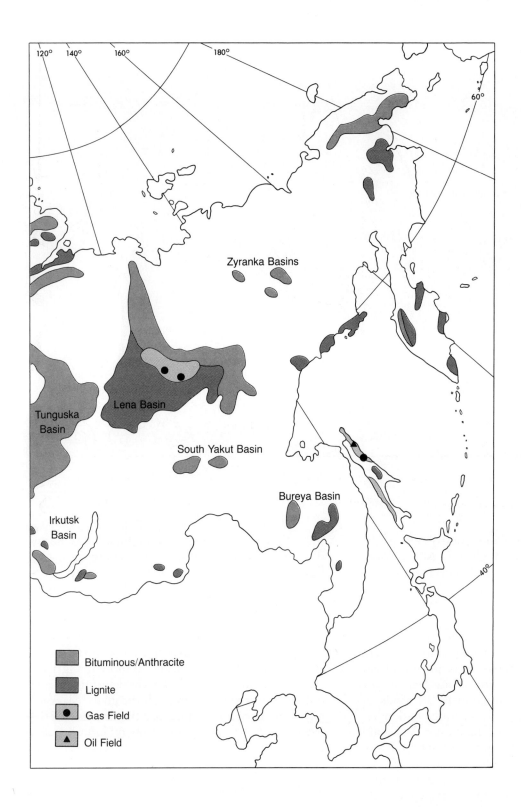

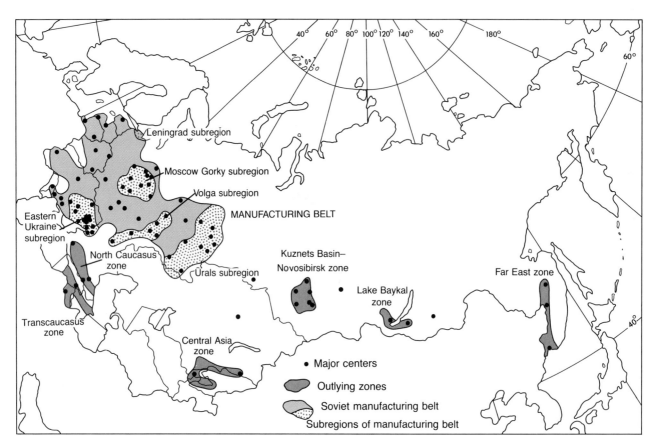

FIGURE 7.13

Manufacturing Regions of the Soviet Union.

power, irrigation, and flood control along the river. The Volga flows through the East European Plain and drops only 650 feet (200 m) from source to mouth without any rapids or falls, so it is navigable along its entire course. The Volga also supplies some fish for local markets. The river is the heart of the region and ties it together.

Besides the Moscow and Volga regions, light manufacturing is carried on in areas noted for their heavy industry. North of the Black Sea, in the Ukraine and including the Donbas, all kinds of manufacturing takes place. This favored region has mining, heavy industry, manufacturing, and good agriculture. It is not surprising that the population density here is one of the highest in the Soviet Union. The Kuznetsk Basin also supports manufacturing, but rather than letting it grow naturally on the basis of the local minerals, the Soviet planners are promoting manufacturing in the region. The Kuzbas has had mining and heavy industry for some time, but manufacturing of lighter-weight commodities is more recent.

Most economic activities in the Soviet Union are owned or controlled by the government, but a huge illegal black market exists, and illegal private production and service firms are periodically exposed. Plans for Soviet industry are set on five-year state planning periods, similar to those for agriculture. Preset production goals are determined, and output is geared to accomplish the goals. Slackening of industrial output during the first half of the 1980s was due to shortages in the oil industry, less steel production, and poor grain harvests.

Transportation

A large country has many advantages, especially with regard to resources, but a major disadvantage

The Volga is the standard Soviet passenger car, but it is priced too high for the average citizen. Photograph by Lawrence Gibbs.

is the cost of developing, maintaining, and using efficient transportation systems. The Soviet Union has great distances to conquer, and because the resources are widely dispersed the distances must be overcome.

Moscow, the hub of all Soviet transportation systems, is the focus of all roads and railroads and the center for air travel and communications networks (see Figure 7.14). West of the Urals, the railroad network of the Soviet Union is well developed, and every major city and many smaller towns are connected. Both passenger and freight movement by rail is heavy. The railroads were started during the time of the czars, and the old 5-foot (1.5 m) wide gauge of track has been retained. The subway system within Moscow itself is clean, modern, and functional and rivals any system in the world.

East of the Urals the railroads thin out considerably. A meager network ties the mining and manufacturing regions together, stretching between the cities of Krasnoyarsk on the east to Karaganda and Sverdlovsk on the south and west. Much of this thin lacework of rails consists of feeder lines sprouting off the main lines. The famous Trans-Siberian Railroad runs east of Irkutsk to Vladivostok. This double-track line has become overloaded, so the government recently completed a second parallel route. The trans-Siberian route is a notably long railroad line. It takes a fast train 10 days to travel from Vladivostok to the Polish border.

Highways are not a major aspect of the Soviet transportation system. Some freight is hauled by trucks, but not nearly as much as in Europe or Anglo-America. Passenger cars are very expensive and not commonly available to the average citizen. A standard Volga, the Soviet counterpart to the Chevrolet, costs the equivalent of five to six years of an ordinary worker's wages. Because only government officials, certain celebrities, factory heads, and other prominent people can afford a car, there is no need for good highways for automobiles, but good highways are becoming more essential for the increased amount of truck traffic and for military movements. Most urban streets are paved, but the pavement often ends at the city limits. A few major highways connect the cities west of the Urals, but on the east only gravel roads are available.

Water transportation in the Soviet Union is important but hampered by the cold winters. As noted above, the Volga River carries large volumes of freight. In fact, it carries more than half the inland waterway tonnage, although the river is frozen solid every year, sometimes for as long as

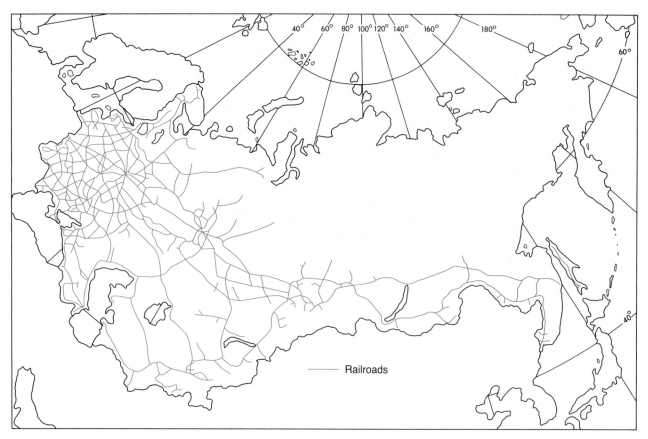

FIGURE 7.14

Railroads of the Soviet Union.

five months. The most important cargo on the Volga is petroleum from the Baku field, but much grain, coal, rock building materials, and timber also are carried. The rivers in Siberia flow into the Arctic Ocean and are frozen for more than six months of the year. Coastal shipping is prominent, but again hampered by the cold. The port at Vladivostok is kept open in the winter with the aid of ice-breaker ships—tedious, costly work, but necessary for most Soviet ports.

The Soviet Union is covered by an airline network. Modern aircraft provide regular services to all major cities. Excessive distances within the country are overcome somewhat by the speed of air travel, but because of the cost, passenger and air cargo traffic is limited to only special people and commodities. The Soviet airlines can be bolstered at any time by about 2,500 military cargo planes.

Communication

Freedom of the press and freedom of speech do not exist in the Soviet Union. All communications media are owned and closely screened by the state. All television and radio broadcasts are monitored, and any even slightly adverse comment made about communism or the state will cause the broadcast to be cut off immediately. All reading materials are subject to censorship; either the undesirable parts are eliminated, or the entire document is destroyed. Television sets, radios, and newspapers are common, however, because those media are effective for government propaganda purposes. About 75 million television sets were in use in 1983, or about one for every four people. Because as late as 1970 only one television set for every 20 people was available, massive production

Travelers in the Soviet Union can stop at stations for refreshments. In this case, vending machines dispense a colored sugar-water, but everyone must use the same cup. Photograph by Lawrence Gibbs.

has occurred recently. Much air time is devoted to government propaganda, but soccer games and the ballet are the most popular television programs.

■ *Political Geography*

Since the 1917 revolution the Soviet Union has had internal stability. Ethnic uprisings and minor revolts were quickly suppressed. The central government has absolute control over every aspect of life in the country. Individual freedoms are ignored in favor of the socialist state concept. Except for the nagging food-shortage problem, the basic needs of the people are being met. At this time there seems to be no internal movement that might challenge the absolute power of the government. The Communists are in firm control and will likely remain so in spite of the loosening of party control suggested by some of the tenets of the recent policy of *perestroika*.

The Soviet Union maintains a very strong military and does not hesitate to assert itself in world affairs. Kremlin leaders have maintained their determination to persuade the non-

These Soviet gentlemen near the city of Kuibyshev provide musical entertainment for highway travelers. Photograph by Lawrence Gibbs.

The Soviet Union

Communist countries of the world to adopt their system. Now that it has the world's largest army equipped with the most modern equipment, Moscow will probably never again be overrun by foreign troops, and Soviet political influence in world affairs will continue to be immense.

Key Words

communism
continental climate
Donbas region
Kuzbas region
permafrost
Povolzhye region
Russian Orthodox church
Slavic languages
state farms
Turkic languages
Ural-Volga region

References and Readings

Allworth, Edward (ed.). *Ethnic Russia in the U.S.S.R.: The Dilemma of Dominance.* Elmsford, N.Y.: Pergamon, 1980.

Ambler, John (ed.). *Soviet and East European Transport Problems.* New York: St. Martin's, 1985.

Andrusz, Gregory D. *Housing and Urban Development in the U.S.S.R.* Albany, N.Y.: State University of New York Press, 1985.

Baransky, N. N. *Economic Geography of the U.S.S.R.* Translated from the Russian by S. Belsky. Moscow: Foreign Language Publishing House, 1956.

Bater, James H. *The Soviet City: Ideal and Reality.* Beverly Hills, Calif.: Sage, 1980.

Bater, James H., and French, Richard A. (eds.). *Studies in Russian Historical Geography.* New York: Academic Press, 1983.

Bergson, Abram, and Levine, Herbert S. (eds.). *The Soviet Economy: Toward the Year 2000.* Winchester, Mass.: Allen & Unwin, 1983.

Bialer, Seweryn. *The Soviet Paradox: External Expansion, Internal Decline.* New York: Knopf, 1986.

Blainey, Geoffrey. *Across a Red World.* New York: St. Martin's, 1968.

Brown, Archie, et al. (eds.). *Cambridge Encyclopedia of Russia and the Soviet Union.* New York: Cambridge University Press, 1982.

Clark, M. G. *Economics of Soviet Steel.* Cambridge, Mass.: Harvard University Press, 1956.

Cole, John P. *Geography of the Soviet Union.* London: Butterworths, 1984.

Crankshaw, Edward. *Russia and the Russians.* New York: Viking, 1948.

Cressey, George B. *How Strong Is Russia? A Geographical Appraisal.* Syracuse, N.Y.: Syracuse University Press, 1954.

Deutsch, Robert. *The Food Revolution in the Soviet Union and Eastern Europe.* Boulder, Colo.: Westview, 1986.

Dewdney, John C. *A Geography of the Soviet Union.* Elmsford, N.Y.: Pergamon, 1979.

———. *U.S.S.R. in Maps.* New York: Holmes & Meier, 1982.

Dohrs, Mary Ellen. *Sketches of the Russian People.* Detroit: Garelick's Gallery, 1959.

Ginsberg, Norton, et al. *The Pattern of Asia.* Englewood Cliffs, N.J.: Prentice-Hall, 1960.

Goodall, George (ed.). *Soviet Union in Maps.* Chicago: Denoyer-Geppert, 1947.

Harris, Stuart A. *The Permafrost Environment.* Totowa, N.J.: Rowman & Littlefield, 1986.

Hedlund, Stefan. *The Crisis in Soviet Agriculture.* New York: St. Martin's, 1984.

Holderness, Mary. *New Russia.* New York: Arno, 1970.

Hooson, David J. M. *The Soviet Union: Peoples and Regions.* Belmont, Calif.: Wadsworth, 1966.

Hopkirk, Peter. *Setting the East Ablaze: Lenin's Dream of an Empire in Asia.* New York: Norton, 1985.

Howe, G. Melvin (ed.). *The Soviet Union: A Geographical Survey.* Plymouth, Eng.: MacDonald & Evans, 1983.

Jensen, Robert G., et al. (eds.). *Soviet Natural Resources in the World Economy.* Chicago: University of Chicago Press, 1983.

Karklins, Rasma. *Ethnic Relations in the U.S.S.R.: The Perspective from Below.* Winchester, Mass.: Allen & Unwin, 1986.

Levine, Irving R. *Travel Guide to Russia.* Garden City, N.Y.: Doubleday, 1960.

Linz, Susan J. (ed.). *The Impact of World War II on the Soviet Union.* Totowa, N.J.: Rowman & Allanheld, 1985.

Lowery, J. H. *Europe and the Soviet Union.* London: Arnold, 1966.

Lydolph, Paul E. *Geography of the U.S.S.R.* New York: Wiley, 1977.

Mehnert, Klaus. *Soviet Man and His World.* New York: Praeger, 1962.

Mellor, Roy E. H. *Geography of the U.S.S.R.* New York: St. Martin's, 1964.

———. *The Soviet Union and Its Geographical Problems.* London: Macmillan, 1982.

Mikhailov, Nikolai N. *Soviet Russia: The Land and Its People.* New York: Sheridan House, 1948.

Mirov, Nicholas T. *Geography of Russia.* New York: Wiley, 1951.

Morton, Henry W., and Stuart, Robert C. (eds.). *The Contemporary Soviet City.* Armonk, N.Y.: Sharpe, 1984.

Nazaroff, Alexander I. *The Land of the Russian People.* New York: Lippincott, 1944.

Nove, Alec. *The Soviet Economic System.* Winchester, Mass.: Allen & Unwin, 1980.

Parker, William H. *The Soviet Union.* New York: Longman, 1983.

Pipes, Richard. *The Formation of the Soviet Union: Communism and Nationalism, 1917–1923.* New York: Atheneum, 1968.

Pitcher, Harvey J. *Understanding the Russians.* London: Allen & Unwin, 1964.

Ro'i, Yaacov (ed.). *The U.S.S.R. and the Muslim World: Issues in Domestic and Foreign Policy.* Winchester, Mass.: Allen & Unwin, 1984.

Sagers, Matthew J., and Green, Milford B. *The Transportation of Soviet Energy Resources.* Totowa, N.J.: Rowman & Littlefield, 1986.

Sanchez, James J. *A Bibliography for Soviet Geography.* Chicago: Council of Planning Librarians, 1985.

Settles, William F. *Life Under Communism.* Chicago: Adams, 1969.

Shabad, Theodore. *Geography of the U.S.S.R.: A Regional Survey.* New York: Columbia University Press, 1951.

Smith, Hedrick. *The Russians.* New York: Times Books, 1983.

Stuart, Robert C. (ed.). *The Soviet Rural Economy.* Totowa, N.J.: Rowman & Littlefield, 1984.

Symons, Leslie. *Russian Agriculture: A Geographic Survey.* New York: Wiley, 1972.

Symons, Leslie, et al. *The Soviet Union: A Systematic Geography.* Totowa, N.J.: Barnes & Noble, 1983.

Taaffe, Robert N. *An Atlas of Soviet Affairs.* New York: Praeger, 1965.

Thayer, Charles W. *Russia.* New York: Time Inc., 1965.

U.S.S.R. Energy Atlas. Washington, D.C.: U.S. Central Intelligence Agency, 1985.

Wagret, P. (ed.). *U.S.S.R.* Geneva: Nagel, 1969.

Wilson, David. *The Demand for Energy in the Soviet Union.* Totowa, N.J.: Rowman & Allanheld, 1983.

Wixman, Ronald. *The Peoples of the U.S.S.R.: An Ethnographic Handbook.* Armonk, N.Y.: Sharpe, 1984.

Zum, Brunnen and Osleeb, Jeffrey P. *The Soviet Iron and Steel Industry.* Totowa, N.J.: Rowman & Littlefield, 1986.

Chapter 8

JAPAN

■ Introduction

It might seem strange that a small island nation deserves an entire chapter in a book about world regions, but Japan is a special country. It is indeed small, especially when compared with the areas of the Soviet Union and Anglo-America, but Japan is definitely part of the world we call "developed." In addition to being small in size, the country has a poverty of natural resources, a rugged surface, and an isolated location, but despite these handicaps, the Japanese have built an economic empire.

Industrialization in Japan seems to be at its zenith, but each year new production records are set. Japanese industries produce every commodity imaginable—from giant oil tankers to the smallest microchips. The industry thrives on trade with other countries, and the United States is Japan's leading trade partner. The United States is also bound by treaty to protect Japan militarily, and the country itself has only a small military budget. This lack of a need to produce aircraft and machine guns releases funds for further development of the industries.

The Japanese islands lie off the east coast of the Asian continent (Figure 8.1). They consist of four closely grouped large islands and hundreds of

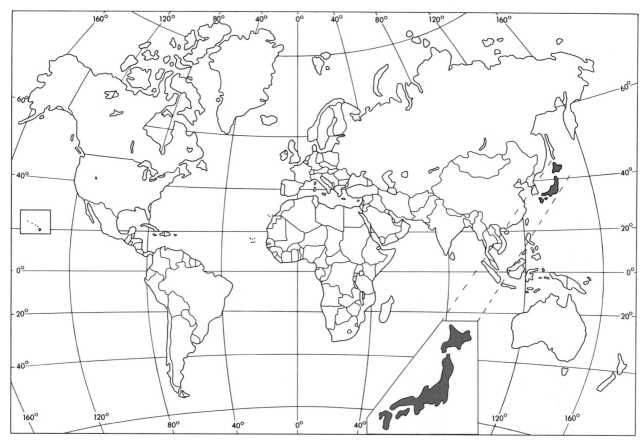

FIGURE 8.1
World Location of Japan.

smaller ones. The large islands from north to south are: Hokkaido, Honshu, Shikoku, and Kyushu, and they constitute about 98 percent of the Japanese territory. The island chain forms an arc, convex toward the Pacific Ocean, and stretches from about 30 degrees north latitude to about 45 degrees north latitude. The latitudinal extent is similar to that of the 48 states of the continental United States. Japan's nearest neighbors are Korea on the west and the Soviet Union on the north. It is about 118 miles (190 km) across the Tsushima Strait to Korea, and about 25 miles (40 km) across the Soya Strait between Hokkaido Island and the Soviet Union's Sakhalin Island. The land area of all the Japanese islands is 147,470 square miles (381,947 km^2), or similar to that of Montana.

Physical Geography of Japan

Landforms

Japan is located in what is known as the *Ring of Fire*—the belt of earthquakes, mountain-building, and volcanic activity that surrounds the Pacific Ocean. In the western Pacific this belt takes the form of a series of island arcs that are all geologically complex. Movements of the earth's crust, such as folding and faulting, have created conspicuous knots of highlands and mountains that form the backbones of the individual islands. Repeated volcanic eruptions also have added to the complexity of the landscape, and short but swiftly moving streams have carved and sculptured the surface into intricate and varied contours.

conurbation A region that contains a large aggregation of urban communities and is usually dominated by one large center, such as the Tokyo conurbation.

The topography of Japan cannot be described according to any large, contiguous physical provinces, such as those found in the Soviet Union and Anglo-America. The landform categories are not regions, but groups of similar types of topography: (1) the hill and mountain country, (2) volcanic cones, (3) the lowlands, and the (4) coastline. (See Figure 8.2.)

Hill and Mountain Country. The hill and mountain country of Japan makes up about 75 percent of the country's territory. The highest elevations and most rugged terrain are found in central Honshu. These highlands, called the *Japanese Alps,* contain more than a dozen peaks over 10,000 feet (3,049 m). Japan's tallest peak, Mount Fuji, is near the Japanese Alps but geologically unrelated to them. The Alps are uplifted mountains, while the 12,461-foot (3,799 m) Fuji is a volcanic cone. Most of the hill and mountain country is made up of steep slopes and sharp divides. Landslides are common, caused by the frequent earthquakes. More than a thousand earth tremors occur each year, but those that cause damage to buildings occur on an average of only once every five years.

The mountains contain short, fast-moving streams that are fed by melting snow in the winter and heavy rains in the summer. These streams seem to be ideal for generating hydroelectricity while their flows are torrential, but they can quickly become placid trickles of water, useless for generating constant power. In spite of the erosional work of the streams, however, the landscape is sharp and angular, with steep slopes and knife like ridges. Rounded surfaces are not common in the Japanese highlands, as one would expect in similar wet locales.

Volcanic Cones. Volcanic cones occur throughout the length of Japan. With their symmetrical slopes, lava flows, and ash plateaus, they add variety to the landscape. Some, like Mount Fuji, have been dormant for hundreds of years, but others remain active. Some 13 active volcanoes are located in the eastern part of Hokkaido, and there are 18 more on the western part, along with the associated region of northern Honshu. About another dozen active volcanoes form a line through central Honshu, and about 14 are grouped on the island of Kyushu. Only Shikoku has no active volcanoes. Besides the volcanoes, there are many hot springs, some of which are used for resorts.

The Lowlands. The most important aspect of the landscape of Japan is the lack of level land. Only 15 percent of the land area is cultivable, and in a country that ranks seventh in the world in total population, this is critical. The small areas of lowland contain not only the cultivable land but also most of the population and industry. The Japanese lowlands are small coastal plains formed from the deposition of river-borne material. These plains are separated from each other by mountains that run directly into the sea. Japan does not have any long rivers, but the short streams can come out of the mountains as a torrent of water after a heavy rainfall. Once the streams reach the coastal plains, they slow down and meander, leaving deposits of alluvium, and finally empty into bays and estuaries.

The largest of Japan's coastal lowlands is the Kanto Plain in eastern Honshu (Figure 8.3 on page 208). It contains about 5,000 square miles (13,000 km^2) of level land and one of the highest concentrations of people anywhere in the world, with the Tokyo-Yokohama **conurbation**. Most of the other major cities of Japan also are located on similar types of coastal plains. The populations of the Kanto, Nobi, and Kansai plains combined make up more than one-third of the people of Japan, and these people produce about half the country's industrial output. None of the alluvial plains is extensive, and many are only small pockets of land hemmed in by steep slopes on one side and by the sea on the other. Every piece of level land is used to the fullest.

The Coastline. Japan's coastline is very long because of the hundreds of inlets, bays, and estuaries.

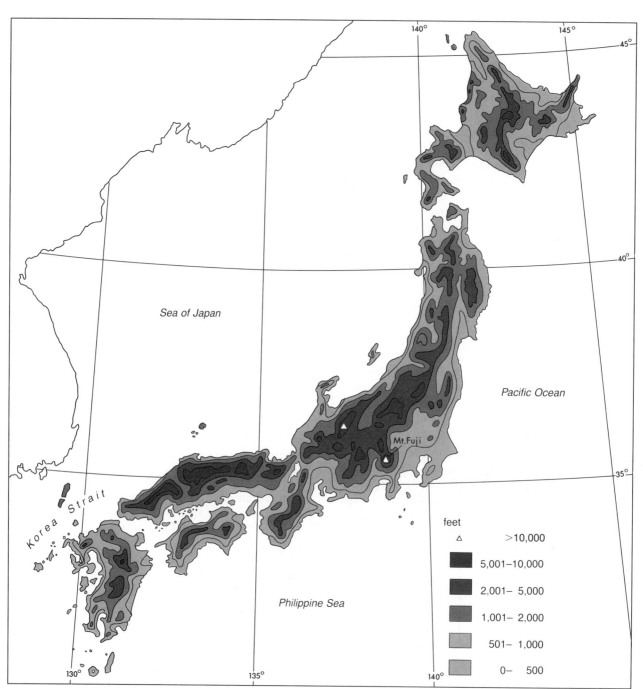

FIGURE 8.2
Land Elevations of Japan.

Siberian High A large, stable high-pressure system that builds up over eastern Asia (Siberia) during the winter months because of the coldness created by the lack of marine influences (continental location).

The four major islands alone have about 17,000 miles (27,353 km) of coastline. It is also a very intricate coast. Major faults cut across the grain of the land on the Pacific side, and these have helped create the larger coastal plains on that side. The Japanese shoreline facing Asia is less diversified than that along the Pacific; the coastal plains there are smaller, and more sand dunes and beach ridges appear.

Japan probably has more of a variety of interesting coastal features than anywhere on earth, and as already noted, most of the Japanese people live along the coast. Consequently, the people of Japan have what is called a "maritime outlook." Not only are they among the world's greatest fishermen, they are superb shipbuilders, and their commerce is oriented toward overseas ports. The waters between the islands are teeming with all sorts of vessels, for both local and overseas transportation.

Climate and Vegetation

The topography of Japan limits economic activity and restricts transportation on land, but the overall climatic conditions are favorable. The greatest control operating to produce the climates of Japan is the monsoonal air movements. The climates of monsoon Asia are characterized by seasonal changes in the direction of the prevailing winds. In the winter the huge land mass of Asia creates what is known as the **Siberian High**. This anticyclone, centered over the Lake Baikal region forces cold, dry air away from it, and the air circles off the mainland to blow over Japan. As it passes over the Japan Sea, the bottom layers of the air masses become warmed by the water and pick up mois-

The bridge over this slow-moving stream is designed in classic Japanese architecture. Note the rugged terrain and the lush vegetation cover. Photograph by Kazuyo Nakao.

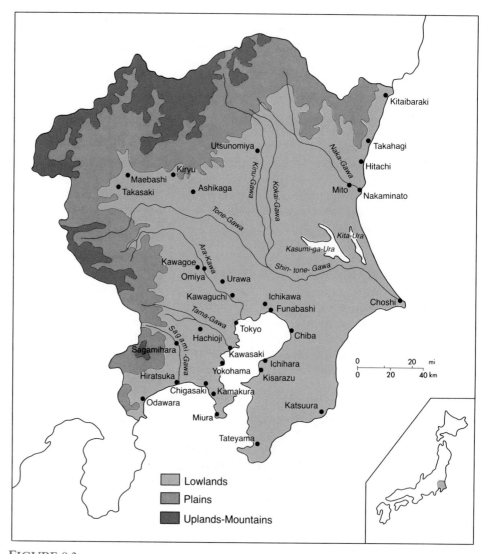

FIGURE 8.3

The Kanto Plain. The coastal plain of central Japan is a heavily populated region containing both intensive agriculture and numerous manufacturing cities.

ture. The air mass continues to move over Japan from the west, and as it rises over the islands it is cooled, which causes snow to fall. The resulting winter weather along the western side of the islands is generally cloudy, with continuous snowfalls that pile up in deep blankets. By contrast, the Pacific side of the islands experiences cold but dry winters, with very few clouds and very little snow. The Pacific coast is influenced by the ocean, especially the warm Kuro Shio current. (See Figure 8.4.)

In the summer the Asian continent becomes warm and a thermally induced low pressure builds up. Low pressure creates a vacuum, causing surface air to be drawn in toward it. The result for Japan is a complete reversal of the winter winds, because in the summer the prevailing air currents move toward the Asian mainland. Throughout the winter months, winds in Japan are from the northwest, but in the summer they are from the southeast. These are called the *winter monsoons* and

double-cropping Planting another crop immediately after one crop has been harvested, so that two crops can be grown during the same season.

summer monsoons, respectively (see Figure 8.5). The summer monsoon draws warm, moist air from off the Pacific Ocean, so Japanese summers are warm and humid but lacking in extremely high temperatures, due to the oceanic influence.

One result of the monsoonal air movement over Japan is that the country has no dry region. In fact, the precipitation range is from wet to extremely wet. The precipitation is normally heaviest during the summer months, but the difference between summer and winter is not great. Many small localities, especially in the highlands of Honshu, receive more than 120 inches (305 cm) of precipitation annually, and some places get as much as 160 inches (406 cm). Hokkaido receives the lowest amounts of total precipitation (about 40 inches annually, or 102 cm) but the largest amounts of snow.

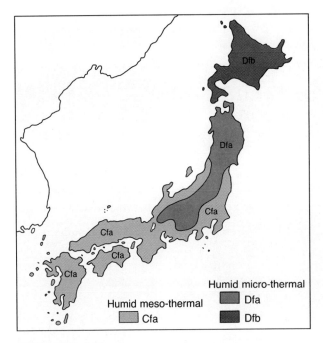

FIGURE 8.4
Climate Regions of Japan.

Some of the large precipitation totals in Japan are caused by late summer *typhoons* that usually scrape across the country. These tropical cyclonic storms, called "hurricanes" in the western hemisphere, can drop as much as a foot of rain in less than a day. The storms often are violent, carrying winds up to 125–130 miles per hour. They originate over tropical waters of the western Pacific near the Mariana Islands and travel northwest, striking Japan from the southeast. The violent winds and beating rains often cause damage, especially to crops and rural villages. Fortunately, these severe storms are not frequent, averaging about one or two a month from late July to early October.

In addition to the contrasting climates between western and eastern Japan, there is a latitudinal change in temperature. The break-point is about 38 degrees north latitude. The average January temperatures north of that line fall below 32° F (0° C), and in northern Hakkaido they often get as low as 20° F (−10° C). These colder conditions in the north prohibit **double-cropping**, or growing two crops in one season, as is done south of 38 degrees north latitude. Southward from the transition line, average January temperatures increase gradually with distance. The January average for Tokyo is about 39° F (4° C), or comparable to cities along the North Carolina coast in the United States. Kagoshima in southern Kyushu averages about 43° F (6° C) in January, or about the same as Birmingham, Alabama.

The overall large amounts of precipitation in Japan allow the growth of forest vegetation. The Japanese are fond of trees and plant them in almost any available space. Natural forests cover about 55 percent of the land of Japan, but the trees grow on slopes too steep for agriculture. Broadleaf trees predominate in about half the forest land; the other half is divided equally between mixed forests and coniferous forests. The forest types depend on the latitude and elevation. The *subtropical forests* grow south of the 38 degrees north latitude line and consist of bamboo, palms, and bananas on the south, and oak, camphor, and wax trees on the north. The *temperate mixed forest* is found north of the 38 degrees north latitude line on northern Honshu and southern Hokkaido. Deciduous trees, such as oak, birch, beech, and poplar, prevail, but

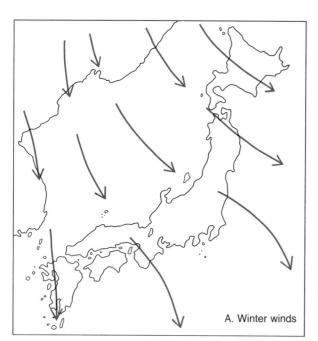 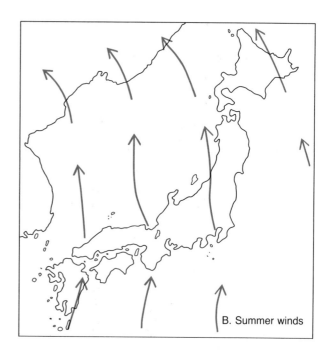

FIGURE 8.5

The Summer and Winter Winds over Japan. The winter winds blow off the Asian mainland (A), the summer winds come from the Pacific Ocean (B).

A thick forest cover surrounds this Shinto shrine. The trees are lush but not tall and straight, so they are not good for lumber. Photograph by Kazuyo Nakao.

> **prefecture** A subdivision of a province for local administration under a governor, assembly, and council. Japan has 47 prefectures.

they are interspersed with the needleleaf types, such as fir, pine, hemlock, and cedar. In northern and eastern Hokkaido, where annual average temperatures are lower, the forests become entirely needleleaf. Generally, the conifers grow at higher latitudes and elevations, while the broadleaf trees prefer lower latitudes and elevations. All the Japanese forests have a dense undergrowth, which reflects the large amounts of rain.

The long occupation of Japan by sedentary people has led to a confusion of vegetation. A great variety of both natural and cultivated species exist, from dwarf trees to giant conifers. Other than the latitudinal differences described above, there are no simple vegetative zones. In addition to the great variety of native plants, many species that are native to much warmer regions grow in Japan. Also, many relict plants from past cultivation help to confuse the botanists.

Human Geography of Japan

Population

Japan is crowded, but the country is not as small as it might seem. It is half again as big as the United Kingdom, and it stretches for 1,300 miles (2,092 km) from north to south. Large parcels of rugged terrain and steep slopes are devoid of people, however, so the flat areas are extremely crowded. The mid-1986 population of Japan was 121.5 million, and with the total land area of 147,470 square miles (381,947 km^2), the overall density was 824 people per square mile. The number of people per square mile of arable land is an incredible 5,498, compared with 481 per square mile for the United States. The population per unit of productive land is by far the highest in the world. In that respect, the country is incredibly overcrowded. (See Figure 8.6.)

The annual rate of natural increase of Japan's population is 0.7 percent, the lowest in Asia. The Japanese are striving to control their population growth, but all methods of birth control are voluntary and are not imposed by the government. The legalization of abortion caused the birth rate to decrease from about 18 per 1,000 in 1975 to less than 13 per 1,000 in 1985. Life expectancy rates in Japan are not only the highest in Asia, but also among the highest in the world: age 73 for males and age 78 for females. This means that the population of the country is getting older. About 10 percent of the people are age 65 or older. That is similar to the ratio of older, dependent people in Anglo-America, but it is much greater than the 0.4 percent for the remainder of Asia.

Urbanization

The areas of population growth in Japan are in the city regions (Figure 8.7). For the last 40 years, the rural population experienced a steady decline as streams of migrants poured into the urban centers seeking employment with the ever-expanding industries. Of the 47 **prefectures** (political subdivisions), more than half now have fewer people than they did in 1960. Every city in the country has experienced growth during the same period. About 76 percent of the people of Japan live in urban areas, and only 8.9 percent of the total labor force is employed in agriculture, forestry, and mining. Essentially, Japan has an urban society.

Ten cities in Japan have more than 1 million people, and 10 others have populations between 500,000 and 1 million (see Table 8.1). This compares with 6 cities of more than 1 million in the United States, and 18 U.S. cities between 500,000 and 1 million. Remember, however, that the United States is 25 times larger than Japan and has nearly twice as many people. Tokyo is by far the largest Japanese city, with about 8.5 million people. The urban area of Tokyo merges into two other millionaire cities of Japan. Kawasaki has a population of 1.2 million, and its near neighbor, Yokohama, is Japan's second largest, with 3 million people. The Tokyo-Yokohama conurbation contains more than 10 percent of the people of the entire country.

Three of Japan's other large cities also are located on the lowland plains of eastern Honshu. These are Nagoya (2.1 million population), Osaka

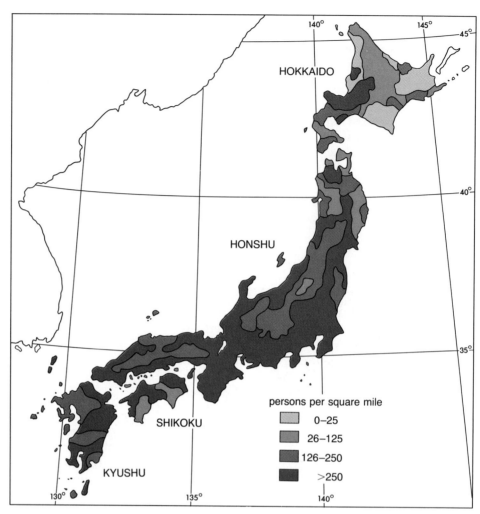

FIGURE 8.6
Population Patterns of Japan.

TABLE 8.1

Japan's Cities by Size

POPULATION	NO. OF CITIES
Over 2 million	4
1 million to 2 million	6
500,000–1,000,000	10
400,000–500,000	16
300,000–400,000	21
200,000–300,000	35

SOURCE: *The Far East and Australia* (London: Europa Publications, 1983), p. 441.

(2.6 million), and Kobe (1.4 million). The ancient capital of Kyoto (1.5 million) lies about 30 miles (48 km) northeast of the Osaka-Kobe conurbation. Two other millionaire cities are located on the northern plains of the island of Kyushu: Kitakyushu (1.1 million) and Fukuoka (1.1 million). Sapporo (1.5 million) is in northern Japan on the island of Hokkaido. All these large cities, with the exceptions of Kyoto and Sapporo, are located on low-lying ground along natural bays.

In many ways, the city of Tokyo represents Japan. It is the center of the national administration, education, culture, and economy of the coun-

FIGURE 8.7
Major Cities of Japan.

try. The largest number of foreign business establishments among Japanese cities are located in Tokyo. Sprawling over nearly 800 square miles (2,071 km^2) of the Kanto Plain and facing Tokyo Bay, the city is a thriving industrial center and truly an international city. It has more neon lights and movie theaters than any other city in the world. Traffic is incredible, and Tokyo may be the noisiest city in the world. The streets and sidewalks are full of vehicles and pedestrians for many hours every day. The population of Tokyo had reached 1.5 million by the year 1800, making it one of the earliest large cities of the world. The city was wrecked by an earthquake in 1855 and, along with Yokohama was destroyed again in 1923. American bombing during World War II caused massive fires that nearly destroyed the city again in 1945.

Race and Culture

The exact origin of the Japanese people is not known, but they are definitely a mixture of racial types. Today, they are a homogeneous people with very few ethnic minorities living on the islands. The Korean minority makes up less than one percent of the total population, and it is the largest. The Japanese are basically of Mongolic stock, but the mixture of Malayan and Polynesian strains is evident. Prehistoric migrations came from the Asian mainland and from the islands to the south of Japan. The first people to arrive were the *Ainu*. They were the ancestors of the few present-day Ainu who live mainly on the northern island of Hokkaido. These people are Caucasoids and are believed to be from Central Asia. The southern island people, called the *Yamato*, were the next to arrive in Japan. They took over the islands from the south, mixing with the Ainu but also driving them northward. The *Mongols* were the last to

Japanese racial mixture is evident in this woman's features. She also is wearing classic attire for Japanese women, which is not commonly seen today. Photograph by Kazuyo Nakao.

A common sight in Japan: the crowded street. This scene is near part of the elevated railroad in Kyoto. Photograph by Kazuyo Nakao.

arrive, and their bloodlines and culture mixed with those of the other groups. They came from Korea and northern China about 2,000 years ago.

The racial mixture of the Japanese today is about 60 percent Malayan-Mongolian, 30 percent Chinese-Mongolian, and 10 percent Ainu, but the mixing is so thorough that the physical traits are very uniform. The larger percentage of Malayan blood accounts for most physical distinctions attributed to the Japanese. Their island background also explains some of their modes of living, such as food and dwellings, which are typical of the rainy tropical climates found in the islands to the south.

Language

The Japanese language is isolated from the other languages of the world. It is related to Korean, Chinese, and possibly some other Altaic tongues

Shinto The ethnic cult and religion of the Japanese in which spirits of ancestors and some deities of nature are revered. Shinto is not regarded as a religion, so it can be followed by members of all faiths—that is, Buddhists and Christians alike.

but is quite distinctive. The spoken language is divided into two styles, the "recording" style and the "conversational" style. The recording style is used in most written documents, such as newspapers, journals, and official papers, but most people do not actually speak that way. There are four forms of the conversational style: the "abrupt," "plain," "polite," or "very polite" forms. Besides these subtle but distinct differences, another usage exists for women. All this means that a simple phrase may be stated in 30 to 40 different ways. Compared with the written form, however, conversational Japanese can be easy to learn.

In writing, the Japanese again use two forms: *kata-kana* and *hira-gana*. Both are from *kanji*, characters derived from written Chinese. The *kata-kana* form of the Japanese language was invented in the eighth century and is a phonetic writing using simplified Chinese characters. It is believed to have been influenced by Sanskrit, an ancient language of northern India. The *hira-gana* form originated in the ninth century and is written in a cursive style. *Kata-gana* is used in transcribing most foreign languages into Japanese and for such things as telegrams and children's books. Most other publications are in *hira-gana*, but mixed with Chinese characters.

Religion

As with their language, the Japanese have been very receptive in absorbing different religions from other cultures. Also, similar to language, the religions introduced to Japan have been combined into a uniquely Japanese faith.

Shinto. The native religious philosophy of Japan is **Shinto**. It is the basic, common belief of most Japanese, but it borders on animism because it has a variety of "gods": the creator, the sun, the

The Japanese make use of every piece of land both in the cities and in the rural areas. Land use intensity is illustrated by this scene in Kyoto. Photograph by Kazuyo Nakao.

Japan 215

moon, and the stars, as well as rivers and mountains, and vegetables and animals, among many others. For years, Shinto was the national faith, and it is still deeply rooted in the minds of most Japanese. About 70 million people in Japan today claim to be adherents of the Shinto religion.

Buddhism. **Buddhism** was first introduced to Japan in the middle of the sixth century. It came from northern India by way of China, and its acceptance was enhanced by the highly developed Chinese culture of the time. Buddhism was viewed by the Japanese as a vehicle of culture too important to be ignored and was therefore supported by powerful families in the government. Eventually, characteristics of Buddhism were modified to closely resemble Shintoism, and the religion became a national cult. Most things in the Japanese culture of today are tinged with Buddhist influences, but Japanese Buddhism is different from the Buddhism found today in India or Southeast Asia. It is no contradiction to the Japanese to believe in both Shinto and Buddhism. When a baby is born, a majority of Japanese parents visit a Shinto shrine, but when a death occurs in the family, a Buddhist priest is called to perform the interment ceremonies.

Buddhism A religion from India based on the doctrine originally taught by Gautama Buddha, consisting of the "Great Enlightenment," which lists the causes of suffering and how salvation can be obtained through suffering.

Christianity. Christianity did not arrive in Japan until the middle of the sixteenth century. Roman Catholic missionaries worked in the country for about 100 years, until they were expelled in 1639, and during that relatively short period about 200,000 Japanese were converted to Christianity. Although the missionaries were forced to leave Japan, Christianity was never completely eradicated. When missionaries were allowed to reenter the country in 1873, they found many faithful

Buddhist monks are a common sight in Japan. The one in this picture remains in this praying position on a busy street corner in Tokyo for hours. Photograph by Kazuyo Nakao.

labor-intensive Describes any economic activity that requires or is done by hand labor with a noted absence of machines.

believers who were descendants of the original converts. Today, there are about 1,750 Roman Catholic churches in Japan, with about 320,000 adherents. Among the Protestant churches in Japan, the United Church of Christ has about the same number of churches as the Roman Catholics, but about half as many members. The Anglican Episcopal church is the third largest Christian church in Japan, followed by the Eastern Orthodox church. The Baptists and Mormons also claim to have recently entered the Christian movement in Japan.

For those who are not accustomed to them, the differences between Shinto and Buddhist shrines are difficult to discern. This small Shinto shrine is in Kyoto, where the greatest concentration of shrines is located. Photograph by Kazuyo Nakao.

Economic Geography of Japan

Agriculture

Japanese farming is characterized by small, fragmented land holdings with very high productivity. Until the end of World War II in 1945, about half the arable land in Japan was owned by absentee landlords. As a result of land reform measures after the war, however, about 95 percent of the cultivated land today is in the possession of owner-farmers. The land reform brought about increased productivity, but it also created extensive fragmentation of the holdings. The small, scattered holdings of the farmers make it difficult to use large, modern equipment, so farming in Japan remains a **labor-intensive** occupation. Farming in the United States requires less than one person per acre, but in Japan about 60 people per acre are used. However, only about 15 percent of the Japanese farmers are engaged exclusively in farming; the remainder have secondary employment in industry.

By far the most important crop raised in Japan is rice. About one-third of the total agricultural output is devoted to that crop. In fact, rice is usually overproduced. The Staple Food Management Law requires the government to set the price paid to the rice producers. The price is adjusted yearly, but it usually gives an advantage to rice growers. The result is that more rice is produced than otherwise would be the case in an unsupported market. The price support for rice is an intricate, unsettled political issue, and the government is now trying to get farmers to grow crops other than rice.

Rice is in many ways an ideal crop for Japan, regardless of the historical momentum to produce it. The subtropical, monsoon climate, abundant summer rainfall, and the easily irrigated alluvial lowlands are suitable and attractive conditions for growing rice. Also, the yield per unit area for rice is highest among the small grains. Furthermore, in spite of the warm and humid conditions, rice will keep under storage. All these factors favor rice production, and a more highly recommended standard food crop would be difficult to find.

Rice is grown on all the major Japanese islands, but it eventually disappears from the

Focus Box

Sports in Japan

The oldest sport in Japan is *sumo*, a form of wrestling that dates to 30 B.C. Primarily because of its antiquity, it is called the "national sport" of Japan. Sumo wrestling is noted for its slowness and complexity of ritual and for the physical attributes of the contestants. The ritual includes demonstrating that no hidden weapons are available, and the wrestlers purify themselves with water and the wrestling ring with salt. The wrestlers are huge, weighing well over 300 pounds, with rolls of fat around their midsections. Before the matches begin, the wrestlers perform an elaborate bobbing and weaving, crouching and stamping activity designed to bring a spiritual element to the proceedings and to scare their opponents. When the match starts, the contestants lunge at each other, trying to throw their opponent from the ring. The end usually happens very quickly—the average match last about 10 seconds. Sumo wrestling is big business in Japan, and the wrestlers enjoy good incomes.

Even though sumo claims to be the national sport of Japan, baseball attracts larger crowds. Both sports are extremely popular, however, and both are shown regularly on national television. Baseball was introduced to Japan by an American schoolteacher in 1873, three years before the founding of baseball's National League in the United States. The Japanese took to the game with great

This three-tiered driving range at the Shiba Golf Center in Tokyo can accommodate 155 players at once. Lack of space is critical in Japan, but the zeal for sports has created this special need, resulting in an interesting use of land. Photograph by Colin Hagarty.

fervor, and it became popular throughout the country. Sandlot baseball is everywhere, and nearly every city has at least one professional team. Japanese players are highly skilled, and many could play for American professional teams. Salaries are high, and some American players, attracted by the high incomes, have chosen to play in Japan. The standard of excellence in baseball is still in America, but enthusiasm for the game provides a bond between the people of Japan and the United States.

Besides the spectator sports, the Japanese are very fond of individual physical activities, such as golf, karate, boxing, swimming, skiing, ping-pong, and gymnastics. Japan has produced some world champions in the individual sports. The Japanese also compete on the world level in some team sports, such as volleyball, but they lack the height to be world-class basketball players, and are not generally large enough to be good at American football. Sports is an important part of the Japanese culture, attracting huge numbers of participants and spectators alike.

---

paddy From the Malay *padi*. In its original meaning, unmilled or rough rice, whether growing or cut, but the word also refers to the field where the rice is grown.

---

landscape in extreme northern and eastern Hokkaido (see Figure 8.8). Most of Japan is too cool in the winter to permit double-cropping of rice, which takes longer to mature than some other crops. Rice is grown in **paddies**—small, enclosed areas where water for irrigation can be retained. The irrigation water comes mostly from the rivers instead of from reservoirs, but small ponds also are important. Irrigation canals are small and never used for transportation. The landscape in the lowlands consists of a multitude of tiny fields with irregular outlines, enclosed by dikes and with standing water in them. The mosaic of fields is interrupted occasionally by tree- and hedge-enclosed villages and raised footpaths and roads. Rice is started in seedbeds during the early spring, then transplanted to the paddies after about two months growth. By midsummer the water-dominated scene has changed to a lush green as the rice plants flourish. The rice changes from green to yellow as it matures and is usually harvested in late October or early November. (See Figure 8.9.)

Other crops produced by Japanese farmers include wheat, barley, soybeans, potatoes, sweet potatoes, other vegetables, and fruit. Cabbages and radishes are found in nearly every garden, and pears, apples, mandarin oranges, and tea are grown for home use, with any surplus sold to local markets.

The Japanese farm labor force totals about 5 million, and these people work on about 4.5 million farms. The average size of the individual farm is about one acre, and less than 10 percent of all farms are larger than two acres.

Much of Japan is covered with forests. In spite of the large number and variety of indigenous trees, more than half the country's timber requirements must be imported. After coal and petroleum, timber is the third most important import item for Japan. One important type of tree is the mulberry, because its leaves are the exclusive diet of silkworms. Japan produces about 14 million metric tons of raw silk each year, but production is decreasing.

The growing of rice in this small field tucked between an elevated road and some apartment buildings indicates the high utilization of land in Japan. Photograph by Kazuyo Nakao.

In other middle-latitude regions of the world where rugged terrain dominates, animal husbandry is a popular way of supplementing the diet. This is not true in Japan. The Japanese have only a small-scale animal industry, conducted primarily by individual farmers. The slopes are too steep for grazing cattle, and very little good pastureland is available elsewhere. Consequently, the Japanese never acquired a taste for dairy products, simply because they have not been available. The Buddhist faith discourages the killing of animals, so meat in the diet also has been uncommon. In spite of these historical influences, milk and meat are becoming more and more a part of the Japanese diet.

The Fishing Industry

The small-scale animal industry in Japan is offset by the country's fishing industry. In most countries, including the United States, fishing is looked at more as a pastime than as an industry. In Japan, however, fishing is a leading enterprise, a genuinely big business, and fish appear regularly on nearly every household's daily menu. Japan is the world's greatest fishing nation, and the annual catch accounts for about 15 percent of the world's total. The average yearly catch is around 10 million metric tons, compared with about 450,000 metric tons of beef and veal produced. About 98 percent of the fish caught come from the Pacific Ocean and the inland waters, but the fishermen roam the Atlantic and Indian oceans as well.

Fish and fish products are the main animal food of the Japanese and are second only to rice in their diet. About 500,000 full-time fishermen catch enough for both home consumption and exporting. The fish caught include every edible species, but sardines, mackerel, and Alaskan pollack account for the most tonnage. Squid, oysters, and various crustaceans also are important food items, and the Japanese use about 600,000 metric tons of aquatic plants (mostly seaweed) each year.

The fishing industry in Japan was hurt by the 1976 international agreement establishing 200-nautical-mile (370 km) offshore fishing zones. Prior to that agreement, about 40 percent of the

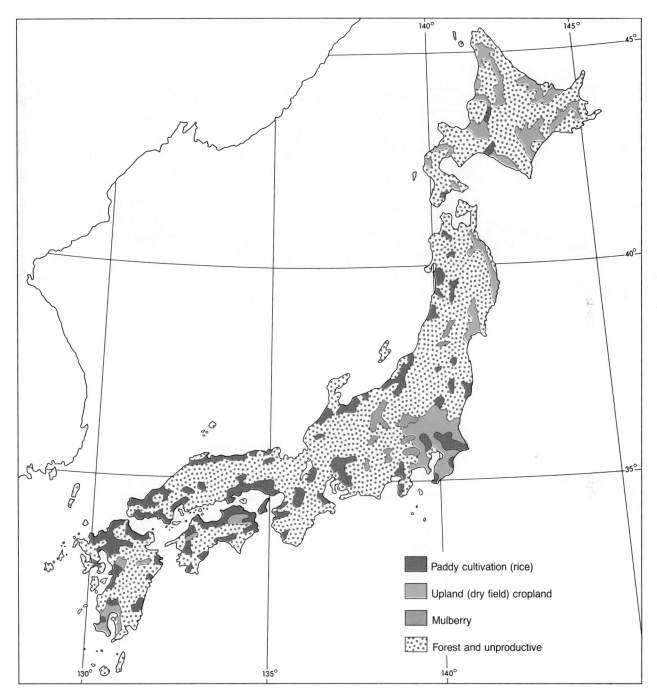

FIGURE 8.8

Land use in Japan. Note the numerous but relatively small areas of rice cultivation.

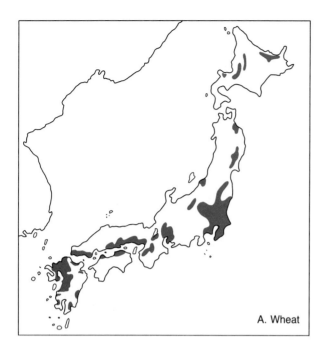

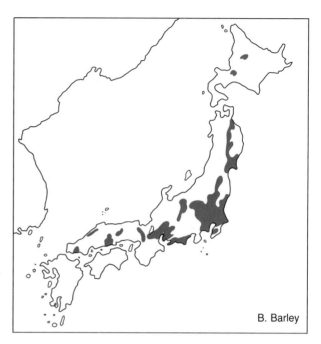

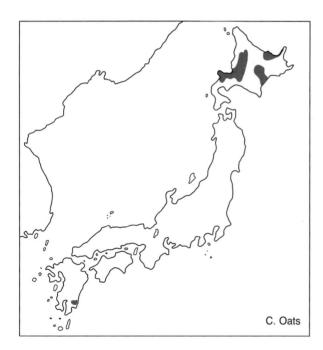

FIGURE 8.9

Wheat, Barley, and Oats Producing Areas of Japan. Other than the extensive cultivation of rice, Japan also produces many other grains and food crops. Wheat (A) is the most common crop after rice, then barley (B), and oats (C).

fish caught by the Japanese came from inside other nations' offshore zones. The United States and Japan have had disagreements over this issue, especially as it applies to whaling. Japan still catches about 5,000 whales a year, in spite of the worldwide whaling moratorium. The whales are for research purposes and come mostly from the Antarctic Ocean.

Mineral Production

Japan has a rich variety of minerals for mining, but the quantities are very small. Only about 100,000 people are employed in mining, and in terms of tonnage, the leading item produced is limestone. Limestone is used for building material, but it is more important for the manufacturing of steel. Enough limestone is produced to satisfy local demands. Japan also produces its own sulfur, chromite, and manganese. One major problem confronting the country, however, is the lack of both essential minerals and fuels. Japan must import nearly all of its bauxite, petroleum, and iron ore, 95 percent of its copper ore, and 90 percent of the coking coal. This means that all the essential ingredients for making steel, except for limestone, must be imported. Other industrial areas of the world have a good supply of either coal or iron ore or both, but Japan has become industrialized without them.

Manufacturing

Japan's remarkable industrial expansion began in the 1930s. The growth of the iron and steel industries followed the government's declared policy of political expansion. The unnatural emphasis on heavy industry was motivated by a desire to make the country self-sufficient in the materials needed for modern warfare. In 1930, Japan produced about 1 million metric tons of steel, but by 1940 the output had increased to more than 5 million metric tons. After World War II, production of steel continued to be emphasized, and today Japan averages about 100 million metric tons each year. Only the Soviet Union exceeds Japan in steel production. The iron ore is imported from Australia, Brazil, and India, and the coking coal comes from Australia, the United States, and Canada. After Japan imports the ore and the fuel for making the steel, approximately one-third of the output is exported. Japan is the world's leading steel-exporting country, accounting for about 30 percent of the world's total.

Industrial production in Japan is concentrated in four major regions (see Figure 8.10). About half the country's total manufacturing output comes from the combined productions of the Tokyo-Yokohama, Osaka-Kobe, Nagoya, and Kitakyushu regions. Japan is the world's leading producer of ships, passenger cars, trucks, and buses, and it ranks second in the manufacture of paper, synthetic fibers and resins, cement, and steel. Japan ranks second only to the United States in overall industrial production.

Manufacturing transportation equipment is the single most important industry in Japan, and passenger cars and ships are the leading items. Japan's nine major car manufacturers build about 7 million cars a year and compete for shares of the

Although much manufacturing in Japan takes place on production lines and by machine, furniture and other household goods are made in thousands of small factories. This small shop in Tokyo produces wall panels for houses. Photograph by Kazuyo Nakao.

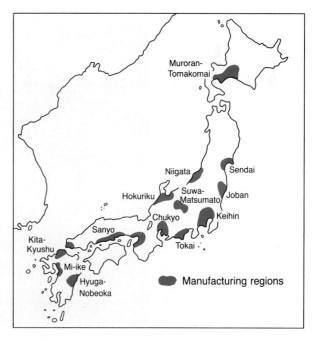

FIGURE 8.10

Manufacturing Regions of Japan.

world market. More than half the cars are exported, and the United States takes the largest share. The domestic market increases each year, and about 65 percent of the families in Japan now own automobiles, nearly all of which are made in Japan. Japan's shipbuilding industry accounts for about half the world's output, but production declined from a high of 18 million gross tons of ships in 1975 to about half that in 1985. The reduction was caused by the fall in the world's need for oil tankers because of the drop in oil prices during that period. Also, Japan's shipbuilding industry has met recent competition from South Korea and Taiwan, where labor costs are lower than in Japan.

Japan is also a major producer of heavy electrical machines, electrical appliances, electronic equipment, cameras, and watches. The country provides a large local market with its demand for consumer goods and still exports about one-third of what is produced. The electronics industry has had the most rapid growth, and Japanese television sets, video recorders, tape recorders, radios, stereos, and calculators can be found in stores throughout the world. Japanese cameras also are known worldwide, and Japan's output of watches surpasses that of Switzerland. Japan has more working robots than any other country and is a leader in manufacturing computers.

Textile manufacturing, timber cutting, and paper production are the only industries in Japan that declined between 1965 and 1985. Other industrial outputs rose and fell, but overall there were increases during that period. Textile manufacturing was the first major industry in Japan. The leading commodity was silk, and until about 1930 Japan led the world in the production of raw silk. It was the country's leading export item until the industrial expansion during the 1930s. Silk is no longer a major trade commodity for Japan, but the country still produces about one-quarter of the world's total. The Japanese seem to think it is cheaper and easier to import wood and wood products than to try to tap their own timber resources. Wood is thus a major import item, and it comes from the United States, Canada, Australia, and various countries of Southeast Asia.

Most Japanese people enjoy rapid ground transportation, epitomized by Japan's railroad system. This bullet train is leaving a station at Kyoto on its way to Tokyo. Photograph by Kazuyo Nakao.

Trade and Transportation

Japan has a large and growing domestic market, but foreign trade is critically important to the country's economy. Japanese government officials and business leaders must aggressively pursue world trade. Both sources of raw materials and markets for manufactured goods must be found and maintained. The Japanese have learned to be efficient at what they need to do. Before World War II, Japan's principal export items were raw silk, cotton fabrics, green tea, ceramics, and cheaply constructed toys, but postwar industrialization changed the structure of exports dramatically. Now, cars, trucks, steel, cameras, and television sets make up the bills of lading for outgoing ships—ships that are also built in Japan.

Japan's foreign trade is dominated by the United States (see Table 8.2), but importance of the United States is declining as Japan strives to open new markets, especially with China. The large trade surplus Japan has with regard to the United States—a surplus of more than $10 billion—is of concern to both countries. In addition, of the 10 major countries from which Japan imports, most of them produce either petroleum or iron ore. Petroleum usually accounts for about one-third of Japan's imports in terms of value.

Transportation for Japan is dominated by shipping, and the country has about 10,000 merchant vessels used primarily in the domestic waters and about 38,000 ships used on the international seas. The total displacement of all these ships is about 375 million tons, which is sufficient for most of Japan's world trade. Most city streets as well as rural roads in Japan are congested, and some streets carry 30 to 40 times more traffic than their design suggests they should. About 25 million cars and 14 million trucks are registered in Japan, and the numbers increase each year. Japan has both national and private railroad systems (Figure 8.11). The national railroad is devoted mostly to carrying freight, while the private railroads carry mostly passengers. The passenger trains in Japan are fast and modern, but most are extremely crowded. Some commuter trains are literally packed with passengers.

Standard of Living

In spite of the crowded conditions in Japan, and even though the country must import most of its raw materials, the standard of living is very high. The per capita gross national product (GNP) is comparable to the GNPs in Europe and just

Table 8.2
Japan's Ten Principal Trading Partners, 1985

IMPORTS		EXPORTS	
Country	%	Country	%
United States	18.0	United States	26.0
Saudi Arabia	16.0	Saudi Arabia	4.8
Indonesia	9.0	Germany	3.6
United Arab Emirates	6.1	South Korea	3.5
Australia	5.3	United Arab Emirates	3.5
China	4.1	Hong Kong	3.4
Canada	3.4	Australia	3.3
South Korea	2.5	Singapore	3.2
Malaysia	2.3	Indonesia	3.1
Brunei	2.0	Taiwan	3.0

SOURCE: Ministry of Finance, *Japan Statistical Yearbook* (Tokyo, 1986).

FOCUS BOX

Japan's American Investments

An imbalance of trade between two countries means that the flow of goods is greater in one direction than in the other—that is, one country purchases from another country more goods than it sells to the other country. A trade imbalance gives an advantage to the country that is selling more than it is buying, and a surplus of money accumulates for that country. A trade imbalance between Japan and the United States has existed for a number of years. The value of commodities produced in Japan and sold in the United States (such as automobiles, cameras, and electronic equipment) far exceeds the value of materials produced in the United States and purchased in Japan. Japan has been able to use the income produced by this imbalance to invest in the United States, and the amount of investment has increased dramatically during the decade of the 1980s.

As of 1987, Japan had invested about $95 billion in corporate stocks, bonds, and government securities in the United States, another $15 billion had been invested in real estate. A large percentage of these totals were purchased since 1980. For example, Japanese-owned factories in the United States increased in number from 190 in 1983 to 640 in 1987, and employment in these factories increased during the same time from 45,000 to 160,000. The number of automobiles and pickup trucks manufactured in the United States by Japanese companies increased from about 50,000 in 1980 to nearly 600,000 in 1987. Honda is the leading producer, followed closely by Nissan. Toyota, however, started manufacturing in the United States in 1986, and Mazda began in 1987. Japanese investments in American real estate have increased from almost nothing in 1980 to the 1987 total of $15 billion. These investments include office buildings, hotels, and residential properties. A large portion of the investments have been made in west coast cities, although New York City is the leading location, with nearly $6 billion worth of Japanese-owned properties. The other leading cities are Los Angeles, Honolulu, Phoenix, San Francisco, Seattle, and San Diego.

Although Japanese investments in the United States are growing rapidly, so far they are only a very small part of America's $4.7 trillion economy. They are, however, beginning to challenge the traditional leaders of the foreign investors in the United States—the British and the Dutch. For example, Japan's 1980 American investments totaled about one-third of the British investments, but by 1987 the Japanese investments had climbed to about half the British total; even faster gains have been made by the Japanese on the Dutch. The U.S. government is seeking ways to bring a balance to the trade with Japan, and if that occurs the Japanese investments will slow down dramatically. Until then, the Japanese will continue to "buy America."

---

slightly lower than those in the United States and Canada. Workers' monthly incomes are good. Consumer products are available and widespread; nearly every household has a refrigerator, a washing machine, and a color television set. Educational attainments are high in Japan, and the number of university graduates climbs each year. Most of the college graduates find jobs in industry, and Japan has one of the lowest unemployment rates in the world.

The major drawbacks to the high standard of living in Japan are less-than-well-developed hous-

FIGURE 8.11
Railroads of Japan.

ing, pollution problems, and the recent rise in consumer prices. Japanese houses are noted for their poor construction and for the weak materials used in construction. A new phenomenon in Japanese housing is the "2 × 4" house, which means construction is in the Anglo-American tradition of using 2-by-4-inch lumber for the frame. These American houses are becoming popular in spite of their very high cost. Land values, especially in the urban areas, also are high, which adds to the overall cost of housing. Pollution control laws have been passed, but the urban crowding and increased use of the automobile do little to help clean the environment. Consumer prices have risen faster than the overall economy, so it is difficult for wage-earners to keep up with prices. The Japanese are thrifty, however, and somehow manage to save a higher percentage of their wages than workers in any other country in the world.

This ancient symbol of the Shinto religion used as a modern entranceway to a park indicates how the Japanese are able to mix the ancient and the modern. Photograph by Kazuyo Nakao.

Political Geography of Japan

For many centuries, Japan was a remote and isolated group of islands, but in the 1930s the country began to enter the modern world. During the early 1940s, however, the country joined forces with Germany and declared war on the United States. The country expanded its political control over vast reaches of the Southwest Pacific region, including both island countries and mainland countries. After years of bitter fighting that included naval battles, jungle warfare, and island-hopping, the United States and its allies defeated the Japanese, and the once vast empire was reduced to its present size.

Japanese industry, which began in the 1930s, was expanded and became a source of pride for the country. Steady and sometimes phenomenal growth occurred, especially in the manufacturing of such consumer commodities as cameras, automobiles, and television sets. The United States became a prime trade partner for these commodities. The United States also agreed to protect Japan militarily, so the country's leaders did not have to worry about money for a military budget. Money that might have gone into making combat airplanes, for example, was released to help the consumer products industries. The Japanese government has subsidized research that has helped the manufacturing industries in many areas.

Japan has become politically isolated again with regard to its near neighbors, but it is tied economically to many countries of the world. Because trade is crucial for Japan's industries to continue, the country's prosperity relies on maintaining good relations with its trading partners. Japan seems to be content with its role as East Asia's industrial giant. As long as Japan can remain prosperous, the country probably will not seek military answers to its potential food shortages.

Key Words

Buddhism	paddy
conurbation	prefecture
double-cropping	Shinto
labor-intensive	Siberian High

References and Readings

Akaha, Tsuneo. *Japan in Global Ocean Politics.* Honolulu: University of Hawaii Press, 1985.

Allen, G. C. *A Short Economic History of Modern Japan.* New York: St. Martin's, 1981.

Burks, Ardath W. *Japan: A Postindustrial Power.* Boulder, Colo.: Westview, 1984.

Christopher, Robert C. *The Japanese Mind: The Goliath Explained.* New York: Linden, 1983.

Hall, Peter. "Tokyo." In *The World of Cities,* pp. 179–197. New York: St. Martin's, 1984.

Hall, Robert. *Japan: Industrial Power of Asia.* Princeton: Van Nostrand, 1973.

Hall, Robert B. *Japanese Geography.* Ann Arbor: University of Michigan Press, 1956.

Hane, Mikiso. *Modern Japan: A Historical Survey.* Boulder, Colo.: Westview, 1986.

Haring, Douglas G. *The Land of Gods and Earthquakes.* New York: Columbia University Press, 1929.

Harris, Chauncy D. "The Urban and Industrial Transformation of Japan." *Geographical Review* 72 (1982): 50–89.

Isenberg, Irwin. *Japan: Asian Power.* New York: Wilson, 1971.

Japan: A Regional Geography of an Island Nation. Tokyo: Teikoku-Shoin, 1985.

Japan: The Official Guide. Tokyo: Japan National Tourist Organization, 1964.

Kornhauser, David H. *Japan: Geographical Background to Urban-Industrial Development.* New York: Longman, 1982.

MacDonald, Donald. *A Geography of Modern Japan.* Ashford, Kent, Eng.: Norbury, 1985.

Murata, Kiyogi, and Ota, Isamu (eds.). *An Industrial Geography of Japan.* New York: St. Martin's, 1980.

Reischauer, E. O. *Japan: The Story of a Nation.* New York: Knopf, 1981.

Roberts, Dorothy E. *A Scholar's Guide to Japan,* Honolulu: East-West Center, 1966.

Seidensticker, Edward. *Japan.* New York: Time Inc., 1965.

Trewartha, Glen T. *Japan: A Physical, Cultural, and Regional Geography.* Madison: University of Wisconsin Press, 1945.

Chapter 9

AUSTRALIA AND NEW ZEALAND

■ Introduction

Australia and New Zealand are located in the Southwest Pacific Ocean and form part the chain of Pacific islands that trail off the Southeast Asia mainland (see Figure 9.1). These two southern hemisphere countries usually are considered neighbors, even though they are separated by 1,300 miles (2,100 km) of ocean. Australia and New Zealand are located within the Asian realm of nations, but they are the only Asian countries inhabited primarily by people of European origin and are part of the world we call "developed." Both cultural isolation and physical isolation have been factors in the development of these two countries.

Although Australia is called a continent, it is basically a very large island. Compact in shape, it has few inlets or peninsulas disrupting its configuration. The land area of Australia is about 3 million square miles (7,773,000 km^2). It is similar in size to the United States, excluding Alaska and Hawaii. More than one-third of the country lies within the Tropics, and overall it extends from 10 degrees south latitude to nearly 45 degrees south.

New Zealand is located farther south than Australia and thus generally has a colder climate.

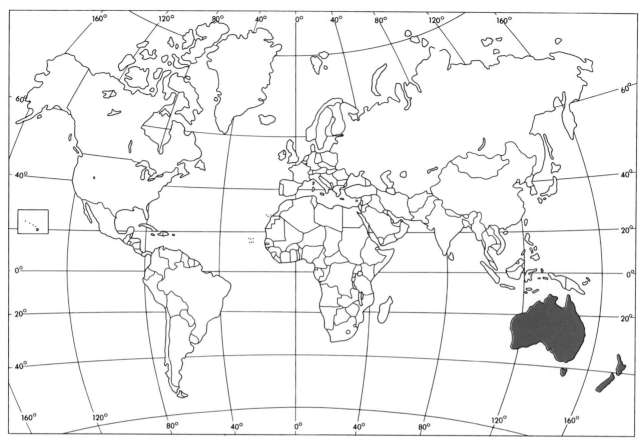

FIGURE 9.1

World Location of Australia and New Zealand.

The country consists of four islands, but only two are of significant size. The two main islands are somewhat linear in shape and stretch for more than 1,000 miles (1,600 km) between 34 degrees and 48 degrees south latitude. The land area of New Zealand is 103,736 square miles (290,470 km²), or about the same as Colorado.

The major portion of New Zealand consists of North Island and South Island, located with respect to each other as their names indicate. The two islands are separated by the Cook Strait, which is 16 miles (26 km) wide at its most narrow point. The two smaller islands of Stewart and Chatham are located off South Island. Stewart Island is a triangular-shaped land mass, 670 square miles (1,736 square km) in area and located 102 miles (164 km) off the south coast of South Island.

The Foveaux Strait flows between Stewart Island and South Island. Chatham Island is actually a group of islands, with one main island, three smaller ones, and many islets and atolls. The total area is 372 square miles (964 square km) but the largest is Chatham Island, consisting of 222,490 acres (90,041 hectares). The Chatham group of islands is located 575 miles (925 km) due east of Christchurch, the largest city on South Island.

Australia and New Zealand are separated by a portion of the South Pacific known as the Tasman Sea. The Coral Sea, another extension of the Pacific Ocean, lies off the northeast coast of Australia. The Coral Sea is noted in American history as the site of a critical 1942 World War II naval battle, where the sea and air forces of the United States defeated Japanese forces. Australia's north-

Southern Alps The chain of mountains mostly on South Island of New Zealand but also extending onto North Island. About half the surface area of South Island is mountainous, while only about one-tenth of North Island is covered. New Zealand's Southern Alps, with their serrated crests, rank among the world's more spectacular mountain ranges.

Great Dividing Range Sometimes called the Eastern Highlands of Australia, the Great Divide is a complex region of uplifted blocks and folds with intervening lowlands. The highland region parallels the eastern and southeastern coast for nearly 2,500 miles (4,023 km). Mt. Kosciusko (7,328 feet, or 2,234 m), in the southeastern corner of the continent, is the highest point in Australia.

Tasmania An island separated from the Australian continent by the Bass Strait, which is about 150 miles (241 km) wide. It is essentially a southern outlier of the Eastern Highlands.

east coast also is fringed by the Great Barrier Reef. The multicolored coral, the stony mass created from cementation of the skeletons of marine animals, extends for more than 1,000 miles (1,600 km) along the coast. The reef has been a hazard to shipping because the coral builds up even with sea level and therefore can be seen only at low tide. Today, however, waste from shipping may be a hazard to the reef, which is retained as a biological preserve. The reef's lagoon is home to thousands of species of tropical fish, as well as sharks, large barracuda, and giant clams.

■ Physical Geography of Australia and New Zealand

The landscapes of Australia and New Zealand are quite different. Australia is generally flat with a few low mountains, while New Zealand is primarily mountainous, with a few scattered flat plains. The tallest mountain on the continent of Australia is Mount Kosciusko, located in the Great Dividing Range about halfway between the two largest cities, Sydney and Melbourne. The mountain rises to 7,328 feet (2,234 m). In contrast, Mount Cook on New Zealand's South Island is 12,349 feet (3,764 m) tall. In addition to Mount Cook, there are 17 other peaks over 10,000 feet tall in New Zealand's **Southern Alps.**

Landforms of Australia

The Great Dividing Range is one of three major landform regions of Australia. The other two physiographic regions are the Central Lowlands and the Western Plateau. The **Great Dividing Range** is the most elevated part of the continent. It extends from the northern tip of the Cape York Peninsula southward along the east coast to the southern border of the continent, a distance of nearly 2,500 miles (4,023 km). It also includes the mountainous island of **Tasmania** off the south coast. Even though the Great Dividing Range is not a chain of lofty alpine peaks, its rugged surface and the absence of natural passes has hampered transportation. The first people to penetrate the eastern coastal valleys found the mountains very difficult to cross. Today, most of the people of Australia live in towns and cities located along the flanks of the mountain range.

The Tasman Glacier descends through the main divide of Mount Cook on New Zealand's South Island. Photograph by Ross Barnett.

The Great Dividing Range was named for its function of dividing watersheds, which determines the direction of flow of many rivers. This eastern highland region consists of block-fault mountains that rise steeply on the east and slope gradually on the west. The rivers on the east, then, are short, swift-flowing, and unnavigable, while those on the west flow more slowly and for longer distances and can be used for navigation (see Figure 9.2). The Great Dividing Range is nowhere wider than 100 miles (161 km), and in some places it is as narrow as 30 miles (48 km). These eastern mountains are the major source of some minerals and most of the country's timber, water, and hydroelectric power. The region is also a prominent grazing area for both beef and dairy cattle, and the less steep slopes are used for mixed farming and fruit growing. The eastern mountains also contain national parks, tourist resorts, and ski resorts.

The Central Lowlands are located along the western flanks of the Great Dividing Range. This linear region runs parallel to the eastern mountains from the Gulf of Carpentaria on the north to the southeastern edge of the Great Australian Bight on the south. The Central Lowlands region consists of a series of basins, low-lying land, lakes, and old

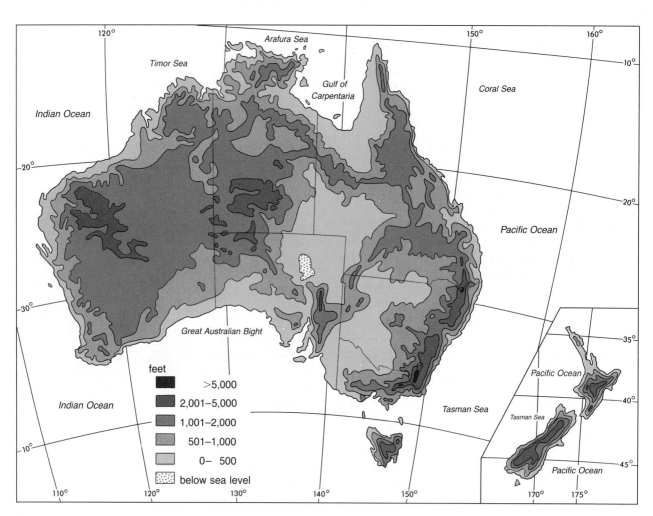

FIGURE 9.2
Landforms of Australia and New Zealand.

lake beds. Most of the land lies below 500 feet (153 m) elevation. The surface of Lake Eyre is the lowest point in the region, at about 40 feet (12 m) below sea level. The region contains two large basins; the largest is the Great Artesian Basin, the other is the Murray Basin. The Great Artesian Basin covers two-thirds of Queensland as well as parts of three other states. This large area is noted for the production of artesian water. The water is usually hot and high in salt and other minerals, and it is under constant pressure. Thus, it flows freely once it is tapped with a well, but the salinity makes it good only for livestock, not for irrigation. About one-quarter of Australia's sheep are entirely dependent on artesian well water. Artesian basins cover about one-third of the land area of Australia, and more than a quarter of a million wells have been bored into the sedimentary rock that holds the water.

Australia's major rivers flow through the Central Lowlands region. The Murray River flows northwest from its source in the Australian Alps near Mount Kosciusko, through the Murray Basin, and into Encounter Bay south of the city of Adelaide. The Darling River is the major tributary of the Murray, and along with other tributaries it drains most of the Great Dividing Range north of the Murray. Numerous lesser rivers flow through the Central Lowlands on the north and drain the region around the southern shore of the Gulf of Carpentaria. Because of the extremely dry climatic conditions, there are no other large rivers in the rest of Australia.

The Western Plateau, Australia's largest structural unit, occupies about 60 percent of the continent (see Figure 9.3). This high tableland extends through most of western Australia and contains the desert regions of the continent. Most of this large desert plateau is riverless and treeless, and large parts of the landscapes of the Great Sandy, Gibson, and Great Victoria deserts consist of sand plains, fixed sand ridges, and permanent sandhills. The fixed (nonmigratory) dunes are held together by tough desert grasses. One section of the Great Sandy is covered with sand ridges up to 60 feet (18 m) high that run parallel to each other for miles and miles in a northwest to southeast direction. The Western Plateau is fringed by low-lying mountain ranges and vast basins along the coastal areas.

The Western Plateau supports only limited agriculture, found along the west coast near Perth. Sheep and cattle grazing is carried on in the steppes surrounding the deserts, but the capacity is limited by the scant rainfall. Consequently, the population density of the region is low. The Western Plateau is composed of a large shield of ancient rocks that contain some valuable mineral resources. Thus, the poverty of agriculture is partly compensated for by the wealth of minerals. Mining activity does not require many people, however, so the population of the region remains small.

Landforms of New Zealand

In New Zealand, about four-fifths of South Island and one-fifth of North Island contain mountain landforms (see Figure 9.2). Most of the remainder of North Island consists of steep hills and dissected plateaus. Plains are not prominent features of either island; those that do exist are small and isolated and are located along the coasts of both islands.

The mountains of the Southern Alps of South Island are the highest of Australasia. The tallest peak, Mount Cook, and numerous other mountains in the range exceed 10,000 feet. These mountains contain permanent snowfields, and glaciers flow down both sides of the range. South of this high country are two other distinct landform regions. The Otago region is also mountainous, but it is different from the lofty alpine region to the north in appearance and structure. The smaller, block mountains of this region are separated by dry, terrace-sided basins, and there are fewer glaciers. It is the driest region of South Island. The third landform region on South Island is the Fiordland, located on the southwestern side of the island and noted for its outward-radiating, vertically walled valleys cut by ancient glaciers. On the seaward side of the mountains, fiords are common, and on the inland side, fresh-water lakes fill the ancient valleys. This region is the wettest part of New Zealand, and yearly rainfall amounts often exceed 250 inches (6.35 m).

The highest point of North Island is Mount Ruapehu, which reaches 9,175 feet (2,797 m). This

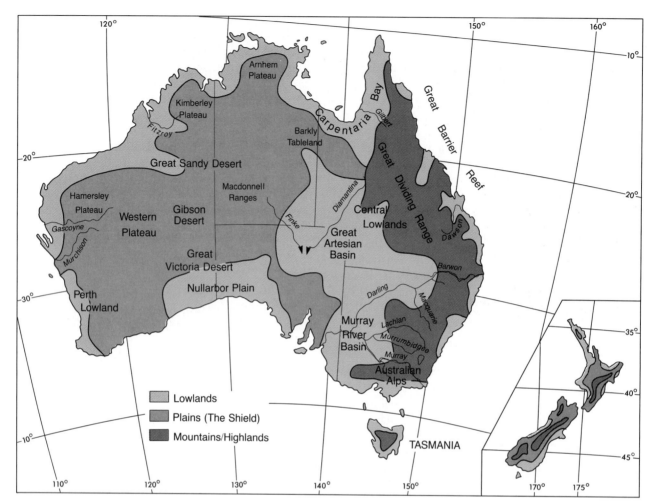

FIGURE 9.3

Physiographic Regions of Australia.

mountain, as well as the other taller peaks on North Island, are volcanic in origin. Most are extinct, but some remain active. The volcanoes have provided New Zealand with areas of rich volcanic soil. The dissected plateaus of North Island also were formed originally by volcanic flows. Thus, a large part of the landforms on North Island are related to volcanism, while those on South Island were created by uplifting.

Climate and Vegetation of Australia

The locations of Australia and New Zealand provide some important features that affect the climates. First, even though Australia is a continent it has no *continental*-type climates. An equivalent northern hemisphere location for Australia would be similar to Mexico. Thus, no severe winter climates nor taiga exist. Second, the southern hemisphere location of the two countries makes the seasons reversed from those in the United States. Christmas comes in mid-summer, and one must travel north to heat and south to cold. It should also be noted that cyclonic winds and ocean currents rotate clockwise, rather than counterclockwise as in the northern hemisphere.

The most outstanding feature of the climates of Australia is the very large portion of the country

The Waimakariri River flows through a mountain valley in Arthur's Pass National Park in New Zealand. Photograph by Ross Barnett.

The 7,515 foot (2,291 m) Mount Ngauruhoe is one of the active volcanic mountains on North Island in New Zealand. Photograph by Ross Barnett.

that receives very little precipitation (see Figure 9.4). Three-fourths of the continent is either arid steppe or desert. The dry region stretches from the Indian Ocean across the continent to the western edge of the Great Dividing Range. The Australian desert is not as dry as some other deserts of the world, but it is a true desert. Most of the region receives at least 5 inches (13 cm) of precipitation a year. The desert is unique in the world because it covers nearly half the Australian continent. All other deserts form only small portions of the continents on which they are located.

Most of Australia is dry, because the ocean waters near the southern and western coasts are

Antarctic Drift Sometimes called the West Wind Drift, the ocean current that extends around the world off the coast of the Antarctic continent. Cold water flows toward the east without interference from continents, and the water movement depends on prevailing winds and the earth's rotation.

cooled by the **Antarctic Drift,** a cold current that is chilled as it passes through the Antarctic region. Because the cold waters do not allow much evaporation to occur, the winds that come onto the continent are dry. The Great Dividing Range on

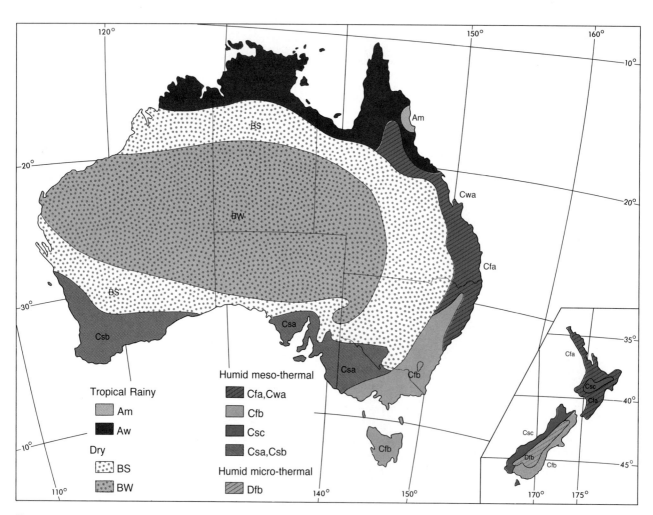

FIGURE 9.4

Climate Regions of Australia and New Zealand.

marsupials An order of species comprising the lowest mammals and having a pouch for carrying the young. It contains the kangaroos, wombats, bandicoots, opossums, and others.

the east works as an effective barrier that blocks the winds that might carry moisture from the east. During the winter months, subtropical high pressure builds up over the continent, which causes the winds to blow outward from the continent rather than in toward it. These winds also are dry.

Adequate amounts of precipitation in Australia fall only in the coastal regions. As the world wind-belts move southward during the summer months, the north coast receives moisture-laden air carried in by the northeast trade winds. A small portion of the northeast coast, along the eastern edge of the Cape York Peninsula, receives more than 80 inches (200 cm) of rain a year. Even though the moisture comes mostly in the summer, this region is the wettest in Australia. Australia's entire north coast gets most of its rain in the summer—that is, it has what is called a *summer maximum* rainfall pattern. Most of the east and southeast coasts are similar to the southeast coast of the United States, where the precipitation falls throughout the year with no seasonal maximum. The southwest coast, near Perth, is similar to the American west coast in that most of the precipitation comes in the winter months (*winter maximum*). Only those three coastal areas receive adequate moisture. The remaining coastal regions are desert. Essentially, Australia is very dry in the interior and humid along all coasts, except on the south and west.

The temperature patterns of Australia are characterized by hot summers and mild winters (see Figure 9.5). The location straddling the Tropic of Capricorn assures much intense sunshine for the entire continent, especially during the high sun period. In the desert areas, summer temperatures are usually above 90° F (32° C) every day, and days where the temperature exceeds 100° F (38° C) are common. The northwest coast is extremely hot and dry, and the northeast coast is hot and humid. The east, south, and southwest coast are milder, and it is cooler in the higher elevations along the Great Dividing Range. Winter temperatures for most of Australia average between 50° and 60° F (10° to 15° C), although it is warmer along the north coast.

Lush vegetation is scarce in most of Australia because of the low amounts of precipitation. The wetter north and northeast coasts are fringed with tropical rainforests, but the lush plant growth changes quickly as one moves inland. Many species of grasses grow in the inland steppe regions, and they give way to the sparse xerophytic plants of the desert. Scattered gum and acacia trees are found where sufficient moisture allows, and many pure stands exist in the wetter areas. More than 600 varieties of gum (eucalyptus) trees are found in Australia, and the tree virtually has become a symbol for the country. The exclusive diet of the native koala bear consists of gum tree leaves. The acacia tree is a relative of the mimosa tree found in the United States, and sketches of its branches adorn the Australian coat of arms. Thus, it *is* a symbol for Australia.

The physical isolation of Australia influenced the development of some peculiar native animals. Australia is the original home of most of the marsupials found on the earth. **Marsupials** are mammals that give birth to their young but carry them in a pouch until they are mature enough to keep pace with the adults. The largest marsupial is the kangaroo. The koala bear also carries its young in a pouch, and wombats are smaller versions of the koala. Wombats burrow in the ground like moles. The Tasmanian devil is a flesh-eating marsupial found only on the island of Tasmania. Australia's exotic nonmarsupials include the duckbill platypus, the dingo, the emu, and the goanna. The platypus is amphibious, lays eggs, and has a duck bill, webbed feet, and a fur coat. The dingo is a wild, yellow dog. The emu is a large ostrichlike bird that does not fly but grows as tall as humans. The goanna is a six-foot-long lizard that barks and hisses. Australia also is noted for large numbers of rabbits, but these animals were not native to the continent.

Climate and Vegetation of New Zealand

Compared with Australia, the climates of New Zealand are not nearly as extreme (see Figure 9.5). Most of New Zealand receives between 40 and 80

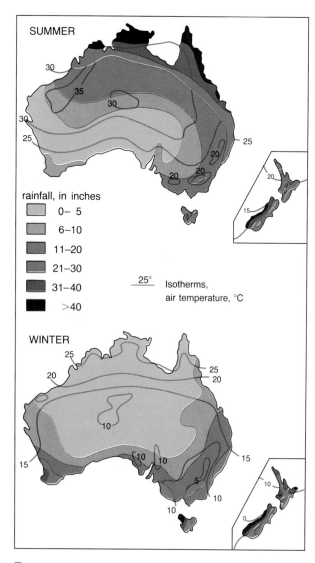

FIGURE 9.5

Summer and Winter Temperature and Precipitation Patterns of Australia and New Zealand.

Variation in vegetation found on either side of the highlands of New Zealand can be seen in the photographs here and below. The lush, almost tropical, vegetation of Pukekura Forest is located on the wetter, western side of the island.

Xerophytic plants and grass of Bannock Byrn are typical of the drier, eastern side of New Zealand. Photographs by Ross Barnett.

inches (100–200 cm) of precipitation annually, with little seasonal variation. The west coast of South Island is the wettest part of the country. The prevailing westerly winds push moist Pacific air onto the Southern Alps, and the resulting uplift caused by the mountains produces precipitation that totals more than 120 inches (300 cm) a year. Much of the moisture comes in the form of snow that helps feed the snowfields and glaciers as well as the numerous mountain lakes.

Aborigines The earliest known inhabitants of Australia, these small, black people, called Australoids, lived in wandering bands and became remarkably adapted to the heat and cold of their desert homeland.

The temperature patterns of New Zealand are cooler than those for Australia and, as with precipitation, there are fewer extremes. New Zealand has no hot regions like Australia's deserts, but the higher island mountains get much colder than anywhere in Australia. The annual range of temperatures is not as great in New Zealand as it is in Australia, however. Summers in New Zealand are relatively short and cool, but the winters in the inhabited areas are mild.

The natural vegetation on New Zealand is about half temperate rainforest and half grassland. The forest includes such trees as conifers, beech, and tree ferns, as well as lianas (vines) and epiphytes. Areas of scrub vegetation are found between the forests and the grasslands.

■ Human Geography of Australia and New Zealand

Settlement

The **Aborigines** are the black, native people of Australia. It is estimated that they arrived on the continent as long ago as 25,000 B.C. For thousands of years they lived under primitive conditions, but now they are being absorbed into Australia's dominant lifestyle. These short, chocolate-brown people have wavy hair, wide noses, and low foreheads, and the men have full beard growth. Originally, the Aborigines were exclusively nomadic hunters and gatherers, with no permanent homes or settlements and few possessions. The women hunt small animals, pick fruit and berries, and dig roots for food. The men hunt larger animals, such as kangaroos and emus.

Prior to the settlement of Australia by whites, there were an estimated 300,000 Aborigines on the continent. Only about 75,000 remain today, and many of those are of mixed blood. The Australian constitution assures protection and aid for the

Tourists visit the glaciers and snowfields in the mountains of New Zealand by landing directly on the snow in small airplanes. Photograph by Ross Barnett.

These Maori boys of Whakarewarewa, New Zealand, are descendants of the original Maori settlers of 1350. Photograph by Ross Barnett.

Maoris The New Zealand aboriginal people, primarily Polynesian with some Melanesian admixture. They are tall in stature and were originally very warlike, but now are citizens of New Zealand.

merino sheep A hardy breed of fine-wooled white sheep originating in Spain. Their wool excels all others in weight and quality, but the breed does not rank high as a mutton producer.

Aborigines, but many still live in their nomadic tribal regions. Schooling and agricultural training are provided, and they are encouraged to move into environments where conditions are better than in the desert wastelands. Some have sought employment in the urban areas, and numerous types of government subsidies are provided.

The **Maoris,** the native people of New Zealand, did not arrive on the islands until after A.D. 1300. Compared with Australia's Aborigines, they are newcomers. The Maori people are Caucasian, not black; they are of the Polynesian race. The original migratory group came in seven canoes in the year 1350, followed later by other groups as the settlement became established. Each of the seven canoes had a name, and the Maoris of today delight in tracing their own names to one of the original canoes. When white settlers began to encroach on Maori land, bitterness resulted, so from the time of the first white settlement in 1840 until the late 1860s, there were sporadic armed clashes between British and New Zealand troops and the Maori forces. In spite of problems created by prejudice, including some bloody racial clashes, the Maori are much more integrated into New Zealand's society than the Aborigines are in Australia's society. The majority of New Zealand's place names are Maori in origin, so daily reminders exist about the country's native people.

The first Europeans to set foot on Australia and New Zealand were the seventeenth- and eighteenth-century explorers. In 1642, a Dutchman, named Abel Tasman, found both Tasmania and New Zealand, as well as many other islands of the Southwest Pacific, but somehow he missed the mainland of Australia. The famous British explorer, Captain James Cook, finally claimed Australia for England in 1770, nearly 130 years after Tasman's voyages. Following Cook's claim, many British explorers criss crossed the interior of Australia.

During the eighteenth century, a common sentence meted out to criminals in England was "transportation" meaning deportation to America—but after Americans gained their independence from England, Australia became the next logical place to ship criminals. Thus, in 1788 the first white settlers arrived in Australia in the form of 750 English convicts. Many of the criminals were women, most of whom were prostitutes. During the next 100 years more than 150,000 convicts arrived from England.

Free colonists began to settle in Australia around the turn of the eighteenth century, and they soon outnumbered both the Aborigines and the convicts. One early settler, John Macarthur, introduced **merino sheep,** a hardy breed with long, fine, silky wool, originally from Spain. The herds developed from these sheep became the foundation of the Australian sheep industry, which added much to the prosperity of the country.

The Australian states existed as English colonies throughout the nineteenth century. During that era, gold was discovered, the large sheep and cattle "stations" were established, and frozen meat was successfully shipped back to England. All these events were significant to the economy of the country. Finally, in 1901, the colonies became states in the Australian Federation, and modern Australia was born.

Settlement of New Zealand was not as easy as settlement of Australia. The Aborigines in Australia merely complained about intrusion, then quietly retreated. The Maori of New Zealand were more militant. Hostilities raged on the islands for more than a decade, and the Maori nearly drove the whites away but eventually were defeated. Today, the Maori make up about 10 percent of the population of New Zealand, totaling about 330,000. Once in control of the land, the New Zealanders established large sheep and cattle ranches similar to those in Australia. And like its larger neighbor, New Zealand began exporting wool, mutton, beef, and dairy products. The country became a "dominion" of the British Empire in 1907 and gained its independence in 1947.

Population

By most world standards, population density in Australia is extremely low. The 15.8 million people in Australia have nearly 3 million square miles (7.8 million km^2) of territory. The overall population density is 5.3 people per square mile, but most of the people live in coastal cities and towns (see Figure 9.6). Most of the interior of the continent is essentially empty of people. The major exception is along the narrow, fertile valley of the Murray River. About 40 percent of the Australians live in the two large coastal cities of Sydney and Melbourne.

Sydney is the largest city in Australia, with about 3.3 million people, followed closely by Melbourne, with about 2.8 million. Sydney is located on the narrow plain between the Great

The unusual architecture of this Maori meeting house in Ohinemutu, New Zealand, is typical of the traditional Maori building methods. Photograph by Ross Barnett.

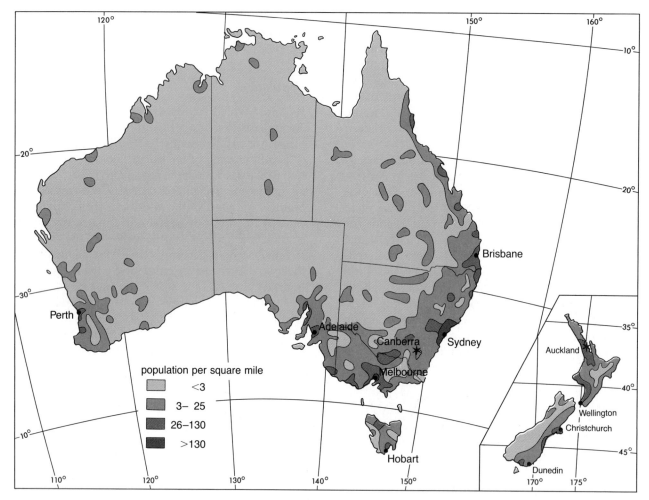

FIGURE 9.6

Population Patterns of Australia and New Zealand.

Dividing Range and the Tasman Sea, on the southeast coast of the continent. Melbourne is located within Port Phillip Bay, on the south coast, about 600 miles (900 km) by water from Sydney. Brisbane is the third largest city, with about 1.1 million people. It is located on the east coast about the same distance north of Sydney as Melbourne is south. Adelaide and Perth are both close to 1 million in population, but next there is Newcastle, with about 400,000, and the remaining cities are quite small: none is over 250,000. Canberra, the national capital, is close to 250,000 and the only major inland city in the country. The 10 largest coastal cities and towns contain about 70 percent of Australia's population.

Most of the people of New Zealand also are confined to the coastal lowlands (see Figure 9.6). The overall population density is higher in New Zealand than in Australia, but still relatively low. The 3.3 million people with 103,736 square miles (270,000 km^2) of territory yields a density of about 32 people per square mile. Auckland and Wellington are the largest cities on North Island, and Christchurch and Dunedin are the largest on South

FOCUS BOX

Australian Immigration

In 1787, Arthur Phillip set sail from Portsmouth, England, with 750 convicts who had been sentenced to "transportation" to Australia. About one-fourth of the convicts were women, many of whom were professional prostitutes. The ships sailed southwest, around South America, and in January 1788 landed on the east coast of Australia near what is now the city of Sydney. Australia's modern history began with these first immigrants. Arthur Phillip became the first governor of the colony. Before "transportation" was abolished in 1868, about 168,000 convicts were sent to Australia.

Among Australia's early settlers was John Macarthur, who introduced Merino sheep to the continent. Large-scale sheep production began around the turn of the century, and sheep eventually became a source of much prosperity for many Australians. One early governor of the colony, William Bligh, is famous for his role as captain of the ship *Bounty*. Another early governor was Lachlan Macquarie, who became known as the "father of Australia" because of the massive public works projects started during his term in office.

Free immigrants began to come to Australia in the 1830s, and by 1850 more than 200,000 Britons had moved "down under." During the 1830s the Australian leaders started a "bounty" system for immigrants in order to attract promising settlers. Such incentives as free transportation and/or free land were offered. In 1851 gold was discovered in New South Wales, and the subsequent gold rush brought settlers from all the world. Chinese laborers brought to help fill the labor void created by the gold rush were the ancestors of the present Chinese minority in Australia.

During the first half of the twentieth century, Australia's population grew steadily, but the real population boom started after World War II. After the war, Australia began a massive immigration policy. The bounty policy was expanded to include not only cheap sea passage but also jobs and living accommodations. In the two decades after the war, more than 2 million people, one-fifth of the country's population, entered Australia as immigrants. Most came from Europe because of Australia's "White Australia Policy." The policy excluded not only blacks from Africa, but also all Asians. Of the 2 million immigrants, about 50 percent came from the United Kingdom, 20 percent from southern Europe (mostly Italy), 15 percent from northern Europe (primarily Germany and the Netherlands), and most of the remainder from the Slavic countries (mostly Poland).

During the second half of the twentieth century, migration to Australia has slowed, but the incentive program is still followed. The requirements have increased, however, so that now a college education or expertise as a skilled laborer are desirable attributes. The "White Australia Policy" has been eased since it came under attack from human rights advocates. Permission to enter the country, however, is arbitrarily granted by the minister for immigration. This means that people from Europe can expect to get permission, but not Afro-Asians. The total non-European population of Australia remains at less than one percent.

Island. Approximately 40 percent of New Zealanders live in Auckland, but the more centrally located Wellington is the capital city.

The rates of natural increase of the populations of Australia and New Zealand are both less than one percent. This means that the birth rates, death rates, and doubling time of the population also are similar. The populations of both countries are about 85 percent urban. The age structures of the populations are similar to each other, and similar to those for Europe. About 25 percent of the people are under 15 years old, and about 10 percent are over age 65. All the population figures are unique for both the Oceania region and the Asian region, in which the two countries are located.

Language and Religion

Nearly all the people in Australia and New Zealand are English-speaking Protestants. The early British settlers brought their language and religions with them, and the traditions have been maintained. The most common churches in New Zealand are Anglican and Presbyterian; in Australia the Anglican and Roman Catholic churches are prevalent. Most of the Catholics are Irish immigrants and their descendants. The Presbyterians and Methodists each claim about 10 percent of the total church membership in Australia. There are no significant regional differences among the four major churches.

Although English is the common language for both Australia and New Zealand, the people of each country have their own distinct accents and use many words differently from other English-speaking people. Despite Australia's large size, very little regional variation in the accent can be noted. This is somewhat ironic because Great Britain itself, the mother country and much smaller than Australia, has many regional and class accents for spoken English. The list of unique terms that are common in Australia is quite long, and non-Australians need a glossary for complete understanding. Many of the terms—such as "willy-willy" used for "cyclonic storm," and "jumbuck" for "sheep"—are old standards, but new slang terms that confound the non-Australian are always being added.

Minorities

Strong immigration laws and physical isolation, have kept the cultures of both New Zealand and Australia from becoming too complex. Besides the Maoris and the Aborigines, few ethnic minorities live in either country. Australia does have a very small minority of Chinese, but they are not recent arrivals. They are descendants of a very few Chinese laborers brought to Australia to work in the gold mines during the 1850s, and they have not significantly increased in number. Other minorities include European immigrants from such countries as Italy and Spain, but these people have learned to speak English and have become part of the Australian society. The least integrated of the Australian minorities are some Southeast Asian "boat people" who arrived in Australia, but these groups are small.

The place names on this roadside signpost near Chateau Tongariro are in the Maori language. Photograph by Ross Barnett.

Health and Welfare

Australia and New Zealand are capitalist democracies. Although free enterprise is encouraged, the

National Health Scheme Australia's socialized medical program, which covers the health costs of all citizens.

government intervenes in the economies of both countries. Taxes are very heavy in both countries, and the monies are used for institutional functions designed to help all citizens. With the revenue, the countries have done much to conquer pain and poverty. Life expectancy rates are among the highest in the world. The **National Health Scheme** in Australia provides free prescription drugs and helps with hospital and other medical expenses. New Zealand's Social Security program is even more extensive than Australia's and may be the most comprehensive in the world. Pension plans in both countries provide for invalids, the blind, the unemployed, and victims of chronic diseases.

The governments of both Australia and New Zealand provide free and compulsory education to age 17 and encourage higher education. Australia has 18 major universities, of which the University of Sydney is the largest. New Zealand has six major universities. Australia and New Zealand both train military personnel, but only a very small part of their national budgets is designated for defense. Except for the Maori conflicts in New Zealand during the nineteenth century, internal warfare in both countries has been nonexistent. Australia is the only large country in the world that has avoided internal war or revolution.

Economic Geography of Australia and New Zealand

Agriculture

Despite the heavy concentrations of urban populations today, both Australia and New Zealand owe their initial economic success to agriculture. Wool and meat have been important products for both countries. Australia still produces about one-third of the world's wool and about half the world's merino wool. The annual amount of wool production is about 1.2 million metric tons, down from about 2.0 million metric tons in 1977. Australia is the world's leading exporter of both beef and lamb. New Zealand is the second leading meat exporter in the world, and most of the meat is mutton. About 75 percent of New Zealand's export dollars come from wool, meat, and dairy products.

Many of the animals in Australia are raised on "stations," which are huge ranches located in the interior, or "outback," of the country. The lack of rainfall produces only sparse vegetation, so foraging animals need more acres to find an adequate amount of food than they would in wetter areas. The stations often extend over hundreds or even thousands of square miles. Most are owned by large meat-exporting companies, but they are operated by locally hired managers. The water problem is partially overcome with artesian wells, but these are located only along the western flank of the Great Dividing Range. Some rangeland water is obtained from springs, rivers, and shallow lakes, but wells with windmills are most common. Despite the huge livestock stations, most of the sheep production is carried on in the wetter, southeastern part of the country. The states of New South Wales and Victoria are the primary sheep producers. Most of the dryland stations, located throughout the periphery of the very driest part of the desert interior, raise cattle. (See Figure 9.7).

In 1982, Australia had about 138 million sheep, so sheep outnumbered the people by nine to one. From these sheep came 230,000 metric tons of mutton, 276,000 metric tons of lamb meat, and 1,153,000 metric tons of wool. The same year, the country had about 25 million head of cattle, from which came 1,573,000 metric tons of beef and veal, and about 2.4 million pigs provided 228,000 metric tons of meat. The livestock industry of Australia produced 9.6 billion Australian dollars worth ($A 9.6 billion) of income from exporting live animals, meat, dairy products, wool, hides, and animal oils. These figures are representative of the livestock production in Australia during the first half of the 1980s. From 1980 to 1985, cattle production declined by about one million head a year, but the numbers of sheep and pigs have remained stable.

Because of the greater amounts of rainfall in New Zealand, compared with Australia, the animal-per-acre ratio is larger. Each acre of land in

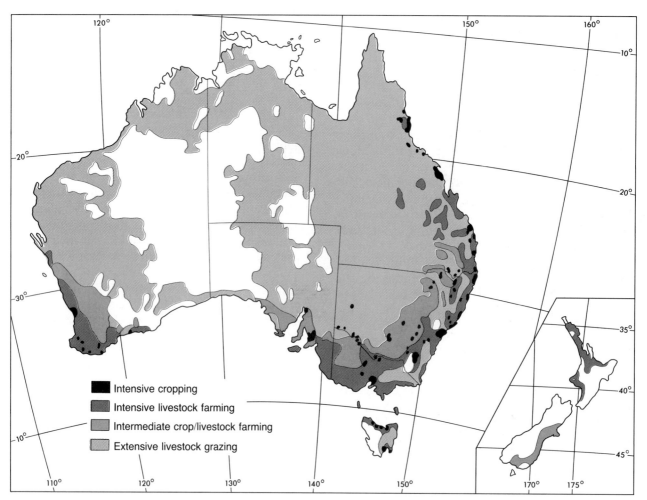

FIGURE 9.7

Agricultural Areas of Australia and New Zealand.

wet areas can support more cattle than acres in dry areas, simply because under wetter conditions more and better grass grows. As a result, the livestock ranches in New Zealand are smaller than those in Australia. Also, a larger percentage of the cattle in New Zealand are raised for dairy use rather than for beef. Dairy cattle must be located close to the milking facilities, so they cannot be allowed to roam over large areas to forage for food. Consequently, they are fed more grain and hay than beef cattle, and they require less grazing space. In 1982, New Zealand had about 8 million head of cattle, about 2 million of them dairy breeds. These animals were used to produce 519,400 metric tons of beef and 7.2 million metric tons of liquid milk, butter, cheese, preserved milk, and casein.

In terms of acres under cultivation and amount of money produced from export, wheat is Australia's most important cultivated crop. The two major wheat-growing regions of Australia lie in relatively narrow strips, one between Perth and Albany in western Australia, another along the Murray River between Adelaide and Sydney in the southeastern part of the country. It is grown in many other areas, however, where annual precip-

The Glentanner sheep station in New Zealand is in the foothills below Mount Cook, which can be seen in the background. Photograph by Ross Barnett.

itation allows. Wheat was introduced to the continent in 1788, when the first seeds were brought from England, but local environmental conditions, especially the dryness, were much harsher in Australia than in England, so new varieties had to be developed. Today, there are about 1,000 different varieties of Australian wheat, but only about 45 of them are grown consistently. Most of the varieties were developed to grow under dry conditions. Spring wheat is more popular than winter wheat, but both are grown. Wheat is often grown in rotation with other crops or in conjunction with sheep farming. Australia usually produces about 400 million bushels of wheat, and more than half of what is produced is exported.

The other grain crops produced in Australia include barley, oats, sorghum, and corn, but in terms of income produced, sugar cane is more important than any of them. Cane to be crushed for sugar is grown along the hot-wet northeast coast, especially between Cairns and Townsville but also as far south as Brisbane. The northeast coastal strip also produces tropical fruit, such as bananas and pineapples. The nontropical fruit varieties—apples, peaches, and pears—are grown in the southeastern agricultural region along the Murray River. Grapes for both table fruit and wine are grown along the southwestern-facing coasts near Perth and Adelaide. These two regions have the Mediterranean-type climate conducive to the production of grapes and most varieties of citrus fruit.

Mineral Production

Australia's mining activities started when gold was discovered in New South Wales in 1851. Soon after that another strike was made in Victoria. The wealth produced from the mines was not great, but the discoveries set the stage for later mining industries in the country. The strikes also lured many people to the mining areas from the stations and towns as well as from overseas. Many became the exploration and mining experts for the country. As the extractive industries grew, the foundation for industrial development was formed. Large deposits of both coal and iron ore—the essential ingredients for steel making, which in turn is the basis for other heavy industries—were discovered. The early interest in mining, the availability of raw materials, and the highly productive agriculture gave Australia the natural elements it needed to become an industrialized country. (See Figure 9.8).

Some of the largest mining discoveries were made during the latter part of the 1960s and the early 1970s. In 1965, a hugh new deposit of iron ore was discovered in western Australia, and the world's largest single deposit of bauxite was unearthed a year later in the same area. In 1969, a new source of nickel was found, and in 1972 uranium deposits were discovered. In addition, new oil and gas fields were found in the 1960s. All these discoveries have made Australia self-sufficient in nearly everything necessary for industrialization, although oil is somewhat short. Australia imports oil from the Middle East, primarily Saudi Arabia.

Industry

All the sophisticated products that modern consumers demand are manufactured in Australia. Industrial output includes all types of machinery, automobiles, television sets, aircraft, electrical equipment, chemicals, and drugs, as well as such basic items as textiles and foodstuffs. The building

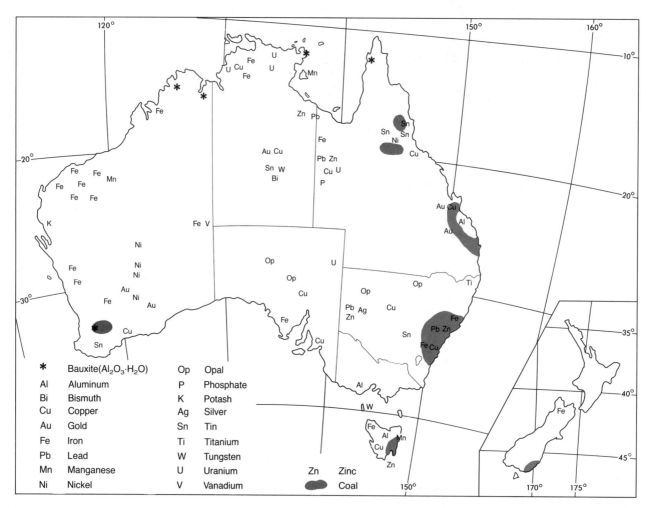

FIGURE 9.8

Mineral Deposits of Australia and New Zealand.

Part of New Zealand's energy supply comes from this geothermal power station at Wairakei. Photograph by Ross Barnett.

industries are supplied with 2.3 million tons of clay bricks and 5.5 million metric tons of cement annually.

Australian steel mills produce about 7.5 million metric tons of steel ingots each year. The steel industry is owned by Australians, but many of the other industries have had to rely on foreign investments. These investments, mostly from Canada, United States, and Great Britain, amount to several billions of dollars. About one-fourth of Australia's industries are owned by overseas investors. Because Australia's domestic market is small, industrial growth must rely on trade. For this reason, all industrial commodities are protected by tariffs, and much effort is given to expanding overseas markets.

Industrial development in New Zealand is also significant, in spite of the general lack of minerals and the overall importance of agriculture. More than 30 percent of the labor force in New Zealand is engaged in manufacturing. Food-processing is a major industry, followed closely by pulp wood and paper-processing. Natural gas has been discovered, and both lignite and bituminous coal are mined. Beyond these resources, hydroelectricity is utilized and seems to have the most potential for future energy needs.

Trade

Australia maintains a fairly balanced foreign-trade ledger: both imports and exports are valued at about $A 20 billion a year. Australia's 10 major trade partners in 1983, in terms of value of commodities, are ranked in Table 9.1.

The United States is a major trade partner, but notice also Australia's ties with its neighbors of the oriental world. These are the markets that Australia is striving to expand. For many years, the United Kingdom was Australia's leading trade partner for both imports and exports, because of the old political ties between the countries. Recently, however, the United Kingdom has slipped in the rankings because of Australia's increased trade with neighbors and because the United Kingdom entered the European Economic Community and gets from European countries commodities that once came from Australia. Most of

TABLE 9.1

Australia's Major Trade Partners, 1983

EXPORTS	IMPORTS
1. Japan	1. United States
2. United States	2. Japan
3. New Zealand	3. United Kingdom
4. United Kingdom	4. West Germany
5. South Korea	5. Saudi Arabia
6. Soviet Union	6. New Zealand
7. China	7. Singapore
8. Singapore	8. Taiwan
9. West Germany	9. France
10. Taiwan	10. Canada

SOURCE: *Handbook of International Trade and Development Statistics,* New York: United Nations Publications, 1986.

Australia's imports consist of mineral fuels and lubricants, heavy machinery, and transportation equipment. The bulk of the exports include foodstuffs and live animals, metal and ores, and crude materials, such as fibers.

New Zealand's major trade partner for both imports and exports is Australia. Most other trade is with Japan, the United States, the United Kingdom, Germany, and the Soviet Union. The primary import items are mineral fuels and lubricants, machinery, transportation equipment, and chemicals. The major export items include meat, hides, butter, and forest products.

Transportation

Large portions of Australia's railroad and road networks are centered on the city of Melbourne in the state of Victoria (see Figure 9.9). The surrounding region is the well-watered and fertile southeastern corner of the continent, and Melbourne is the gateway for the area. Numerous towns and sheep and cattle stations are located in the region, all connected by road and rail. The network of routes extends northwest of Melbourne to Adelaide and northeast to Sydney, but not far beyond those cities both railways and roads become single routes.

The only railroad that penetrates the interior of the continent runs north of Adelaide for about 900 miles (1,450 km) to Alice Springs. The single transcontinental route runs from Adelaide westward to Perth on the west coast. The route between the two cities is 1,600 miles (2,560 km) over flat and monotonous land. One stretch of the track along this route runs for 300 miles (485 km) perfectly straight and is the longest length of flat, straight track in the world. Fighting sleep is an occupational hazard for train engineers that drive across this route.

A small local network of rails exists in the vicinity of Perth, but from there northward and around the northern part of the continent to Cape York Peninsula only a few miles of track have been constructed. The routes in this large expanse are short lines running from coastal ports to interior resource sites. On the eastern side of the continent, northward from Sydney, a single railroad line follows the coast north through the sugar cane fields to Cape York. From this coastal route, these treelike networks extend over the Great Dividing Range into the interior. At one time each state in Australia maintained its own railroad, and the gauge (width) of the tracks varied from state to state. Single trains could not travel between states, and all commodities and passengers had to be transferred at the state boundaries. This problem has now been corrected. About 412 million passengers journey over Australia's railroads annually, and about 128 million metric tons of goods and livestock.

The road pattern in Australia is similar to the railroad network, but roads connect many areas not served by railroads. Most new road construction occurs around and in the urban areas, however, so the vast interior is still essentially devoid of major highways. As of June 1984, some 8.2 million motor vehicles were licensed in Australia, of which about 65 percent were passenger cars. An important means of transportation for the interior is by aircraft.

Passengers, freight, and mail are carried by Australia's domestic and international airlines. In addition to the major airlines, numerous feeder lines and private aircraft provide important air links within the country. The remotely located

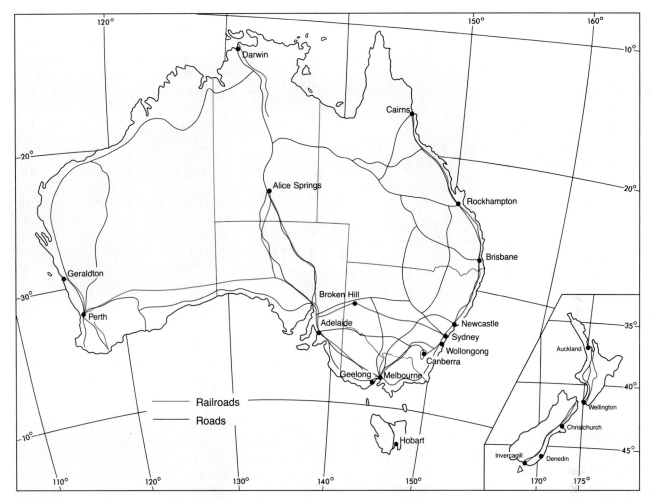

FIGURE 9.9

Roads and Railroads of Australia and New Zealand.

cattle and sheep stations all have runways to accommodate aircraft for medical emergencies and business trips. Many station operators own their own aircraft. Australia's international carrier is Qantas. The name was derived from "Queensland and Northern Territory Aerial Service," which started as a domestic airline servicing those areas. About 12 million passengers move each year on Australia's domestic airlines, and about 2 million fly with the international carrier.

New Zealand's road and railroad networks are restricted by the mountainous terrain, but all the major towns on North Island are connected by both roads and railroads. The road and railroad networks on South Island are much more limited and are found primarily on the more populated eastern half of the island. New Zealand's major internal airline (NAC) is used extensively because of the terrain and the great length of the islands. About 1.2 million passengers fly on the domestic airlines annually. Small feeder lines and private aircraft also are important carriers in New Zealand. Mount Cook Air Service Ltd., for example, makes regular tourist flights to the glacier country, where the small aircraft land directly on the glaciers.

Queenstown, New Zealand, located on the shores of Wakatipu Lake in the mountains, is somewhat isolated because of the difficulty of building roads in the mountainous terrain. Photograph by Ross Barnett.

Nearly every coastal city and town in Australia and New Zealand has fishing fleets. These well-rigged, modern fishing boats are anchored at Milford Sound in New Zealand. Photograph by Ross Barnett.

Both Australia and New Zealand have extensive external and coastal shipping lines. Shipping is of considerable importance because the economies of each country rely heavily on foreign trade and because all the major cities of both countries lie along the coasts. Vessels for both international and coastal shipping are built in Australian shipyards.

■ Political Geography of Australia and New Zealand

Modern, stable economies, peaceful existence, pleasant climates, and friendly people combine to create very good living conditions in both Australia and New Zealand. The poor and hungry masses of humanity in the near north on the Asian mainland may be the only threat to the people of Australia and New Zealand, but so far the two countries continue to thrive. Both countries maintain an active military, but they could hardly defend themselves against a large external attack. They look to the United States and the United Kingdom for military support, although New Zealand does not want nuclear-armed American ships to dock at any of its ports.

The somewhat ideal living conditions along the southeast coast of Australia are tempered by the harshness of the interior. The rapidly growing cities and thriving industries are countered by the raw solitude and decreasing populations of the outback. Most Australians live along the coasts but are sentimentally attached to the difficult life of the dry and remote interior. These contrasts give Australians a unique outlook on the remainder of the world and provide them with insights they might not otherwise have. The people of both Australia and New Zealand tend to be individualists who are isolationists at heart. They would much rather be left alone than enter into the squabbles of the remainder of the world, but they certainly are not afraid to fight. Troops from "down under" helped the American cause in World War I and World War II, as well as in the Korean and Vietnam wars. They fought not only courageously but also with utmost distinction.

■ Key Words

Aborigines
Antarctic Drift
Great Dividing
   Range
Maoris
marsupials
merino sheep
National Health
   Scheme
Southern Alps
Tasmania

■ References and Readings

Alley, Roderic, (ed.). *New Zealand and the Pacific.* Boulder, Colo.: Westview, 1983.

Barrett, Raes D., and Ford, Roslyn A. *Patterns in the Human Geography of Australia.* Melbourne: Macmillan, 1987.

Bernardi, Debra (ed.). *Fodor's Australia, New Zealand, and the South Pacific.* New York: Fodor's, 1985.

Bolton, G. *Spoils and Spoilers: Australians Make Their Environment, 1788–1980.* Boston: Allen & Unwin, 1981.

Brander, Bruce. *Australia.* Washington, D.C.: National Geographic Society, 1968.

Burnley, Ian H. *Population, Society, and Environment in Australia.* Melbourne: Shillington House, 1982.

Burnley, Ian H., and Forrest, James (eds.). *Living in Cities: Urbanism and Society in Metropolitan Australia.* Winchester, Mass.: Allen & Unwin, 1985.

Carter, Jeff. *Outback in Focus.* London: Angus & Robertson, 1968.

Courtney, Percy P. *Northern Australia: Patterns and Problems of Tropical Development in an Advanced Country.* New York: Longman, 1983.

Daly, M. *Sydney Boom, Sydney Bust.* Sydney, Aust.: Allen & Unwin, 1982.

de Blij, Harm J. *Wine Regions of the Southern Hemisphere.* Totowa, N.J.: Rowman & Allanheld, 1985.

Eliade, Mircea. *Australian Religions.* Ithaca, N.Y.: Cornell University Press, 1973.

Gentilli, J. (ed.). *Climates of Australia and New Zealand.* New York: Elsevier, 1971.

Hanley, Wayne, and Cooper, Malcolm (eds.). *Man and the Australian Environment: Current Issues and Viewpoints.* Sydney, Aust.: McGraw-Hill, 1982.

Jeans, Dennis N. (ed.). *Australia: A Geography.* New York: St. Martin's, 1978.

MacInnes, Colin. *Australia and New Zealand.* New York: Time Inc., 1966.

McKnight, Thomas L. *Australia's Corner of the World.* Englewood Cliffs, N.J.: Prentice-Hall, 1970.

McLeod, Alan L. *The Pattern of Australian Culture.* Ithaca, N.Y.: Cornell University Press, 1963.

Parkes, Don (ed.). *Northern Australia: The Arenas of Life and Ecosystems on Half a Continent.* Orlando, Fla.: Academic Press, 1984.

Rees, Henry. *Australasia: Australia, New Zealand, and the Pacific Islands.* London: Macdonald & Evans, 1962.

Rich, David C. *The Industrial Geography of Australia.* North Ryde, Aust.: Methuen, 1985.

Roberts, John E. (ed.). *Bold Atlas of Australia.* Sydney, Aust.: Ashton Scholastic, 1983.

Saddler, H. *Energy in Australia: Politics and Economics.* Sydney, Aust.: Allen & Unwin, 1981.

Simpson, Colin. *The New Australia.* Sydney, Aust.: Angus & Robertson, 1971.

Spate, O. H. K. *Australia.* New York: Praeger, 1968.

Taylor, M. J., and Thrift, Nigel. *The Geography of Australian Corporate Power.* Beckenham, Eng.: Croom Helm, 1984.

Trollope, Anthony. *Australia.* St. Lucia, Brisbane: University of Queensland Press, 1967.

White, Richard. *Inventing Australia.* Boston: Allen & Unwin, 1981.

Chapter 10

MIDDLE AMERICA

Introduction

Numerous terms are used to describe the land areas of the western hemisphere lying south of the United States. The general term describing all these countries is *Latin America,* which gives reference to the Latin cultural characteristics of the people that settled the land. These characteristics include the common heritage of the people, their languages, religions, customs, and economic ideas, as well as numerous other cultural traits. The heritage they share came from the Latin countries of Spain, France, Italy, and Portugal.

Middle America (sometimes "Meso-America") is the area between South America and the United States (see Figure 10.1). It includes Mexico but may not always include the Caribbean Islands. *Central America* is the group of countries between Mexico and South America. The Caribbean Islands are usually called the *West Indies,* a historical phrase that came into use because Columbus thought he had sailed to India. Little confusion exists with the term *South America,* the large landmass south of the border between Panama and Colombia, but Mexico and Central America are part of the landmass of North America. See Figure 10.2.

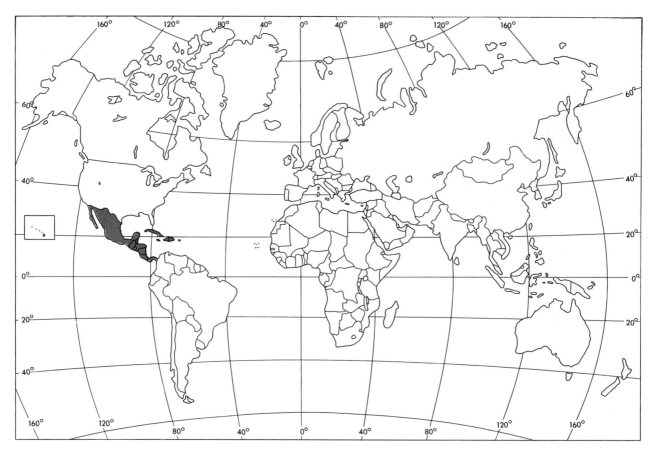

FIGURE 10.1

World Location of Middle America.

In spite of the various divisions of Latin America, a broad unity does exist, but for practical purposes a division had to be made in order to discuss this large area of the Developing World. The division used here is between South America and the remainder of Latin America. This chapter focuses on the lands between the United States and South America, and the following chapter will cover South America.

Mexico

The Republic of Mexico is approximately one-fourth the size of the United States and has about one-third as many people. The triangularly shaped country consists of 758,278 square miles (1,963,940 km²) and lies between the southern border of the United States and about 14 degrees north latitude. The westernmost point of Mexico is directly south of San Diego, California, and the easternmost point is due south of Mobile, Alabama. The Tropic of Cancer, the line marking the northernmost limits of the direct rays of the sun, bisects the country. Thus, much of Mexico is tropical, receiving two seasons of direct, intense heat from the sun. The remainder of the country also is quite warm, with no real winter season.

Mexico's location is important to its social and economic development. The latitudinal location and the corresponding warm climates have shaped some social and economic institutions. The relative location, with respect to its rich and powerful northern neighbor, also has affected the economic and political positions of the country. The boundary with the United States is more than a 1,600-mile-long border (2,577 km) between two independent republics (see Figure 10.3). It is a frontier zone separating Latin America from Anglo-America—a barrier yet a zone of contact, because the Latin influence is strong in the southwestern part of the United States. Mexico lost nearly half its original national territory to the United States, so it still looks with caution toward its neighbor.

Central America

The countries of Central America include Belize, Costa Rica, El Salvador, Guatemala, Honduras, Nicaragua, and Panama (see Figure 10.4). These seven countries are all generally small, poor, and weak, with rapidly expanding populations. The total land area (208,725 square miles (540,738 km^2)) of the seven countries is less than that of Texas, and the total population (26.7 million) is about the same as California's. Most of the people speak Spanish and are Roman Catholics.

The economies of all the Central American countries are based primarily on agricultural commodities, especially the exportation of coffee and bananas. Guatemala is the richest country in the region in terms of gross national product (GNP), but Costa Ricans are richer in terms of per capita income. Belize has the smallest GNP, but El Salvador has the lowest per capita income. Nicaragua is the largest country, El Salvador is the smallest; Guatemala has the most people, Belize the least. The countries of Central America are remarkably similar.

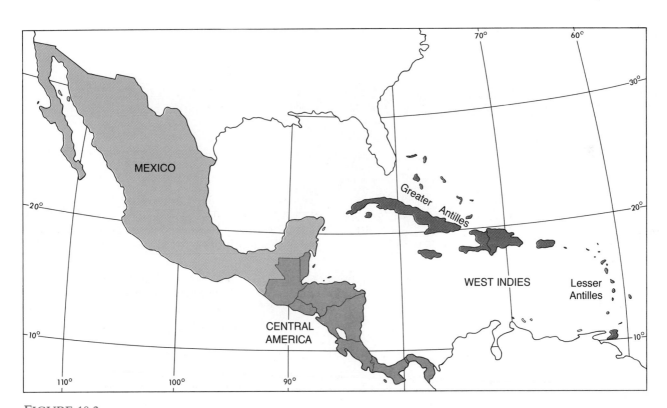

FIGURE 10.2

The Countries of Middle America. Divisions are according to the discussion in the text.

The Caribbean Islands

The islands of the West Indies form a 2,000-mile-long chain (3,221 km) from the southern tip of Florida to the northern coast of Venezuela (see Figure 10.5). The chain contains hundreds of islands, some quite large but most very small and uninhabited. In contrast to the similarities within Central America, this vast region contains great varieties in both the physical and the human aspects of the countries.

Lesser Antilles A north-south trending arc of islands that separates the Caribbean Sea from the western portion of the North Atlantic Ocean. It extends from Sombrero in the north to Grenada in the south.

The West Indies usually are divided into four main groups of islands: the Bahamas, the Greater Antilles, the Lesser Antilles, and the Venezuelan coastal group. The **Lesser Antilles** are further

FIGURE 10.3

The States of Mexico. Mexico, like the United States, is divided into political units. The subdivisions are called "states," and each state has a capital city. The national capital of Mexico City is in a "federal district" similar to Washington, D.C.

Greater Antilles The large east-west trending islands between the Gulf of Mexico and the Caribbean Sea. Composed of Cuba, Hispaniola, Jamaica, Puerto Rico, and the Virgin Islands and comprising 90 percent of the land area of the West Indies.

divided into the Windward Islands and the Leeward Islands. The countries of the **Greater Antilles** are probably the most familiar to Americans. They include Cuba, Jamaica, Puerto Rico, each on its own island, and Haiti and the Dominican Republic sharing the island of Hispaniola. The United States has leased territory on Cuba near Guantanomo Bay for a naval base, and Puerto Rico is a U.S. territorial possession.

The West Indies has less than half the land area of Central America but about 25 percent more people. Although many of the islands are uninhabited, some are very crowded. None of these island countries is considered wealthy, and Haiti is among the poorest countries in the world. Local resources are scarce, and in many places agriculture is nearly impossible. Many islands survive economically on tourist money alone. The white sand beaches and warm tropical sun, then, can be considered resources.

■ Physical Geography of Middle America

Landforms of Mexico

Only about one-third of the land in Mexico can be classified as level, and most of that is too dry for nonirrigated agriculture. Most of the territory is mountainous, and more than half the country lies above 3,000 feet (900 m) elevation. The mountains are of various types—some were formed through

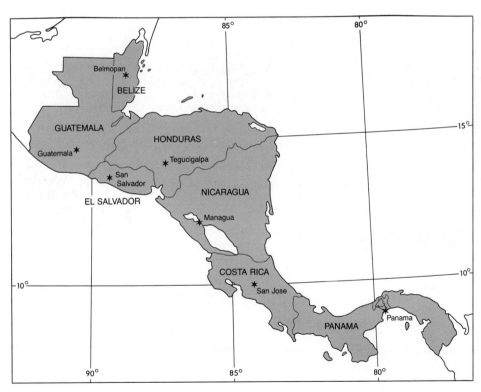

FIGURE 10.4
The Countries of Central America.

FIGURE 10.5

The Greater Antilles and Lesser Antilles. The Greater Antilles are the large islands of the Caribbean, and the Lesser Antilles are the small islands. The Lesser Antilles can be divided into (A) the low islands, (B) the high islands (volcanic), and (C) the islands that are geologically associated with South America.

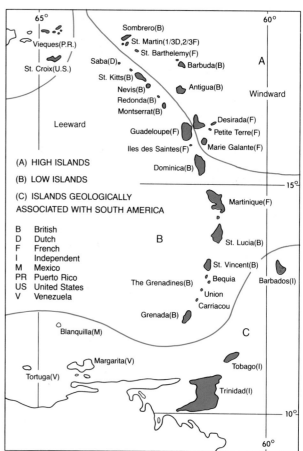

erosion, some through block-faulting, and others are of volcanic origin. These contorted rock structures have created barriers to transportation, micro-climatic variations, and some of the most spectacular scenery anywhere in the world.

The land surface features of Mexico can be divided into two chief parts: the *Central Plateau* and the rugged *Dissected Borders*. The central plateau is a high table land that is dry on the north near the border with the United States but gets progressively wetter toward the south. Few flat plains areas exist on the central plateau, which is cut by canyons and interrupted by scattered block mountains and volcanic peaks. There are numerous interior drainage basins, or **bolsons,** most of which contain the towns and cities of central Mexico. Mexico City is located in such a basin.

The dissected borders of the Central Plateau are made up of three major mountain ranges. The *Sierra Madre Occidental* run along the western edge of the Central Plateau from northwestern Mexico near the United States border southward, where

bolson A flat-floored desert valley that drains into a playa. A playa is the bottom of an undrained desert basin that becomes at times a shallow lake, which when it evaporates leaves deposits of salt or gypsum.

Ring of Fire The region of active volcanoes along the rim of the Pacific Ocean from the southern tip of South America northward through North America and west and south through East and Southeast Asia.

---

they extend into the Pacific Ocean near the city of Acapulco. These mountains are the highest of the three ranges and have created a transportation barrier between northwestern Mexico (the peninsula of Lower California and the state of Sonora) and the Central Plateau region. At their southern edge, the Sierra Madre Occidental turn eastward and become the *Sierra Madre del Sur* without interruption to the mountainous terrain. Thus, land transportation is difficult all along the west coast and in the south. The Sierra Madre del Sur run west to east along the southern part of Mexico and join the *Sierra Madre Oriental* on the east. The Sierra Madre Oriental mountains are not as tall as those in the other two ranges, but they are complex, and they too constitute a nearly impassable barrier between the east coastal plain and the Central Plateau. (See Figure 10.6.)

Southeast of the Central Plateau the tangle of complex mountains begins to smooth out. The flatter, lower land is pinched into the Isthmus of Tehuantepec, the most narrow part of Mexico. The low, flat land continues eastward into the limestone plains of the Yucatán Peninsula, but on the south the elevations rise again. The Highlands of Chiapas, along the southern border of Mexico, are the northwestern end of the mountainous region that extends into Guatemala and on into the remainder of Central America.

The mountainous land of Mexico has few places where the gradients are gentle, and mixed in with the complexly folded mountains are high volcanic peaks. The peaks include, both active volcanoes and volcanoes that are dormant and extinct. So many volcanoes exist that no accurate count has been made, and only the largest or most active are well known. The highest volcanic peak in Mexico is Orizaba (or Citlaltépetl) at 18,700 feet (5,700 m). It is located on the border between the states of Vera Cruz and Puebla, where the Sierra Madre del Sur meets the Sierra Madre Oriental. One of the most famous volcanoes in Mexico is Parícutin. It first appeared in 1943, when it ruptured the surface of a cornfield west of Mexico City in the state of Michoacán. A terrified farmer was working in the field when the ground split open and the first gaseous belches exploded out of the broken surface. Lava and ashes followed the gases and continued to spew out of the opening for months. The resulting lava flow reached a height of 1,700 feet (518 m), covered the nearby town of San Juan Parangaricutiro, and spread outward for nearly 20 miles.

Associated with the volcanic activity in Mexico are the numerous earth tremors. Mexico and Central America are part of the earth's most active earthquake and volcanic zone, the so-called **Ring of Fire** that surrounds the Pacific Ocean. Small tremors are recorded daily, and fairly frequent larger, destructive shakes occur. During the summer of 1957, Mexico City was rattled, about 60 people were killed, and destruction was heavy. A series of more violent tremors shook the city again in September 1985, when more than 300 buildings were destroyed and an estimated 5,000 lives were lost. Tens of thousands of people were left homeless and forced to live and sleep in the open. The most severe earthquake and volcanic zone in Mexico covers an area of about 100,000 square miles (259,067 square km) and runs across the middle of the country from Banderas Bay on the Pacific to Veracruz on the Gulf Coast.

Climate and Vegetation of Mexico

The rugged land surfaces in Mexico create contrasts in climatic conditions that occur within short distances (see Figure 10.7). In turn, these contrasts lead to variations in the natural vegetation as well as in the types of crops that can be grown. The vertical variations within short horizontal distances are called "vertical zonations." In Mexico, as well as in the other mountainous areas of Latin

America, the vertical zones are identified by names.

The lowest climatic and vegetation zone in Latin America is known as the **tierra caliente,** or hot country (see Figure 10.8). It is the lowland area, usually found along the coasts, and could be called the "tropical crops zone." This is where most of the banana and sugar plantations are found. The zone lies between sea level and about 3,000 feet (915 m) elevation. The second zone upward from the tierra caliente is the **tierra templada,** or temperate country. The average temper-

---

tierra caliente The "hot zone," or lower elevations, of Latin America where crop specialization (bananas and sugar cane) is based on the hot and humid conditions.

tierra templada The "temperate zone," or elevations of about 2,000 to 5,000 feet (610 to 1,524 m), in the vertical zonation of climates in Latin America.

---

atures in the tierra caliente are between 75° and 80° F, but in the tierra templada they only average

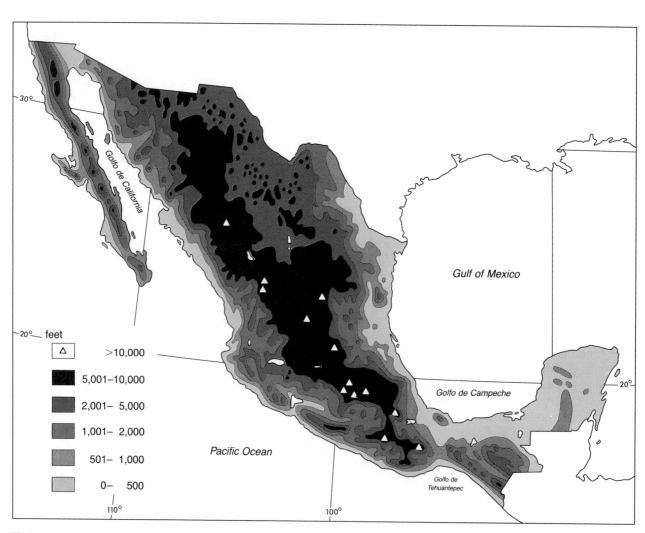

FIGURE 10.6

Land Elevations and Water Bodies of Mexico.

tierra fria The "cold zone," or elevations above about 5,000 feet (1,524 m), in the general altitude zones of Latin America that create local climates and are based on the decrease in temperature with increasing elevation.

paramos A mountain grassland that consists of tall bunch grasses and shrubs and includes some areas above the snowline.

between 65° and 75° F. The tierra templada, the "coffee zone," lies between 3,000 and 6,000 feet (915–1,829 m) above sea level.

The **tierra fria,** or cold country, lies between 6,000 and 10,000 feet (1,829–3,049 m) elevation. In this zone the crops grown are grains at the lower elevations and root crops, such as onions and potatoes, at the higher elevations. The average annual temperatures found in the tierra fria range between 55° and 65° F. Neither forests nor agricultural crops grow above 10,000 feet, but some hardy grasses do exist. That zone, between 10,000 and about 15,500 feet (4,726 m), is the zone of alpine meadows and is called the **paramos.** Above the paramos is the zone of permanent snow and glaciers. The upper limits of the vertical zones vary throughout Middle America according to latitude and whether the slopes face into or away from the prevailing wind and the sun.

The highest temperatures in Mexico occur at low altitudes in the deserts of the northwest and along the eastern coast of the Gulf of California. Summertime temperatures of 100° F are common, and each summer there are usually about 15 to 20 days when the temperature exceeds 110° F. These are the hottest temperatures not only in Mexico but also in any place in Central America and the Caribbean Islands. The coldest average

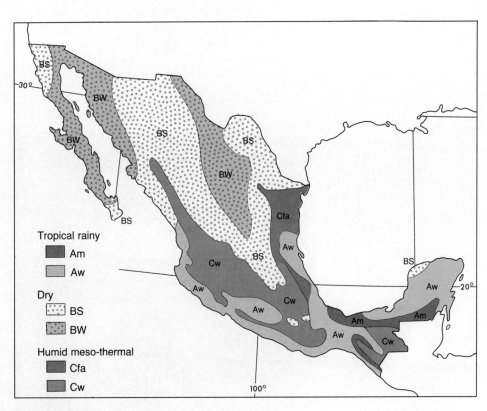

FIGURE 10.7
Climate Regions of Mexico.

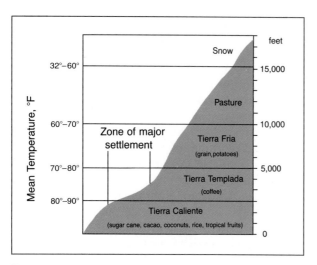

FIGURE 10.8

Altitudinal Zones of Middle America.

temperatures in Mexico are found in the highlands, but cold air masses from the Arctic reach all the way to Mexico City and the Yucatán Peninsula. The movement of cold air is more common along the east coast than along the west coast. The Pacific coast is generally frost-free throughout the year, but as far south as Tampico on the Caribbean side of the country, frosts occur nearly every year.

About half the territory of Mexico is deficient in moisture and is classified as either arid or semiarid. The arid region extends southward from the northern border to the city of San Luis Potosí and across the country from the Pacific Ocean to the mouth of the Rio Grande. The semiarid region lies to the south of the arid area and includes all of the Central Plateau except the part south of Mexico City. Areas where precipitation amounts are adequate in all seasons of the year make up only 15 percent of the territory of Mexico. The two chief rainy areas are found along the east and west coasts of the Isthmus of Tehuantepec. The region along the east coast is the larger of the two and runs from Tampico on the north, through the isthmus, and into the Yucatán. The remainder of Mexico has climates with rainy seasons in the winter that are not very wet, and with extremely dry summers. Thus, most of Mexico is deficient in moisture, either all year round or during the summer growing season.

Coffee is grown in the *tierra templada* region of Central America. This Costa Rican coffee plantation is in the foothills of the mountains near Paos.

Potatoes are produced in the upper reaches of the *tierra templada* and the lower levels of the *tierra fria,* where it is too cold for coffee. Notice the steepness of the slopes where these potatoes are growing and that the clouds are lower than the field.

Landforms of Central America

The structural makeup of the landforms of Central America is distinct from those of North and South America. On the surface, the mountains that form the backbones of all the countries of Central America run from Mexico on the north to Colombia on the south. The underlying structure of these mountains, however, is dominated by an east-west trending system that includes not only Central America but also the islands of the Caribbean. Central America and the West Indies make up a small, unified plate of the earth's crust that is separate from the plates that make up North and South America. This means that although North and South America are connected on the surface by Central America, they are disconnected geologically by the underlying Central American plate. Furthermore, Central America and the West Indies are separated on the surface by the Caribbean Sea but are connected geologically because they are parts of the same underlying plate.

Central America can be divided into three basic physiographic provinces (see Figure 10.9). Most of the region is dominated by the Mountainous Core. These folded and faulted mountains are interspersed with volcanic peaks as well as high, intermontane basins and plateaus. The rugged terrain has been a barrier to transportation as well as to the cooling and moisture-laden northeast trade winds. The second physical region of Central America is the Eastern Lowlands, which stretch along the Caribbean Sea from the Yucatán in Mexico to the Gulf of Darien off Colombia. The plain is widest along the Honduran and Nicaraguan coasts, where it extends inland as much as 150 miles (241 km) and is most narrow along the coast of Panama. The natural vegetation of these coastal lowlands is tropical rainforest, but much of the forest has been removed to make way for the tropical commercial crops, such as a cocoa, bananas, rubber, and coconuts.

The third physical region of Central America is the Pacific Coastlands. On the west, the mountains of Central America descend abruptly into the Pacific Ocean. The coastal plains are generally very narrow along the entire length of the coast from Mexico to Colombia. The only wide coastal plains are between the mainland and the peninsulas of De Nicoya in Costa Rica and De Azuero in Panama.

Middle America

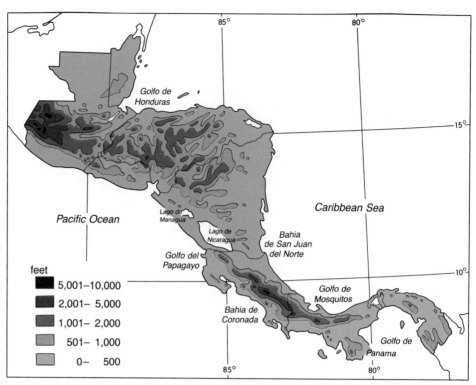

FIGURE 10.9

Land Elevations and Water Bodies of Central America. Note the higher elevations running through the middle of the countries.

The peninsulas themselves, however, are the remnants of old, worn-down mountains and are somewhat rugged.

Climate and Vegetation of Central America

All of Central America is tropical, but climatic conditions vary with the differences in precipitation and elevation (see Figure 10.10). Precipitation varies in amount from place to place as well as seasonally at any particular place (Figure 10.11). The whole region is affected by a two-season rhythm in precipitation. Maximum amounts fall during the summer, or from May to October, and some locations receive more than 100 inches (254 cm) in that period. Rainfall during the six winter months usually totals about one-fourth to one-third the amounts for the summer months. These seasonal differences are caused by the movement of the intertropical front, which in turn follows the general movement of the world wind-system. In other words, the seasonal variations are similar to those caused by the monsoons of Southeast Asia.

Besides the seasonal changes, the east coast receives much more rain than the west coast. For example, Greytown, along the Caribbean coast of southern Nicaragua, averages more than 250 inches (635 cm) of rain a year, but Puntarenas, on the Pacific coast of Costa Rica, receives only about 60 inches (152 cm) a year. As noted above, the Caribbean shore is exposed to the rain-bearing northeast trade winds, whereas the mountains block these winds from reaching the western shores. The eastern coastal lowlands have tropical rainforest (Köppen's Af) climates, while the western coastal regions have tropical savanna (Köppen's Aw) climates. These differences in rainfall amounts are reflected in the types of natural vegetation found in the two regions as well as in the types of agriculture that can be

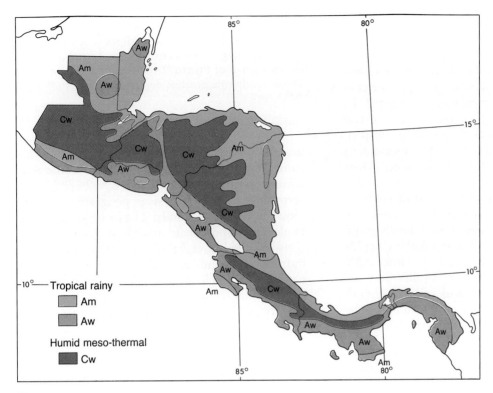

FIGURE 10.10

Climate Regions of Central America.

conducted. Broadleaf, evergreen forests and banana plantations are on the east coast; whereas, grasslands and cattle grazing are prominent on the west coast.

Variations in the climate, the natural vegetation, and agriculture also occur because of the differences in elevation. The tierras caliente, templada, and fria cause the patterns of temperature, precipitation, vegetation, and land use for crops to vary with altitude throughout Central America. Therefore, the mountains not only act as barriers for the prevailing wind, causing the eastern coast to be wet and the western coast to be dry, they also cause temperatures to decrease and precipitation to increase with higher elevations. As a consequence, both the natural vegetation and types of crops that are grown change with altitude.

Landforms of the Caribbean Islands

The islands of the Caribbean Sea were formed as the crests of submerged mountains. New growth through faulting, uplifting, and vulcanism has created some rugged landscapes on some of the islands; other islands have been eroded away so they are nearly flat and barely above sea level. Numerous mountain chains exist, but there are structural similarities within groups of islands. One way the hundreds of islands can be grouped according to geologic similarity is as follows: (1) the Bahamas, (2) the Greater Antilles, (3) the Lesser Antilles, and (4) the South American continental islands (see Figure 10.12).

The Bahama Islands are limestone platforms and are structurally related to the peninsula of Florida, the main part of Cuba, and the Yucatán Peninsula in Mexico. These irregularly shaped islands, along with the Caicos and Turk groups, spread for 900 miles (1448 km) south and east of Miami, Florida. All these islands are low and flat. The highest is Cat Island in the Bahamas, at slightly over 200 feet (61 m) elevation.

The Central American–Antillean mountain system is the name of the main axes of the Greater

Middle America **271**

Antilles. All the upland regions of Puerto Rico, Haiti, the Dominican Republic, Jamaica, and southeastern Cuba, as well as the mountains of Central America, are structurally related. The system is not one great mountain range, but a series of folded and block-fault mountains separated by plains and valleys. The highest point in the system is Pico Duarte, located near the center of Hispaniola in the Dominican Republic. The mountain's crest is at 10,417 feet (3176 m) elevation. Deep submarine troughs run parallel to the mountain system, and the **Brownson depression** north of Puerto Rico is one of the deepest in the world.

The Lesser Antilles consist of a north-south trending arc of islands that is about 450 miles (724 km) long. This group of islands is structurally distinct from both the Greater Antilles and the mainland of South America, and submarine fault troughs delimit the island chain at each end. The Lesser Antilles have been divided administratively into the Windward and Leeward groups, but these names have no significance in relation to the actual wind patterns. The islands also have been grouped according to their physical geography into the "low" and "high" islands. The low islands are low in elevation and consist mainly of level limestone reefs. All the Lesser Antilles are of volcanic origin, but the low islands are old worn-down cones that have subsided below sea level. The high islands all have similar landscapes created by the volcanic activity and characterized by high, steep mountain cones, geysers, craters, and fast-flowing streams.

Brownson depression Sometimes called the Brownson deep or the Puerto Rican Trench, this is the area north of Puerto Rico where the Atlantic Ocean reaches its greatest depth: 27,972 feet (8,526 m).

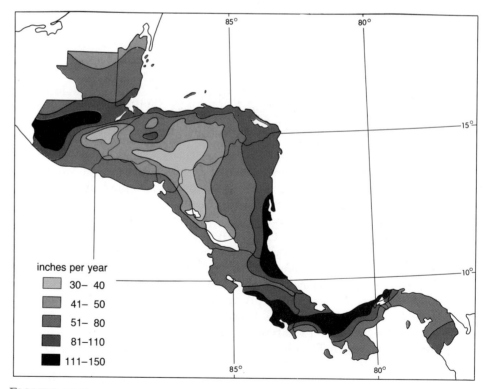

FIGURE 10.11

Annual Precipitation of Central America. Compare the precipitation pattern with the elevations shown in Figure 10.9.

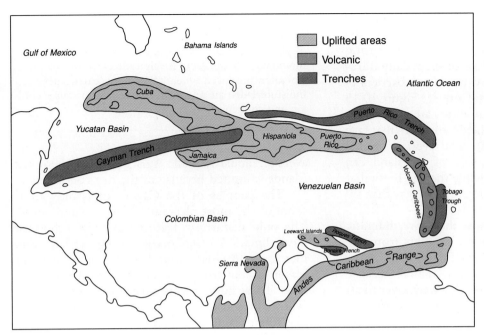

FIGURE 10.12
Geology of the Caribbean. The geologic units of the Caribbean are distinct entities sometimes separated by deep trenches under the water.

The volcanic activity that created many of the islands of the Caribbean has worked in favor of the people as well as to their disadvantage. Generally, soils that develop from volcanic rock are very fertile. These volcanic soils, combined with the warm, humid climates, promote the potential for good agriculture. On the other hand, the volcanoes, along with the associated earthquakes, also have the potential for being very dangerous to humans. For example, Mount Soufrière on Guadeloupe erupted on May 7, 1902, killing more than 2,000 people. The next day an even worse disaster occurred on Martinique. Mount Pelée's eruption spread lava, hot gas, and ash over large parts of the island. The city of Saint Pierre was destroyed, and 40,000 people died. (See Figure 10.13 on page 276.)

The continental island group is an east-west trending chain that stretches for about 700 miles (1,126 km) from Aruba on the west to Barbados on the east. The island chain lies adjacent to the north coast of South America and is associated structurally with the northern highlands of Venezuela. The islands are uplifted sections of the earth's crust and are not volcanic in origin.

Climate of the Caribbean Islands

All the islands of the Caribbean have warm, tropical climates. Temperatures are high throughout the year and are enhanced by warm ocean currents that wash the islands from the east. The high temperatures are moderated by the persistent breezes of the trade winds and, on the high islands, by elevation. The normal summer temperature for the entire region is about 80° F (27° C), with a yearly range of only about 6 or 7 degrees. The yearly range of temperature is greatest in Cuba because the winters are slightly cooler there than in the more southerly islands.

The prevailing winds and the local relief work together to create dramatic rainfall patterns for the Caribbean Islands. The trade winds, flowing from the northeast, pick up loads of moisture from the readily evaporated warm waters of the Caribbean

Middle America

(see Figure 10.14 on page 277.). The waterlogged, constantly moving air is forced to rise over the numerous islands. The rising air is cooled, causing condensation, and heavy rains of short daily duration result. Rain caused by mountain barriers is called **orographic precipitation** (see Figure 10.15 on page 278). The rainfall amounts are greatest on the northeast sides of the islands, the sides that face into the wind (windward). The average annual rainfall on the windward side of an island can exceed 200 inches (60 m), while the yearly total at stations on the leeward side averages less than 30 inches (9.2 m).

The taller the mountains, the more dramatic the differences in precipitation amounts from the windward to the leeward side of the islands. Also, the low islands receive very little rainfall because the moisture-laden winds merely sweep over them without releasing their loads. The low islands also are noted for the porosity of their limestone surfaces. The rainwater rapidly soaks through the soil and the underlying limestone and disappears. Thus, there are no streams, and extreme aridity results in spite of the 45 inches (13.7 m) or more of average annual rainfall. Agriculture on these low islands is limited because of the dry conditions.

---

orographic precipitation Orography is the branch of physical geography concerned with the study of mountains. Orographic precipitation pertains to precipitation produced when a mountain barrier deflects moisture-laden air upward, which in turn causes cooling and condensation.

The peoples of the Caribbean Islands enjoy some of the most pleasant climates in the world. The only disturbing features of these otherwise ideal climatic settings occur during the late summer, from August through October. That is the season of the tropical cyclones, known as *hurricanes*. These huge storms cover hundreds of square miles and consist of tremendous volumes of air that rotates counterclockwise (see Figure 10.16 on page 279). Near the surface the air moves toward the center of the storm (the "eye"), then ascends. The entire mass of whirling air originates over the warm waters of the Atlantic Ocean off the west

The 11,260 foot (3,432 m) Mount Irazú is a volcanic peak and one of the highest points in the mountain core of Costa Rica.

> **mestizo** The racial mixture in Latin America composed of American Indian and European ancestry.
>
> **mulatto** The racial mixture in Latin America composed of Black African and European ancestry.
>
> **zambo** The racial mixture in Latin America composed of Black African and American Indian ancestry.

coast of Africa. From there the storms move westward, eventually ripping across the Caribbean Islands, then they turn in an arc back toward the northeast. They tend to follow similar paths but are known to stray considerably. One storm might go directly west, striking the Yucatán; another may turn sharply and hit the west coast of Florida.

To be classified as a hurricane, a tropical storm must contain minimum surface wind speeds of 75 miles per hour. Some storms have winds much greater than that, so the destruction of homes and crops can be considerable. The winds also produce massive waves that can destroy coastal installations. Destruction also occurs from the huge amounts of rain these storms can produce. World record rainfalls for a 24-hour period are all associated with hurricanes or their Pacific Ocean counterparts, typhoons. As much as 4 *feet* (48 inches) of rain has been recorded coming from one such storm.

■ Human Geography of Middle America

Population Patterns

The people of Middle America reflect the racial intermingling that has occurred in the region. About half the 139 million people of Mexico, Central America, and the Caribbean Islands are of mixed racial ancestry. The three main elements are Indian, black, and white, but each includes a variety of kinds of people. The mixing of bloodlines between the native Indians and the white Europeans has produced what has been called **mestizos.** The mixture of Black Africans and whites is called a **mulatto,** and the black and Indian mixture is called a **zambo.**

The Pacific, or western, side of Central America is much drier than the mountains and the eastern side, but the rainfall on the west is sufficient for growing certain crops as shown in this scene of western Costa Rica.

Middle America 275

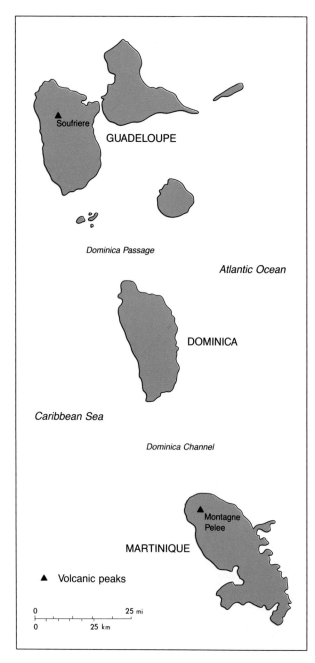

FIGURE 10.13

Guadeloupe, Dominica, and Martinique. The major volcanic peaks of the region are Mount Soufrière on Guadeloupe and Mount Pelée on Martinique.

About one-third of the people of Mexico are Indian; the remaining two-thirds are mestizo, with heavy Indian lineage. In Guatemala the majority are Indians, but the population of Costa Rica is almost entirely white. The remainder of Central America is occupied primarily by mestizos, with Indians found in the higher parts of the tierra fria and in the jungles of the coastal lowlands. On the islands of the Caribbean, the native Indians were virtually eliminated either through battles with the white settlers or from diseases brought by the whites. The most prominent group today is composed of descendants of African slaves imported for labor. The Spanish, who colonized the islands, soon moved to the mainland and left a population consisting overwhelmingly of blacks and mulattoes. Only in Cuba, Puerto Rico, and the Dominican Republic are Caucasoids in a majority today.

People of Asian ancestry live in Middle America, and there have been many mixtures. For example, 36 percent of the people in Trinidad are descendants of laborers imported from India, another 21 percent are Chinese, Lebanese, Syrian, and European, and the remaining 43 percent are black. Trinidad is unusual in the high proportion of Asians, but it does demonstrate other dimensions of the racial makeup of Latin America. In all of Middle America, the Latin cultural influence is strong, but the other imported cultures from Africa and Asia also have contributed to the complex and varied cultural landscape.

Population Distribution

The rugged physical landscape of most of Middle America causes an uneven distribution of population within most countries (see Figure 10.17 on page 280). There is a varied distribution between countries too. For example, Nicaragua has only about 67 people per square mile of territory, while Barbados and Puerto Rico have more than 1,000 people per square mile. Barbados is one of the most densely settled rural nations of the world, at 1,526 people per square mile. Only 42 percent of the people of Barbados live in what can be called an urban area. The capital and largest city, Bridgetown, has a population of less than 8,000.

The most densely settled parts of the mainland of Middle America are on the high plateaus of

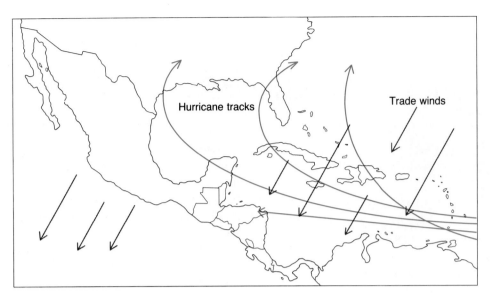

FIGURE 10.14

Tropical Storm (Hurricane) Tracks of the Caribbean. Note that the storms tend to turn northeastward once they leave the zone of the northeast trade winds.

Mexico. About 82 million people live in Mexico, which has the largest population by far of the countries of Middle America. The overall population density is 108 per square mile, but 70 percent of the Mexicans live in the urban areas, leaving vast regions that are nearly devoid of people. Mexico City is now the largest city in the world, with a metropolitan population of about 15 million. It also is one of the fastest-growing cities in the world because the rate of growth of the Mexican population is high and each year more people leave the rural areas looking for a better life in the capital city. The regional industrial centers of Guadalajara and Monterrey also are growing rapidly, and each now exceeds 2 million in population. Along with the continued growth of the older cities of Mexico, new oil towns have sprung up along the Gulf Coast, and new factory districts have appeared in the smaller cities of every state.

The total population of Central America is 26.3 million. Guatemala has the most people, with 8.6 million, and Belize has the least, with about 200,000. As with Mexico, most of these people live in the highlands, either on the plateaus between the mountains or along the mountain

The native Indian influence can be detected in these people of Mexico. Notice their facial features and the types of dresses they are wearing. Photograph by Richard Hecock.

Middle America 277

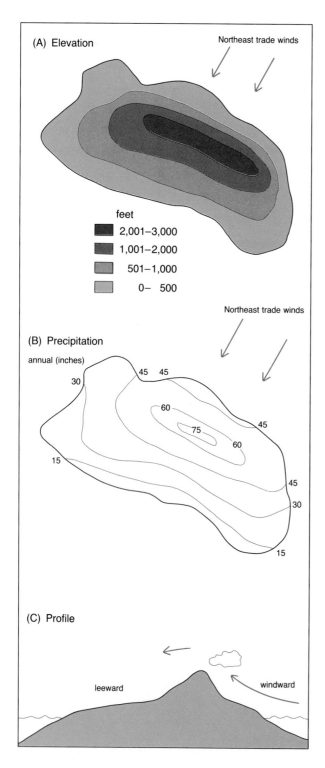

FIGURE 10.15

Precipitation Patterns of a Typical Caribbean Island. Elevations in feet (meters) above sea level (A), annual precipitation amounts in inches (centimeters) (B), and a profile of the island (C). The upper slopes of the windward side of the mountain receive large amounts of precipitation, and the leeward slopes are drier.

Panama the lowest. Mexico's rate of increase (2.6 percent) is about the same as the average for the other mainland countries of Middle America.

The total population for all the islands of the Caribbean is 31 million, but most of these people live on the four largest islands of the Greater Antilles. Cuba alone accounts for one-third of the population, while slightly more than one-third live in Haiti and the Dominican Republic on the island of Hispaniola. Puerto Rico, with 3.3 million, and Jamaica, with 2.3 million, account for another 18 percent of the total. The only other large concentration of population is on the island of Trinidad, where about 1.2 million people live. Unlike the mainland, most of the island people live on the low plains and along the coasts. Another major difference between mainland and island demographics is in the rate of growth. The average rate of annual increase for all the Caribbean Islands is 1.7 percent, the lowest for all of Latin America. The highest growth rate is 2.5 percent in the Dominican Republic, and the lowest is 0.9 percent for Barbados.

Only about 54 percent of the people of the islands of the Caribbean live in the urban areas (see Figure 10.18 on page 281). Urbanization figures vary from 75 percent of the population in the Bahamas to 26 percent for Haiti. The largest city of the region is Havana, Cuba, with nearly 2 million people. The regional centers of Cuba are Santiago de Cuba and Camagüey, each with about 500,000 people. Other large cities of the region are Kingston, Jamaica; San Juan, Puerto Rico; and Port-of-Spain, Trinidad.

Population Structure

As with most rapidly growing areas, the population of Middle America is young—that is, the number of people less than 15 years old is much greater than the number of people older than 65. The average proportion for Central America and Mexico are 42 percent of the population age 15 and

ridges. The average natural rate of population increase for the region is 2.7 percent, with 3.4 percent in Nicaragua the highest and 2.1 percent in

papiamento The jargon spoken by the black people of Curaçao with a vocabulary of Spanish, Portuguese, Dutch, English, Carib, and native African.

younger, with only 4 percent in the 65 and older category. By comparison, the figures for the United States are 22 percent of the population age 15 and younger, and 12 percent for age 65 and older. The youngest countries in Middle America are Nicaragua and Honduras, each with 48 percent of their populations in the young category and 3 percent in the oldest groups. The oldest countries of the region are the Caribbean Island nations of Cuba and Martinique, each with 28 percent of their population age 15 or younger, and 8 percent at 65 or older. See Figure 10.19 on page 282.

The infant mortality rate in Middle America is much greater than in North America, but not as high as in South America. The average rate for Mexico and Central America is 57 deaths per 1,000 live births, and the average for the Caribbean Island is 54 deaths per 1,000. By comparison, the average for the United States and Canada is 9.3 deaths per 1,000 live births, while that for South America is 70 deaths per 1,000. High infant death rates are related to malnutrition and poverty. The lack of resources and the resulting lack of health knowledge and facilities create miserable conditions for saving lives, especially the lives of children in their first few months of life. For example, Haiti has the lowest per capita gross national product in the region and the highest infant mortality rate—108 deaths per 1,000 live births. The more wealthy countries have much lower infant mortality rates. Martinique, for example, has the lowest infant death rate and the second highest per capita income for the region.

Language and Religion

The diversity that separates Middle American countries would be much greater if it were not for the somewhat unifying effect of language. Spanish, the tongue of the original European colonizers, is the official language of the mainland states from Mexico to Panama. In the islands it prevails in Cuba, the Dominican Republic, Puerto Rico, and some of the Lesser Antilles. Thus, Spanish is the language of a vast majority of Middle Americans.

English is spoken in Belize, the Bahamas, Jamaica, the Cayman Islands, Trinidad and Tobago, and most of the Lesser Antilles. French is the national language in Haiti and is common in the central part of the Lesser Antilles from Saint Lucia to Guadeloupe. The only other language found in Middle America is **papiamento,** which is spoken in the Netherlands Antilles—Aruba, Curaçao, Bonaire, and some small islands in the Leewards group. This language is based on Spanish but is much corrupted. The variety of languages reflects the colonial history of the islands.

In Middle America, official languages are seldom standard, and many of the dialects spoken are not intelligible to people in other areas who officially speak the same language. Also, there are regional pockets in which a language dominates but is not the official language of the country. For example, in the region near the eastern coastal town of Limón in Costa Rica, English is common.

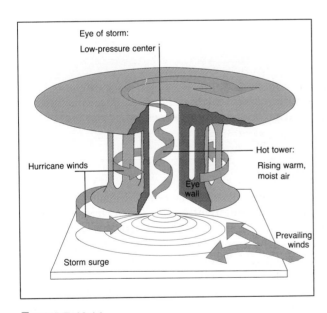

FIGURE 10.16

Profile of a Hurricane. The air movement associated with hurricanes changes at the various elevations. The surface wind rotates in large spirals, but near the center of the storm (the eye) the air rises rapidly in a tightly rotating spiral, then spreads out again at the higher elevations and begins to rotate in a different direction. The storm surge is the dome of water pulled up by the low pressure in the eye of the storm.

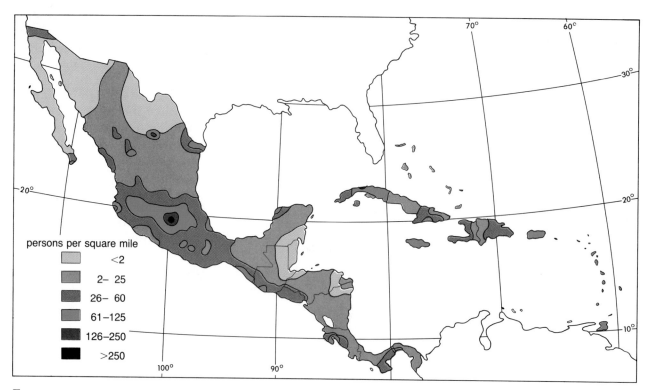

FIGURE 10.17

Population Patterns of Middle America.

It is the language of the black laborers brought from Jamaica to work on the banana plantations. Other examples of regional pockets of language include the more than 50 surviving Indian languages and dialects in Mexico alone. More than one million Indians in Mexico do not speak any Spanish whatsoever, and some cannot understand their neighbors in nearby villages.

While language gives Middle America some measure of self-identity, religion is a more effective unifier. The zeal of the early Spanish priests left most of the countries of Middle America with a strong Roman Catholic heritage. Even in Mexico, where a large Indian minority exists, 95 percent of the people are Catholic. Thus, not only the people with Latin Ancestry are Catholic, but the native peoples also were converted to that faith. In Belize, where the British ruled until 1981, Roman Catholics make up about 66 percent of the population, the lowest percentage of the mainland countries. When the majority of people have similar religious beliefs, it helps shape similar attitudes and promotes common values.

Ethnicity

Some of the ethnic makeup of Middle America has already been mentioned. The primary factor to note is the complexity. Variation among the countries is extreme as far as the percentages of each racial type. The figures in Table 10.1 show the extent of the variation. In spite of the problems associated with obtaining demographic statistics for these countries, the estimates local officials provide are considered to be fairly accurate. Things to note in Table 10.1 are the large percentage of mixed-blood people in all the countries except Belize and Costa Rica, the large percentage of Indians in Guatemala, the large percentage of whites in Costa Rica, and the large percentage of blacks in Belize.

TABLE 10.1.
Racial Composition in Mexico and Central America

COUNTRY	TOTAL POPULATION (MILLIONS)	% WHITE	% BLACK	% INDIAN	% MESTIZO
Belize	0.2	10	55	32	3
Costa Rica	2.7	80	2	1	17
El Salvador	5.1	11	—	11	78
Guatemala	8.6	5	—	60	35
Honduras	4.6	2	2	10	86
Mexico	81.7	15	1	29	55
Nicaragua	3.3	17	9	5	69
Panama	2.2	11	13	10	65

Note: The percentages are estimates, and some do not total 100 because other ethnic groups (e.g., as Asians) live in some of the countries.

SOURCE: Primarily "World Population Data Sheet," Population Reference Bureau, Washington, D.C., 1986; Preston E. James, *Latin America* (New York: Odyssey, 1978); and *World Almanac* (New York: Newspaper Enterprises, 1986).

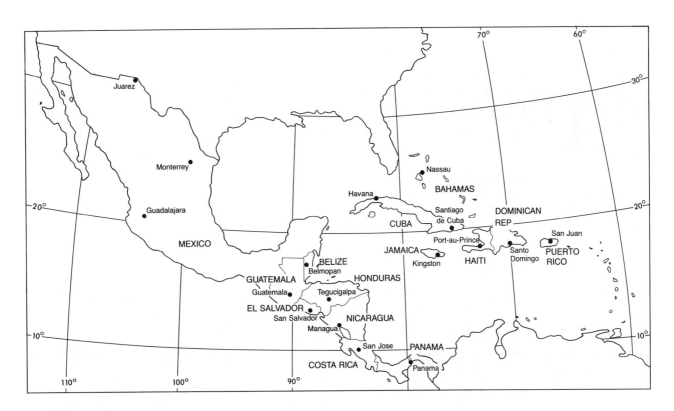

FIGURE 10.18
Major Cities of Middle America.

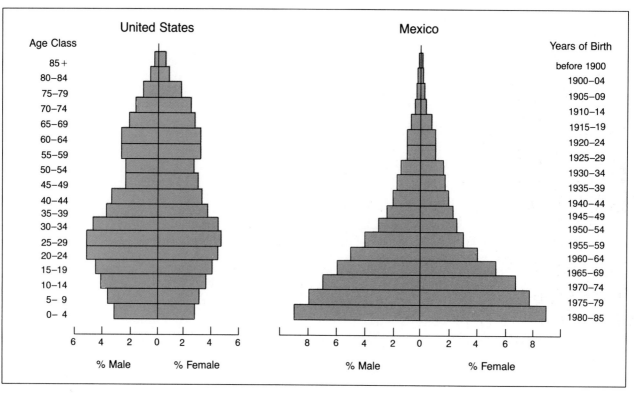

FIGURE 10.19

Population Pyramids for the United States and Mexico. The shapes of the two pyramids indicate slow growth for the United States and fast growth for Mexico. The growth of the population in the United States has slowed since the "baby boom" of the 1950s, but Mexico's population has continued to increase.

TABLE 10.2

Racial Composition in the Major Caribbean Islands

COUNTRY	TOTAL POPULATION (MILLIONS)	% WHITE	% BLACK	% MULATTO
Cuba	10.2	73	12	14
Dominican Republic	6.4	16	11	73
Haiti	5.9	—	95	5
Jamaica	2.3	8	76	15
Puerto Rico	3.3	99	—	—
Trinidad	1.2	49	40	14

Note: The percentages are estimates, and some of the percentages do not total 100 because other ethnic groups (such as Asians) live in some of the countries.

SOURCE: Primarily "World Population Data Sheet," Population Reference Bureau, Washington, D.C., 1986; Preston E. James, *Latin America* (New York: Odyssey, 1978); and *World Almanac, 1986* (New York: Newspaper Enterprises, 1986).

This large Roman Catholic Church is in the central part of Mexico City. The city is the largest in the world and growing larger every day, partly because of the Catholic position on birth control. Photograph by Richard Hecock.

Similar statistics for the six largest countries of the Caribbean are in Table 10.2. Very few Indians or mestizos live in the Caribbean Islands. Note also the larger percentage of blacks on the islands than on the mainland. The mixed-blood groups in the islands, thus, are the mulattoes, instead of the mestizos. Nearly all (99 percent) the Puerto Ricans are of European origin, and most (95 percent) of the Haitians are of African origin. (See Figure 10.20 for the distribution of ethnic groups in the Caribbean.)

■ Economic Geography of Middle America

Agriculture

Most of the people of Middle America earn their living from farming, but only about 10 percent of the land area is tillable. This lack of flat land puts a strain on the ability of the countries to feed themselves. Approximately half the people of Mexico and two-thirds of the people of the other countries of the area eke out a living on small subsistence plots of land. The volcanic soils, which are sometimes very fertile, are usually heavily eroded, primitive tools are still used, malnutrition and disease are prevalent, and grinding poverty prevails. Without a productive form of agriculture, it is difficult for developing countries to modernize in other ways. Before industrial development can occur, a state must be able to feed its people.

The most common agricultural crops in Middle America are beans, corn (maize), and vegetables in the highlands, and manioc, corn, yams, and plantains in the low elevations. These are the subsistence crops grown by the farmers who eat their own products. The major cash crops are coffee, bananas, and sugar. Coffee is grown throughout the highlands of Mexico and Central America, sometimes on small farms and sometimes on plantations, but everywhere in the mountains it is a leading source of income. Coffee is the leading export for Costa Rica, El Salvador, and Guatemala and is usually Mexico's second leading agricultural export after cotton. (Agricultural regions are shown on the map in Figure 10.21.)

In the lowland tierra caliente zone, the leading commercial crops are bananas and sugar. Both are grown on plantations, and production is characterized by the use of seasonal labor, imported capital, one-crop production, and an exported product. The plantations are carved out of the tropical rainforests, and much labor is required to subdue the forest and keep it subdued from year to year. Mechanization is used in harvesting both bananas and sugar cane, but the actual cutting of the cane and the removal of a bunch of bananas from a banana tree still must be done by hand. Much hand labor, then, is required for the harvesting of these two crops. Although the labor requirements are seasonal, the areas with plantation agriculture generally enjoy higher standards of living than similar areas that do not have commercialized agriculture. (See Figure 10.22 for the location of banana production areas.)

United Fruit Company A North American company formed in 1899 to develop banana production in Costa Rica. See Focus Box on page 287.

The banana trade did not develop until late in the nineteenth century, when American companies, most notably the **United Fruit Company,** stimulated the industry along the east coast of the mainland from Mexico to Panama. Bananas are the leading export item today for Honduras, Nicaragua, and Panama. At one time the United Fruit Company was a powerful force in both the economies and the politics of most Central American countries. The phrase "Banana republics" has been used to describe these countries, and Latin Americans still call the United Fruit Company simply "the banana company." The company no longer meddles in Latin American politics, but political

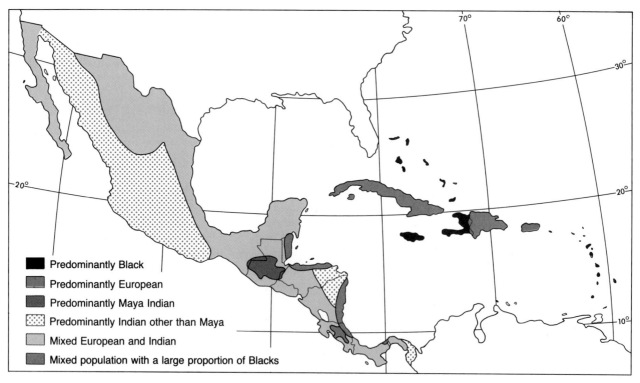

FIGURE 10.20

Racial Composition of Middle America.

Many farm products in Mexico go directly to outdoor markets like this one in Mexico City, where housewives purchase fresh produce daily. Photograph by Richard Hecock.

Much of the farming in Middle America is done without the help of mechanization. In this case, an ox pulls a cultivator plow through a potato field in Costa Rica.

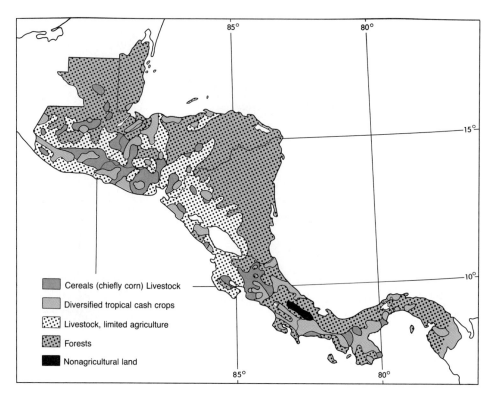

FIGURE 10.21

Agricultural Regions of Central America.

leftists often label it as "el pulpo" (the octopus) that represents the United States.

The major crop of the Caribbean Islands is sugar cane. About 83 percent of Cuba's exports is refined sugar. In Middle America, Mexico is the second leading producer, and it is the leading crop for Belize. The total Central American sugar cane crop, however, averages only about 20 percent of Cuba's production. Sugar is the leading crop for all the major islands of the Caribbean except Puerto Rico. Rum is an island by-product of sugar production and the leading agricultural export for Haiti, Jamaica, and Trinidad. Like bananas, sugar cane is a plantation crop, and sugar's traditional market has been the United States. Since Fidel Castro's Communist takeover in Cuba, however, 65 percent of that country's exports go to the Soviet Union. (See Figure 10.23 on sugar producing areas of Cuba.)

Several other crops are produced on a commercial basis in Middle America. Henequen, sisal, and cotton are grown for fibers. Mexico exports more than half the world's henequen and sisal, yet cotton is its leading agricultural export. Balsam,

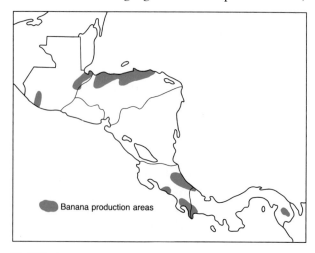

FIGURE 10.22

Banana Production Areas of Central America.

FOCUS BOX

The United Fruit Company

The founders of the United Fruit Company were not interested in gaining power, they were motivated strictly by profit. But the company did become an octopus (*el pulpo*) that reached deeply into the national budgets of Central American countries and altered governments for its own purposes. The legacy of hatred survives, even though the company no longer meddles in politics.

The foundation for the United Fruit Company was laid in 1864, when Carl and Otto Frank started the first banana trade between Central America and the United States. Carl developed a taste for bananas while working as a steward on a ship that made regular stops along the north coast of South America. The German-born brothers decided they could make a profit by shipping the somewhat exotic fruit to the United States. Their business never became very large.

In 1870, Captain Lorenzo Baker took 160 bunches of bananas from Colón in Panama to Jersey City. Baker took the bananas as a gamble and to complete the load on his small schooner the *Telegraph*. After an 11-day voyage from Panama, Baker sold all the bananas in one day at the Jersey City dock. He made a profit of $2 a bunch and entered the banana trade full-time. Later that same year, Andrew Preston, from Boston, was visiting Jersey City and came across some of Baker's bananas. Preston thought the fruit was delicious and made a deal with Baker. He convinced Baker to bypass Jersey City and bring all his bananas to Boston, where Preston would sell them. The deal was made, and the banana trade began to increase.

The true founder of the United Fruit Company was Minor Keith, a young Brooklyn native who in 1870 was invited by his older brother Henry to help build a railroad in Costa Rica. The railroad was to go from the eastern port town of Limón to the highland capital of San José. The lowland coast was a rainy swamp swarming with mosquitoes, and hundreds of American laborers died from malaria and yellow fever. Henry Keith also died, and Minor took over the operation of constructing the railroad. He brought in Jamaican laborers, who were relatively immune to malaria. After about 50 miles of track had been laid, however, the railroad company was near bankruptcy. They needed to begin hauling freight, but the swampy jungle provided little of value.

Minor Keith had the ingenuity and ambition to start growing bananas, a crop suited to the rainy climate. He went to Colón, obtained thousands of banana shoots, brought them back to the Limón area, and planted them. As the bananas began to make money, Keith used the profits to continue building the railroad. The Jamaican laborers turned to the banana harvest and became permanent settlers. Banana production began to spread throughout Central America, and Keith's railroad followed the diffusion of production.

In 1899, Minor Keith was the leading banana grower in Central America, and in the same year he formed a partnership with Captain Baker and Andrew Preston. Keith grew the bananas and moved them to the coastal ports on his railroads, Baker transported the bananas by sea from Central America to the United States, and Preston sold them in the United States. From the beginning, the partnership was called the United Fruit Company.

After the sugar cane is cut by hand, the stalks containing the sugar are loaded onto wagons, which are pulled by tractors to weigh stations (shown here) and then taken to nearby mills for processing.

cacao, palm oil, and tobacco are commercially successful crops for several countries, both for export and for the domestic market. Various tropical fruits also are grown, as well as coconuts and pineapples.

The acute shortage of usable agricultural land in Middle America does not allow for a well-developed animal husbandry industry. Commercial cattle ranches exist in Mexico, especially in the dry areas along the northern border with the United States. And although dairying is not a noted tropical activity, Puerto Rico's dairy product income has now passed that for sugar. Other than these two cases, very little livestock production occurs in Middle America. A few domestic animals can be seen around most subsistence farms, however.

Mineral Production

While mineral deposits are not lacking in Middle America, they are highly localized. (The mineral producing areas of Mexico are shown on the map in Figure 10.24.) Mexico is in the most enviable position because it has good mineral resources and a location close to the world's major trade routes. Mexico ranks with the United States in leading the world in silver production, it is rapidly developing into a world leader in the production of oil, and it has significant amounts of gold, copper, zinc, antimony, mercury, sulfur, molybdenum, graphite, arsenic, opal, iron, and coal. Although only a small percentage of the population is engaged in the production of minerals, because of its highly mechanized nature, the products from Mexico's mines and wells play a vital part in the country's export trade and industrial development. Of the Central American countries, none has found any good deposits of valuable minerals to exploit, although along the east coast there is some oil production.

Among the island states, Cuba is best endowed with minerals, having good deposits of iron, copper, manganese, nickel, and salt. Trinidad is unusual in that about half its income is from mining and manufacturing instead of from agricultural products. The country imports crude oil from Africa, especially Nigeria, and then exports the refined oil to the United States. Oil revenue makes Trinidad one of the more prosperous countries in the region. Jamaican bauxite contributes roughly one-third of that country's income, and

Alliance for Progress A largely defunct organization of Latin American states that once attempted to enhance establishment of a continental free-trade area.

most of the bauxite is exported to the United States. Cuba used to export chromium, manganese, nickel, and iron to the United States, but because Cuba's present main trade partner, the Soviet Union, does not need these items, Cuban production has been greatly restricted.

Industry

Industry in Middle America is poorly developed, and there are few indications that rapid growth will come in the future. Large amounts of capital are required to develop industry, and most people of the area are poor. Those that do have money to invest usually avoid industry. Private capital from outside is available, but unpredictable profits and unstable political systems that often confiscate

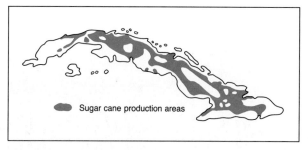

FIGURE 10.23
Sugar Cane Production Areas of Cuba.

foreign investments discourage most potential outside investors. The best hope appears to be projects financed by governments outside the region that are prepared for losses—for example, the U.S.-financed **Alliance for Progress,** inaugurated in 1961. Such programs, however, have made woefully slow progress.

Most of the industry in Middle America is consumer oriented—that is, it produces items for a local market that will be consumed without further production. This type of industry contrasts with industries that produce items that contribute

The Kaiser dock in Jamaica is an important loading point for much of the bauxite ore exported from the country. Photograph by Richard Hecock.

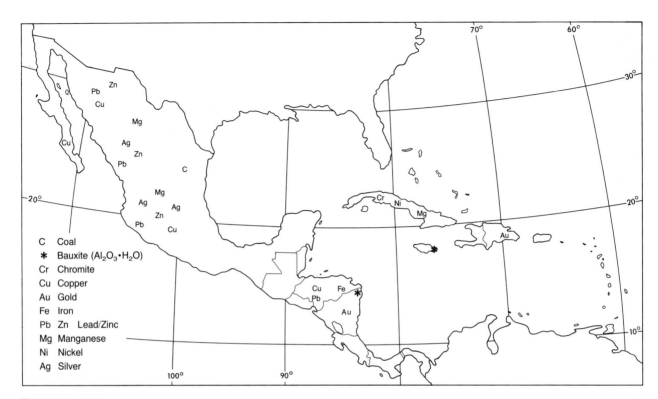

FIGURE 10.24
Mineral Deposits of Middle America.

toward the manufacture of other items. Examples of consumer goods are textiles, clothing, shoes, tobacco products, beverages, and canned food—all items that are manufactured throughout Middle America. In areas of special agricultural production, the processing of the local product creates industrial activity. Common examples of this are the roasting and grinding of coffee, stamping sugar cane, and scraping henequen.

While most industry is light and consumer oriented, some heavy industry is beginning to develop in the region. One factory often found in emerging industrial areas is the cement plant, which utilizes local labor, shale, and limestone and sells to local markets. Such plants are found in all mainland countries of Middle America and on the major islands. Guatemala is the leading producer of cement in Central America, but Mexico produces seven times the combined cement output of Central America. Among the island countries, Cuba is the leading producer.

For overall industrialization, Mexico is the giant of Middle America (see Figure 10.25 on page 294). About 18 percent of the country's labor force is engaged in industry. Monterrey is the iron and steel center for Mexico, although some iron, steel, and numerous related industries are centered around Mexico City. Mexico has a large domestic market, a strong labor source, and good minerals, including plenty of coal and petroleum for energy. Besides textiles and foodstuffs, Mexico produces steel, chemicals, electric goods, rubber, and petroleum products. Many U.S. companies have established assembly plants in northern Mexico to take advantage of the cheap Mexican labor. Parts are made in the United States and shipped to these *maquiladora* plants for assembly, then the finished products are returned to the United States for sale. As many as 1,000 of these plants are now operating along the border between the United States and Mexico.

In Puerto Rico, the main income is from manufacturing. Puerto Rico's industries produce

FOCUS BOX

The Five Nations of Mexico

With a population of about 82 million, Mexico is nearly four times larger than Canada. Half the population of Mexico is under the age of 19, compared with a median age in the United States of 31. A whole range of complex issues—from illegal migration and drug trafficking, to Mexico's foreign debt and the debate over Central American problems—demands that the people of the United States pay more attention to Mexico than they have in the past. Some American leaders now say that improving our relations with Mexico has become a major challenge, and second in importance only to our dealings with the Soviet Union.

The American view of modern Mexico probably should not be based on the existence of the 31 states of Mexico, or on the topographic and climatic regions, or even on ethnic regions, but it should take into consideration the five nations described by Louis Casagrande: (1) Metromex, (2) Mexamerica, (3) South Mexico, (4) New Spain, and (5) Club Mex (see Figure 10.A).

Metromex includes Mexico City, the entire Valley of Mexico, and the industrial corridors from Querétaro on the northwest to Cuernavaca on the south. Before the Spanish Conquest, the Valley of Mexico was a great, shallow lake. The dried-up bed of Lake Texcoco now provides the dust that mingles with the smoke of 100,000 factories and the exhaust of 3 million vehicles, creating one of the most toxic environments known. Half of all the industry in Mexico is located in Metromex, and the region accounts for 38 percent of the country's gross national product. The population of Metromex has doubled in the last decade, despite the unhealthy environment. Nearly 3,000 immigrants arrive each day, and almost one of every four Mexicans lives in the region. The Metromex is the economic, religious, cultural, and political center of Mexico.

nortenos Literally translated from Spanish as "northerner" and used by Mexicans to refer to people from the United States.

Mexamerica is a rival to Metromex. The region straddles the U.S.-Mexico border and has a population of nearly 38 million. Mexamerica historically has been beyond the full control of Metromex. The **nortenos** developed into independent and self-reliant people, mostly free from authority of the central government. Today, the middle-class *nortenos* look to the north for their role models, rather than to the Metromex. The economy of the region is diversified and relatively strong. Heavy industry in the northeast centered on Monterrey, large irrigated commercial farms on the west, and industrial development along the border all serve the region with economic growth potential.

South Mexico is the most impoverished region of Mexico and the most resistant to change. The population of the region, about 15 million, is more than half Indian and mostly poor. There are only a few medium-sized cities and very little industry. Most of the people live in small towns and villages. The economy is predominately agricultural, small-scale, and familial. Crafts are a main source of income for many people, and their pottery, colorful hand-woven textiles, and wooden toys are famous. South Mexico is somewhat ignored by the officials in the Metromex, and although the people of South Mexico would like to initiate independent development, they lack the economic clout to gain self-control.

New Spain is the densely populated colonial center of "Old" Mexico. The region features an elaborate web of large, medium-

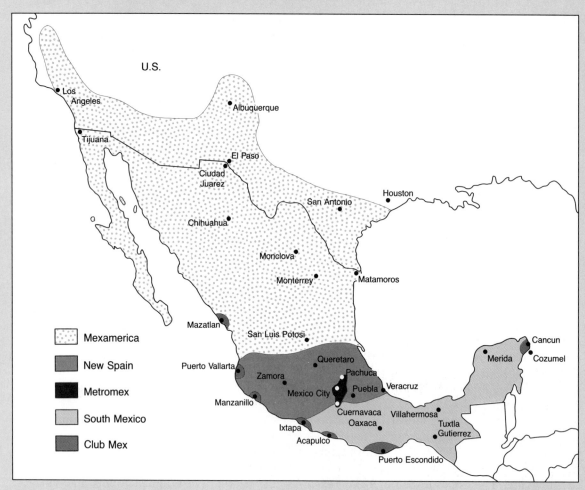

FIGURE 10.A

The Five Nations of Mexico.

sized, and small cities, each with its own hinterland of dependent villages. Most of the communities in New Spain are centuries old. The numerous mountain valleys are well watered and arable, and the region feeds both itself and the Metromex. In return, the Metromex offers New Spain industrial goods. The dual capitals of the region are Guadalajara in the west and Veracruz in the east. The region is overpopulated, and about 80 percent of the young men that go north to the United States in search of work come from New Spain.

Club Mex is the newest and smallest of the Mexican regions. The region consists of converted port towns (Acapulaco, Mazatlán,

Manzanillo), once-sleepy fishing villages (Puerto Vallarta, Zihuátanejo, Puerto Escondido, and Cozumel), and brand-new planned tourist centers (Cancun and Ixtapa). Club Mex is a region because each of the divergent towns reflects a coherent plan to create resort enclaves dedicated to tourists who bring millions of dollars in search of sun and relaxation. Metromex has invested billions of dollars in Club Mex, as have private developers. Club Mex achieved international status in the late 1960s, but the 1.5 million residents of the region have not been greatly affected by the changes brought about by the demands of international tourism.

Despite their common heritage, the five regions of modern Mexico have significantly different histories, economies, and evolving identities. A perception of these five regions based on the foregoing information rather than on older images should help to penetrate the mystique of America's southern neighbor.

SOURCE: Louis B. Casagrande, "The Five Nations of Mexico," *Focus* 37 (Spring 1987): 2–9.

---

electrical equipment, plastics, and chemicals—products that indicate a degree of industrial development beyond the usual consumer-oriented goods. Textiles too are still an important item of production. About 19 percent of Puerto Rico's labor force is engaged in manufacturing, with another 19 percent in trade. Puerto Rico has the lowest percentage of its population engaged in agriculture of any country in Middle America.

Cuba, the largest and most populous of the islands, has an industrial potential because the labor, market, and minerals are available, but it has not developed much beyond the sugar-processing industry. Capital for this industry was secured in

This large market in Mexico was built near an older outdoor market. It now supplies local goods but also caters to the tourist trade. Photograph by Richard Hecock.

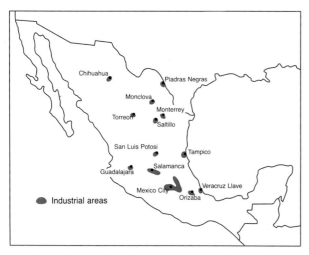

FIGURE 10.25
Industrial Areas of Mexico.

1960 when Cuba confiscated foreign-owned sugar mills worth millions of dollars. Other important products of Cuban industry include refined oil, cement, textiles, and chemicals. Growth or change in Cuban industry is difficult to evaluate because Cuba does not report production figures to the United Nations. Cuba maintains a tight trade relationship with the Soviet Union.

Petroleum refining is an important industry in the Netherlands Antilles. On the islands of Aruba and Curaçao, there are huge refineries erected by the Dutch Shell Oil Company to process crude oil originating in Venezuela and destined for markets in the United States and Europe. Another refinery is located at Freeport in the Bahamas. The New England Petroleum Company and Standard Oil Company of California invested in the facilities that receive oil from the Middle East, Venezuela, and Nigeria. The refined products go on to the United States and Europe. The facilities can accommodate supertankers and process as much as 250,000 barrels of oil a day.

Another major economic activity in the Caribbean is the tourist trade. Balmy weather, warm water, sandy shores, and somewhat different customs entice visitors from temperate latitudes in increasing numbers each year. Such tourist attractions as gambling, horseracing, colorful ceremonials, and sports events are developed almost entirely to attract the visiting currency. The tourist income is a major source of revenue in many island states.

A busy dock in Mexico catering to the import-export business. Notice the numerous small containers of different sizes. Photograph by Richard Hecock.

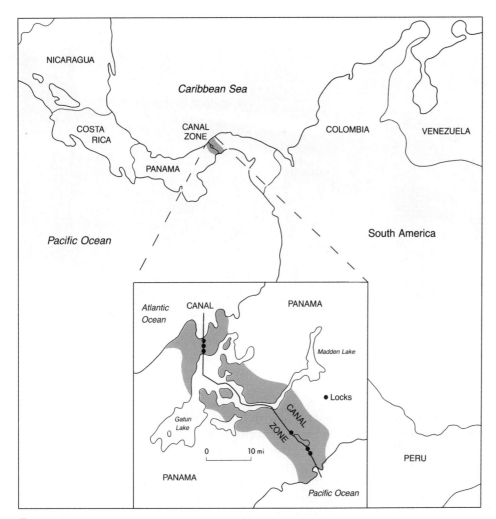

FIGURE 10.26

The Panama Canal. The canal contains six locks for raising and lowering ships. It averages about 40 feet (19 meters) in depth, but the waterway includes parts of lakes. The canal itself varies between 100 (30.48 meters) and 300 (91.44 meters) feet in width.

Most of the tourists visiting Middle America are from the United States, and naturally the places that are the most accessible gain the most revenue from travelers. Puerto Rico, with close political, social, and economic ties to the United States, received $660 million in 1984 from out-of-state visitors—nearly $200 for every person in the country. Cuba was once the mecca of the Caribbean for attracting tourists from the United States, but few Americans have visited there during the nearly 30 years since the Communist takeover. The loss to Cuba was a gain for the Bahamas, where tourism is a leading source of income. Lying within 60 miles (96 km) of the Florida coast, the Bahama Islands enjoy visitors that come on regular excursion boats and daily plane flights from Miami.

Mexico leads the region in tourist income. It lies adjacent to the United States and benefits greatly from the spirited tourism of the *Norteamericanos*. The very long common border, and the fact that Americans can cross the border in their automobiles, gives a decided advantage to the

Middle America 295

This dock is at Limón in Costa Rica. Compare the shipping methods here with those shown in the previous photograph from Mexico. Here the containers are all large and of the same size (note the line of trucks waiting to unload their containers onto the shop).

Mexican tourist trade. Among Mexico's many tourist attractions are the historic ruins of the Mayan, Toltec, and Aztec cultures, as well as the recent commercial developments at Cancun and Acapulco.

Trade and Transportation

The major trade partner for most of the countries of Middle America is the United States. More than half the import and export dollars for Mexico, Honduras, Panama, the Dominican Republic, Haiti, Puerto Rico, and Trinidad are the result of trade with the United States. Mexico is the leading commodity trader of the region, and the leader in amount traded with the United States. About 64 percent of Mexico's imports and 55 percent of its exports are with the United States. Mexico and Cuba are the only countries in Middle America that have a trade surplus—that is, the export revenue exceeds the cost of imports. Numerous countries in Middle America trade with Japan, West Germany, and each other. Other trade ties are more specific and based on historical conditions. For example, Haiti trades with France, Mexico still trades with Spain, and Belize, Jamaica, and Trinidad do business with England.

The focal point of transportation in Middle America is the Panama Canal (Figure 10.26). Completed in 1914 this canal not only connects the coastal commerce of the United States, but also is a major route for Latin American nations and worldwide shipping. Along with the Suez Canal, it is one of the world's most strategic waterways. This vital passage is 50 miles (80 km) long from deep water to deep water and averages 40 feet (12 m) in depth. It is a lock and lake canal that ships traverse in approximately 12 hours.

One would expect that an area where water transportation has been very important would have developed a network of coastal transportation, but this is not the case in Latin America. Coordination between the various countries and long-range planning would be required to establish regular ship or ferry service between the ports of Middle America. The countries have not worked together enough to make long-term plans. Not only is shipping between countries haphazard, but most port facilities in Middle America are extremely poor. Most of the great cities of the

United States are port cities, but in Middle America the great cities are in the highlands.

Air transportation in Middle America is much better coordinated than shipping. All countries except Cuba have connecting lines, and they tie into international systems with South America, Europe, and the United States. Internal systems usually are quite good, and most airlines fly used but well-maintained commercial jets made in the United States. Mexico has 40 local airlines servicing every part of the country. Every Central American state has its own internal airline system. For example, the national airline of Costa Rica is Taca.

Railroad and highway systems in Middle America are not as well developed as air routes. Mexico is the only mainland country with a rail network, and it has laid down about 12,000 miles (20,000 km) of track. The Mexican railroad lines connect with the United States at many points along the border, but the network is most dense in the Mexico City area. Most of the other countries have one railroad line that extends from the highland capital down to a coastal port city. These lines were built at great cost, in terms of both money and human lives, because of the rugged terrain and tropical diseases. Among the Caribbean Islands, only Cuba has any railroad line. About 3,000 miles (4,800 km) of track extends the length of the island with a few feeder lines.

Highways are being constructed throughout Middle America, and bus transportation is replacing some of the railroad carriers. The most promising road, the Inter-American Highway, will someday connect the United States to South America. With the easing of tensions in Nicaragua and El Salvador, the road could become an important tourist route. Nearly complete from Alaska to Colombia, with only a few critical miles remaining in Panama, it already is a strategic road. The road is hard surface and fairly well maintained, and it passes through some of the most spectacular scenery in the world. (See Figure 10.27.)

Most of the cars, trucks, and buses used in Middle America are imported. Many of the cars come from Japan, especially the cheaper, high-gas-mileage vehicles. They are shipped to the west coast port cities of the mainland directly across the Pacific from Japan. Fewer automobiles are found on the Caribbean Islands, but those that are there usually come from the United States, France, or West Germany. Mexico, however, does have automobile assembly plants where they make the older "beetle" version of the German Volkswagen. Used cars from the United States also are popular in Mexico but are rarely seen in any of the other states in the area.

Political Geography of Middle America

As of mid-1986, Middle America had 139 million people, and some countries have the greatest rate of population increase of any in the world. The rates of population increase are much higher for Mexico and Central America than for the islands of the Caribbean. Historically, the Middle American population has grown faster than the population of any other underdeveloped area of the world. This rapid growth has created a young population and a population with pressure on it. The pressure comes not from the localized pockets of very high density, but from the inability of many countries to feed and educate their people.

Middle America remains essentially an agrarian society, although some inroads into industrialization have been made. Puerto Rico, for example, has changed from a poor farming country to one of the wealthiest countries in Middle America, largely through industrial expansion. Mexico has the most potential for industrial expansion in Middle America because it has the best resources. Transportation in Middle America is as varied as the landscape. Modern commercial jets fly over poverty-ridden peasants pulling two-wheeled carts by hand. The streets of Mexico City and Monterrey are choked with traffic, and smog is a critical problem, while the only means of movement along the Mosquito Coast of Nicaragua is by dugout canoe. These contrasts often create political problems both for the local leaders and for the leaders of the countries. When poor, underfed people are able to see the modern luxuries of the rich, demands for land reform and other measures that act to spread the wealth are understandable. When demands are not met soon enough, some countries turn to communism, hoping that a socialist form of government will meet the needs of the people.

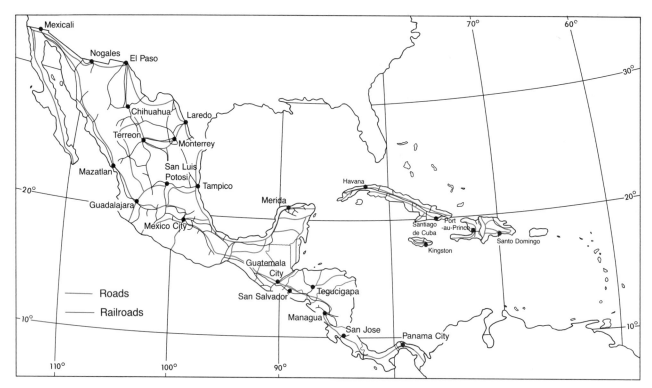

FIGURE 10.27
Roads and Railroads of Mainland Middle America.

Key Words

- Alliance for Progress
- bolsons
- Brownson depression
- Greater Antilles
- Lesser Antilles
- mestizos
- mulatto
- nortenos
- orographic precipitation
- papiamento
- paramos
- Ring of Fire
- tierra caliente
- tierra fria
- tierra templada
- United Fruit Company
- zambo

References and Readings

Anderson, Thomas D. *Geopolitics of the Caribbean: Ministates in a Wider World.* Westport, Conn.: Praeger, 1984.

Benjamin, Thomas, and McNellie, William (eds.). *Other Mexicos: Essays on Regional Mexican History.* Albuquerque: University of New Mexico Press, 1984.

Blakemore, Harold, and Smith, Clifford T. (eds.). *Latin America: Geographical Perspectives.* New York: Methuen, 1983.

Blakemore, Harold, et al. (eds.). *The Cambridge Encyclopedia of Latin America and the Caribbean.* London: Cambridge University Press, 1985.

Blouet, Brian W., and Blouet, Olwyn M. (eds.). *Latin America: An Introductory Survey.* New York: Wiley, 1981.

Boehm, Richard G., and Visser, Sent (eds.). *Latin America: Case Studies.* Dubuque, Iowa: Kendall/Hunt, 1984.

Bolland, O. Nigel. *Belize: A New Nation in Central America.* Boulder, Colo.: Westview, 1986.

Brown, Peter G., and Shue, Henry (eds.). *The Border That Joins: Mexican Migrants and U.S. Responsibility.* Totowa, N.J.: Rowman & Littlefield, 1983.

Butland, Gilbert J. *Latin America: A Regional Geography.* 3d ed. New York: Wiley, 1972.

Calvert, Peter. *Guatemala: A Nation in Turmoil.* Boulder, Colo.: Westview, 1986.

Carim, Enver (ed.). *Latin America and the Caribbean, 1983.* Essex, Engl.: World Information, 1983.

Carlson, Fred A. *Geography of Latin America.* New York: Prentice-Hall, 1943.

Central Intelligence Agency. *The World Factbook.* Washington, D.C.: Government Printing Office, 1985.

Christian, Shirley. *Nicaragua: Revolution in the Family.* New York: Random House, 1985.

Cole, John P. *Development and Underdevelopment: A Profile of the Third World.* New York: Methuen, 1987.

———. *Latin America: An Economic and Social Geography.* London: Butterworths, 1965.

Crassweller, Robert D. *The Caribbean Community: Changing Societies and U.S. Policy.* New York: Praeger, 1972.

Crow, Ben, and Thomas, Allen (eds.). *Third World Atlas.* Philadelphia: Taylor & Francis, 1984.

Cumberland, Charles C. *Mexico: The Struggle of Modernity.* New York: Oxford University Press, 1968.

Davidson, William V., and Parsons, James J. (eds.). *Historical Geography of Latin America.* Baton Rouge: Louisiana State University Press, 1980.

Diagram Group. *Atlas of Central America and the Caribbean.* New York: Macmillan, 1985.

Dickenson, John P., et al. *Geography of the Third World.* New York: Methuen, 1983.

Dominguez, Jorge I. (ed.). *Cuba: Internal and International Affairs.* Beverly Hills, Calif.: Sage, 1982.

Dozier, Craig L. *Nicaragua's Mosquito Shore: The Years of British and American Presence.* Tuscaloosa: University of Alabama Press, 1985.

Falk, Pamela S. *Petroleum and Mexico's Future.* Boulder, Colo.: Westview, 1986.

Feinberg, Richard E. (ed.). *Central America: International Dimensions of the Crisis.* New York: Holmes & Meier, 1982.

Flagg, John Edwin. *Cuba, Haiti, and the Dominican Republic.* Englewood Cliffs, N.J.: Prentice-Hall, 1965.

Gibson, Lay James, and Renteria, Alfonso Corona (eds.). *The U.S. and Mexico: Borderland Development and the National Economies.* Boulder, Colo.: Westview, 1985.

Greenbie, Sydney. *The Central Five: Guatemala, Honduras, El Salvador, Nicaragua, and Costa Rica.* New York: Row, Peterson, 1943.

Hall, Carolyn. *Costa Rica: A Geographical Interpretation in Historical Perspective.* Boulder, Colo.: Westview, 1985.

Hamilton, Nora, and Harding, T. F. (eds.). *Modern Mexico: State, Economy, and the Social Conflict.* Beverly Hills, Calif.: Sage, 1986.

Harman, Carter. *The West Indies.* New York: Time Inc., 1966.

James, Preston E., and Minkel, Clarence W. *Latin America.* 5th ed. New York: Wiley, 1986.

Johnson, William Weber. *Mexico.* New York: Time Inc., 1966.

Latin America. New York: Americana, 1943.

Lavine, Harold. *Central America.* New York: Time Inc., 1964.

Levy, Daniel C., and Szekely, Gabriel. *Mexico: Paradoxes of Stability and Change.* Boulder, Colo.: Westview, 1987.

MacPherson, John. *Caribbean Lands.* New York: Longman, 1980.

Meso-Lago, Carmelo. *The Economy of Socialist Cuba.* Albuquerque: University of New Mexico Press, 1981.

Mintz, Sidney W., and Price, Sally (eds.). *Caribbean Contours.* Baltimore: Johns Hopkins University Press, 1985.

Morris, Arthur S. *Latin America: Economic Development and Regional Differentiation.* Totowa, N.J.: Barnes & Noble, 1981.

Needler, Martin C. (ed.) *Political Systems of Latin America,* New York: Van Nostrand Reinhold, 1970.

Odell, Peter R., and Preston, David A. *Economies and Societies in Latin America: A Geographical Interpretation.* New York: Wiley, 1973.

Parker, Franklin. *The Central American Republics.* London: Oxford University Press, 1964.

Peckenham, Nancy, and Street, Annie. *Honduras: Portrait of a Captive Nation.* Westport, Conn.: Praeger, 1985.

Riding, Alan. *Distant Neighbors: A Portrait of the Mexicans.* New York: Knopf, 1985.

Sanderson, Susan W. *Land Reform in Mexico, 1910–1980.* Orlando, Fla.: Academic Press, 1984.

Scott, Ian. *Urban and Spatial Development in Mexico.* Baltimore: Johns Hopkins University Press, 1982.

Sealey, Neil. *Tourism in the Caribbean.* London: Hodder & Stoughton, 1982.

Siemans, A. H. "Wetland Agriculture in Pre-Hispanic Mesoamerica," *Geographical Review* 73 (1983): 166–181.

Tata, Robert J. *Haiti: Land of Poverty.* Washington, D.C.: University Press of America, 1982.

Veliz, Claudio (ed.). *Latin America and the Caribbean.* New York: Praeger, 1968.

Walker, Thomas W. *Nicaragua: The Land of the Sandino.* Boulder, Colo.: Westview, 1982.

———. (ed.). *Nicaragua: The First Five Years.* Westport, Conn.: Praeger, 1985.

West, Robert, and Augelli, John. *Middle America: Its Lands and People.* Englewood Cliffs, N.J.: Prentice-Hall, 1966.

Chapter 11

SOUTH AMERICA

Introduction

South America is a continental block, somewhat triangular in shape, consisting of about 6,875,000 square miles (17,875,000 km^2) of land area. The continent is bulky in the north, with the widest part between the equator and 10 degrees south latitude, and then tapers to a point at Cape Horn in the extreme south. South America is not directly south of North America; most of the continent lies east of a line drawn due south of Miami, Florida (see Figure 11.1).

Another north-south line has historical significance with regard to the settlement of South America. The **Treaty of Tordesillas** between Portugal and Spain (1494) gave Portugal the right to all lands discovered east of a line drawn north and south 370 leagues west of the Cape Verde Islands, just off the west coast of northern Africa. This line was approximately the same as 50 degrees west longitude. As a result, it was the Portuguese who explored and colonized the eastern coast of South America, most notably the area that is now Brazil. The Spanish moved into the western part of South America from Middle America. Spain and Portugal were neighbors on the Iberian Peninsula in Europe, and their colonies became neighbors on the continent of South America.

The countries of South America consist of 12 republics and one European "department" (see Figure 11.2). These countries are: Argentina, Brazil, Bolivia, Chile, Colombia, Ecuador, French Guiana (Department of France), Guyana, Paraguay, Peru, Surinam, Uruguay, and Venezuela. These political units are today roughly the same in size and shape as the subdivisions of the original Spanish colonial governments. After the successful nineteenth-century revolts against Spain and Portugal, the young political entities were established, and there have been only a few boundary adjustments since. The most dramatic boundary changes affected Bolivia, which lost half its original territory through wars with its neighbors—Peru, Chile, Argentina, Paraguay, and Brazil.

The Spanish colony of New Granada—which consisted of present-day Colombia, Panama, Venezuela, and Ecuador—won independence from Spain in 1819 after 300 years of Spanish rule. Brazil won independence from Portugal in 1822, and by 1830 most of the other countries of South America had become independent. Venezuela and Ecuador broke away from New Granada in 1829, but Panama did not withdraw until 1903. The countries of South America that became independent most recently are the small nations of Guyana (1966) and Surinam (1975) located on the northeast coast of the continent. French Guiana, located on the same coast and adjacent to Surinam, is the only South American country today that is not an independent nation. French Guiana is administered by a prefect and sends one senator and one deputy to the French Parliament in Paris.

Peru and Bolivia are the heart of the old **Inca Empire.** The Inca state originated on the shores of Lake Titicaca and had its capital 200 miles (320 km) northwest of Cuzco. The Spanish conquerors built the new city of Lima near the ocean and made it the administrative headquarters of the western Andes region. Following independence, Lima became the capital for Peru, and Bolivia developed around the population nucleus on the high plateau south of Lake Titicaca. Chile developed in the narrow valley and western slopes between the Andes and the Pacific. It is a long, narrow country stretching for 1,500 miles (2,400 km) between Peru and the Drake Passage.

Uruguay was established as a buffer state between the Portuguese-held Brazil and the Span-

Treaty of Tordesillas After Columbus's first voyage of discovery, and to prevent a clash between Portugal and Spain, the pope drew an imaginary line through the Atlantic Ocean from north to south. He awarded Portugal the lands on the African side of the line, and Spain the lands on the western side. The line was drawn 100 leagues (one league is about 3 miles) west of the Azores and the Cape Verde Islands. In 1494, with the Treaty of Tordesillas, the line was redrawn at 370 leagues west of the Cape Verde Islands. Portugal retained rights to the land east of the line, and Spain retained the land to the west. Spain and Portugal maintained their obligations faithfully, and no serious conflicts arose between them. The Protestant countries of Northern Europe, however, did not recognize the pope's right to make gifts of the New World lands, so British and Dutch explorers ignored the "line of demarcation."

Inca Empire The Inca state had reached and passed its zenith of development before the arrival of the Europeans, but at one time it extended along the west coast of South America from central Ecuador, through Peru, and as far south as the present location of Santiago, Chile. The nucleus of the empire was in the Cuzco Basin in southern Peru. The empire was formed through the conquest and assimilation of numerous separate Indian groups and eventually had political control over a large land area.

ish colony of Argentina. Paraguay was based on early missionary developments along the Paraguay River and became one of the most isolated nations in South America. Argentine colonists won independence from Spain in 1819, when a long period of disorder in Argentina finally came to an end and a strong central government was established. Large-scale European immigration, especially from Italy, Spain, and Germany, occurred after 1880 and helped make Argentina one of the most modern, prosperous, educated, and industrialized nations of Latin America.

■ *Physical Geography of South America*
Landforms

The landscape of western South America (Figure 11.3) is dominated by the Andes Mountains, a

This Peruvian Indian woman is a descendant of the ancient Incas and lives much like her ancestors did. Photograph by Colin Hagarty.

population of Argentina and Uruguay (80–90 percent) and are an influential minority everywhere else. The mestizos also are widely scattered, but they tend to inhabit the *templada* zones of the highlands and the central valley of Chile. The population of Chile is about 25 percent of Spanish descent and 75 percent mestizo. Mestizos also make up the majority in Venezuela and Colombia. Indians predominate in Ecuador and comprise 80 percent of the populations of both Bolivia and Peru.

Many blacks occupy the tropical zones of South America. Brazil is strongly negroid, with about 11 percent pure black and about 41 percent mulatto and zambo. The Guianas have a large number of blacks and mulattoes, but the dominant group in Surinam and Guyana are the descendants of the indentured servants brought from India. French Guiana is the site of the infamous French penal colony Devil's Island, but the prison was closed in 1944 and its 2,800 inmates were returned to France. Pure Frenchmen are therefore not common in the country.

The population density pattern of South America (Figure 11.5) is fairly simple. Most of the people live along the rim of the continent, while the interior is essentially empty. In the west, the main centers of settlement are in the highlands from Venezuela to Bolivia. In Chile, where temperatures become cooler, the people live in the low central valleys between the Andes and the coast. On the east side of the continent, habitation is near the ocean, where onshore breezes cool the air. From Venezuela to the Amazon River, only scattered clusters of people are found, but from the Amazon south to the pampas of Argentina, relatively dense populations occupy a broad, continuous strip of land. South of the pampas, arid climates and sparse vegetation again limit settlement to small clusters.

The rate of annual increase for the population of South America is 2.4 percent. This varies from 2.8 percent for Bolivia, Ecuador, and Paraguay to 0.9 percent for Uruguay. The tropical countries generally have greater rates of increase than the temperate countries. The birth rate in Bolivia is the greatest for the continent, at 43 live births each year for every 1,000 people. However, the infant mortality rate—that is, the number of children under the age of one year that die each year—is 119 out of every 1,000. This reduces the life expectancy for the country to only 51 years. Again, life expectancies in the temperate countries of Argentina, Chile, and Uruguay are longer than they are in the tropical regions.

The population of Brazil is by far the largest on the continent, at 143.3 million. On the other end of the scale is French Guiana, with only 66,800 people in the entire country. Table 11.1 contains some population statistics for all of South America.

Urban Centers, Towns, and Villages

South Americans are mostly urban-dwelling people. They insist on congregating in the cities despite governmental inducements to keep them on the land. Although the urban slums are noted for their squalor and poverty, they are more appealing than the even worse conditions in the rural areas, because the potential for economic improvement is perceived to be better in the urban areas.

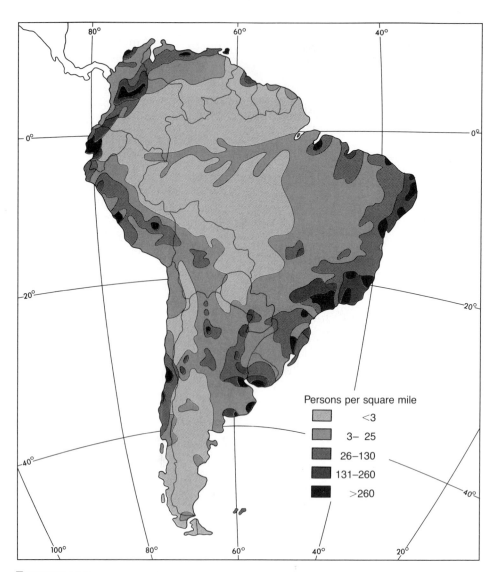

FIGURE 11.5

Population Patterns of South America.

In the mid-1920s the population of South America was one-third urban, and by 1965 more than half the people lived in urban areas. The rural to urban migration trend continues today, but at a more rapid rate. In 1975 some 57 percent of the people were urban-dwellers, and by 1985 the figure had increased to 64 percent. This migration tendency, coupled with the rapid population growth, has produced some very large cities in South America.

For many years, Buenos Aires, Argentina, was the largest city in South America, but during the 1970s and early 1980s a number of other cities surpassed it. The population growth rate of Buenos Aires has stagnated compared with other Latin cities because the rural to urban migration in Argentina is less than in other countries and because the rate of population increase is less (see Table 11.1). Currently, the two largest cities in Brazil are the largest in South America: São Paulo

TABLE 11.1

Population Statistics for South America, 1986

COUNTRY	POPULATION	BIRTH	DEATH	RATE OF INCREASE
Argentina	3.10	24	8	1.60
Bolivia	6.40	43	15	2.80
Brazil	143.30	31	8	2.30
Chile	12.30	22	6	1.60
Colombia	30.00	28	7	2.10
Ecuador	9.60	36	8	2.80
French Guiana	0.07	29	9	2.50
Guyana	0.80	28	6	2.20
Paraguay	4.10	35	7	2.80
Peru	20.20	35	10	2.50
Surinam	0.40	28	8	2.00
Uruguay	3.00	18	9	0.90
Venezuela	17.80	33	6	2.70

SOURCE: *World Population Data Sheet, 1986* (Washington, D.C.: Population Reference Bureau, 1986).

Most Latin American cities have a central plaza area surrounded by government buildings, a Catholic church, and some shops. This is the Plaza des Armes in Arequipa, Peru. Photograph by Colin Hagarty.

South America 315

(7,033,529) and Rio de Janeiro (5,093,232). Santiago, Chile (4,085,300); Bogotá, Colombia (3,800,000); and Lima, Peru (3,158,417) also now rank higher than Buenos Aires (2,908,000) in total metropolitan population, but Argentina remains the most urbanized (86.3 percent) country in South America.

Not all the people live in the large cities of South America (Figure 11.6). Tens of thousands of small towns and even smaller villages dot the landscape, especially in agricultural areas. Latin Americans tend to prefer to live in groups, and even the farmers live in villages. Most towns have the same general plan because during the Spanish colonial period the laws of the Indies were very specific about how towns were to be laid out. The center of the Latin town is a square (see Figure 11.7). On one side of the square is the church, on

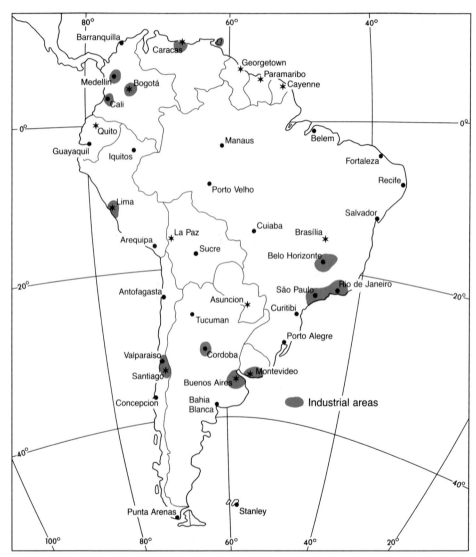

FIGURE 11.6

Industrial Areas and Major Cities of South America.

316   Chapter 11

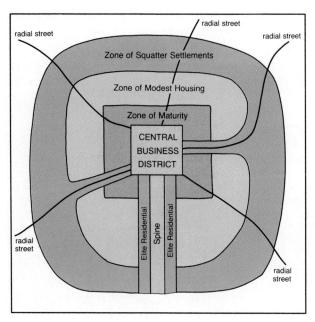

FIGURE 11.7

The Latin American City. Many cities of Latin America have similar internal land-use structures that differ from other cities of the world. The commercial strip (spine) leading from the central business district (CBD) and surrounded by the elite residential sector is the major unique characteristic. The zone of older houses (maturity) around the CBD usually contains the second-best housing, after the elite residential sector. The next zone outward contains the lower-class housing. The final zone outward from the CBD contains the slums of the squatters, which usually penetrate the other zones toward the CBD along some streets.

another side are the government buildings. The square is the gathering place for the people of the community, and in some cases the roads radiate from the square like spokes on a wheel. People meet in the square to visit, and the square is the focus of activities. Nearly every Saturday afternoon and evening, young people promenade in a dignified precourtship ritual around the square. One day a week is designated as market day, when farmers bring surplus produce to the square and display it in colorful arrangements. Socializing is as much a part of market day as buying and selling. The church is usually open for religious activities, and local law enforcement officers are visually prominent. For the local people, the town square is the center of the political, social, economic, and religious lives.

The general movement of the population is from the farm villages to the towns and from the towns to the cities. With people arriving in the cities at such a rapid rate, both by birth and by immigration, the pressure on the urban areas is immense. All cities are ringed with blocks and blocks of slum areas, where squatters have built their own hovels. Housing in these areas consists of shacks erected from paper boxes, tin, wood scraps, or other throw away materials. The percentage of family dwellings that consist of only one room is a measure of the habitation pressure. In Canada and the United States, this percentage is less than one percent, but in Paraguay and Ecuador, for example, it is 45 percent of all the homes in the two countries. Only 6 percent of the homes in Paraguay have water piped to them. These conditions are typical of the urban slums too, and the awful conditions keep social and political unrest constantly festering.

Language and Religion

The Spanish and Portuguese came to the Americas firmly convinced of the rightness of their cultures, and they did not shrink from leaving their imprints wherever they ruled. Officially, South America is Spanish-speaking everywhere except in Brazil, which is Portuguese-speaking, and in the Guianas, where the 50 percent Asian population speaks an Indo-Aryan tongue from India. English and French also have been introduced and are widely spoken in Guyana and in French Guiana.

On an unofficial basis, the language pattern is more complicated. More than 100 Indian languages are spoken, mainly in the tropical rainforests and in the Andean mountain communities. As in Middle America, the official language of Spanish is not always the most common among the natives, and accents vary so much that communication between people who live in different areas can be difficult.

The religious pattern of South America is similar to that for language. The Iberian conquerors were Roman Catholic and tolerated no other

Because of the high rates of illiteracy, letter writers can make a living in the streets of Cuzco, Peru. Photograph by Colin Hagarty.

religion. The French too brought the Catholic religion to French Guiana. Thus, the prominent religion in the entire continent is Roman Catholic. The only exception occurs in Surinam and Guyana, where some religious mixtures exist, including among the Asian groups in Guyana that are Hindu and Muslim.

But the worship of South American Roman Catholics is not the same everywhere. The Roman Catholic church incorporated many of the existing religious activities of the Indians into Christian customs and holidays. As a result, many practices dating back to the Incas, such as religious rites held at planting time, are continued as Christian ceremonies. The Indians also incorporated the new Roman Catholic ceremonies into their older rituals. For example, some Indians still maintain houses for the dead and conduct ceremonies in which they carry mummies and supply them with food, but also regularly attend Catholic church services. In other places, such as Brazil, Roman Catholicism is combined with various forms of voodoo rituals, such as walking on hot coals. Thus, while there appears to be religious uniformity in South America, there is actually much variation from place to place.

■ Economic Geography of South America

Agriculture

About one-third of the labor force in South America is engaged in agriculture. The highest ratios are in Ecuador, Bolivia, Paraguay, and Peru. Brazil, the giant of agriculture, reported a 50 percent agricultural labor force in 1965, but by 1975 the figure had dropped to 43 percent and in 1985 the country reported a figure of 30 percent. Other South American countries exhibit similar declines in the proportion of farm workers (see Table 11.2). Chile has the lowest percentage of farmers on the continent—only 9 percent of workers in that country are farm laborers. These figures tend to confirm the urbanization impulse discussed earlier. It should be noted, however, that families in rural areas are seldom dedicated entirely to farming,

TABLE 11.2
Percentage of Labor Force Engaged in Agriculture, 1975–1985

COUNTRY	% IN AGRICULTURE, 1985	% CHANGE SINCE 1975
Argentina	19	−4
Bolivia	47	−7
Brazil	30	−9
Chile	9	−5
Colombia	26	−2
Ecuador	52	−8
French Guiana	18	−5
Guyana	34	−8
Paraguay	44	−4
Peru	40	−9
Surinam*	—	—
Uruguay	16	−4
Venezuela	16	−9
Total South America	33	−8

*Data for Surinam not available.
SOURCE: Central Intelligence Agency, *The World Factbook, 1985* (Washington, D.C.: Government Printing Office, 1986).

or hybrid seeds, so to minimize the possibility of crop failure, which would mean starvation, they usually plant a variety of crops, often in the same field. Many of the crops had been domesticated by the precolonial farmers of South America. Potatoes and corn (maize) are both native to the Americas, and the peasant farmers still use many varieties.

The traditional farming areas are located in the countries where the percentage of the labor force engaged in agriculture is high (see Table 11.2), generally in the Andes from southern Colombia to northern Chile. But another large area of subsistence farming is the tropical forest of the lowlands, where native Indians clear small plots along the

except in areas where large commercial farms exist. It is common for farmers also to engage in weaving or trade or to work at other jobs during parts of each year.

Although the percentage of the total labor force engaged in agriculture is declining in South America, the total agricultural output continues to increase. Recent United Nations reports indicate that every country of South America except Argentina has increased its agricultural output over the last 10 years. The greatest increase has been for Venezuela, followed closely by Bolivia, Brazil, and Colombia. The agricultural areas of South America are shown in Figure 11.8.

The primary concern of most farmers in South America is not with producing a surplus for trade or for sale. Although farmers do sometimes produce some crops for trade at the local markets, their first concern is to produce enough food for the family to eat. The methods of production these small-scale peasant, subsistence farmers use are based on long experience and tradition. Most such farmers cannot afford to buy commercial fertilizers

Coffee is the premier crop of the highlands of South America. These plants are growing in the mountains of Colombia.

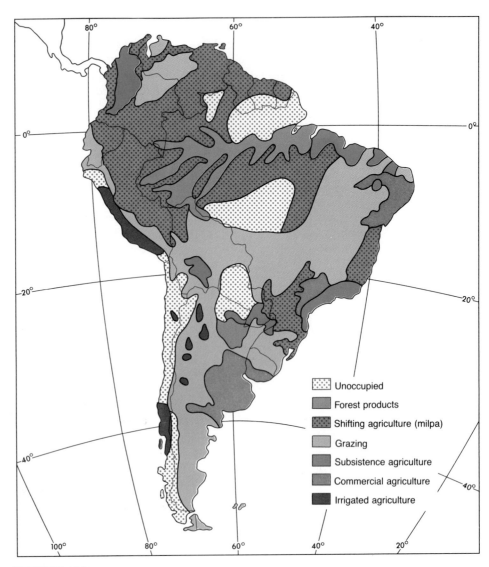

FIGURE 11.8

Agricultural Regions and Land Use of South America.

rivers in a type of agriculture referred to as *milpa* in Mexico and *roza* or *purma* in South America. They use a "slash and burn" system, where the trees are cut down (slashed) and burned (see Figure 11.9). Crude tools are then used to plant the land, and meager yields of corn, beans, and such root crops as cassava (manioc) are produced. Because the constant rainfall leaches the soil nutrients out quickly, the fields must be abandoned within two to three years, and new land must be cleared. It takes the tropical rainforest anywhere from 15 to 100 years to return to the condition it was in before the tall trees were cut down. Thus, a particular field may be used only once during two or three generations of the Indians. This type of agriculture is very destructive to the natural environment, but it has been going on for thousands of years and can work quite well if population densities remain low.

Besides the traditional types of subsistence agriculture, South America also has large areas of commercial agriculture. In the tropical, northern

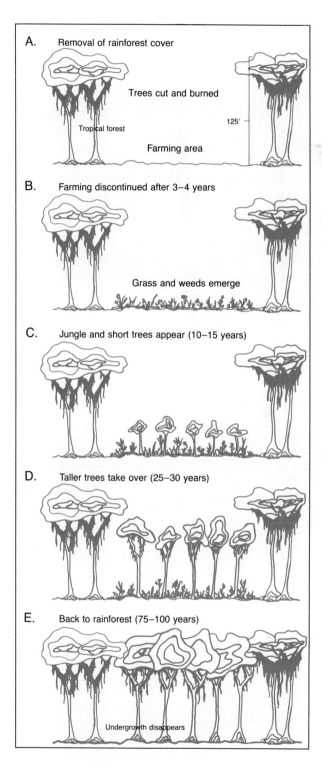

FIGURE 11.9

Forest Sequence After Slash-and-Burn Agriculture. The trees of the tropical rainforest are cut down and burned (A). After three or four years of farming, the plot is abandoned and weeds and grasses emerge (B). After 10 to 15 years, jungle vegetation and some trees appear (C). After 25 to 30 years, the taller trees take over and the jungle declines (D). Finally, after 75 to 100 years, the rainforest is back to its original state, and the undergrowth disappears (E).

the temperate climate regions of Europe and North America. The plantations are located near the coast in Venezuela, Colombia, Brazil, and the Guianas. These agricultural establishments are usually financed by outside capital and imported technology, administration, equipment, and sometimes food. They use local labor, and the trend is to utilize local management.

Typical plantation crops found in tropical South America are cotton, cacao, tobacco, sugar cane, and bananas. Coffee, which is different from the other crops in that it is a highland crop, is produced partly on plantations and partly on small farms. It is the major crop in the mountains of Colombia, Venezuela, and the Brazilian Plateau. For many years, coffee was the number-one export item for Brazil in terms of volume, but it has recently been replaced by soybeans. However, coffee still accounts for 54 percent of the value of Colombia's exports.

Bananas are a consistent source of income for the tropical coastal areas, but the crop is subject to the vicissitudes of weather, labor shortages, disease, and market prices. As a consequence, income from this crop fluctuates from year to year—for example, bananas accounted for about 50 percent of Ecuador's exports in 1975, but by 1985 the percentage had dropped to 9 percent. Sugar cane is grown primarily along the east coast, but all tropical countries in South America produce sugar cane. Sugar accounts for 35 percent of Guyana's exports, but Brazil produces more in terms of total volume. Because most of the cotton produced in South America is used in the local textile industries, it does not appear as a major export commodity. Brazil and Colombia are the major producers of cotton.

The second type of agriculture is located in the temperate zone south of the Tropic of Capricorn,

part of the continent, the commercial agriculture is a plantation type, where the product is usually one crop. The primary markets for the products are in

FOCUS BOX

Cocaine

South America is the primary source for most of the cocaine consumed in the United States and in Europe. The consumer population in the United States, estimated at more than 20 million people, spends $35 *billion* a year on the illicit narcotic. The international cocaine industry depends on the harvest of huge amounts of coca leaves cultivated in Bolivia, Brazil, Colombia, and Peru (see Figure 11.A). In numerous regions, coca cultivation has become dominant over other local cash crops and in some places even over food crops.

The primary coca cultivation areas are in the Chaparé and Yungas regions of Bolivia and the upper Huallaga valley of Peru. These three areas alone provide the coca leaves for about 80 percent of the cocaine consumed in the United States. Agriculture of the regions is completely dominated by coca cultivation, and the crop provides local farmers with their major source of income.

Secondary sources of coca leaves are (1) the Cuzco region of Peru, (2) scattered valleys adjacent to the Marañón River near Cajamarca Department in northern Peru, (3) isolated parts of the Llanos region of Colombia, (4) mountain pockets throughout southern Colombia, (5) the eastern-facing Andean slopes and adjacent lowlands near Santa Cruz, Bolivia, and (6) the upper Amazon Basin of Brazil. As antinarcotic law enforcement authorities eradicate the coca plants from one area, production increased in other areas. Eradication campaigns are very unpopular because coca cultivation and leaf-chewing are deeply embedded in the Indian culture of the eastern Andean region and because farmers can make as much as 10 times more for a cocaine crop than for any other crop.

The coca plant is one of the oldest cultivated plants in South America. The *Erythroxylon coca,* the most primitive of the cultivated cocas, is grown extensively in the moist, tropical valleys of the eastern and central slopes of the Andes from Ecuador to Bolivia. Grown at altitudes up to 9,000 feet, the bushy perennial has a productive life of 20 years. The *Erythroxylon novogranatense,* commonly called "Colombian coca," thrives with less moisture and at lower elevations than the primitive coca. Found originally in Central America and along the northern coast of South America, it is cultivated today in the dry mountain valleys of southern Colombia.

Coca leaves are harvested two to four times a year, depending on the age of the plants and the growing conditions. Small plots of less than a quarter-acre in size are common, but in the primary growing areas plantations of 50 to 100 acres have been discovered. It takes about 500 pounds of coca leaves to make one pound of processed cocaine ready for smuggling into the United States.

After the leaves are picked, they are air-dried in small sheds before the numerous processing steps begin. In processing, the dried coca leaves are soaked in small pits filled with kerosene and sodium carbonate. The solution is agitated, sometimes by stomping with bare feet, until a milky, sludgy residue is produced. The mixture is then treated with sulfuric acid and dried to form a paste; a good paste contains about 75 percent cocaine. Sometimes the dried paste is smoked in pipes or cigarettes by the local people in the producing areas, but most of it is treated with ether, acetone, and potassium permanganate to make a base. The final product, cocaine, is made by dissolving the base in ether, filtering it, and treating it again with a mixture of ether, acetone, and hydrochloric acid. The precipitate is dried in ovens, and the resulting white crystalline salt, which is 95 percent pure cocaine, is packaged in one-kilo bricks.

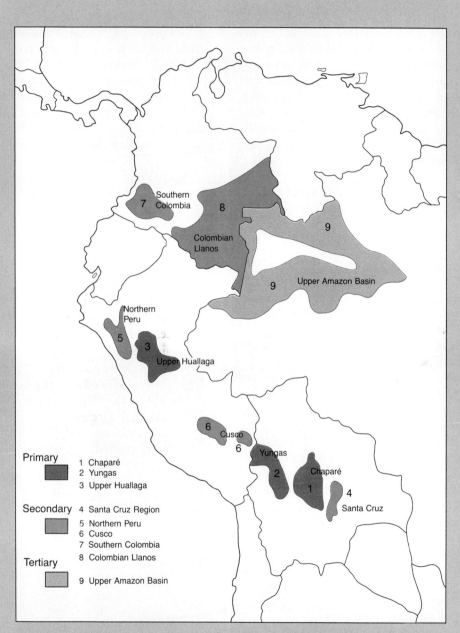

FIGURE 11.A

Coca Cultivation Areas of South America.
SOURCE: Map adapted from an original by Tim Hudson. Used by permission of the American Geographical Society.

Most of the refined cocaine is carried to the United States on private airplanes and ocean ships, but sometimes "mules" (human couriers) use commercial aircraft. Once in the United States, the cocaine is diluted to about 12 percent purity by adding sugar or laxatives. One gram of the diluted cocaine sells for about $150, depending on the supply, but rarely for less than $100. A one-kilo brick of pure cocaine has a street value of more than $1 million after it has been diluted.

Cocaine trafficking is both illegal and highly risky. Smugglers that are caught can expect long prison terms, and narcotic kings kill in order to protect their supplies and markets. With so much money at stake, however, the international cocaine industry is a thriving business.

SOURCE: Tim Hudson, "South American High: A Geography of Cocaine," *Focus* 35 (January 1985): 22–29.

---

where the emphasis is on cattle and sheep in combination with grass and grains, especially corn, wheat, and alfalfa. The area of the pampas around Buenos Aires and in Uruguay is the best example, but this type of agriculture extends from the grasslands of *campos* in southern Brazil to the central valley of Chile.

The Europeans who came to the temperate grassland area of South America found few native Indians. The grasslands were not conducive to hunting, which was the main occupation of the natives, but the Spaniards and the Portuguese found the area ideal for the horses, cattle, and sheep they brought with them. The immigrants soon began large commercial operations and never became involved in subsistence farming. They prospered, and today Argentina, Brazil, and Chile form what are known as the "ABC countries" of South America—the three political and economic powers of the continent. Uruguay, in the same area, is too small to be a major power, but the country's prosperity and standard of living are as high as those of the ABC countries.

Agricultural production in Argentina is located in concentric zones outward from Buenos Aires and the Río de la Plata (see von Thunen, Chapter 4 and Figure 11.10). Immediately surrounding the city, corn is the main crop. Beyond the corn, wheat production forms a semicircle that stretches from the Plata to the Atlantic. The drier land beyond the wheat region is left to grass. Here the prairie slopes for miles westward toward the Andes, and cattle are raised on large *estancias*.

Cocao pods grow on trees and contain the seeds from which cocoa is produced. The seeds are dried, exported, and eventually become a variety of chocolate products.

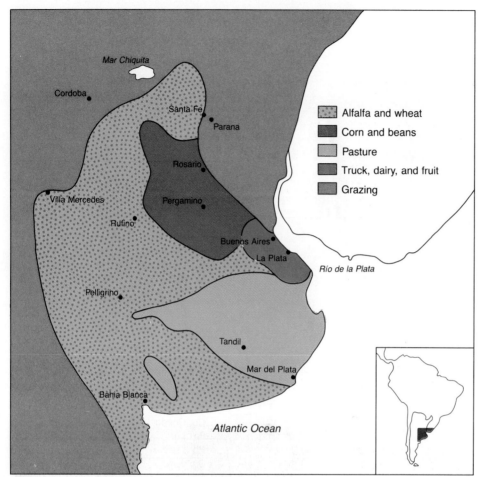

FIGURE 11.10

Agricultural Regions of Argentine Pampa. Argentina has the best-developed agriculture in South America. The crops grown vary with distance from Buenos Aires.

Although many types of animals are produced on these commercial farms, an overwhelming 95 percent of the meat production is beef. Beef is Argentina's leading export item, followed closely by corn. In Uruguay, which is located in a more humid area, sheep are more prevalent. Mutton accounts for 15 percent of that country's meat production, and wool is its leading export item (20 percent of the value of its exports).

The final type of agriculture in South America is not commercial and not completely subsistence. It is carried on mainly by mestizo farmers who are in transition between subsistence farming and commercial farming. This type of agriculture is found north of the commercial grain and meat animal economic region, but it does not reach into the subsistence areas of the high Andes or the tropical rainforest. These farmers grow such crops as corn, beans, and root crops for their own use, but they also grow commodities for sale, such as coffee. The farms are somewhat similar to the old mixed farming types of the North American Midwest, except that the crops grown are more tropical. Most of these farmers own their own

FIGURE 11.11
Mineral Deposits of South America.

land, much of which they received as grants from the government in a manner similar to homesteading in the United States.

Mineral Production

South America has long been a storehouse of mineral resources and is a world leader in the supply of some minerals today (see Figure 11.11). Venezuela has been a noted petroleum producer for years, and virtually 95 percent of that country's export revenue comes from crude oil and petroleum products. This income makes Venezuela one of the richest countries on the continent. Colombia is also a major producer of crude oil, and it comes from the same pool as that for Venezuela—the Lake Maracaibo region. Colombia pipes the crude oil across the Andes to a point near Barranquilla for export. There is also some oil production along the east slope of

World Levels of
ENERGY USE

IN ADVANCED COUNTRIES, the use of electricity is high. Not only do we light our homes and streets with electrical energy, we also use hundreds of types of electric household appliances for such varied things as sharpening pencils to mowing grass. Some homes are heated by electric furnaces, and some people cook and heat water with electricity. Also, some industries require huge amounts of electric energy (such as in refining bauxite into aluminum). On the other hand, people in the economically underdeveloped countries use very little electricity. Nomadic herders in the Sudan, for example, may never use an electrical appliance anytime during their lives. Thus, the production and use of electric energy is a measure of economic development.

World Levels of Energy Use

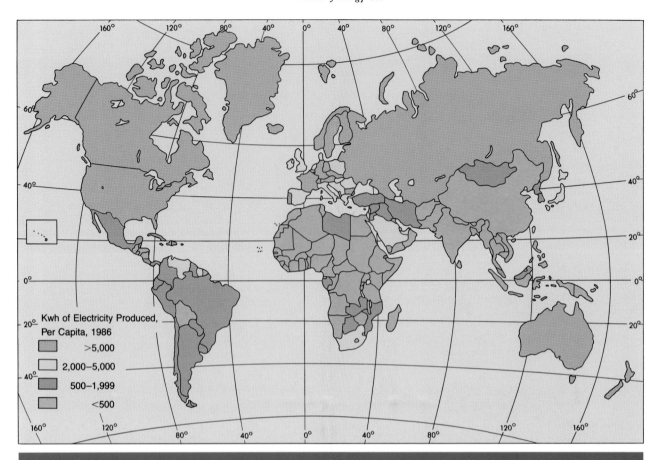

The above map contains one of three criteria used to determine the levels of economic development for the countries of the world, the amount of electricity produced. The categories for the map were computed by dividing the kilowatt hours produced in 1987 in each country by the total population of the country. Thus, the per capita kilowatt hours are shown on the map. The categories are: (1) countries producing more than 5,000 kilowatt hours per person, (2) countries producing between 2,000 and 5,000 kilowatt hours per person, (3) countries producing 500 to 1,999 kilowatt hours per person, and (4) countries producing less than 500 kilowatt hours per person. The range varies from 1 kilowatt hour per person in Lesotho to 25,594 kilowatt hours per person in Norway. Many countries of the **highly developed world** produce over 10,000 kilowatt hours per capita (the United States produces 10,251) but some are surprisingly low. North European countries that rank among the **highly developed** in other categories, but only among the **moderately developed** as far as electricity produced include Ireland and the United Kingdom, The Netherlands, Denmark, and Poland.

World Levels of Energy Use

(Left) The very high levels of energy use in the **highly developed world** is depicted in this photograph of the night skyline of Seattle, Washington. Although the 605-foot-tall Space Needle is very prominent in this view, the Columbia Center (954 ft.) and Seattle 1st National Bank Building (609 ft.) in the background are both taller. Besides all the electricity consumed to light the buildings, thousands of electrical appliances also use electric energy.

(Below) In the **moderately developed world**, one common use of electricity is for public transportation such as for these electric street cars in Poland. Fewer lights and appliances are found here than in the highly developed world.

World Levels of Energy Use

(Above) Many of the people of the **developing world** are without electricity. In this working class district of Mexico City, electricity is a rare commodity. In some cases, squatters illegally tap into nearby electric wires in order to steal energy for their household use.

(Right) In countries of the **underdeveloped world**, most of the people are totally without electricity from either home generators or commercial supplies. The people in this farm village in the Jodphur Rajasthan District of India have no electricity at all. Their night light is entirely from candles and campfires, and they have no electrical appliances to help with their work.

the Andes as far south as Argentina, which is the continent's third largest producer, after Venezuela and Colombia.

Iron ore is a leading product of Brazil, Venezuela, and Chile. Each of these countries uses some ore for local iron and steel production but also exports much of it. Brazilian ore is found in the Minas Gerais hills of the plateau, Chilean ore comes from the Tofo mines in the Atacama Desert, and Venezuela obtains its ore from the Orinoco Basin on the edge of the Guiana highlands near Ciudad Bolívar. Iron ore deposits also have been found in most of the other countries, but most notably in Bolivia and Peru.

Copper is an important mineral in both Chile and Peru. In Chile, copper was discovered in the mid-nineteenth century on the plateau of the Atacama, and that country has been a heavy producer of copper ever since. The main market for the ore is in the United States, and its sale furnishes Chile with 45 percent of its export revenue. During the 1970s, when Chile was experiencing domestic political problems, copper production slowed dramatically, copper dropped to less than 5 percent of the country's total value of merchandise exports in 1972. As copper was making a comeback in Chile in the early 1980s, it was decreasing in value to Peru. In 1975, for example, copper comprised 22 percent of Peru's export revenue, but by 1985 it had dropped to 17 percent. The change occurred because of increased petroleum exporting. Copper, however, is still the major source of export income for both Chile and Peru.

Bolivia is another country that has been dependent on a single mineral for the major part of its exports. Bolivia is the only source of American tin closer than Southeast Asia, and tin amounts to 37 percent of the country's exports. As recently as 1972, tin accounted for nearly three-fourths of the country's exports, but recent increases in the production of petroleum products, especially natural gas, have taken some of the financial burden off tin production. In 1942, when Southeast Asian tin was denied to the United States because Japan controlled the sea lanes of the area, Bolivian tin suddenly became the only source of supply. One

This mining town in mountains outside Cuzco, Peru, indicates that miners are not at all wealthy. They do have entertainment, however, in the form of a local soccer team (note the playing field in the foreground). Photograph by Colin Hagarty.

problem, however, did have to be overcome. No tin-smelting industry existed in the western hemisphere, so a smelter was quickly constructed in the United States to process the Bolivian ore.

Bauxite is mined almost exclusively in Surinam and in Guyana and is marketed in the United States. As the source of aluminum, it is a valuable product. Aluminum is a lightweight, durable, strong, and relatively inexpensive metal used extensively in automobile and aircraft manufacturing as well as in numerous other products. It is the most abundant element in the earth's crust, but concentrated sources of bauxite rich enough for mining are somewhat rare. Besides Surinam and Guyana, Jamaica and Haiti are other major Latin American producers of bauxite, and these four countries produce 22 percent of the world's bauxite.

Other important minerals produced in South America are nitrates and coal. Huge deposits of natural sodium nitrate were discovered in the Atacama Desert in the middle of the nineteenth century. The discovery brought on the War of the Pacific in 1879 between Chile on the one hand and Peru and Bolivia on the other. Victorious Chile took territory from each country and robbed Bolivia of its access to the sea. For many years the Atacama was almost the only source of sodium nitrate, from which nitric acid (used for making dyes and explosives) was made. The nitrate industry thrived for many years, but it declined after cheap synthetic nitric acid was developed. Today, sodium nitrate is not as valuable as it once was, although it is still important as a fertilizer and provides the world with about half its iodine. Coal, an important source of energy, is mostly of medium to poor quality in South America. The best supplies are in Chile and Brazil, where it is used in the local iron and steel industries. Neither pure nitrate nor coal is an important export commodity for any South American country.

Most metals mined in South America are still destined for export, and metals make a larger contribution to the external trade of South America than they do to its internal economy. The uneven distribution of metals means that the mining industry is highly significant in the economic life of some countries and insignificant for others.

Industry

South America is only slightly ahead of Middle America in its industrial development. Both areas have similar handicaps—there is a shortage of skilled labor, a need for capital investments, especially foreign investments, and a decreasing need for manual laborers. Private capital from the outside is difficult to obtain, and trained workers are not eager to migrate to the region. The scarcity of coal and the uneven distribution of other minerals are also barriers to industrial expansion. In spite of these handicaps, however, some countries have expanded their industrial outputs.

Most of the industrial production in South America (see Figure 11.6) is of consumer goods, such as textiles, clothing, beverages, and tobacco products and in food-processing, but there are signs of a potential for development of heavy industry. Five countries have sufficient area, resources, and populations to support an industrial society, and each has high energy production—a good index to industrial capacity. By this index, Argentina leads and is followed in order by Colombia, Brazil, Chile, and Venezuela. Keep in mind that this is only an indication of potential and that these countries still lag far behind the truly industrial states of the world. For example, in 1983 Argentina produced 2.9 million metric tons of steel, while Canada, with a similar population, produced 11.9 million metric tons.

The capacity to produce steel is the basis of an industrial society, and each of the five South American countries mentioned produces some steel. The leader is Brazil, with 12.8 million metric tons produced in 1982. Brazil's Volta Redonda steel plant is the largest such production unit on the continent. It is located on the plateau 60 miles (96 km) north and west of Rio de Janeiro, close to Minas Gerais iron ore and the Brazilian market. The third leading steel producer after Brazil and Argentina is Venezuela, with about 2.0 million metric tons a year. Production is located near Ciudad Bolívar, the navigable river port on the Orinoco. Venezuela's iron ore deposits are located near the city, south of the river. The Chilean iron and steel center is Concepción, located along the coast, where ships can bring coal and take the refined metal to market. Chilean production is about 0.5 million metric tons annually.

import substitution A method, common in Third World countries, of substituting expensive imported finished products, such as automobiles, with locally produced "substitutes" made in assembly plants utilizing inexpensive local labor and imported parts.

Latin American Free Trade Association (LAFTA) An organization of the countries of South America and Mexico begun in 1960 to promote trade among the countries, instead of having each country trade commodities on its own with North America and Europe.

Central American Common Market (CACM) An organization of Central American countries started in 1960 to promote local industrialization and trade among the countries.

Much of the heavy manufacturing in South America is done in assembly plants. This type of industrialization is called **import substitution** because it is a way for Third World countries to make their own consumer goods and substitute them for expensive import items. For example, Chrysler, Fiat, Ford, General Motors, and Mercedes Benz all have motor vehicle assembly plants in Argentina, Brazil, and Venezuela. The largest assembly-type production, however, is the Brazilian Volkswagen industry. Other companies, such as Nissan, Peugeot/Citroen, and Renault, have assembly plants in some of the countries already mentioned, as well as in Chile, Colombia, and Peru (and Mexico). More and more of the components for assembly are being made in the countries where the plants are located. Initially, the motor vehicle assembly plant idea was very expensive and the cost of production was high. For example, locally produced cars in Argentina during the late 1970s were twice the cost of similar cars on the international market level. However, the per-unit cost of production decreased as production increased, because the fixed costs came to be spread over more units. Thus, import substitution works best in countries with large local markets.

The disadvantages of small market size for import substitution industrialization were incentives for the development of economic integration in South America. In 1960 the **Latin American Free Trade Association (LAFTA)** was created. It came to include most of the major countries of South America (plus Mexico) but was unable to stimulate either industrialization or trade. In the same year, 1960, the **Central American Common Market (CACM)** was created, consisting of Guatemala, Honduras, El Salvador, Nicaragua, and Costa Rica, and started with a strong commitment to close economic integration among these countries. Each CACM country was to have a major industrial plant that would serve the markets of all five countries. But industrial expansion was uneven and by 1971 the CACM began to crumble. In 1969, Chile, Bolivia, Peru, Ecuador, and Colombia formed the Andean Pact, which made similar commitments to economic integration. Chile was the main motivator for the formation of the pact, but by 1977 Chile became the first country to leave it, mostly because the member countries registered an overall decline in manufacturing during the first eight years of operation.

Trade and Transportation

In general, trade between countries of South America has expanded to include manufactured items, but the region's world trade is still mostly in primary commodities. Also, trade among the Latin countries has increased, while trade with the United States has decreased. In 1965, for example, 35 percent of South America's imports came from the United States, but by 1981 the percentage had dropped to 27 percent. Other recent tendencies in South American trade have been the decline in both imports from and exports to the European Economic Community (EEC) countries and an increase in the volume of trade with Japan and the Middle East (see Table 11.3). In 1965, five countries in South America received more than 40 percent of their imports from the United States, but by 1985 there was only one country in that category. In 1965, six countries received more than 30 percent of their imports from the EEC countries, but by 1985 none did. During the same 25-year period, however, trade among the countries of South America and with the Middle East and Japan increased significantly.

Containers awaiting transshipment outside the port of Santa Marta, Colombia. Many goods are now shipped by means of such containers.

The destinations for South American exports did not change as much over the 1965–85 period as the origins of the region's imports did. The United States is by far the single leading recipient of commodities from South America, with 35 percent of the 1985 total, in 1965 the United States received 37 percent of the total. In 1985, the EEC countries received about 21 percent of the goods shipped from South America, down from 29 percent in 1965. As noted above, trade among the South American countries has increased, and the change in the export figure is similar to that for imports. Exports to the Middle East and Japan did not change dramatically from 1965 to 1985.

The types of goods shipped from South America have not changed over the last 25 years. Only a few traditional commodities account for most of the exports from South America today, and the same commodities were important in 1965. Some of these commodities and their proportion of the total exports from South America are as follows: petroleum (13 percent), coffee (12 percent), copper

TABLE 11.3

Origin of South American Imports, 1965 and 1985 (in percent of imports)

South American Country	U.S.		EEC		JAPAN		MIDDLE EAST		LATIN AMERICA	
	1965	1985	1965	1985	1965	1985	1965	1985	1965	1985
Argentina	27	18	40	28	4	12	0	3	13	22
Bolivia	44	25	28	17	7	14	0	0	15	13
Brazil	32	25	25	25	5	9	5	17	15	12
Chile	38	23	30	22	3	4	1	6	19	30
Colombia	52	42	26	23	3	9	0	1	6	13
Ecuador	44	37	34	23	4	14	0	1	7	14
Guyana	21	27	46	32	2	4	0	1	13	25
French Guiana*	—	—	—	—	—	—	—	—	—	—
Paraguay	25	12	25	21	8	4	0	1	22	47
Peru	40	31	31	24	5	10	0	1	13	18
Surinam	28	33	26	24	0	5	0	0	15	21
Uruguay	19	9	38	19	2	2	5	20	22	34
Venezuela	54	47	28	26	4	9	0	1	3	7
Total South America	35	27	31	26	4	8	1	4	14	23

*Data for French Guiana are not available.
Note: Percentages do not total 100 because countries are involved in trade with a few other countries not represented on the table.
SOURCE: Direction of Trade, Washington, D.C.: International Monetary Fund, December, 1986. (Used with permission)

(7 percent), sugar (4 percent), beef (4 percent), iron ore (3 percent), cotton (3 percent), bananas (3 percent), corn (2 percent). These nine raw materials and foods account for more than half of all the goods shipped from South America. Table 11.4 contains the first and second most important exports for the countries of South America as of 1982.

Transportation in South America has not been developed to facilitate communication on the continent. The systems that did develop were responses to meet specific needs as they arose. Most of the transportation facilities were constructed from the source of a raw material to the nearest port. The urban centers, with their wide and busy streets, favorably impress the visitor, but the boulevards do not penetrate very far into the countryside. It is also easy to fly from one urban center to another, but small feeder airlines into the interior are inadequate or missing entirely.

In the lowland areas, with the broad rivers, transportation could be cheap and plentiful, but most of the major rivers flow through uninhabited areas. Only the Magdalena and the lower Paraná-Paraguay system flow through populated areas, and they do play important roles in the transportation systems of Colombia and the Plata countries. The Amazon has a 20-foot draft capacity as far upstream as Manaus and is navigated by small ships as far as Iquitos, in Peru. But traffic on the Amazon is limited because of the lack of people in the region. Both the Orinoco and the São Francisco rivers flow through remote areas too. Most of the other rivers of the east coast of Brazil and of the Andean countries are short, broken by rapids, and do not provide access to the interior.

For years the best transport systems in South America were the railroads, but they have declined from their former importance. The only true railroad network was built in Argentina, where lines fan out from Buenos Aires (see Figure 11.12). The rails stretch across the pampas, connecting with Santiago, Chile, on the west and La Paz, Bolivia, on the north. In Argentina, railroad passenger and freight traffic have both declined steadily from a high point in the 1950s. For example, Argentina's railroad traffic in 1950 was 14.5 million passenger-miles and 17.2 million ton-miles of freight, but by 1980 the totals had

TABLE 11.4

Primary and Secondary Export Commodities and Percentage of Total Exports, 1982

COUNTRY	PRIMARY PRODUCT	%	SECONDARY PRODUCT	%
Argentina	Beef	11	Corn	10
Bolivia	Tin	37	Natural gas	35
Brazil	Soybeans	12	Coffee	9
Chile	Copper	45	—	
Colombia	Coffee	54	—	
Ecuador	Petroleum	58	Bananas	9
Guyana	Sugar	36	Bauxite	32
French Guiana*	—		—	
Paraguay	Cotton	38	Soybeans	20
Peru	Copper	17	Petroleum	16
Surinam	Alumina†	55	Bauxite	14
Uruguay	Wool	20	Beef	19
Venezuela	Petroleum	65	Petroleum products	30

*Data for French Guiana not available.
†Alumina is the oxide of aluminum and comes from both bauxite and corundum. Most alumina from Surinam comes from bauxite.
SOURCE: International Monetary Fund, *International Financial Statistics,* November 1983.

FIGURE 11.12
Railroads of South America.

dropped to 11.5 million passenger-miles and 10.8 million ton-miles.

Along the west coast of South America some longitudinal railroad tracks run between Puerto Monte in Chile to Caracas, Venezuela, but the line is not completely connected. Only short spurs feed this railroad, however, and some extensions of the line pass through the capital cities of Santiago, Lima, Quito, Bogotá, and Caracas. La Paz connects to an intercontinental line that crosses the Andes from Buenos Aires. Most railways, however, are short lines from the coast to an interior city or mineral deposit.

The highways of South America are scarce, inadequate, and serve the same areas as the railroads (see Figure 11.13). However, new roads are being built, and the decline in railway traffic has been offset by the increase in motor vehicle traffic, for both passengers and freight. The Pan American Highway has served as a stimulus to the ten

Avianca The national airline of Colombia, which began flying regular passenger routes in 1919. It is the oldest airline in the western hemisphere.

countries along its route to develop better road systems, but road-building is expensive, especially for these poor countries. Typically, travel between towns takes place on buses and trucks, which bounce and careen along the uneven surface with passengers, cargo, and domestic animals all in one enclosure. Dust, noise, and stops to clear the road of rocks or trees are to be expected until the vehicle reaches the outskirts of a city, where smooth going on concrete streets is common.

Air transportation developed rapidly in South America after World War I. Commercial air services started in 1919 in Colombia, where **Avianca** claims to be the oldest such service in the western hemisphere. By 1927, both Bolivia and Brazil had airlines, and other countries have followed. Foreign airlines, such as Pan American, also helped provide air service to many Latin cities, and during World War II the United States built many airstrips in northern South America that have since provided airports for civilian use. Air transport provides both speed and the ability to overcome the huge distances and the barriers of the mountains and rainforests. In addition to the development of scheduled flights between the major cities, airplanes have provided access to remote areas that were impossible to reach by land or by water. In the past three decades the airplane has done much to reduce isolation in South America.

Political Geography of South America

Coups d'état, dictatorships, and government by military junta unfortunately have been an all-too-common way of life in much of South America. Making democracy work is a major problem in many countries. Peru has had 12 constitutions, Bolivia 14, and Ecuador 17. Most of the military activity in South America has been fashioned to contain interior uprisings rather than to engage in

This modern, hard-surface road was built through the tropical rainforest of northern Colombia. The rain, shown in the picture, occurs nearly every day.

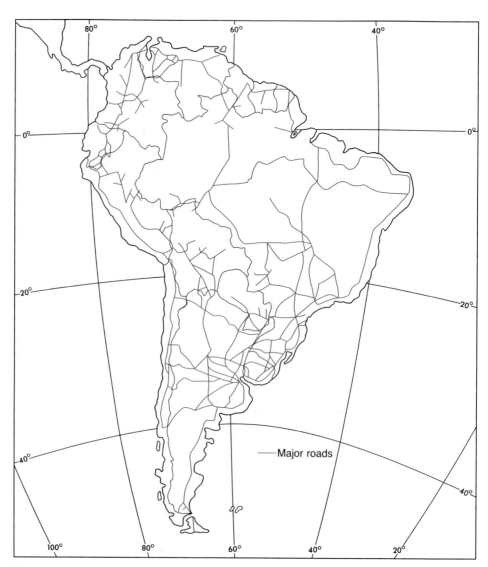

FIGURE 11.13

International Highways of South America.

hostilities with other nations. However, there have been boundary quarrels, especially in the Andean countries, ever since independence from Spain. Peru and Ecuador have fought over their boundary, Chile took territory from both Peru and Ecuador, Bolivia has lost land to Brazil and Paraguay, and Ecuador has been reduced in size by cessions to Brazil, Colombia, and Peru.

The political climate in South America has been a major factor in keeping the continent in the Developing World. As long as the unstable political conditions exist internally, the economic conditions cannot attract help from outside. In addition, the historical factors and exploitations have created an authoritarian and highly polarized society, and there are few indications that this will change in the near future. Most countries will continue to have a few wealthy people with political clout, and many very poor people who are politically impotent. With all the internal problems, none of the countries of South America is a major force in world politics.

Key Words

Avianca
Central American Common Market (CACM)
Cerro Aconcagua
import substitution
Inca Empire
isohyet
Latin American Free Trade Association (LAFTA)
tierra caliente
tierra fria
tierra templada
Treaty of Tordesillas

References and Readings

Aguilar, Luis E. *Latin America, 1984.* Washington, D.C.: Stryker-Post, 1984.

Bishop, Elizabeth. *Brazil.* New York: Time Inc., 1967.

Blakemore, Harold, and Smith, Clifford T. (eds.). *Latin America: Geographical Perspectives.* New York: Methuen, 1983.

Blouet, Brian W., and Blouet, Olwyn M. (eds.). *Latin America: An Introductory Survey.* New York: Wiley, 1982.

Boehm, Richard G., and Visser, Sent (eds.). *Latin America: Case Studies.* Dubuque, Iowa: Kendall/Hunt, 1984.

Braveboy-Wagner, Jacqueline A. *The Venezuela-Guyana Border Dispute: Britain's Colonial Legacy in Latin America.* Boulder, Colo.: Westview, 1984.

Bromley, Rosemary D. F., and Bromley, R. *South American Development: A Geographical Introduction.* New York: Cambridge University Press, 1982.

Butland, Gilbert J. *Latin America: A Regional Geography.* 3d ed. New York: Wiley, 1972.

Carlson, Fred A. *Geography of Latin America.* New York: Prentice-Hall, 1943.

Central Intelligence Agency. *The World Factbook, 1985.* Washington, D.C.: Government Printing Office, 1985.

Child, Jack. *Geopolitics and Conflict in South America: Quarrels Among Neighbors.* New York: Praeger, 1985.

Cole, John P. *Latin America: An Economic and Social Geography.* London: Butterworths, 1965.

Dickenson, John P. *Brazil.* New York: Longman, 1983.

Enders, Thomas O., and Mattione, Richard P. *Latin America: The Crisis of Debt and Growth.* Washington, D.C.: Brookings Institute, 1984.

Fearnside, Philip M. *Human Carrying Capacity of the Brazilian Rainforest.* New York: Columbia University Press, 1986.

Ferguson, J. Halcro. *The River Plate Republics: Argentina, Paraguay, Uruguay.* New York: Time Inc., 1968.

Forbes, Dean K. *The Geography of Underdevelopment.* Baltimore: Johns Hopkins University Press, 1984.

Gauhar, Atlaf (ed.). *Regional Integration: The Latin American Experience.* Boulder, Colo.: Westview, 1986.

Gilbert, A., et al. (eds.). *Urbanization in Contemporary Latin America.* New York: Wiley, 1982.

Griffin, Ernst C., and Ford, Larry R. "A Model of Latin American City Structure." *Geographical Review* 70 (October 1980): 397–422.

Gunther, John. *Inside South America.* New York: Harper & Row, 1967.

Gwynne, Robert N. *Industrialization and Urbanization in Latin America.* Baltimore: Johns Hopkins University Press, 1986.

Hames, Raymond B., and Vickers, William T. (eds.). *Adaptive Responses of Native Amazonians.* New York: Academic Press, 1983.

International Financial Statistics, 1982. New York: International Monetary Fund, 1983.

James, Preston E., and Minkel, Clarence W. *Latin America.* 5th ed. New York: Wiley, 1986.

Johnson, William Weber. *The Andean Republics: Bolivia, Chile, Ecuador, Peru.* New York: Time Inc., 1965.

Latin America. New York: Americana Corp., 1943.

Lombardi, John V. *Venezuela: The Search for Order, the Dream of Progress.* New York: Oxford University Press, 1982.

MacEoin, Gary. *Colombia and Venezuela and the Guianas.* New York: Time Inc., 1965.

Maos, Jacob O. *The Spatial Organization of New Land Settlement in Latin America.* Boulder, Colo.: Westview, 1984.

Mertz, John D., and Meyers, David J. (eds.). *Venezuela: The Democratic Experience.* New York: Praeger, 1986.

Miller, E. Willard, and Miller, Ruby M. *South America: A Bibliography on the Third World.* Monticello, Ill.: Vance Bibliographies, 1982.

Miller, Tom. *The Panama Hat Trail: A Journey from South America.* New York: Morrow, 1986.

Moran, Emilio F. *Developing the Amazon.* Bloomington: Indiana University Press, 1981.

———. *The Dilemma of Amazonian Development.* Boulder, Colo.: Westview, 1983.

Morner, Magnus. *The Andean Past: Land, Societies, and Conflict.* New York: Columbia University Press, 1985.

Morris, Arthur S. *Latin America: Economic Development and Regional Differentiation.* Totowa, N.J.: Barnes & Noble, 1981.

―――. *South America.* Totowa, N.J.: Barnes & Noble, 1987.

Needler, Martin C. (ed.). *Political Systems of Latin America.* New York: Van Nostrand Reinhold, 1970.

Odell, Peter R., and Preston, David A. *Economies and Societies in Latin America: A Geographical Interpretation.* New York: Wiley, 1973.

Pereira, Luiz B. *Development and Crises in Brazil, 1930–1983.* Boulder, Colo.: Westview, 1984.

Smith, Nigel J. H. *Rainforest Corridors: The Transamazonian Colonization Scheme.* Berkeley, Calif.: University of California Press, 1982.

Stevenhagen, Rudolfo (ed.). *Agrarian Problems and Peasant Movements in Latin America.* Garden City, N.Y.: Anchor Books, 1970.

Steward, Julian H., and Faron, Louis C. (eds.). *Native Peoples of South America.* New York: McGraw-Hill, 1959.

Super, John C., and Wright, Thomas C. (eds.). *Food, Politics, and Society in Latin America.* Lincoln: University of Nebraska Press, 1985.

U.N. Statistical Yearbook, 1988. New York: United Nations, 1985.

Uys, Errol L. *Brazil.* New York: Simon & Schuster, 1986.

Veliz, Claudio (ed.). *Latin America and the Caribbean.* New York: Praeger, 1968.

Webb, Kempton. *Geography of Latin America.* Englewood Cliffs, N.J.: Prentice-Hall, 1972.

Weil, Connie H. (ed.). *Medical Geographic Research in Latin America.* Elmsford, N.Y.: Pergamon, 1982.

Wilkie, Richard W. *Latin American Population and Urbanization Analysis: Maps and Statistics, 1950–1982.* Los Angeles: UCLA Latin American Center, 1984.

World Population Data Sheet, 1986. Washington, D.C.: Population Reference Bureau, 1986.

Chapter 12

AFRICA

Introduction

Africa, the second largest continent, is bisected by the equator. The continent extends from approximately 35 degrees north latitude to 35 degrees south latitude. More of the continent lies north of the equator than south of it, because of its prominent northwestern "hump."

Africa is situated southwest of Asia and south of Europe. On the northeast, Africa is separated from Asia only by the Suez Canal. On the northwest, it is separated from Europe by the narrow Strait of Gibraltar. The Mediterranean Sea borders the continent along the north from Gibraltar to Suez, a distance of about 2,000 miles (3,218 km). See Figure 12.1.

The Red Sea lies along the northeast coast of Africa and separates the continent from the Arabian Peninsula. This long, straight, narrow sea stretches directly southeast of the Suez for nearly 1,300 miles (2,092 km) to the Bab al-Mandab, the Red Sea entrance to the Gulf of Aden. On the east, the Gulf of Aden and the Indian Ocean form another 4,000 miles (6,436 km) of African coastline. Then, on the west, waves from the Atlantic Ocean roll onto the African shores from Cape Town to Gilbraltar, a distance of nearly 7,000

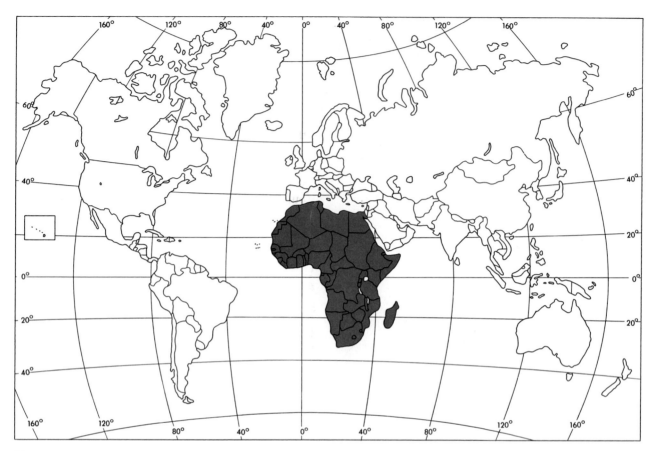

FIGURE 12.1
World Location of Africa.

miles (11,263 km). Actually, about 1,000 miles (1,609 km) of this coastline have a southern exposure because of the western hump of the continent.

The very long shoreline circumference of Africa outlines a massive land area of 11,688,000 square miles (30,300,000 km²), which is 20.4 percent of the earth's total. Much of the huge area is devoid of people, and the 583 million Africans make up only about 11.4 percent of the world's population. About 77 percent of Africa lies within the Tropics, and more than 25 percent is made up of the great wasteland of the Sahara Desert. Because life is difficult in the tropical areas, and especially in the desert, population densities are low in both areas.

The long shoreline and the continent's location affected Africa's human development, both economically and politically. For many years, Africa was looked on as a huge barrier jutting southward and presenting an obstacle to early sailing vessels plying the waters between Europe and Asia. Later, almost the entire continent became a political appendage to Europe. For nearly 80 years, during the later part of the nineteenth and early twentieth century, Africa was under European colonial rule. As colonies, the countries of the continent could not develop economically or politically. During the colonial period, the continent became a supplier of raw materials for the industrialized nations of Europe and the Americas, an extremely economically suppressing condition. Today, most Africans are still struggling to emerge from the legacies associated with the history of their continent.

▪ Physical Geography of Africa

Africa's overall geology may be the least complicated factor related to the geography of the continent (see Figure 12.2). No vast and complex mountain ranges, like the Rockies of North America or the Himalayas of Asia, exist in Africa. The entire continent is basically a single block of rock that was uplifted intact nearly a quarter of a billion years ago. The gigantic block was tilted slightly, causing higher elevations on the east than on the west. Africa is primarily a very large plateau fringed by a narrow coastal plain. The plateau is divided into a series of steps, higher on the east and gradually declining toward the west. The generally flat landscape is relieved by large but shallow basins and their associated rivers, by the deep cut of the Rift Valley, and by occasional volcanoes and fault-block mountains.

The temperature and rainfall patterns of Africa may also be generalized, but local variations are much more pronounced for the climatic elements than for the geologic elements. Except for the few highland areas, all of Africa is generally hot. This is attributable to its location, with the equator dividing the continent. As a consequence, throughout the year the noon sun is directly overhead in some part of Africa. In January, Africa is warm and dry on the north and hot and dry on the south. In July, as the direct rays of the sun move into the northern hemisphere, it is hot in the north and warm in the south. The midsection of the continent is continually rainy, but the northern and southern sections are dry. The rainfall patterns shift north and south with the changing seasons, but the shift is back and forth across the equator. Thus, the large rainfall

The small town of Matmata on the edge of the Sahara Desert is noted for its underground homes. The original Berbers dug caves to hide from their enemies, but because the underground dwellings provide relief from the desert heat they are still used today.

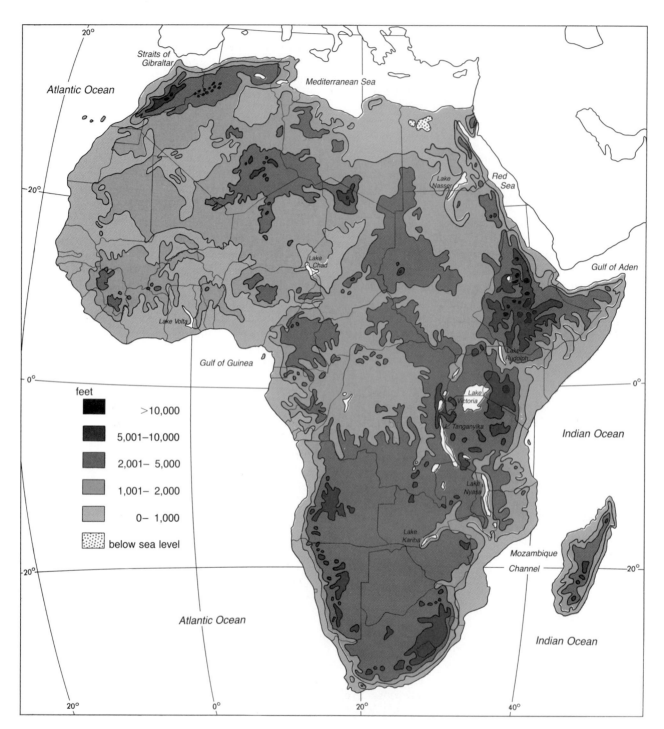

FIGURE 12.2

Land Elevations and Water Bodies of Africa.

Hamites Members of the native races of North Africa, which include the Berber peoples north of the Sahara Desert as well as the Fula, the Tibbu, and the Tuaregs of the Sudan, the ancient Egyptians and their ancestors, the tribes of Ethiopia, and the Galles and Somalies to the south.

Berbers The Hamites of North Africa west of Tripoli who are closely related to southern Europeans, Egyptians, and Ethiopians and who vary from blue-eyed blonds in the Atlas Mountain region to black-skinned inhabitants of the oases.

Bedouins The nomadic Arabs of the Arabian peninsula and North African deserts.

Pygmies The dwarf people of Central Africa, generally under 5 feet (1.5 m) in height and dark-skinned, but lighter in color than the true Negroid.

Watusi A branch of the Bahima people of Central Africa living near Uganda who are Hamite-Negro mixtures.

---

amounts generally occur no farther north or south than the 15-degree parallels.

Natural vegetation patterns correspond to the rainfall and elevations. The rainy areas that straddle the equator are covered with tropical rainforests, and the tall trees can be dense enough to make canopies over the ground below. The areas farther away from the equator, where the rain is seasonal, have savanna grass. Some trees grow in the savanna areas, but they usually are found along the rivers and in bunches where the ground water is near the surface. The dry areas have steppe and desert conditions where few trees are found. Grasses are tall and lush in the savanna, but gradually become shorter and eventually change to sparse bunch grasses. Finally, the grasses give way to the desert shrubs, which in turn give way to no vegetation at all. The extreme northern and southern edges of the continent, however, are located in the westerly wind-belts that provide moisture from cyclonic storms. The rain is concentrated in the winter months, producing Mediterranean (Köppen's Csa) type climate. In these areas the landscape is green, a welcome relief from the nearby brown desert conditions.

■ Cultural Geography of Africa

The origin of human life in Africa is bathed in obscurity. Some theorize that *Homo sapiens* originated in Africa, others maintain that humans crossed the Suez land-bridge from Asia in more recent times. The fact remains that humans have been in Africa for several million years.

Much of the history of Africa was never recorded, so little is known about human activities there except what can be deduced from legends and archaeological digs. It has been established that considerable migration has occurred over the centuries. A group of people, known collectively as the **Hamites,** migrated from Arabia at various times during the past. The Hamite group includes the **Berbers,** the **Bedouins,** and the Arabs, who settled most of northern and eastern Africa. Black people are probably indigenous to Africa, but they too have a history of movement. The Bantus moved from their original home in the Cameroon area, migrating long distances both east and south. The Bushmen spread north and south from the Lake Tanganyika area, and the Pygmies diffused throughout the Zaire (Congo) Basin from their original home near Lake Victoria.

The migrations and the resultant racial admixtures that developed created an extremely complex cultural base. Today, Africa's people vary from the world's shortest (**Pygmies**) to the tallest (**Watusi**). Skin colors range from pure white, through pink, yellow, tan, and brown, to coal black. The blackness of Central Africans is well known, but it is not common knowledge that the descendants of some Berbers have blond hair and blue eyes. The Dutch and British settlers are, of course, white, but Africa also has large populations of South and East Asians. Thus, all the human races can be found on the continent.

Recent African history has been dominated by the European colonial influence and the political independence that followed. The most prominent countries of Europe that held control over large areas of Africa were Spain, Belgium, Portugal, France, Great Britain, Italy and Germany. During the 1960s, when the idea of independence spread throughout the continent, political action in Africa was largely African-motivated. Momentum for independence reached its zenith in 1960 and 1961, when 19 new countries sprang into being during that two-year period. Today, the continent is divided into more than 53 countries, including the island nation of Madagascar.

The early migrations, the mixing of races, and the European domination all left marks on Africa. The essence of the marks is one of variation, but one cultural factor that is unvaried throughout much of Africa can be summed up in the term "poverty." One would think that average incomes would be high because population densities are generally low and natural resources are abundant. But the average per capita gross national product for the continent is $740, ranging from a low of $120 in Ethiopia to a high of $8,460 in Libya. The GNP figure is a close approximation of the average yearly income per person. Besides lack of income, poverty also includes low health standards. The life expectancy for the entire continent is 50 years, and much lower for some countries. For example, the average person in Gambia or Sierra Leone can expect to live only to the age of 35. Poverty also means low standards of living and poor housing but high birth rates, high infant mortality rates, and high disease rates. Wealth is not common in Africa.

In the previous chapter, the entire continent of South America was treated as a unit, but that is impossible for Africa because of the much larger number of countries involved. Traditionally, geographers have identified three to five subregions on the African continent. Such divisions are somewhat arbitrary, but they are necessary for a meaningful discussion. Here we will use three subregions: Northern Africa, Tropical Africa, and Southern Africa (see Figure 12.3). For some parts of the discussion, Tropical (or Middle) Africa will be divided into its eastern, western and central parts.

NORTHERN AFRICA

■ Introduction

Northern Africa is dominated by the Sahara Desert, one of the world's best-known physical features. The desert is a unifying element that affects each of the 11 countries of the region. The

The Romans controlled most of North Africa during the zenith of their empire about 2,000 years ago. The ruins of this Roman colosseum are located in El Djem, Tunisia.

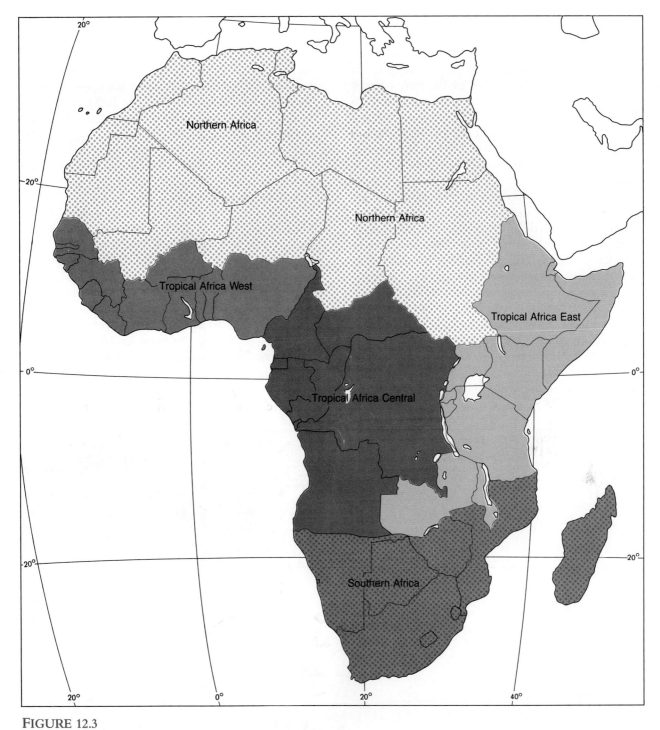

FIGURE 12.3

Regional Divisions of Africa. The five regions shown correspond to the discussion in the text and have no other significance.

line that separates arid Northern Africa from the humid middle of the continent is approximately 15 degrees north latitude. Most of Africa north of that latitude receives less than 10 inches (25 cm) of rainfall annually, while south of the line, toward the equator, annual rainfall amounts increase rapidly. The only exception to the dryness is a narrow strip of land along the north coast from Gibraltar to Banghazi. Much of the human activity of Northern Africa is related to the persistence of the Sahara Desert.

The boundaries of the Northern African states do not correspond exactly to the boundary of the desert. Five countries overlap into the more rainy parts of Tropical Africa, but these countries—Mauritania, Mali, Niger, Chad, and Sudan—are tied to the northern region historically, linguistically, and through their common bond with the desert. The remaining countries of Northern Africa are Egypt, Libya, Tunisia, Algeria, Morocco, and Western Sahara (formerly Spanish Sahara).

Most of Northern Africa has been settled for thousands of years. The remainder has been occupied, but only in the sense that nomads "occupy" an area. Sedentary people have lived along the fertile Mediterranean rim of the north coast and inland along the Nile Valley since prehistoric times. The Berbers were the original inhabitants of most of the region, followed by the Carthaginians, the Romans, the Vandals, and the Turks, but the entire region was overrun by the Arabs in the later part of the seventh century. Most of the region came under the political control of various European countries during the nineteenth century. Finally, independence came to the countries of the region in the late 1950s and early 1960s. See Table 12.1.

The European influence in Northern Africa remains strong, even after 30 years of independence. The newer parts of the cities look very European in both layout and building types. In Tunis, Algiers, and Casablanca, French automobiles and buses are prominent, and most of the people in those cities speak both French and Arabic. The older parts of the cities retain much of their Arabic quality. Narrow, crooked streets are lined with small shops, and the eaves overhang the streets to form canopies for protection from the bright African sun. Two-wheeled carts pulled by

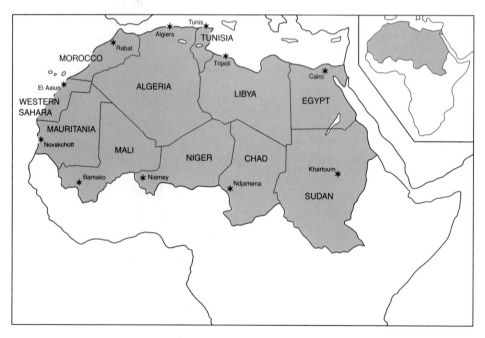

FIGURE 12.4
The Countries of Northern Africa.

TABLE 12.1

Countries of North Africa, with Former Rulers and Date of Independence

COUNTRY	FORMER RULER	INDEPENDENCE
Algeria	France	Sept. 1, 1962
Chad	France	Aug. 11, 1960
Egypt	Great Britain	1936★
Libya	Italy	Jan. 2, 1952
Mali	France	June 20, 1960
Mauritania	France	Nov. 28, 1960
Morocco	France	Mar. 2, 1956
Niger	France	Aug. 3, 1960
Sudan	Great Britain	Jan. 1, 1956
Tunisia	France	Mar. 20, 1956
Western Sahara†	Spain (Morocco and Mauritania)	

★Egypt secured independence from Great Britain in 1922, but British military occupation continued until the end of 1936.
†Most of the cities of Western Sahara are occupied by Moroccon troops, but the open country is under the control of the Polisario Front, an independent guerrilla movement.

donkeys replace the French automobiles. The North African cities are a study in cultural contrasts.

■ Physical Geography of Northern Africa

The three major physical features of Northern Africa are the Nile River on the northeast, the Atlas Mountains on the northwest, and the great expanse of the Sahara Desert between. The Nile, one of the great rivers of the world, flows northward from the rainy highlands of Tropical Africa through the dry areas of Sudan and Egypt. Once it leaves the upper reaches, it picks up very little water from tributary streams. After the Blue Nile and the White Nile join near Khartoum, no other permanent stream flows into the river until it reaches the delta, a distance of more than 1,200 miles (1,931 km). The Aswan High Dam in Egypt now controls the flow of the great river. The dam was completed in 1971 with financial aid from the Soviet Union. Lake Nasser backs up behind the dam for about 250 miles (402 km) and extends into the Sudan. Historically, the Nile has been the lifeblood of Egypt, and it remains so today. About 95 percent of Egyptians live within a few miles of its banks.

The Atlas Mountains are divided into five separate ranges that run parallel to each other and to the northwest African coast. The highest and most prominent range in the group is the Great Atlas Mountains, located almost entirely in Morocco. The Anti-Atlas and Middle Atlas ranges fuse into the southwest and northeast sections respectively of the Great Atlases. In Algeria, toward the east, the Great Atlas range gives way to the Saharan Atlas and the Maritime Atlas. The names indicate the locations of the ranges with respect to the desert and the sea. The tallest single peak in Northern Africa is Toubkal, at 13,664 feet (4,165 m), located in the Great Atlases approximately 200 miles (322 km) south of Casablanca. The peak can be seen from the nearby city of Marrakech. (The major features of the landforms of Northern Africa can be seen on the map in Figure 12.5.)

The world's greatest desert, the Sahara, spreads over 3.5 million square miles (9.1 km^2) of Northern Africa. It covers nearly one-third of the entire continent. The desert covers about the same land area

Africa

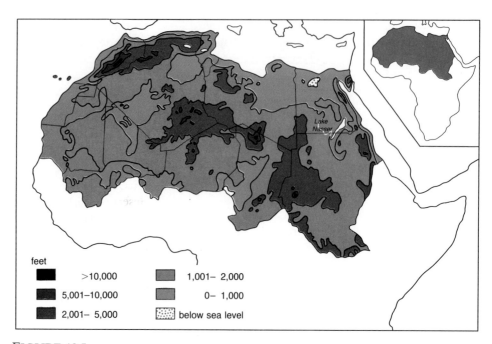

FIGURE 12.5

Land Elevations and Water Bodies of Northern Africa.

as all 50 of the United States, and Algeria is more than three times larger than Texas. The desert landscape is not without variety, but there is also mile after monotonous mile of nothing but bare rocks or desiccated dunes. Individual sand dunes reach windblown heights of 600 to 700 feet (183–213 m) and extend for miles. Land completely covered with sand dunes is called an *erg,* but sand cover extends over only about 10 percent of the Sahara. The remainder consists of barren rocky type landscape (*reg*) or has a sparse cover of highly specialized xerophytic-type vegetation.

The mostly flat plateau of the desert is interrupted by two mountain ranges located deep in the interior. The Hoggar Range in southeastern Algeria rises to a height of 9,850 feet (3,002 m) at Mount Tahat, and the Tibesti Range in northern Chad is topped by Emi Koussi (11,200 feet, or 3,414 m). Both mountain ranges run east to west. The Hoggar Range is approximately 400 miles (644 km) long by 200 miles (322 km) wide, and the Tibesti Range is about half as large, but buried as they are in the vastness of the Sahara, they seem insignificant in size.

Desert areas are defined as areas that receive less than 10 inches (25.4 cm) of annual precipitation, but most of the Sahara receives less than half that amount. All of Northern Africa, except the northern coastal and southern fringes, falls into the desert category. (See Figure 12.6 for rainfall distribution.) Rainfall along the coast usually exceeds 20 inches (51 cm) each year, and as with other Mediterranean-type climates, the rain comes mostly in the winter months. Precipitation averages can be misleading, however, especially in the arid regions of the world. Some deserts go for years, even decades, without a trace of rain. Then a violent storm can drench the area with two or three years' supply in a few hours. The moisture supply from precipitation in Northern Africa is not only scarce but also variable and unpredictable.

The lack of precipitation, lack of moisture in the air, lack of cloud cover, and lack of vegetation all contribute to the high temperatures of the desert. The sun penetrates the clear atmosphere and reflects off the barren surface, causing intense heat. Summer daytime temperatures can exceed 120° F (49° C) day after day. The world's record high temperature was recorded on September 13, 1922, at Azizia, a small town south of Tripoli in Libya. The official high temperature for that day and place was 136.4° F (58° C), recorded in the shade, and ground temperatures in the sun soared to 60° F (16° C) above that. Because of the lack of atmospheric moisture and clouds, which ordinarily would hold the heat near the ground, the daytime heat in the desert dissipates fairly quickly at night. Temperatures may drop 50° F (10° C) or more from the daytime highs. The day to night temperature change is called the *diurnal range* and is less in the winter because the days are not as hot.

■ Human Geography of Northern Africa

The 1986 population for the 11 countries in Northern Africa totaled 153.6 million. The population density for the area is about 20 people per square mile (7.72 per km^2), but the distribution is most uneven (see Figure 12.7). The people are located in clusters along the north coast and along the Nile River, the same places they have been for thousands of years. For example, the population densities for the coastal states of Morocco and Egypt are about 125 people per square mile, but those for Chad and Mali in the interior are about 12–13 people per square mile. Furthermore, nearly half the people in Morocco and Egypt live in urban areas, whereas only about 10 percent of those in Chad and Mali live in cities. The sparsely populated interior can be attributed directly to the great desert.

The northern population clusters include numerous cities that have a population of more than 1 million. Cairo, the capital of Egypt and the largest city of the region, has slightly over 5 million people. It is also the only large city of the region not located along the coast. Egypt's coastal city of Alexandria is the second largest of the region and has about 2.3 million people. Westward along the Mediterranean coast from Alexandria are Banghazi and Tripoli in Libya. Both cities are nearing the 1 million population mark. Tunis, the capital of Tunisia, is over 1 million, and Algiers, the capital of Algeria, is over 2 million, as is

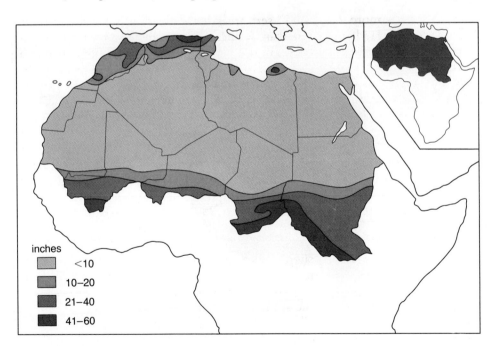

FIGURE 12.6

Precipitation Patterns of Northern Africa. Precipitation amounts decrease rapidly from the Mediterranean coast southward into the Sahara Desert.

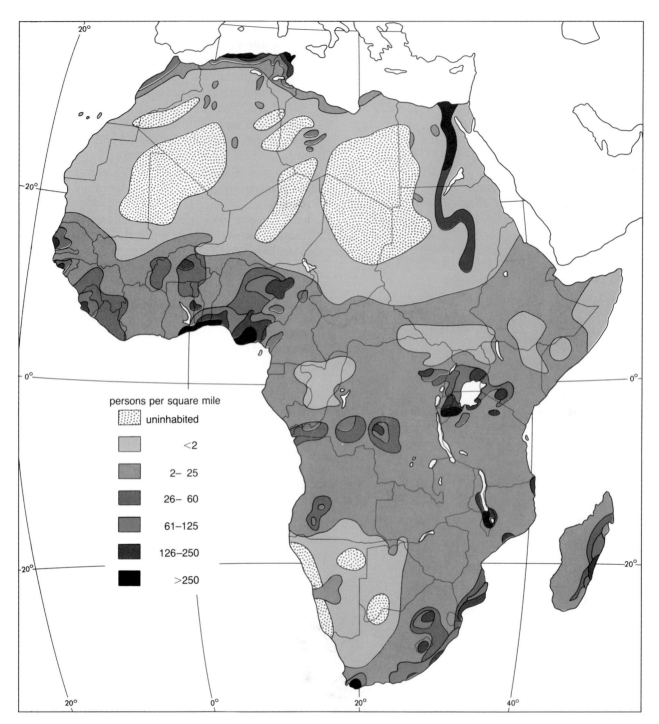

FIGURE 12.7

Population Distribution of Africa.

Casablanca in Morocco. All these great cities, except Cairo, are major ports for their countries. They are mixtures of the old and the new, the ancient and the modern, and of different cultures. All have pleasant year-round climates, and because they are fascinating to visit they draw much money from European and American tourists.

No cities have grown up in the desert, but some small population clusters can be found at the few oases that do exist. Some nomadic tribes still range across the region, but most of the time the desert is devoid of people. Farther south, however, in the southern parts of Mauritania, Mali, Niger, Chad, and the Sudan, small villages appear along the fringes of the desert, some of which are beginning to grow rapidly. For example, Nouakchott, the capital of Mauritania, has grown from 15,000 in 1970 to 250,000 in 1986. And Khartoum, the capital of Sudan, has tripled in population over the last 20 years.

Income and Life Expectancy

Unyielding and harsh as the desert may seem, it has become the financial salvation for many North Africans. The average per capita gross national product for the region is $952 a year. This average excludes Libya, however, where the figure is about $8,500 a year. In the early 1960s, Libya's per capita GNP was about $150 per year, but only a decade later it had risen dramatically to more than $2,000 a year and has continued to increase over the last few years. The change is due to the discovery of oil in the Sahara. A desert resource has brought some prosperity and an improved standard of living to the Libyans, but poverty prevails in most of the remainder of Northern Africa. The three landlocked countries of Chad, Mali, and Niger are among the poorest nations in the world. Natural resources are scarce, and the region's livestock industry is crippled by recurring droughts associated with the vagaries of the desert. The Sahara goes through periods of expansion and contraction and has been expanding in recent years. Thus, while the desert helps some people financially, it adversely affects others.

The average life expectancy for a newborn child in Northern Africa is 51.5 years. The range is from 61 years in Tunisia to 42 years in Mali. The figures for all the countries in the region have

The architecture on Habib Bourguiba Avenue in central Tunis indicates the influence of the French, who ruled Tunisia for many years, as well as the local Arabic influence.

improved over the last two decades but still remain low by world standards. The average life expectancy for the world is 62 years. Low life expectancy rates are related to low incomes. Poverty contributes to conditions that cause people to die at an early age, but the rates also are related to the lack of access to medical facilities. Egypt, for example, is not a wealthy state ($690 per capita GNP), yet it has a relatively high life expectancy rate (58 years). Egypt is the most urbanized state in Africa, so most Egyptians have nearby medical facilities that are not available to the nomads and tribesmen of the nonurbanized countries. (See Figure 12.8.)

Low life expectancy rates are also related to high infant mortality rates, and the poor countries of Northern Africa have very high infant death rates. For every 1,000 births in the world, an average of 82 children die before their first birthday. In Mali, however, the rate is 180 deaths per 1,000 live births, one of the highest in the world. And Chad (142 per 1,000), Mauritania (137 per 1,000), and Niger (140 per 1,000) are not much better off. The infant mortality rate for every country in Northern Africa is higher than the world average. The average for the region is 118.4 deaths in infancy per 1,000 live births.

The low life expectancy rates and high infant mortality rates would lead one to believe that the overall rate of population increase would be low, but the opposite is true because the birth rates are very high. Many infants die, but the fertile women continue to have more babies, probably to help compensate for their losses. Lack of birth control knowledge is also a problem. The world average is 27 births annually per every 1,000 people. The rate for Northern Africa is 46.6 per 1,000, and it is over 50 for Mali, Mauritania, and Niger. These high birth rates offset the high infant mortality rates, so

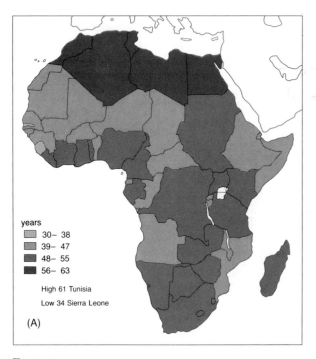
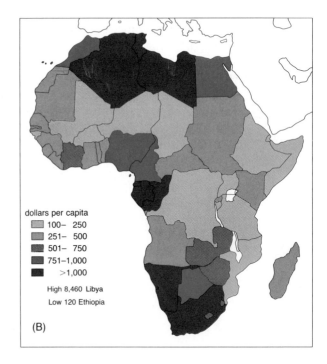

FIGURE 12.8

Life Expectancy (A) and Per Capita Gross National Product Rates (B) for Africa. The rates indicated on Map A are the average number of years a newborn child can be expected to live under current mortality levels. Tunisia has the highest life expectancy (61 years) in Africa, and Sierra Leone has the lowest (34 years). The gross national product (GNP) per person by country is shown in U.S. dollars on Map B. Libya has the highest ($8,460), Ethiopia has the lowest ($120).

that the overall rate of population increase is high. The average annual rate of increase for the region is 2.78 percent, compared with 1.7 percent for the world and 0.7 percent for the United States.

Culture, Language, and Religion

Northern Africa is part of a much larger cultural region known as the "Arab World," which is a subset of a larger region known as the "Muslim World." The Muslim region, includes most of the lands conquered in the seventh century by the followers of Muhammad in their quest to spread their religion, known as Islam, worldwide. The Muslim World includes Northern Africa, the Middle East, and the part of South Asia up to the Pakistan-India border, part of the Soviet Union, and most of Indonesia. It is a large and varied region tied together by similarities in ethnicity, language, and religion. (See Figure 12.9.)

The religious patterns of Northern Africa correspond to the early settlement and migration patterns. In the countries along the north coast, most of the people are Muslims, adherents of the Islamic faith. In Egypt, for example, numerous ethnic elements are present, but more than 90 percent of the population is Muslim. The faith predominates in all the states along the Mediterranean Sea. In the countries to the south, however, Islam gives way to other religions. Chad, for example, is 44 percent Muslim and 33 percent Christian. Another 23 percent of the people of Chad believe in *animism,* which holds that inanimate objects possess spirits. Similar transitional religious patterns exist in Mali, Mauritania, and Niger. The faiths other than Islam become stronger in the most southern part of the region, the area closest to the equator.

The language patterns of Northern Africa are somewhat similar to the religious patterns. Many dialects exist in the region, but four major language groups predominate. All four are among the Hamito-Semitic family of languages and are present-day representations of the ancient settlement patterns. The Semitic group, including Arabic, Hebrew, and Phoenician, is the most com-

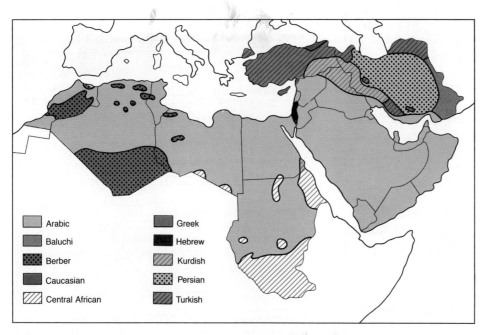

FIGURE 12.9

The Arab World. Arab ethnicity, religion, and languages are found in a region stretching across Northern Africa and southwest Asia. The locations of people who speak the related Arabic languages are shown on this map.

Africa 353

mon and is found throughout Northern Africa. The second most common language is Berber, the predominant language in Niger and also found in pockets along the north coast. The two remaining languages of Northern Africa—Kanuri, spoken in Chad, and Kyshit, spoken by some people along the east coast of Sudan—are less significant.

Superimposed on the indigenous languages of Northern Africa are the languages of the European colonial countries. Many people speak either English or French as well as their native language. And in most of the countries in the southern part of the region, such as Chad, Mali, Mauritania, and Niger, the official language is still French. In the popular tourist areas along the north coast, it is not uncommon to find multilingual people who speak Berber, Arabic, French, and English.

■ Economic Geography of Northern Africa
Agriculture

Agricultural production in Northern Africa is limited by the lack of precipitation in most areas. The only producing regions are along the north coast, along the Nile River in Egypt, and at a few oases, the same places where most of the people are located (see Figure 12.10). Agriculture along the north coast is of the Mediterranean type, and crops include small grains, citrus fruits, figs, dates, grapes, olives, vegetables, and pomegranates. The chief agricultural exports of the region include fruits, wines, grapes, and olive oil. The commercial olive groves cover the most land area, and some extend for miles inland from the coast. The largest single olive tree grove in the world surrounds the city of Sfax on the east coast of Tunisia.

In Egypt, agriculture depends primarily on irrigation from the Nile River. Thus, only about 4 percent of the land area of the country is arable. The crops grown under irrigation are not the same as those grown along the north coast. The most important irrigated crops are cotton (Egypt is one of the world's leading cotton producers), rice, sugar cane, grains, beans, and corn. Fruits and vegetables also are important.

> **nomadic pastoralism** "Pastoralism" pertains to caring for livestock (shepherding), and "nomadic" means roaming around looking for pastures, so "nomadic pastoralism" refers to the type of economic activity carried on by nomadic herders.

Other cash crops of Northern Africa include barley, tobacco, oats, and wheat. Some of these crops are grown on the fringes of the rainy regions under nonirrigated conditions, but crop failure is common and agriculture becomes more and more risky the closer it is to the Sahara. Nothing is grown in the desert itself, except a few date palms located at the scattered oases. On the south side of the region, where the rainfall increases enough to produce grass, the people subsist on nomadic herding. They raise cattle, sheep, goats, camels, and donkeys and live almost exclusively from the animals and animal products. The herders must wander in constant search for both water supplies and forage for their animals. This type of occupation is called **nomadic pastoralism.** It is the

Written symbols of the Arabic language can be seen on this sign advertising a movie under a theater marquee in Sfax, Tunisia. Note also how the Arab men stand very close to each other during a normal conversation.

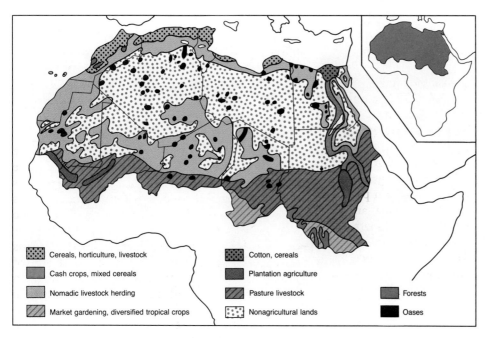

FIGURE 12.10

Agriculture of Northern Africa.

reason that Mauritania, Mali, Niger, and Chad have few urban centers and low gross national products.

Mineral Production and Industry

Northern Africa is not without mineral resources (see Figure 12.11). Although Chad, Niger, and Mali have little or no mineral production, the other countries produce significant amounts. Both Libya and Algeria usually rank among the top 10 oil producers of the world. Egypt is one of the few world sources of titanium, and Mauritania has large deposits of iron ore and copper. Phosphate rock is mined in all the north coast countries, and Morocco leads the world in its exportation. Morocco is also a major world source of cobalt. Most of the north coast countries produce iron ore, zinc, and manganese too.

The colonial influence of the European countries retarded industrialization in Northern Africa for many years. The Europeans wanted their colonial countries to remain undeveloped so they would be markets for the goods manufactured in Europe. Since the 1960s, however, the African nations have started to develop more than the light industries they have had. Algeria and Libya have constructed oil refineries and iron and steel plants. Egypt has chemical, steel, and fertilizer plants, as well as a film industry that supplies the entire continent. Most of the coastal countries have phosphate-processing plants, but very little manufacturing of consumer goods, such as automobiles and household appliances, occurs in Northern Africa. These items are imported from the industrial countries of Europe.

Other industries of Northern Africa include tourism, and Tunisia and Morocco lead the region in accruing tourist money. The Roman and Carthaginian ruins, Mediterranean beaches, mild climates, friendly people, and native crafts such as blankets and carpets draw European and American tourists. Sudan is the world leader in the production of gum arabic, a product from a type of mimosa tree used for making dyes. Meat and leather products are produced for exportation by

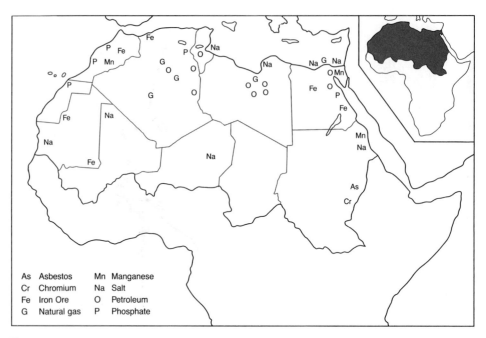

FIGURE 12.11

Mineral Production Areas of Northern Africa.

Chad, some rubber is produced in Mali, and peanuts are the leading export item for Niger. However, modern industrialization, especially in the heavy industries, is still in the future for most North African countries. While Libya has made giant strides toward economic viability, the other countries still have trouble merely feeding their people. Egypt has good mineral resources and adequate agriculture, but the population growth has been too rapid for the economy to keep pace. Potential mineral production, tapping the vast Sahara, may be the economic hope for the future of the region.

Transportation

Railroads are scarce in Northern Africa (see Figure 12.12). One major route from Casablanca to Tripoli parallels the coast for nearly 1,500 miles (2,414 km), and a few feeder lines lead to it from the interior. The feeder lines are 150 to 200 miles (241—322 km) apart and extend about the same distance inland. Another major route runs from Alexandria up the Nile Valley to the Aswan High Dam in Egypt. Sudan has the nearest thing to a railroad network in Northern Africa. Lines run from Khartoum in all directions, up and down the Nile Valley, westward to the interior, and eastward to the Red Sea port of Port Sudan. The remainder of Northern Africa has meager railroad facilities. No railroads cross the Sahara from north to south, and Western Sahara, Chad, and Niger have no railroad tracks at all.

Roads are not much more prevalent than railroads in Northern Africa, but it is possible to drive on good highways for long distances (see Figure 12.12). For example, one can drive from Casablanca, along the north coast, through Algeria, Tunisia, Libya, and into Alexandria, Egypt. The drive could continue south from Alexandria along the Nile Valley to Khartoum, back westward through Chad and Niger, and take one of the two trans-Sahara routes back to Casablanca. Following this route, the traveler would traverse about 90 percent of all the roads in Northern Africa. The traveler should not expect to find rest stops, motels, or even many gas stations along this 7,500

These farmers are selling artichokes from their wagon on a street in Sousse, Tunisia.

mile (12,068 km) route. The traffic is light, with only an occasional truck or two-wheeled donkey cart on the road. In and around the urban areas, however, traffic congestion is common, as is the accompanying smog and noise. The variety of vehicles is much greater in the North African cities than in European or American cities. Traffic congestion includes trucks, buses, and cars, but also many bicycles, scooters, and two-wheeled carts drawn either by donkeys or by hand. Occasionally,

The coastal towns of Northern Africa all have port facilities, but most are not well developed. The port shown here is at Sousse, Tunisia.

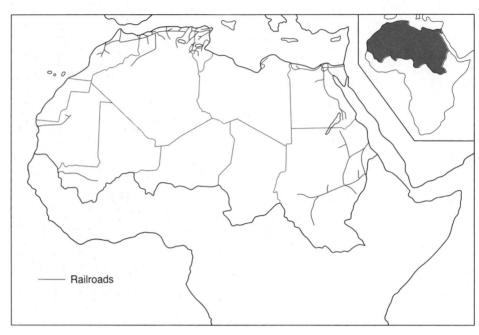

FIGURE 12.12
Railroads of Northern Africa.

a Berber on a camel will be seen in the middle of a traffic jam.

The large cities of Northern Africa all have airports, and much passenger traffic moves by air. Tourists from Europe land daily, and every coastal country has a major international airport. Even the smaller settlements have air strips, and local airlines, such as Air Mali, fly established internal routes.

Television stations are not common in Northern Africa, nor are telephone and telegraph wires. However, the number of radio stations is growing quite rapidly, thanks to the cheapness of transistor radio receivers. According to the government licensing figures for Morocco for 1983, there were 800,000 television receivers in the country but nearly 3 million radios. The ratios are similar in the other north coast countries. Loud radio music adds to the din of the local marketplaces in all areas of Northern Africa, but radios are also a means of getting instant information about world events.

■ Political Geography of Northern Africa

Northern Africa is a region of the world that has recently been realizing its political independence. Except for Egypt, all the states have been granted full independence since 1950, and half of them since 1960. They are still adjusting to the withdrawal of the colonial powers. Having shared so long the political interests of Europeans, their present international posture reflects this interest modified by African nationalism, the Islamic religion, and petroleum profits. Stability in the framework of world politics will depend on how these three regional interests are adjusted to international realities.

Relations among the Northern African countries are only slightly more stable than relationships with areas outside Africa. While the Arabic language and the Muslim religion give a certain amount of cohesion to the region, many factors tend to disturb amicable relations. Regional leadership is one such factor, with the Egyptian population and cultural maturity giving Egypt a historic leading role. But the oil wealth and nationalistic fervor of Libya are attractions to many Muslim countries looking for leadership. Such fervor, however, has led Libya to attack its neighbors and cause international problems as well.

TROPICAL AFRICA

■ Introduction

Tropical or Middle Africa is the most diverse of the three African subregions. In addition to much physical and cultural diversity, a complicated political fabric exists in the region. In areal extent, this region is not as large as Northern Africa, but it contains three times as many political entities. The 35 countries of the region (see Table 12.2 and Figure 12.13) range in size and complexity from the tiny island republic of the Seychelles (108 square miles, or 280 km^2) to Zaire, the largest in area at 905,328 square miles (2,344,800 km^2).

Six of the countries of Tropical Africa are island nations. On the other hand, Burkina Faso, Central African Republic, Uganda, Rwanda, and Burundi do not have any sea coasts. Zaire's boundary is about 4,000 miles (6,436 km) long and touches nine other countries. In spite of the long boundary, Zaire's sea coast is only about 45 miles (72 km) long. The name of the largest town on that particular coast is "Banana," indicating what is grown there.

Tropical Africa has been occupied for millions of years, probably even several million years. The Olduvai Gorge, where fragments of human bones have been dated as being 2 million years old, is located in northern Tanzania. The 14-million-year-old remains of a humanlike creature also were discovered in Kenya. Both countries are in the eastern part of Tropical Africa. The region also is the ancient homeland of Bantus, Pygmies, and Bushmen, as well as the Sudanese Negroes. Not much is

TABLE 12.2

Countries of Tropical (Middle) Africa

WESTERN	MIDDLE	EASTERN
Benin (Dahomey)	Angola and Cabinda	Burundi
Burkina Faso (Upper Volta)	Cameroon	Comoros (I)
Cape Verde Islands (I)	Central African Republic	Djibouti (Afars & Issas)
Gambia	Congo	Ethiopia
Ghana	Equatorial Guinea	Kenya
Guinea	Gabon	Malawi
Guinea-Bissau (Portuguese Guinea)	Sao Tome & Principe (I)	Mauritius (I)
	Zarie (Belgian Congo)	Tanzania
Ivory Coast		Réunion (I)
Liberia		Rwanda
Nigeria		Seychelles (I)
Senegal		Somalia
Sierra Leone		Uganda
Togo		Zambia (Northern Rhodesia)

Note: Former names of the countries appear in parentheses, and the six island republics are indicated with an (I).

Africa

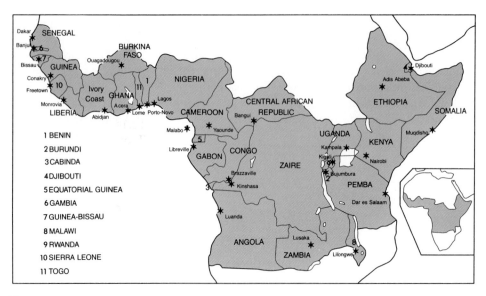

FIGURE 12.13

Countries of Tropical Africa. The divisions correspond to the discussion in the text and have no other significance.

known about the history of these people, except that they migrated from place to place throughout the region. Forced migrations in the form of slave trade caused many to leave the region entirely.

The idea of slavery seems to have originated in Africa with the Arabs about 2,500 years ago. Slaves were commercial items for the flourishing trans-Saharan trade that existed for many hundreds of years. So slavery was not started by the Europeans, as many believe, and it actually ended when the Europeans gained political control of Tropical Africa. Europeans did purchase thousands of Black Africans, however, and moved them from their homelands. Much of the movement was to the Americas. Today, more descendants of Black Africans live in New York City than in the two largest cities of Tropical Africa combined.

Trade with Tropical Africa (sometimes called Middle, Central, or Black Africa) was originally opened by Europeans along the west coast. The Portuguese were looking for access to slaves without going through the Arab (North African) middlemen. For many years after that, the region was not penetrated much beyond the coast, partly because slaves could be obtained from dealers located at coastal outposts and partly because the tropical rainforest was so forbidding to the Europeans. Finally, missionaries went into the area looking for converts, and they were followed closely by political and economic colonialism. Today, all the countries of Tropical Africa that were colonies are now sovereign states, and many of the people enjoy political freedom. They all also have severe economic problems.

■ Physical Geography of Tropical Africa

No outstanding landscape feature ties Tropical Africa together, as the Sahara does for Northern Africa, although the tropical rainforest covers much of the region. Swampy lowlands are found along the west coast, large lakes and mountains on the east, and mile after mile of savanna grasslands surround the rainforest on all sides.

Probably the most notable features of Tropical Africa are the great rivers, but some individual mountains and lakes are impressive. The Zaire

River (previously known as the Congo River) and its lesser known tributaries flow generally westward from the eastern highlands to the Atlantic coast. The sources of the Nile, flowing northward to the Mediterranean Sea, and the Zaire are very near each other in the eastern highland region. The Ubangi River, the largest tributary of the Zaire, begins less than 100 miles (161 km) from the source of the White Nile. The Zaire itself begins near Lake Tanganyika, makes a large loop northward through the country of Zaire, and eventually empties into the Atlantic Ocean near Cabinda. It is one of the largest rivers in the world, both in length and in amount of discharge. (See Figure 12.14 for the location of landforms and rivers of Tropical Africa.)

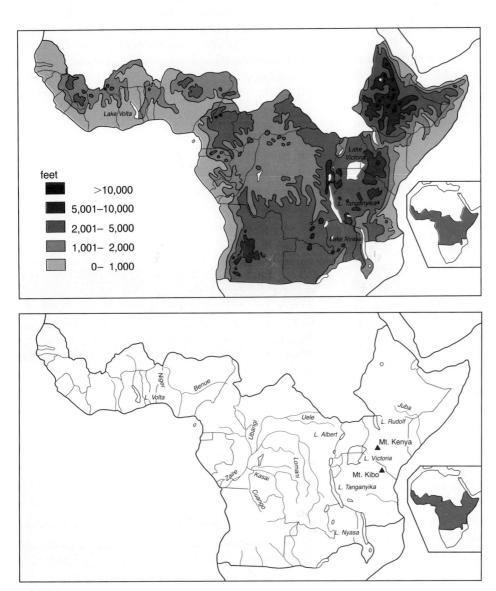

FIGURE 12.14
Land Elevations and Water Bodies of Tropical Africa.

The Niger is another great river of Tropical Africa. It may not be as well known as the Zaire, but it is nearly as long and carries nearly as much water. The Niger begins near the Atlantic coast in Sierra Leone, flows northeast through Guinea and Mali, turns southeast and flows through Niger and Nigeria, and finally empties into the Gulf of Guinea. The two largest rivers of Tropical Africa begin thousands of miles apart, and both make large northern loops, flow generally toward each other, and empty into the Atlantic only a few hundred miles apart.

The tallest mountain peak on the African continent is located in the eastern highlands of Tropical Africa. Mount Kibo, probably better known by its former name of Kilimanjaro, stands in northeastern Tanzania. The mountain is near the border with Kenya and rises to 19,340 feet (5,895 m). Kibo is a unique mountain because it is situated only 200 miles (322 km) from the equator but is snowcapped the entire year. Also, the mountain is different from other tall peaks because it stands alone; it is not part of a range of tall mountains. Mount Kenya, the second tallest peak (17,058 feet, or 5,199 m) in Africa, is located in Kenya directly north of Mount Kibo. It straddles the equator and also is snow- and glacier-covered near the top. Both these peaks are volcanic in origin, which explains their large circular base and isolation from mountain chains.

A series of large fresh-water lakes is situated in the eastern highland region. Most of the lakes fill parts of *rift valleys* between the mountains and therefore are long, narrow, and deep. Lake Victoria, the largest of the lakes and not in a rift valley, is more circular in shape and much more shallow than the others. About the size of the state of West Virginia, it is shared by Kenya, Uganda, and Tanzania and is the third largest lake in the world (following the Caspian Sea and Lake Superior). Other smaller lakes near Victoria reflect the names of the House of Hanover: Rudolf (now Turkana), Albert, George, and Edward. Stretching southward from Lake Victoria are Lakes Tanganyika and Nyasa, the second and third largest of the region. Lake Tanganyika is 400 miles (644 km) long, or roughly the distance between Los Angeles and San Francisco, yet it is never more than 50 miles (81 km) wide.

The highlands of eastern Tropical Africa contain some of the most spectacular scenery in Africa. The beauty of the high mountain peaks and large lakes is reinforced by active volcanoes in the Mufumbiro Range (also known as the Virunga Mountains), immense grasslands containing great herds, and huge flocks of birds along the lakes. In contrast, the western portion of Tropical Africa contains little noteworthy scenery. The land is very flat and covered with trees, so no open areas and few high places are available for viewing the landscape. The only exception is Cameroon Peak, located about 25 miles from the Atlantic coast in Cameroon. The mountain top is 13,350 feet (4,069 m) above sea level, but the peak is like a lonely

A lion pauses for a drink while hunting for antelope on the Serengeti Plain of northern Tanzania. Photograph by Gordon Matzke.

> **orographic precipitation** "Orography" refers to mountains, so orographic precipitation is produced when mountain barriers deflect moisture-laden air upward, causing cooling and condensation, which in turn causes the moisture to fall to the ground.

sentinel standing guard over the swampy coastal lowlands.

All climates in the central region of Africa are tropical, and most are rainy. The entire area of Tropical Africa lies within 15 degrees latitude of the equator. Thus, the region has only slight seasonal rhythms with regard to temperature changes. Seasons are evident, however, by when the most rain falls. Near the equator, the two rainy seasons correspond to the time of the equinoxes. North and south of the equator, the rainy seasons come once a year, at the time of the high sun (summer). Generally, the amount of rainfall received decreases with distance from the equator, but there are two major exceptions to this—on the extreme east coast of the continent and along the south coast of the hump. Somalia, on the east, is dry with steppe and desert conditions. At a similar latitude on the west coast, Cameroon, Nigeria, Liberia, and Sierre Leone receive huge amounts of precipitation. These conditions are created by west coast trade winds that blow landward over warm ocean water, while the winds in the east pass over land masses before reaching Somalia. (See map in Figure 12.15 for distribution of rainfall in Tropical Africa.)

The wettest place in Africa is along the slopes of Mount Cameroon. Air heavily laden with moisture moves inland off the warm Gulf of Guinea. The moving air is forced to rise over the mountain barrier, thus causing **orographic precipitation.** Along the mountain slopes, rainfall amounts exceed 30 feet (360 inches, or 9.1 m) a year. In eastern Africa the rain is less predictable, especially in Ethiopia and Somalia. Some years have no rainy season at all, resulting in severe droughts. The very dry conditions not only kill many of the large animals on the game preserves, but also threaten the people with starvation.

Aside from the extremely rainy west and the dry but variable east, most of the remainder of Tropical Africa receives sufficient moisture to support a tree cover. Much of the forest is tropical rainforest, where the monthly temperature never averages below 64.4° F (18° C) and usually about 80 inches (203 cm) of rain falls each year. The warm, humid conditions create dense forest (selva) cover, where many species of trees are mixed together within a small area. The evergreen forests occupy the hearts of the large basins but give way as the land rises to high savanna type of forest, where the trees lose their leaves during the dry season. Evergreen forests also stretch like long tongues up the valleys that have been eroded into the higher lands. Much of the tropical rainforest has been destroyed for cultivation, and nearly everywhere in the region there is evidence of past or present human settlement. (See Figure 12.16.)

■ *Human Geography of Tropical Africa*
Population

In 1986, the total population of Tropical Africa, including the island nations, was 359.0 million. Nigeria, with 105.4 million, had the most people, and the island countries of Seychelles and Sao Tome and Principe each had about 100,000—the fewest.

The population density for Tropical Africa is about twice that of Northern Africa. Nigeria is the most populous state on the African continent. It has twice as many people as Egypt, although the two countries are similar in size. Two general areas of Tropical Africa have the heaviest concentrations of people. The first is along the south coast of the African hump, the second is around the lakes in the eastern highlands. Nigeria, on the west, averages 265 people per square mile, while the two small lake countries of Rwanda and Burundi have population densities exceeding 550 per square mile. The most sparsely populated areas are found in the heart of the tropical rainforest. Gabon and Congo, for example, average less than 10 people per square mile.

The population densities in the states of Tropical Africa are generally much higher than those in Northern Africa, but the percentage of people living in urban areas is much lower. Thus, the

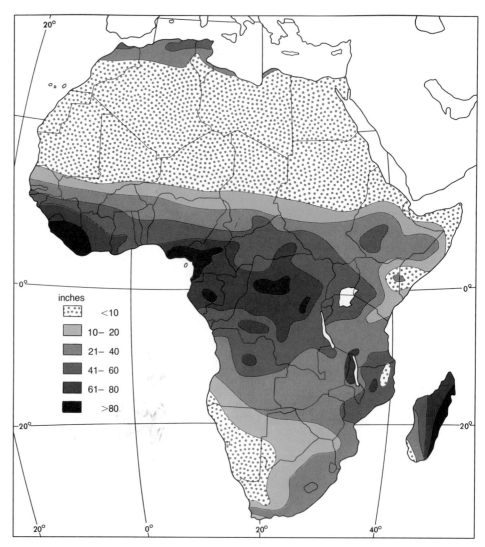

FIGURE 12.15

Annual Precipitation of Africa. The distinctly high amounts of precipitation are received in the tropical midsection of the continent.

urban centers in Tropical Africa are smaller. Again, comparing Nigeria and Egypt, the total population of the five largest cities of Nigeria (Lagos, Ibadan, Kano, Ogbomosho, and Port Harcourt) is less than the population of the single city of Cairo. The largest city of Tropical Africa is Kinshasa, the capital of Zaire (see Table 12.3).

As in other areas of the world, migration from rural areas to urban areas is occurring in Tropical Africa, so most cities are growing rapidly. Many of the new arrivals, however, are unemployed squatters, like those in the slums around Latin American cities. These people exist in pitiful conditions, many living in one-room huts made of mud and sticks. All the large urban centers contain pleasant residential areas comparable to those of Western cities, so the less well off can see that there can be a better life. The cities draw the rural people

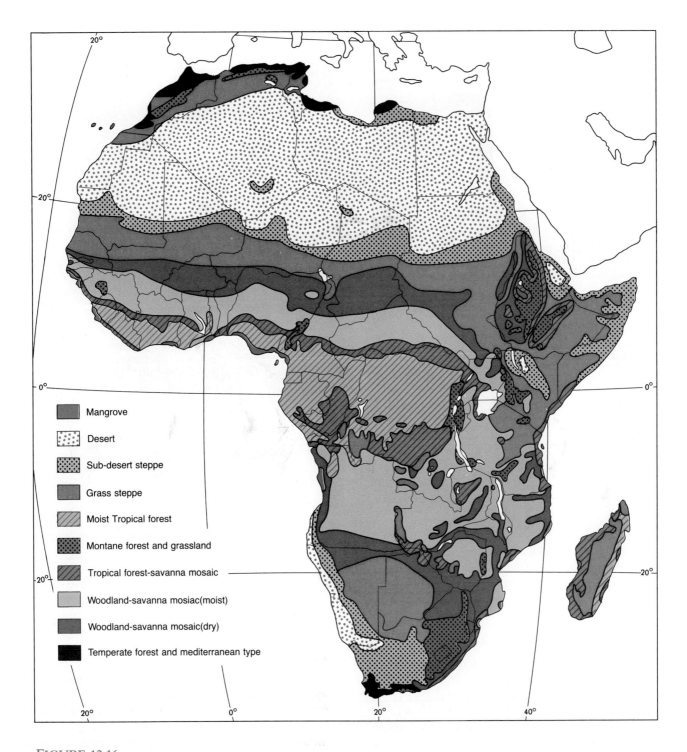

FIGURE 12.16

Natural Vegetation Regions of Africa. Compare the vegetated areas with the precipitation patterns shown in Figure 12.15.

Part of the skyline of central Nairobi, Kenya. Notice the new, modern buildings—a sign of recent growth. Photograph by Gordon Matzke.

TABLE 12.3

Cities of Tropical Africa with Populations of One Million or More, 1986 (estimates in millions)

CITY	COUNTRY	POPULATION
Kinshasa	Zaire	3.0
Abidjan	Ivory Coast	1.7
Addis Ababa	Ethiopia	1.4
Dar-es-Salaam	Tanzania	1.4
Dakar	Senegal	1.3
Luanda	Angola	1.1
Lagos	Nigeria	1.1
Accra	Ghana	1.0
Ibadan	Nigeria	1.0

Note: All the cities above are capital cities, with the exception of Ibandan.

like magnets, but only to live under conditions that are probably much worse.

All the countries of Tropical Africa have high infant mortality rates, but the birth rates are high enough to produce overall population increases. In fact, the percentage increases are among the highest in the world. Kenya, with an annual natural population increase of 4.2 percent, is the most rapidly growing country in the world, but for 13 other countries of the region the rate of increase is more than 3.0 percent, and it averages 2.6 percent for the entire region. These figures can be compared with a 1.7 percent rate for the world and a 0.6 percent rate for the more developed countries (see Figure 12.17a on page 369). The infant mortality rates are highest in the western part of the region (129 deaths per 1,000 live births) and lowest in the eastern part (99

Focus Box

Burkina Faso

Burkina Faso (formerly Upper Volta) is one of the eight countries that make up the geographic region known as the Sahel, a semiarid area just south of the Sahara Desert (see Figure 12.A). The annual rainfall is slight and variable, the area has experienced periodic droughts for centuries, and the soil is poor. In spite of these fragile environmental conditions, nearly 90 percent of the people make their living by traditional agriculture and/or livestock-raising. The 1988 gross national product per person was about $180, making the country one of the poorest in the world.

The ethnic composition of the Burkinabe roughly follows regional divisions. The Bobo tribe inhabits the southwestern part of the country, the Mossi are in the north, the Goruma in the east, and the nomadic Fulani live along the northern areas that border Mali and Niger. Until the end of the nineteenth century, the Mossi tribe ruled the area and their empire was centered on what is now the capital city, Ouagadougou. The French overthrew the Mossi in 1896. After that, the Islamic religion, which was resisted by the Mossi, became more popular. Currently,

FIGURE 12.A

Burkina Faso. The absence of roads, railroads, and cities in Burkina Faso is evident from this map.

Africa

about half the Burkinabe practice native religions, while the remainder are divided between Islam and Christianity.

Since independence in 1960, political power has alternated between civilian and military governments. The current military regime came to power in a 1983 coup. The left-leaning government has links with the Soviet Union, Cuba, and Libya, but other nations, including the United States, contribute economic aid. In 1984 the current government leaders changed the name of the country from Upper Volta to Burkina Faso. Roughly translated, the new name means "Land of the Upright Men."

According to the World Bank, Burkina Faso is the sixth poorest country in the world in terms of gross national product per capita. As a result, a substantial number of working-age Burkinabe males migrate in search of work. Many go to Ivory Coast and Senegal, where the economies are much better. The contribution that such migrants make to the Burkina Faso's GNP is estimated at as much as 8 percent of the total. The literacy rate is Burkina Faso is only 11.4 percent for males and 3.6 percent for females, and only about one-quarter of the school-age children are enrolled in school. Thus, workers can obtain only the most menial jobs. Health standards are low, and the birth rate is high, so the country faces many economic obstacles. Burkina Faso's immediate future seems to depend on the vagaries of the Sahel rainfall.

SOURCE: Nancy V. Yinger, "Spotlight: Burkina Faso," *Population Today* 14 (May 1986): 12.

---

per 1,000). The countries with the highest rates are Sierra Leone (180 per 1,000) and The Gambia (174 per 1,000); those with the lowest are the island nations of Seychelles and Reunion, with less than 15 per 1,000 births. These figures can be compared with the world average of 82 per 1,000, and 17 per 1,000 for the more developed areas (see Figure 12.17b). The average birth rate for all of Tropical Africa is 44.1 births annually for every 1,000 people. This compares with an average of 27 for the world and 15 for the more developed areas. The highest birth rates are in Kenya (54 per 1,000), Malawi (53 per 1,000), and Benin (51 per 1,000), and the lowest are for the island countries, and the states in the very wet areas, such as Gabon, Zaire, and Equatorial Guinea (see Figure 12.17c).

Ethnicity

Tropical Africa is a tribal region, and most of the people are black. There are literally hundreds of tribes, but most of them have Bantu origins. Some of the smallest countries have the greatest tribal purity. For example, in Rwanda and Burundi, about 90 percent of the people belong to the Hutu tribe, with about 9 percent Tutsi and the remainder Twa (Pygmy). In Equatorial Guinea about 80 percent of the people are Fangs, but in the other countries no tribal group constitutes more than one-third of the population. Over 200 tribes live in Cameroon, even though about 30 percent of the people are Bamileke. At one time, all the tribal groups of Tropical Africa had established area boundaries, which were later superimposed by European boundaries. Thus, tribal boundaries now overlap the international boundaries in hundreds of places.

Poverty

The rapid population growth in Tropical Africa is occurring in a region that has been very poor, so

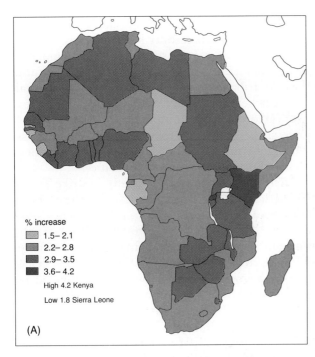

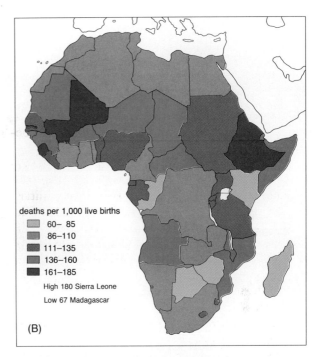

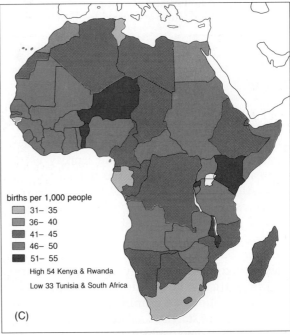

FIGURE 12.17

Infant Mortality (A), Birth Rates (B), and Rate of Population Increase (C) for Africa. The infant mortality rates are the annual number of deaths of children under the age of one for every 1,000 live births. The high is for Sierra Leone (180), the low is in Madagascar (67). The crude birth rates indicate the annual number of births for each 1,000 people; the high is for Kenya and Rwanda (54), the low is for Tunisia and South Africa (33). The annual rate of population increase is a percentage of the total population and is calculated as the birth rate minus the death rate, disregarding migration. The highest rate of population increase is for Kenya (4.2), the lowest is for Sierra Leone.

the crunch of additional people is causing severe economic problems. The average gross national product per capita for the entire region is $693 per year. It is difficult for North Americans to visualize such poverty when most of us spend that much for a color television set or a wedding dress to be used once. The oil-rich countries of the middle part of the region have the highest per capita gross national products—for Gabon, for example, it is $3,430—the highest in Tropical Africa. Ethiopia, on the other hand, has the lowest with $120. Ethiopia has become known for its poverty, which along with the droughts helps to explain the high death and infant mortality rates for the country.

Africa

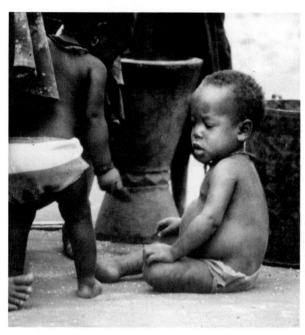

Poverty in parts of Africa is epitomized by this malnourished child in Tanzania. Notice the bloated stomach, a sign of food shortage. Photograph by Gordon Matzke.

Health

Only within the last 25 to 30 years have the terrible epidemic diseases been brought under control in Tropical Africa. Typhus, yellow fever, plague, smallpox, sleeping sickness, and leprosy used to decimate large segments of the population regularly. Although these mass killers have been controlled, sickness, pain, and starvation remain a normal part of life for most tribal Africans. Dietary deficiencies are caused in part by the droughts but also in part by the low fertility of the soils. Where rainfall is heavy, the soils are leached of their nutrients. Poor soils, combined with such cultural traditions as primitive farming methods and devotion to starchy foodstuffs, generally result in inadequate diets, which in turn lead to poor health and run-down physical conditions that are detrimental in fighting infections.

Primitive medical practices are still prevalent in many tribes, but local healers and curers (witch doctors) are of little value in fighting fungus and parasitic infections. Some exotic and localized habits, such as the clitorectomy initiation rites of girls, mutilation of faces in scarred designs, and extended lips and earlobes, remain simply because they are tribal customs. Many people in Tropical Africa believe that daily events are controlled by magical forces. This, combined with high illiteracy rates, makes it difficult to relieve pain and sickness when governments are occupied with maintaining their political and economic viabilities.

The most recent health tragedy to strike Tropical Africa is the geometric explosion in the number of cases of AIDS (acquired immunodeficiency syndrome). The AIDS virus is transmitted primarily through sexual contact and first appeared among African prostitutes in the late 1970s. Since then it has become a worldwide killer, but it is most prevalent in Tropical Africa. The countries hit worst by AIDS are Uganda, Rwanda, Burundi, Tanzania, Zaire, and Zambia (see Figure 12.18). Estimates in early 1987 indicated that in some cities of the region, such as Lusaka, the capital of Zambia, as many as one-third of the adult population is affected. At least 50,000 people in Tropical Africa have already died of AIDS, and if no cure is found another 2 million will die within the next 10 years. Most of the victims have been and will be between the ages of 19 and 40—the sexually active, and the ones the countries

These tribal women live in Kenya. Notice the mutilation of the ear for holding ornaments. Photograph by Gordon Matzke.

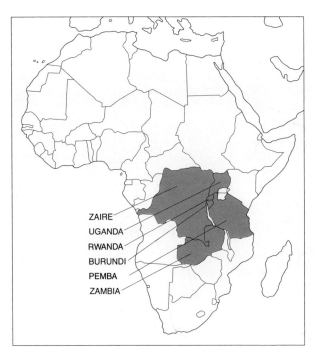

FIGURE 12.18

AIDS in Tropical Africa. In the six countries Zaire, Uganda, Rwanda, Burundi, Tanzania, and Zambia, the percentage of people who have been affected by the AIDS virus is the highest in the world.

can ill afford to lose. AIDS could have a major impact on the development of Tropical Africa.

Religion

The most prominent religious groups in Tropical Africa are the various tribal sects. Each group has its own set of beliefs, customs, and ceremonies, but the religions are all classed as *animistic*. Animists believe that such natural phenomena as water, trees, and wind are alive and have souls, or spirits. In addition, many believe in demons and spirits, and most tribes believe in a supreme being and a type of life after death. Reincarnation is another aspect of animism. Some tribes incorporate all these facets into their beliefs, and then include magic, sorcerism, and even certain aspects of Christianity. As a result, daily life becomes a complexity of activities geared to placate, appease, and apologize to various gods and spirits.

Colonial and missionary influences have made Christianity common in many parts of Tropical Africa. The largest proportion of Christians are Roman Catholic. In many cases, the dilution of Christian beliefs by animism is so complete that the religion appears to be Christian in name only. Liberia is 90 percent Christian, at least in name, and other countries where half or more of the people profess to be Christian are Uganda, Rwanda, Burundi, and Ethiopia.

Many Muslims live in Tropical Africa, especially in the countries that lie closest to the Sahara and along the coast. In Nigeria, for example, Islam predominates in the northern part of the country, while in the southern part Christianity prevails. In Senegal and Somalia, 80 percent of the people are Muslim, and in Gambia about 50 percent are Muslim. Because the Muslim religion, Islam, does not interfere with the treasured customs of the tribal Africans as much as Christianity does, it has enjoyed more success in gaining converts in Tropical Africa.

Language

Language patterns in Tropical Africa are among the most complex of anywhere in the world. One reason for this is that, while the languages spoken by the most people are indigenous, many of the official state languages have been imported. English has been the official language for Gambia, Sierra Leone, Liberia, Uganda, Somalia, and Kenya. French is the official language for Senegal, Ivory Coast, Zaire, Rwanda, and Burundi. In every case except Liberia, however, more people speak the tribal languages than the official language of their country. Some countries, such as Kenya, have changed their official language to a tribal language (Swahili for Kenya), and others are adding a tribal language as their second official language.

The most common language in Tropical Africa is one of the many dialects of Bantu, although in the northern part of the region the Semitic influence is strong. The tribal languages commonly are mixtures, such as Berber-Bantu or Arabic-Bantu, and even new forms of tribal languages continue to develop. The complexity of the

This Venda tribal chief's home carries messages in the designs inscribed on the outer walls. Photograph by Gordon Matzke.

mixtures of languages and the numerous dialects makes it impossible to make a meaningful map of languages for Tropical Africa. Only generalized regions where language families, such as Sudanese, Bantu, Semitic, and Hamitic are found, can be shown (see Figure 12.19).

Economic Geography of Tropical Africa

Agriculture

The majority of the people in Tropical Africa farm, raise livestock, or do both. In many countries, agriculture accounts for 90 percent of the occupations of the people, and most of these people are subsistence farmers. Plantation agriculture is common, however, and some products have made an appreciable contribution to world commerce.

Commercial products grown for trade and export include peanuts, coffee, tea, and cacao. Peanuts are grown extensively along the south coast of the African hump, and the crop is the leading export for Senegal, The Gambia, and Guinea-Bissau. Coffee is grown in all of Tropical Africa, but the major exporters are the eastern highland countries: Burundi, Ethiopia, Kenya, Tanzania, and Uganda. It is the main export for each of these countries. Tea also is grown most extensively in the eastern highlands, and producers are enjoying a recent upturn in trade, but the product is not yet the leading export of any country. Cacao trees, the source of cacao beans used in making cocoa and chocolate, are grown in the rainy areas along the south coast of the hump. Cacao is the leading export for Cameroon, Equatorial Guinea, Ghana, and Sao Tome and Principe. In fact, Ghana, the world's leading producer of cacao beans, provides the world market with more than 25 percent of its total.

Some commercial income is derived from the trees of the vast forested areas. Exotic woods, such as mahogany, ebony, and teak, have commercial value, and Congo, Gabon, and Zaire are leading exporters of timber and wood products. The palm trees also provide products for exchange and cash,

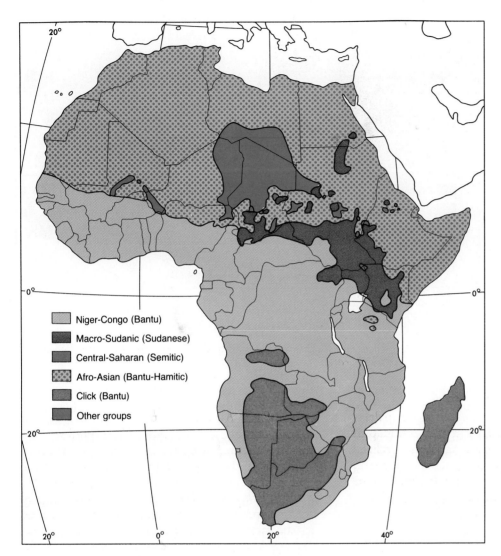

FIGURE 12.19
Language Regions of Africa.

including kernels, nuts, and palm oil. Palm products are found in the hot-wet areas and are the major export for Benin and Guinea-Bissau. One somewhat exotic commercial crop that should be mentioned is cloves. The island of Zanzibar (now part of Tanzania) produces most of the world's supply of that spice. (See Figure 12.20 for the location of the agricultural regions of Africa.)

In spite of the thousands of large commercial plantations, most agriculture in Tropical Africa is at the subsistence level. Most rural Africans live where their ancestors lived and farm as they farmed—for survival. Patches of the rainforest are burned off, and seeds are planted by poking sticks in the ground. The crops these farmers usually produce include yams, cassava, millet, maize, and rice, but many other crops also are grown—in fact, about 90 percent of all cultivated plants grown in the world can be found in Africa. Subsistence farmers often grow small amounts of commercial produce, such as peanuts, for exchange at their local markets. Any surplus, no matter how mea-

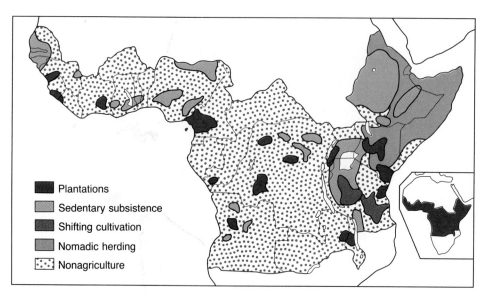

FIGURE 12.20
Agricultural Regions of Tropical Africa.

ger, is hand carried to a periodic local market for trade. Such markets serve both social and economic needs.

After farming, the most common occupation in Tropical Africa is animal husbandry. In many cases, the ownership of animals is a symbol of wealth. Cattle are a medium of exchange, and in some tribes, brides are purchased with a certain number of cattle. Other herd animals include donkeys, pigs, sheep, goats, and some horses. In most areas, animals are used strictly for subsistence, although in the eastern highlands they are commercial items. Hides and skins are among the major exports for Burundi, Djibouti, Somalia, and Burkina Faso.

Mineral Production and Industry

Because the continent of Africa is a huge plateau, Tropical Africa shows great promise as a source of energy. Rivers must flow over the edge of the plateau on their way to the ocean, so Africa has a potential for hydroelectric power which surpasses that of North America, South America, Europe, and Australia combined. Only a small fraction of this potential has been developed, but it means that Africa could become one of the world's great industrial areas.

With the exception of the fossil fuels used for energy, Tropical Africa is also blessed with great mineral resources. Large deposits of bauxite have been found in Guinea, Ghana, and Somalia, and near Boké in Guinea the bauxite deposit is one of the world's richest. Ghana has capitalized on both its bauxite and its hydroelectric power, and the plant at the Volta River project converts the ore to aluminum before it is exported.

Diamonds, gold, platinum, and silver are other commercially important minerals found in Tropical Africa. Most of the diamonds are of the industrial type, but gems also are found. Diamonds are among the chief exports of Central African Republic, Liberia, Sierra Leone, and Tanzania and account for more than half the income for the exports of Sierra Leone and Central Africa Republic. Gold is important in the eastern highland countries, especially Ethiopia, Rwanda, and Tanzania, and both gold and silver are found in the southeastern part of Zaire. (See Figure 12.21 for the location of the mineral producing areas of Tropical Africa.)

Most farmers in Tropical Africa still use primitive methods. These farmers in Zambia are using oxen to pull their plow. Photograph by Gordon Matzke.

Iron ore is found throughout Tropical Africa, but because coal is not in abundance anywhere in the region the potential for a steel industry does not exist. Iron ore is a major export for Liberia and Sierra Leone and accounts for about 70 percent of Liberia's trade income. Nigeria is one of the world's leading oil producers, so fossil fuels are not entirely nonexistent. Crude oil accounts for about 85 percent of Nigeria's income from exports and more than half of Gabon's exports. Other than in those two countries, very little oil or natural gas is produced in Tropical Africa. Exploration for oil is being conducted, however, especially in the eastern highland countries.

Other minerals produced in Tropical Africa are important to individual countries. Zaire, for example, produces about 45 percent of the world's cobalt and about 6 percent of the world's copper, but copper is the leading income producer and consists of about 65 percent of Zaire's exports. The Katanga region of southern Zaire is the rich mineral area. Copper is also the major export commodity for Zambia. Nigeria produces most of the world's columbium, a mineral used as an alloy to add hardness to steel; manganese comes from Gabon, and tungsten comes from Rwanda. Thus, Tropical Africa produces a variety of minerals, but they come from individual countries. No single area has a large variety of minerals.

Heavy industry in Tropical Africa is notably absent. The light industries that do exist are associated with agriculture. Tanning, textiles, and food-processing are common. Other industries include shoe production, a few bicycle and radio manufacturers, and fishing. Tourism is important in the eastern highlands, where the magnificent scenery and wild animals are attractions. Tourism is promoted in all areas, but few major attractions exist in the west. One exception should be noted. The Nobel Prize winning physician Albert Schweitzer worked in Gabon, and the medical facilities and programs he initiated, as well as his grave, are visited regularly by tourists.

Transportation and Communications

No transcontinental railroad lines exist in Tropical Africa, and no regions have what could be called a

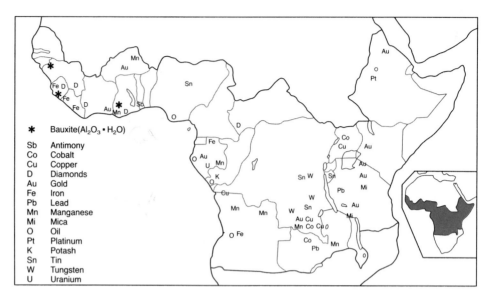

FIGURE 12.21

Mineral Deposits of Tropical Africa.

network of lines. Most railroads run only from the coastal ports to the interior, so few lines cross the international boundaries. Most of these single lines have very few feeder lines. The purpose of these railroads it to tap some interior resource and bring the resulting commodity to the coast for exporting. The railroad mileage is expanding in most countries, but some of the lines, remnants of the colonial era, are in much need of repair. Gambia, Guinea-Bissau, Guinea, Sierra Leone, Liberia, Rwanda, and Burundi do not have a single mile of railroad track.

Tropical Africa is in slightly better shape with respect to roads than with railroads, and trucks carry things that normally would be hauled by rail. It is possible to travel between most countries on paved highways, but many of the roads are gravel. In Nigeria the road system actually approaches what might be called a network. Motor transportation outside the urban areas is not common, and most of the traffic consists of trucks, military vehicles, police cars, and a few government cars. The roads are used extensively, however, by pedestrians, cyclists, and two-wheeled carts. Goods, even commercial items, are often transported by human power over local trails and soft-surface roads.

Principal airports are scattered throughout Tropical Africa. Most of the capital cities have major airports, and a few are at such unlikely places as Bukavu in eastern Zaire. The two largest airports in Tropical Africa are located at Dakar in Senegal and at Nairobi, Kenya. From either place, international flights arrive and depart regularly. Although airplanes are common in the region, they certainly do not carry most of the African people that travel.

Communications are hampered by the large number of languages that are spoken in Tropical Africa. Even if funds were available for governments or independent agencies to pursue communications development, the language barrier is nearly insurmountable. Nevertheless, radios and radio stations are becoming common, especially in the urban areas. Television is not common, but Nigeria does have about 500,000 licensed receiving sets. Most countries still have no television transmitters, and many people have never seen a television. Nigeria, the leading country of the region, with a population of 105.4 million, is served by

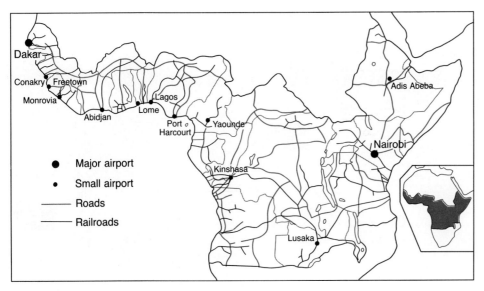

FIGURE 12.22
Roads, Railroads, and Airports of Tropical Africa.

only 154,200 telephones. There are local newspapers, but nearly 80 percent of the people of Tropical Africa are illiterate. Newspapers are found only in the urban area.

Political Geography of Tropical Africa

For many reasons, and especially because of the scattered resources, political integration would benefit nearly all the countries of Tropical Africa. Most Africans agree with this idea in theory but differ sharply on how to achieve unity. Two approaches have been suggested. One is to federate all governments immediately and press hard for complete political unification. The other approach is to begin with economic and social cooperation, modeled after the European Economic Community, and then build toward political unity through such ties. But political unity would mean lost world prestige, money, and power for many government officials, not to mention the number of votes in the United Nations. Also, tribesmen and the common people have little awareness of or concern with the problem of statehood. Most African countries will therefore remain weak, separate political entities.

Many boundary problems exist in Tropical Africa. The colonial empires drew boundary lines on a map according to their agreements with the Europeans and not according to the location of African tribes. Thus, many tribes united by culture and language were separated by political lines, and the old colonial boundaries were retained when independence occurred. Tribes remain divided, yet most people have a fierce loyalty to their tribe. Some have pressed for regional autonomy, but that only serves to divide the states even more. Carried to the ultimate, tribal autonomy could fracture Tropical Africa into hundreds of political units.

For the immediate future, few drastic changes may be expected. Each government wants to maintain its present position. Tribal loyalties will be constrained, by force if necessary, to conform to national programs. Any combination of states will be opposed by the present leadership. Central governments, strong in a local sense, will continue. Many African leaders today recognize the

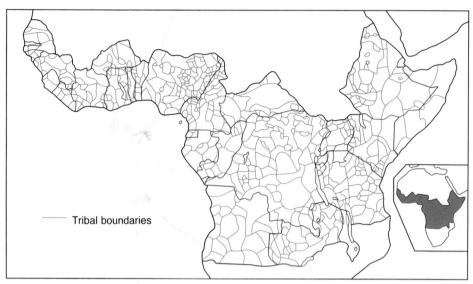

FIGURE 12.23

Tribal and National Boundaries of Central Africa.

SOURCE: This map is an adaption of "Peoples and Political Boundaries in Africa" from Paul W. English. *World Regional Geography: A Question of Place,* New York: John Wiley and Sons Publisher, 1984. Used by permission.

problems, however, and most seek viable answers. The future of Tropical Africa rests with these leaders. (See Figure 12.23.)

SOUTHERN AFRICA

▪ Introduction

The countries of Southern Africa include Botswana, Lesotho, Madagascar, Mozambique, Namibia, South Africa, Swaziland, and Zimbabwe. They lie generally south of the 15th parallel. Only parts of Mozambique and the island nation of Madagascar are north of that line. No single physical feature separates the region from the remainder of the continent, but a human factor does tend to give the area a uniform aspect—the political strength of European settlers. This political strength is at its utmost in the Republic of South Africa, where the system of apartheid assures white control. Until recently, Mozambique was controlled by Portugal, and Namibia by South Africa, and Zimbabwe (formerly Rho-

Bantus The large family of Negroid tribes occupying Tropical and Southern Africa.

desia) was a completely white-controlled state until the late 1970s. The political strength of the whites has declined in all countries except the Republic of South Africa, but its influence remains. (See Figure 12.24 for location of countries of the realm.)

The region was settled originally by Hottentots and Bushmen. The **Bantu** tribes came at a later time during their migrations from the Cameroon region. Moving eastward through the rainforest and then into the highlands, they followed the lakes southward into Southern Africa. The later stages of the migration took place about the time of the birth of Christ and continued until modern times. It is considered to be one of the greatest migrations conducted by humans.

As the Bantus moved through the continent, they displaced and absorbed other, more placid tribes. The Pygmies, for example, were once great

The Kariba Dam on the Zambezi River forms Lake Kariba, which is part of the international boundary between Zambia and Zimbabwe. Photograph by Gordon Matzke.

Bushmen A race of nomadic hunters of Southern Africa chiefly confined to the region of the Kalahari Desert.

Hottentots A race of people of Southern Africa thought to be a mixture of Bushmen and Bantus with a yellowish-brown complexion. Noted for the clicks in their speech, which make them appear to stutter.

Zulu wars The conflicts between the Boers and under later kings of the British with the great Bantu nations of the Natal region of South Africa.

Boer A rural descendant of the early Dutch settlers of South Africa.

in number, but they have dwindled to only about 150,000 today. These small, pleasant people average only about 4.5 feet (1.37 meters) tall. The **Bushmen** were pushed by the Bantus into the harsh arid climate regions of Botswana, Namibia, and Angola, and only about 50,000 remain today. The **Hottentots** faired only slightly better, but they still found themselves and their herds of fat-tailed sheep forced into the drier areas while the more fierce Bantus took over the humid regions. When the Dutch farmers moved away from the south coast and onto the plateau of the High Veld, they met the still-expanding Bantus. This confrontation, known as the "**Zulu wars,**" still furnishes substance for the legends of both groups.

The Great Trek of the **Boers** from the area near the Cape of Good Hope to the High Veld is a major event in the history of South Africa. Dutch in origin and espousing a fundamentalist Calvinistic religion, these farmers resented the British manners, morals, and interference. Loading their possessions into large two-wheeled carts, they left the land they had settled and crossed hundreds of dusty miles to the high country, where grass and water nurtured their crops and livestock. There, however, they met the fighting Bantus. The bloody struggle between these groups was settled by Dutch dominance and the establishment of Transvaal and the Orange Free State. At the conclusion of the Boer War (1899–1902), the hated British rule moved northward and the British victors incorporated the two Dutch states into the political area presently known as the Republic of South Africa.

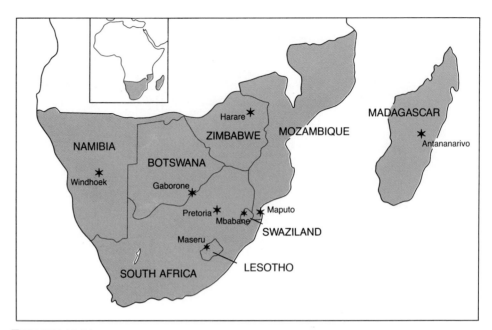

FIGURE 12.24
Countries of Southern Africa.

Southern Africa today is the most prosperous part of the continent, and the Republic of South Africa is its most affluent state. The region's outstanding mineral deposits, pleasant climate, and relatively good farm land have made it an attractive place to live. These seemingly utopian conditions are mitigated by the political disruptions in South Africa created by the complete political control that a white population of 18 percent has over the black population of 82 percent.

Physical Geography of Southern Africa

Southern Africa is mostly a high plateau. The narrow coastal plain widens somewhat on the east in Mozambique, but all the other parts of the region are above 3,000 feet (914 m) in elevation and about one-fourth of the region is above 4,500 feet (1,372 m). Mountains are not common, but the Drakensberg Range, near the southeastern coast in South Africa, is significant. The tallest peak in the range is Thabana Ntlenyana at 11,425 feet (3,482 m). The highlands of Namibia drop to the Atlantic Ocean on the west in only a few miles, and the coastal plain is very narrow. On the east, in Mozambique, the coastal plain averages about 150 miles (240 km) wide.

The island of Madagascar has a low ridge running through its middle and for nearly its entire length. The ridge is approximately 4,500 feet (1,372 m) in elevation and is fringed by coastal plains. Numerous rivers flow in all directions off the highlands and into the Indian Ocean. (See Figure 12.25.)

The two largest rivers in Southern Africa are the Zambezi and the Orange. The Zambezi begins near the Katanga region of southern Zaire, within 50 miles of the headwaters of the Zaire River, loops southward through Zambia, and then turns east and is used as the border between Zambia and Zimbabwe. Located along this stretch of the river are two interesting features. The first is the famous Victoria Falls, where the river plunges over the edge of the plateau, forming a waterfall that is three times wider than Niagara Falls and more than twice as high (it has a drop of 355 feet, or 108 m).

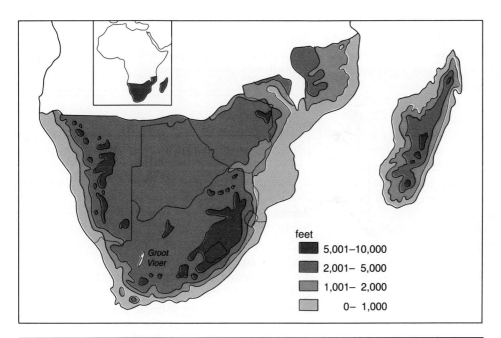
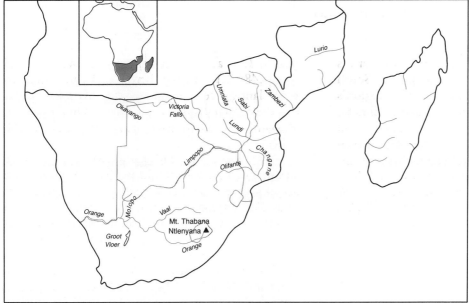

FIGURE 12.25

Land Elevations and Water Bodies of Southern Africa.

The falls are one of the great natural sights of the world. A mist cloud formed by the crashing water is seen for miles around, and the roar is heard from 10 miles away. The second noteworthy feature along the Zambezi is the Kariba Dam, built in the Kariba Gorge in 1959. The dam is 420 feet (128 m) high, and Lake Kariba backs up for 150 miles (240 km) behind it. After the dam, the river continues on through the plains of Mozambique, eventually emptying into the Indian Ocean. The Orange

River and the Vaal, its major tributary, begin in the Drakensberg Mountains and flow westward through South Africa. The Orange, used as the boundary between South Africa and Namibia, empties into the Atlantic Ocean.

Southern Africa has four distinct climatic regions. The deserts are in the southwest, the middle is steppe, in the north is the forest, and in the far south and southeast the climate is marine with the dry season in the winter. The rainfall on Madagascar decreases drastically from east to west across the island; the eastern half is always wet, while rain is seasonal on the western half.

Much of Southern Africa is covered with steppe and desert. Rainfall amounts decrease from the eastern highlands of Tropical Africa south and westward to the Atlantic coast (see Figure 12.26). The Namib Desert, located along the coast, is often foggy, but very little moisture falls to the ground. The result is one of the driest deserts in the world. The Benguela Current, a cold ocean current, flows northward in the Atlantic Ocean just off the coast. Winds that blow over this current pick up very little moisture because the cold water does not evaporate. Thus, the winds carry hardly any moisture onto the land. The conditions are very similar to those in the Atacama desert on the west side of South America, and for similar reasons.

Inland toward the east from the Namib, another great desert area begins. The Kalahari Desert covers 220,000 square miles (551,980 km^2) of Namibia and Botswana. Surrounding the Kalahari on all sides are the steppe grasslands. North of the grasslands, the thorn forests and dry woodlands begin. Most of the rain in the region falls during the southern hemisphere summer, or from November to April.

■ Human Geography of Southern Africa

Population and Urbanization

The overall density of population in Southern Africa is similar to that in Northern Africa, but the Kalahari and the Namib do not influence the entire region, as the Sahara does in the north. The

Coloreds Strictly meaning "having color," this word pertains to the people of South Africa that are Caucasian-Negroid mixtures.

population density ranges from about 128 people per square mile in Lesotho to less than 3 people per square mile in Botswana. The total population of the region was 71 million in mid-1986, but nearly half the population lives in the Republic of South Africa (33.2 million). The rate of natural increase for the region is 2.8 percent, varying from a high of 3.5 percent in Zimbabwe to 2.3 percent in the Republic of South Africa.

The birth rates and death rates for the whites and Asians in the Republic of South Africa are similar to those for the more developed areas of the world, but the figures for the blacks and the remaining countries are similar to the figures for Tropical Africa. The average birth rate for the region is 45 per 1,000 people, excluding South Africa, where it is 33 per 1,000. The overall death rate is 16 per 1,000 for the region, but only 10 per 1,000 in South Africa. Infant mortality rates are high in all countries of the region, including South Africa. The average for the region is 102 per 1,000 live births, compared with 82 per 1,000 for the world.

The percentage of the population living in urban areas in Southern Africa is relatively low. Most people live on farms and in small villages and towns, but some great cities have developed. Cape Town, on the southwest tip of the continent in South Africa, has a metropolitan population of about a quarter of a million, but another million live in the urban area. The majority of the people in Cape Town are "Cape Coloreds," usually just called "**Coloreds**," and are descendants of white and black parents. Cape Town is a port city as well as an industrial city. From Cape Town northward along the west coast of the continent, there are no large cities until Luanda, Angola, nearly 2,000 miles (3,220 km) away. This scarcity of cities is caused by the barren interior—the desert does not produce enough of anything to be shipped out—and the lack of any natural harbors.

No major harbors and no large cities exist east of Cape Town until Port Elizabeth, a distance of

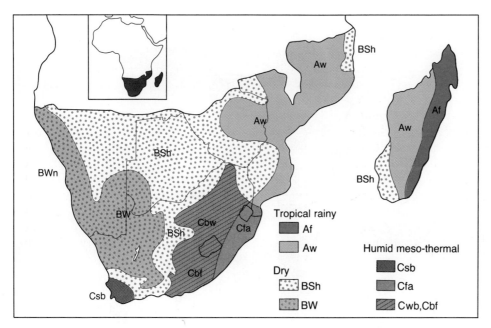

FIGURE 12.26
Climate Regions of Southern Africa.

500 miles (800 km). Except for the small port of East London and a few lesser ports, not many people live between Port Elizabeth and Durban, another 500 miles. Durban, with more than half a million people, is by far the largest of these three port cities. A large portion of the people in these east coast cities are of Asian descent. Between East London and Durban, however, there is a heavy concentration of Bantu tribes, but no cities of any size. Farther north along the Mozambique coast, the largest city is the capital of Maputo (about 785,000 and once called Lourenço Marques).

The largest city in Southern Africa is Johannesburg, South Africa, an inland city located in the northern fringe of the Drakensberg Mountain range (see Figure 12.27). The elevation of the city is similar to that of Denver, and the population is about 1.5 million. Another 500,000 people live in Pretoria, just to the north of Johannesburg, and there are many other smaller cities and towns in the Johannesburg vicinity. The reasons for this heavy concentration of people will be discussed later.

Other urban places in Southern Africa include Harare (once called Salisbury), the capital of Zimbabwe (population 656,000), and Antananarivo (population 650,000), the capital of Madagascar. The countries of Botswana, Lesotho, Namibia, and Swaziland have no large cities. For example, Mbabane, the capital of Swaziland, has only 39,000 people.

Ethnicity

The vast majority of people in Southern Africa are black and are descendants of various indigenous African tribes. The country with the most white Europeans is the Republic of South Africa. Some 18 percent (about 6 million) of that country's 33.2 million people are white. In all the other countries, however, less than 5 percent of the population is white, and in most cases less than 1 percent. Most of the blacks are Bantu, but other tribes, such as the Ndebele, the Swazi, the Shona, and the Sotho, are represented. Madagascar is populated with 18 different Malayan tribes, a few Arabs, and a small number of Black Africans. About 3 percent of the people in the Republic of South Africa are Asian, and about 10 percent are Colored (black and white mixture).

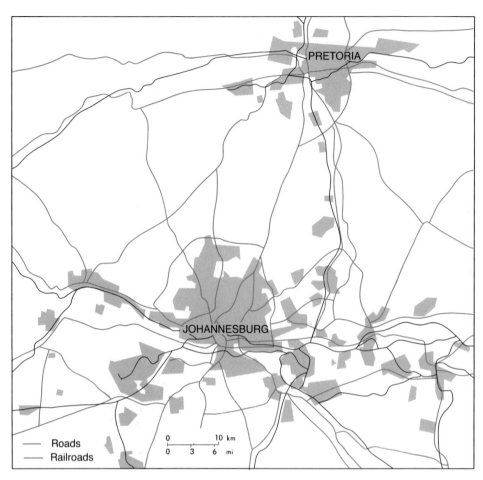

FIGURE 12.27

Johannesburg and Pretoria. Johannesburg is the largest city in Southern Africa, with more than 1.5 million people. Pretoria is the administrative capital of South Africa.

Religion and Language

A surprising number of the people in Southern Africa are Christians. They are predominately Protestant: Dutch Reformed, Anglican, and Methodist, although 43 percent of the people in Lesotho are Roman Catholic. The heaviest concentrations of Christians are related to the white settlements in Zimbabwe, Botswana, and the Republic of South Africa, although missionaries have converted large numbers of the blacks. About 90 percent of the people in Lesotho and 80 percent in Swaziland are Christian, as well as nearly half the people in Mozambique and Madagascar.

Among the other predominant faiths in Southern Africa are the various tribal beliefs. These are found within all the tribal groups throughout Southern Africa, but they prevail along the southeast coast and north of the Zambezi River. Most of the tribal beliefs include aspects of animism. About 10 percent of the people of Mozambique and 2 percent of those in Madagascar are Muslim, and a few Hindus also live in Southern Africa.

Of the many languages spoken in Southern Africa, various forms of Bantu are the most common. These are scattered throughout, but concentrated in the northern and eastern parts of the region. The Hottentots and the Bushmen have

Afrikaners Another name for the Boers of South Africa.

---

their own languages, as do the smaller countries. Sotho, for example, is the language of Lesotho, and siSwati is spoken in Swaziland. The language of Madagascar is Malagasy, of Malayan origin, but French is also common on the island.

English is the official language of Botswana, the Republic of South Africa (along with Afrikaans) Swaziland (along with siSwati), and Zimbabwe, even though most of the people of these countries do not speak it. **Afrikaners** (pronounced Af-ri-con-ers) are South Africans who are descendants of the seventeenth-century Dutch settlers, and their language is also of Dutch origin. The original Dutch has been diluted with English and Bantu and differs considerably from the parent language.

Economic Geography of Southern Africa

Economic Conditions

The poorest country in Southern Africa is Mozambique, but the people in Madagascar are not much better off. The average gross national product per capita for the region is $928, and $150 for Mozambique. Economic problems became severe in Mozambique shortly after the country obtained its independence from Portugal. A Marxist, one-party state was created, and private schools were closed, private homes were nationalized, and collective farms were organized. As a result, most of the country's 160,000 whites emigrated, leading to chaos in both the farming and the administrative systems. Only about 4 percent of the country's land area is considered arable, and because the Bantu people are not expert farmers, agriculture is limited. In addition, very few minerals have been found that might supplement the country's meager income.

The more wealthy countries of Southern Africa have good supplies of mineral resources, and some have favorable environments for crops and livestock. The richest country of the region, the Republic of South Africa, has both minerals and some good farming areas. It also has one of the highest GNPs per capita in Africa ($2,240). About 30 percent of the people work in each sector of the economy: agriculture, industry, and the services. The small remainder work in mining, even though mining is a major industry and provides the country with much of its income.

Agriculture

The livestock raising and agriculture regions of Southern Africa correspond closely to the rainfall pattern. Because of the dryness of the Kalahari and the Namib, a large part of the region's northwest is devoid of any agricultural activities. On the northern edge of the Kalahari in Botswana, Namibia, and Zimbabwe, rainfall is sufficient to support grass. The major occupation of that area is nomadic herding. Farther north, where it is wetter, the grassland gives way to what was once forests of broadleaf, deciduous trees. The forests have now been replaced with farms. A similar sequence of rainfall and natural vegetation is found south and east of the Kalahari—first grassland, then the forests. The grasslands, however, are used for ranching, and the once-forested parts are now mixed farming regions. (See Figure 12.28.)

The agricultural economy of each country of Southern Africa may be predicted from the country's location with respect to the overall rainfall pattern of the region. For example, the areas of Botswana and the Republic of South Africa that are not desert are grassland, and the predominant industries of both states are cattle raising and sheep raising. Farmlands in Zimbabwe are rich, and the seasonal rainfall is sufficient to grow such products as tobacco, sugar, and cotton; cattle also are important. Thus, meat, tobacco, and sugar are Zimbabwe's major exports, and cotton-textile production is the major industry. Mozambique is in the wettest part of Southern Africa, and the country exports copra (dried coconut meat from palm trees) and sisal (rope fiber from hemp), both of which require large amounts of moisture. Cotton is also produced in Mozambique, but as in Zimbabwe the textile industry is for local production

FOCUS BOX

Madagascar

Madagascar is a poor, underdeveloped island country that has undergone the downward spiral in its per capita gross national product within the last few years. A decade of state control has not yet delivered the promised improvement in material well-being of the island people, known as Malagasy. Frustration, disillusionment, and recrimination are part of the international response to the failed expectations of progress. Madagascar is a poor country that has gotten poorer, and the economic dilemma has no ready solution. (See Figure 12.B.)

Unlike other countries of the African region, Madagascar boasts a singularly hybrid cultural heritage. Ancient mariners from Indonesia first arrived on the island about 1,500 years ago. Their contributions were rice as the preferred staple, hilltop villages, rectangular dwellings, and the worship of ancestors. Later, Africans reached the island and introduced zebu cattle, the groundnut, wood sculpture, and toga clothing. As settlement spread from the coast to the interior, the people differentiated into 18 tribes. All islanders, however, share one fundamental culture and speak a language in the Malayo-Polynesian family called Malagasy.

In the 1820s the British missionaries became the first Europeans to live in Madagascar, but the French eventually gained political control of the island. Madagascar became a colony of France in 1896. A revolt against French rule in 1947 eventually led to independence in 1960. The first government was a constitutional republic, but it was replaced by a leftist regime in 1972. Eventually all major industries, banks, and insurance companies came under government ownership or control. The Communist bloc has furnished weapons, military advisers, ideology, and scholarships to the Soviet Union.

To finance its socialist agenda of nationalizations and state investment, Madagascar accumulated a public debt to foreign creditors that surpassed $3 billion in 1985. The yearly

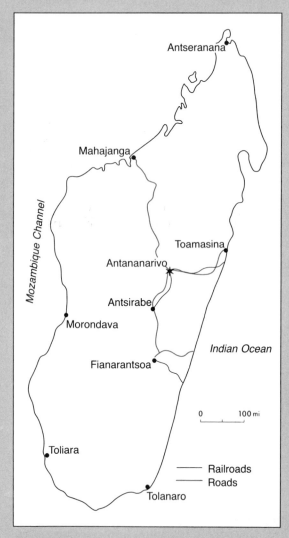

FIGURE 12.B

Madagascar. Most of the cities in Madagascar are located along the coast of this island nation. Few roads penetrate the interior.

repayment schedule is more than three-fourths of the country's export revenues and has recently been abandoned. Everyday life reflects the consequences of an economy in disarray. Most consumer goods are no longer available, and shortages of most basic food needs occur in the cities. Economic conditions steadily decline, and large infusions of outside aid have not saved Madagascar from being one of the poorest countries in the world.

SOURCE: Daniel W. Gade, "Madagascar and Nondevelopment," *Focus* 35 (October 1985): 14–21.

only. The leading industry in Lesotho is livestock raising, and its important exports are wool and mohair (Angora goat hair). Swaziland's agriculture is largely subsistence, but the state does export sugar and citrus fruits. Rubber grown in the wet eastern half of Madagascar is that country's leading export, but rice, coconuts, and tapioca are also grown.

The Republic of South Africa produces many types of agricultural commodities but has little natural agricultural land. Only about 12 percent of the country is arable. Generally, the climate is too dry, and the typically poor African soils are not fertile. In some places, however, conditions are good for farming: (1) the Natal, part of the coastal plain north of Durban, (2) an area of the Orange Free State between the Vaal and Orange rivers, and (3) the coastal lowlands surrounding Cape Town. The three areas were originally settled by the Afrikaners, and most of the farmers in

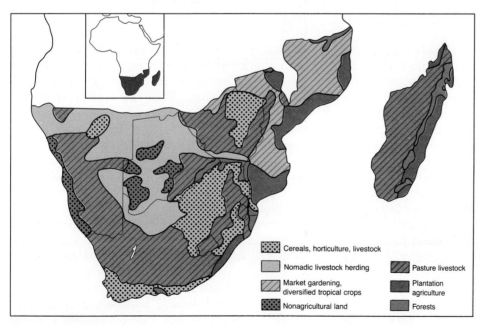

FIGURE 12.28
Agricultural Regions of Southern Africa.

these areas today are their descendants. The most notable farm products are corn, wheat, peanuts, and sugar, but much wine is produced in the vineyards along the south coast. Livestock raising is also important. Wool and mutton come from the numerous sheep ranches, and butter, cheese, and fresh milk are produced on the many dairy farms. Much of the beef that is eaten is imported from Botswana.

In some ways, the Republic of South Africa and California are comparable. South Africa is three times bigger than California, but their populations are similar in size. Both places have large farming industries with little space in which to farm, and the economies of both are dominated by industries other than farming. California and South Africa have regions with favorable climates and attractive coastlines, and tourism is a major industry. The production of wine is an important industry in each. And California and South Africa are among the few places in the world that had nineteenth-century gold rushes and that mine gold now.

Mineral Production and Industry

The Republic of South Africa is the industrial giant of Africa, in large part because of the rich mineral deposits with which the country was endowed. The country does lack adequate petroleum deposits, but more than 50 other minerals are produced, and the 1983 gold output alone was $8.9 billion. South Africa leads the world in the production of gold, gem diamonds, and antimony (used as an alloy to harden other metals, as well as in medicines and pigments). The country is the only African state with large coal and iron ore deposits, which makes steel production possible. Thus, the manufacturing of steel is one of the major industries of South Africa. Other important industries include the production of tires, electric motors, plastics, textiles, and furniture. Financially, however, the capture of minerals is the most important industry in the Republic of South Africa.

Mineral production in the Republic of South Africa began in 1868 with the discovery of diamonds. Diamonds were found first along the banks of the Vaal River. Dry digging, however, began at a place southeast of the river, and many people rushed to the site, hoping to cash in on the discovery. What had been a barren spot soon became the city of Kimberley. Today, because of its central location, Kimberley is the transportation hub of the state, and it still leads the world in the production of diamonds. (See Figure 12.29).

About 20 years after diamonds were found, gold was discovered north of the Vaal. The area is now a province of the Republic of South Africa known as the Transvaal, or "on the other side of the Vaal." The gold was found on a ridge called Witwatersrand, which means "ridge of white water." Again, people rushed to the site of the discovery, many more than those that came to the diamond discovery. People came from all over the world, and what had been a desolate place became another city. The original mining town is now called Johannesburg, one of the largest cities in Africa. The Witwatersrand today is dotted with towns and is one of the most famous and productive mining areas of the world.

Shortly after the gold discovery, coal too was found in the Witwatersrand. Industrially, the coal became more important than the gold. The coal was used originally to power the gold-mining machinery, but it soon became the power source for many other industries. Iron ore was discovered in the same general area, and the steel industry followed soon after. As if these discoveries were not enough, still other rich veins of minerals were discovered, including asbestos, copper, manganese, and platinum. Even the tailings from the gold mines were found to contain a high concentration of uranium. Mineral production in the region remains high today.

Aside from South Africa, other states in Southern Africa also have mineral resources. In Swaziland, mines produce gold, tin, mica, asbestos, and coal. Zimbabwe exports asbestos, copper, and chrome, and nickel comes from deposits in Botswana. Nearly all the minerals known are produced in the Southern African states, and in some cases a large share of the world's known deposits are located within the region. In spite of this vast mineral wealth, however, only the Republic of South Africa can be classified as an industrial state.

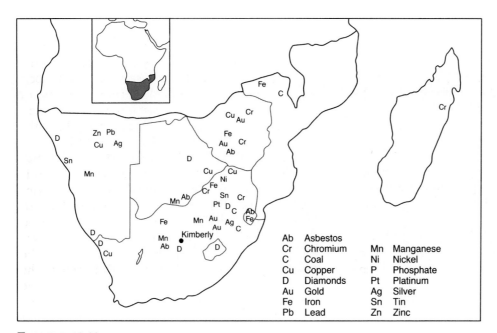

FIGURE 12.29
Mineral Deposits of Southern Africa.

Transportation and Communications

The Republic of South Africa is the only country of Africa with a railroad network. In addition to the routes that lead to other states in the region, all its major cities, ports, and resource regions are connected by rail (see Figure 12.30). Most of the countries of Southern Africa have only one or two routes running from interior resource areas to coastal ports. Of the eight South African states, four are landlocked and another is an island. Railroads are especially critical to the landlocked states for both exports and imports. Both Swaziland and Zimbabwe, for example, rely on routes through Mozambique. And trains from Botswana must go through Namibia, South Africa, or Zimbabwe to get to a port.

Major highways of the region follow a pattern similar to that of the railroads. South Africa has a complete network, and the other countries have one or two major roads. Most roads are tarred rather than paved, and it is possible to drive on these all-weather roads for great distances and at good speed. Of the 142,665 miles (229,690 km) of highways in South Africa, about half are constructed of crushed stone or gravel.

Air service is good in most of Southern Africa. The Republic of South Africa alone has 106 permanent-surface runways, and 12 that will handle large commercial jets. Passage can be made to every city of the region, and the country has about 20 international airline connections.

The telecommunications system of the Republic of South Africa is the best developed, most modern, and has the highest capacity of any state in Africa. The country has 13.1 telephones for every 100 people and supports 67 main television stations and 286 FM radio stations. Zimbabwe, on the other hand, has allowed its once excellent system to deteriorate from lack of maintenance. That country has only 3.1 telephones for every 100 people, and only 8 television stations. All the countries of the region have at least one television station, and tiny Swaziland has 11.

Political Geography of Southern Africa

With the exception of Botswana, all countries of Southern Africa have in recent years been the subject of strong protest and debate in the United Nations and in the media worldwide. The countries themselves also have experienced much violence. The major political problem stems from the fact that the majority people, the blacks, have been controlled completely by the minority whites. The region contains most of the white population of the continent, and they have been in political control since statehood. Only recently did the Portuguese disengage themselves from Mozambique, and the blacks win political control of Zimbabwe. The Republic of South Africa remains in the hands of the white officials. Under the country's **apartheid** system, blacks are restricted to certain living places and occupations and are paid much less than whites for similar work. In addition, blacks cannot vote or run for public office. In 1976, in protests over restrictions, 600 Bantus were killed in riots, and the protests continue today. Agitation from outside the country has increased, but peaceful racial harmony seems remote as long as the whites control the economy and the political system.

apartheid Afrikaans for "apartness," which means a strict racial segregation as practiced in South Africa.

The Republic of South Africa also holds political control of Namibia. Although sparsely populated, Namibia is nearly two-thirds the size of South Africa and contains many valuable mineral deposits. The South Africans won Namibia from the Germans in 1914, but other African nations have charged that South Africa has exploited Namibia and built military bases and imposed apartheid on the region. Finally, in 1971, the International Court of Justice declared that South Africa was occupying Namibia illegally. South Africa has since fought the decision as well as any military attempts to enforce it. Although Namibia is supposed to be independent, South Africa retains direct control of the territory today.

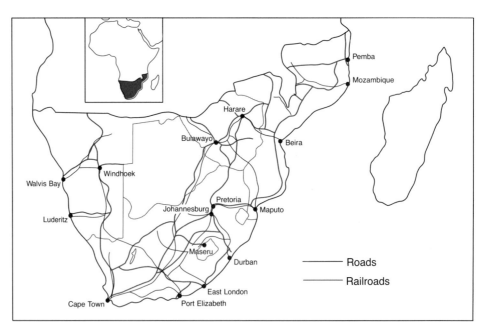

FIGURE 12.30
Roads and Railroads of Southern Africa.

FOCUS BOX

Apartheid in South Africa

"Apartheid" is the name for the policy of racial segregation practiced in the Republic of South Africa. Separate development of the races became official government policy with the 1948 election of Daniel Malan's National Party. The policy called for separate development, separate residential areas, and ultimate political independence for the whites, Bantus, Asians, and "Coloreds." In 1959 the government passed laws providing for the eventual creation of several Bantu nations (Bantustans) on 13 percent of the country's land area.

Of the 33.2 million people in South Africa today, 67 percent are black, 18 percent are white, 10 percent are Colored, and about 3 percent are Asian. The tricameral parliamentary government has one chamber each for whites, Coloreds, and Asians; the Colored and Asian chambers were added in 1983. The country is divided into 4 provinces and 12 "homelands" for the blacks. The first homeland, established in 1963, is called the Transkei and is made up of three separate land areas located in the southeastern part of the country. Transkei is an independent, self-governing area. The other homelands are Ciskei, Bophuthatswana (six separate units), and Venda (two discontinuous unit areas). None of the homelands has received international recognition.

Under apartheid, blacks also are severely restricted to certain occupations and are paid much lower wages than whites for similar work. In addition, only whites may vote or run for public office. In the 1980s, however, some liberalization measures were taken. As a result of internal unrest created by antiapartheid factions and in response to international sanctions imposed on the country. Laws banning interracial sex and marriage were repealed in 1983. In 1986, the nation's system of racial pass laws was ended, so that blacks are no longer required to carry identification passes to enter and leave certain areas. Blacks now also have an advisory role in the government.

For much of the world, the battle against apartheid in the Republic of South Africa is symbolized by a few black leaders. The best known of these is Desmond Tutu, who is the Anglican archbishop of Cape Town, the spiritual leader of 1.3 million Anglican South Africans, both black and white, and winner of the Nobel Peace Prize. This religious leader does not have a huge political following, but he is very popular both inside and outside South Africa. Antiapartheid leader Nelson Mandela, on the other hand, does have a large political following, but he has been imprisoned for many years for his antigovernment stand. Another important black leader in South Africa is Mangosuthu Buthelezi, chief of the 6-million-member Zulu tribe. The more the South African government condemns these men, the more recognition and honor they receive, both locally and worldwide.

In 1976, at least 600 people, mostly Bantus, were killed in riots protesting apartheid, and protests by blacks continued through the 1980s. Police reactions have caused several hundred deaths. Blacks also have organized work stoppages that created chaos in the mines and in the manufacturing and retail sectors of the country. The European Economic Community, Japan, and the United States have all imposed economic sanctions against South Africa, none of which seems to have had much impact on the government. A nationwide state of emergency exists, but as of early 1989, apartheid continues to be national policy.

SOURCE: Mark S. Hoffman, ed., "South Africa," *World Almanac, 1987* (New York: Pharos Books, 1987), pp. 612–613; Jill Smolowe, "South Africa: Battle at the Burial Grounds," *Time*, September 15, 1986, pp. 39–40.

Key Words

Afrikaners	Hamites
apartheid	Hottentots
Bantus	nomadic pastoralism
Bedouins	orographic
Berbers	precipitation
Boers	Pygmies
Bushmen	Watusi
Coloreds	Zulu wars

References and Readings

Abu-Lughod, J. *Rabat: Urban Apartheid in Morocco.* Princeton: Princeton University Press, 1980.

Adalenro, I. A. *Marketplaces in a Developing Country: The Case of Western Nigeria.* Ann Arbor: University of Michigan Press, 1981.

Africa at a Glance. Pretoria: Africa Institute of South Africa, 1978.

Ajayi, J. F., and Crowder, Michael (eds.). *Historical Atlas of Africa.* New York: Cambridge University Press, 1985.

Altschul, D. R. "Transportation in African Development." *Journal of Geography* 79 (1980): 44–56.

Barbour, Kenneth M. *The Republic of the Sudan: A Regional Geography.* London: University of London Press, 1961.

Barbour, Nevill (ed.). *A Survey of North-West Africa: The Maghrib.* London: Oxford University Press, 1962.

Bell, Morag. *Contemporary Africa: Development, Culture, and the State.* White Plains, N.Y.: Longman, 1986.

Berg, Robert J., and Whitaker, Jennifer S. *Strategies for African Development.* Berkeley, Calif.: University of California Press, 1986.

Beyer, J. L. "Africa." In *World Systems of Traditional Resource Management,* ed. G. A. Klee. London: Arnold, 1980.

Blumefeld, Jesmond (ed.). *South Africa in Crisis.* Beckenham, Eng.: Croom Helm, 1987.

Boateng, Ernest A. *A Geography of Ghana.* Cambridge: Cambridge University Press, 1966.

Carter, Gwendolen M. *Which Way Is South Africa Going?* Bloomington: Indiana University Press, 1980.

Central Intelligence Agency. *The World Factbook.* Washington, D.C.: Government Printing Office, 1985.

Christopher, Anthony J. *Colonial Africa: An Historical Geography.* Totowa, N.J.: Barnes & Noble, 1984.

———. *South Africa.* New York: Longman, 1982.

Clark, John I. et al. (eds.). *Population and Development Projects in Africa.* New York: Cambridge University Press, 1985.

Clark, John I., and Kosinski, Leszek A. (eds.). *Redistribution of Population in Africa.* Exeter, N.H.: Heinemann, 1982.

Cloudsley-Thompson, J. L. (ed.). *Sahara Desert.* Oxford: Pergamon, 1984.

Cole, Monica M. *South Africa.* London: Methuen, 1966.

Davenport, T. R. H. *South Africa: A Modern History.* Toronto: University of Toronto Press, 1982.

Davidson, Basil. *Africa: History of a Continent.* New York: Macmillan, 1966.

de Blij, Harm J. *Africa South.* Evanston, Ill.: Northwestern University Press, 1962.

de Blij, Harm J. and Martin, E. (eds.). *African Perspectives: An Exchange of Essays on the Economic Geography of Nine African States.* New York: Methuen, 1981.

Fetter, Bruce. *Colonial Rule and Regional Imbalance in Central Africa.* Boulder, Colo.: Westview, 1983.

Frank, Richard, and Chasin, Barbara H. *Seeds of Famine: Ecological Destruction and the Development Dilemma in the West African Sahel.* Montclair, N.J.: Allanheld-Osmun, 1980.

Gautier, Emile F. *Sahara: The Great Desert.* Translated by Dorothy F. Mayhew. New York: Columbia University Press, 1935.

Gibbs, James L. (ed.). *Peoples of Africa.* New York: Holt, Rinehart & Winston, 1965.

Good, Charles M. *Ethnomedical Systems in Africa: Patterns of Traditional Medicine in Rural and Urban Kenya.* New York: Guilford, 1987.

Green, L. P., and Thomas J. Denis. *Development in Africa: A Study in Regional Analysis with Special Reference to Southern Africa.* Johannesburg: Witwatersrand University Press, 1962.

Griffiths, Ieuan L. L. (ed.). *An Atlas of African Affairs.* New York: Methuen, 1984.

Grove, A. T. *Africa South of the Sahara.* London: Oxford University Press, 1970.

Gulliver, Philip H. *The Family Herds: A Study of Two Pastoral Tribes in East Africa, the Jie and Turkana.* London: Routledge & Kegan Paul, 1955.

Hance, William A. *African Economic Development.* New York: Praeger, 1967.

———. *The Geography of Modern Africa.* New York: Columbia University Press, 1975.

Harrison, Paul. *The Greening of Africa.* New York: Viking, 1987.

Harrison-Church, Ronald J. *West Africa: A Study of the Environment and Man's Use of It.* London: Longman, 1980.

Hibbert, Christopher. *Africa Explored: Europeans in the Dark Continent, 1769–1889.* London: Lane, 1982.

Hoyle, Brian S. *Seaports and Development: The Experiences of Kenya and Tanzania.* New York: Gordon & Breach, 1983.

Kimble, George H. T. *Tropical Africa.* New York: Twentieth-Century Fund, 1960.

King, Lester C. *South African Scenery: A Textbook of Geomorphology.* New York: Hafner, 1963.

Knight, C. G., and Newman, J. (eds.). *Contemporary Africa: Geography and Change.* Englewood Cliffs, N.J.: Prentice-Hall, 1976.

Lamb, David. *The Africans.* New York: Random House, 1983.

Leach, Graham. *South Africa: No Easy Path to Peace.* New York: Methuen, 1987.

Legum, Colin (ed.). *Africa: A Handbook to the Continent.* New York: Praeger, 1966.

Mabogunje, Akin L. *Urbanization in Nigeria.* London: University of London Press, 1968.

Mandy, Nigel. *A City Divided: Johannesburg and Soweto.* New York: St. Martin's, 1985.

Meredith, Martin. *The First Dance of Freedom: Black Africa in the Postwar Era.* New York: Harper & Row, 1985.

Miller, E. Willard, and Miller, Ruby M. *Northern and Western Africa: A Bibliography on the Third World.* Monticello, Ill.: Vance, 1981.

———. *Tropical, Eastern, and Southern Africa: A Bibliography on the Third World.* Monticello, Ill.: Vance, 1981.

Minns, W. J. *A Geography of Africa.* London: Macmillan, 1984.

Morgan, William T. W. *Nigeria.* White Plains, N.Y.: Longman, 1985.

Morrison, Donald, et al. *Black Africa: A Comparative Handbook.* New York: Free Press, 1972.

Mountjoy, Alan, and Hilling, David. *Africa: Geography and Development.* Totowa, N.J.: Barnes & Noble, 1987.

Murdock, George P. *Africa: Its People and Their Culture and History.* New York: McGraw-Hill, 1959.

Murray, Jocelyn (ed.). *Cultural Atlas of Africa.* Oxford: Phaidon, 1981.

O'Connor, Anthony M. *The African City.* New York: Holmes & Meier, 1983.

Oliver, Roland, and Crowder, Michael (eds.). *The Cambridge Encyclopedia of Africa.* Cambridge: Cambridge University Press, 1981.

Osei-Kwame, P. A. *A New Conceptual Model of Political Integration in Africa.* Laham, Md.: University Press of America, 1980.

Parker, Richard B. *North Africa: Regional Tensions and Strategic Concerns.* Westport, Conn.: Praeger, 1984.

Pollock, Norman C., and Swanzie, Agnew. *An Historical Geography of South Africa.* London: Longmans, 1963.

Poltholm, Christian, and Dale, Richard (eds.). *Southern Africa in Perspective.* New York: Free Press, 1972.

Pritchard, John M. *Africa: A Study Geography for Advanced Students.* London: Longmans, 1979.

Prothero, R. Mansell (ed.). *A Geography of Africa.* New York: Praeger, 1969.

Rogge, John R. *Too Many, Too Long: Sudan's Twenty-Year Refugee Dilemma.* Totowa, N.J.: Rowman & Allenheld, 1985.

Senior, Michael, and Okunrotifa, P. *A Regional Geography of Africa.* New York: Longman, 1983.

Singleton, F. Seth, and Shingler, John. *Africa in Perspective.* New York: Hayden, 1967.

Smith, David M. *Apartheid in South Africa.* Cambridge: Cambridge University Press, 1987.

——— (ed.). *Living Under Apartheid.* Winchester, Mass.: Allen & Unwin, 1982.

Soja, Edward W. *The Geography of Modernization in Kenya: A Spatial Analysis of Social, Economic, and Political Change.* Syracuse, N.Y.: Syracuse University Press, 1968.

Stamp, Lawrence Dudley. *Africa: A Study in Tropical Development.* New York: Wiley, 1959.

Thompson, Virginia, and Adloff, Richard. *The Western Saharans: Background to Conflict.* Totowa, N.J.: Barnes & Noble, 1980.

Trimmingham, J. Spencer. *The Influences of Islam upon Africa.* New York: Longman, 1980.

Udo, Reuben, K. *The Human Geography of Tropical Africa.* Exeter, N.H.: Heinemann, 1982.

Ungar, Sanford J. *Africa: The People and Politics of an Emerging Continent.* New York: Simon & Schuster, 1985.

Wellington, John H. *South West Africa and Its Human Issue.* Oxford: Clarendon, 1967.

Wubneh, Mulatu, and Abate, Yohannis. *Ethiopia: Transition and Development in the Horn of Africa.* Boulder, Colo.: Westview, 1987.

Yeager, Rodger. *Tanzania: An African Experiment.* Boulder, Colo: Westview, 1983.

Chapter 13

THE MIDDLE EAST

Introduction

A group of nation-states located in southwestern Asia in the general vicinity of the Persian Gulf make up a world region commonly known as the Middle East. The term "Middle East" was first used by the British to identify a military and administrative area. The Balkan countries of Eastern Europe were labeled the "Near East," and East Asia was called the "Far East," so the area between them became the Middle East (see Figure 13.1). Boundaries between these regions are vague, and the terms make locational sense only when the regions are viewed from Great Britain. A more appropriate name to describe the location of the Middle East would be "Southwest Asia," but the name Middle East has gained worldwide acceptance and will be used here.

The Persian Gulf is an arm of the Arabian Sea, which in turn is part of the Indian Ocean. The region of the Middle East surrounds the Persian Gulf and is bordered on the north by the Caspian Sea and the Black Sea and on the west by the Mediterranean Sea and the Red Sea. The territorial waters of the Middle East also include the Gulfs of Aden, Aqaba, and Oman.

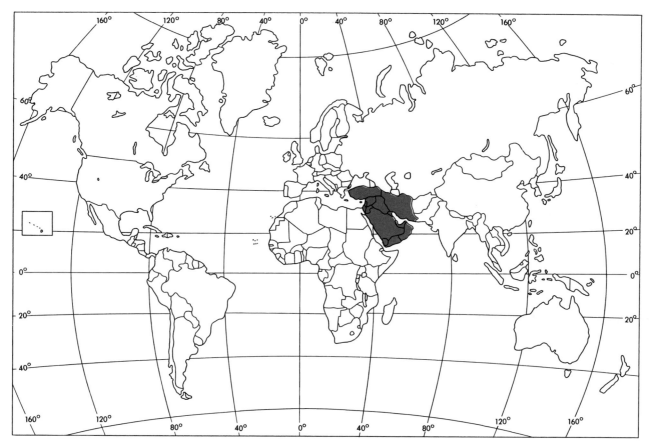

FIGURE 13.1
World Location of the Middle East.

The Middle East is a large but fairly compact region. It stretches east-west for more than 2,000 miles (3,218 km), from Bulgaria and Greece on the west, to Afghanistan and Pakistan on the east. The region borders the Soviet Union on the north and extends for 1,500 miles (2,414 km) to the Gulf of Aden on the south. The most significant factor about the location of the Middle East is what geographers call its *relative location*. The location is important relative to the location of other things. The region is the land connection between Africa, Asia, and Europe, and since the discovery of oil, it has played a major role in world affairs.

As a crossroads for three continents, the Middle East, has served for centuries as the bridge for the movement of ideas, people, and goods between large portions of the world. In addition, and equally as important, the region is the source area for ideas that have had a lasting and profound impact on people all over the world. Some ideas that have originated and diffused outward from this "cradle of civilization" include the wheel, agriculture, urbanization, a number system, and probably most important, three major world religions: Christianity, Islam, and Judaism. Moreover, the region contains a large portion of the world's known petroleum reserves. The importance of the Middle East in world affairs is therefore generally related to three factors: location, religion, and petroleum.

The 16 states of the Middle East (Figure 13.2) include 7 located on the Arabian Peninsula: Saudi Arabia, North Yemen, South Yemen, Oman, United Arab Emirates, Qatar, and Kuwait. An-

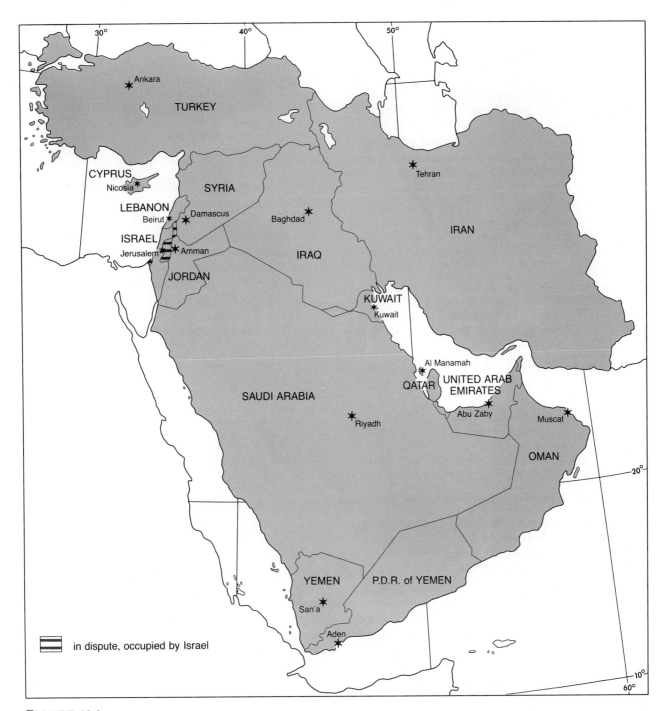

FIGURE 13.2
The Countries of the Middle East.

other 7 states are located north of the Arabian Peninsula but close to it: Turkey, Syria, Lebanon, Israel, Jordan, Iraq, and Iran. The final 2 countries are island nations: Cyprus, located in the Eastern Mediterranean, and Bahrain, in the middle of the Persian Gulf. Saudi Arabia, the largest of these 16 states, comprises more then one-third (36 percent) of the area. The most populous country is Turkey, with 52.3 million people, or about one-third of the region's 162.6 million total population.

Egypt is historically tied to the Middle East and has been a political and economic leader in the region, but Egypt is on the continent of Africa and was considered in Chapter 12 as part of that region. In fact, all of Northern Africa is more closely associated with the Middle East than with the rest of Africa, at least in terms of cultural geography. And although the Middle East is the heart of the Arab World, it is in the continent of Asia. Regional divisions on a world scale do not always conform to cultural divisions.

■ *Physical Geography of the Middle East*

The Middle East may be divided into two major components based on surface configuration. The northern and eastern parts of the region, essentially Turkey and Iran, are dominated by high mountains and intermontane plateaus. The remainder of the region is characterized by tableland with local hills but fringed by small mountain ranges.

From the Aegean Sea through Turkey and Iran to the Arabian Sea, the area is continuously mountainous. The outlines of the various mountain ranges correspond closely to the international boundaries of Turkey and Iran and form the shape on the map of a drooping bow tie. The imaginary tie's knot, known as the **Armenian Knot,** is located near the border between Iran and Turkey (see Figure 13.3). To complete the analogy, four mountain ranges extend outward east and west from the knot; two of these are in Turkey, and two are in Iran. In Turkey, the Taurus Mountains run southwest of the knot and border the Mediterranean Sea. The Pontus Mountains run west of the Armenian Knot and parallel the Black Sea. East-

Armenian Knot The region immediately west of the Caspian Sea in the Armenian state of the Soviet Union. Also known as the Armenian Plateau, the region contains many faults with folded ridges, but many volcanic intrusions are also evident. The region is called a "knot" because it is a focal point for numerous mountain ranges that radiate outward from it.

ward from the knot into Iran, the Elburz Mountains extend along the Caspian Sea, and the Zagros range stretches for more than 900 miles (1,448 km) southeast along the Persian Gulf. The Zagros Mountains cover the most land area, but the Elburz are the highest and most rugged. The tallest mountain peak in the Middle East is Mount Damavand, at 18,934 feet (5,771 m). It is located in Iran in the Elburz Range about 50 miles (80 km) northeast of the capital city of Tehran. (See Figure 13.4.)

The second major landform region of the Middle East includes most of the land area south of Turkey and Iran and all of the Arabian Peninsula. The entire region is a large, tilted plateau. The slope is from the higher southwest corner of the Arabian Peninsula downward toward the low-lying Tigris and Euphrates river valley. The highlands are an L-shaped mountain range centered in the Yemens adjacent to the East African highlands and paralleling the Red Sea and the Gulf of Aden. The mountains are known as the Al-Hijáz Asir, and the tallest peak is J Hadur Shu-ayb at 12,336 feet (3,760 m). The highlands extend along the Red Sea from the Gulf of Aden northward to the Gulf of Aqaba. The entire Arabian Peninsula slopes downward from the Red Sea on the west to the Persian Gulf on the east. The Red Sea itself is in a large rift valley, part of which extends into Africa and holds the lakes of Africa's eastern highlands. This great rift separates the continents of Africa and Asia.

Rivers are not common in a large part of the Middle East because of the arid climatic conditions, and the Arabian Peninsula has no permanent streams anywhere. Some of the rivers that do exist, however, are among the most famous in the world. Their fame is for their political and historical importance, however, not for their size or geographical significance.

Babylon The ancient city, now in ruins, that grew up on the Euphrates River about 55 miles (89 km) south of Baghdad, Iraq, near the modern city of Hilla. The city existed from about 2200 B.C. to about 280 B.C. and was the home of the famous Babyonian kings Hammurabi and Nebuchadnezzar. It is also the place where Alexander the Great died, in 323 B.C.

The Tigris and Euphrates rivers flow through Mesopotamia, and the lowland area in Iraq that some consider to be the birthplace of civilization. **Babylon**, the world's first large city, was located along the Euphrates in the center of Mesopotamia. The two rivers begin in the mountains of Turkey and flow roughly parallel to each other toward the southeast. The Euphrates cuts through Syria, both rivers flow through Iraq, and they eventually empty into the Persian Gulf. The Tigris is actually a tributary of the Euphrates, and it joins the larger river about 100 miles (161 km) north of the gulf. The lower part of the river has been used as the international boundary between Iraq and Iran, and the region has been hotly contested in a bitter war between the two countries.

Baghdad, the capital and largest city of Iraq, is located on the banks of the Tigris River not far from the ruins of the ancient city of Babylon. The Jordan River, which is only 150 miles (240 km) long, is one of the world's best-known rivers because of its location in the Holy Land and significance for Christianity (see Figure 13.5). Essentially, the river flows from the Sea of Galilee southward into the Dead Sea. The surface of the Dead Sea lies 1,299 feet (375 m) below mean sea level. It is the lowest place on earth. Other rivers of the Middle East are located in the mountainous north. They are all short streams that carry the mountain moisture to nearby bodies of water. Some rivers flow into the intermontane basins, but because they disappear into the arid bolsons they do not flow for great distances.

Although ringed by the waters of many seas and gulfs, the Middle East is generally hot and dry. Very little of the moisture from the surrounding water bodies is carried onto the parched land. Subtropical high pressure dominates the region throughout the year, keeping storms from entering. The high pressure ridge is broken only slightly during the winter months, allowing some cyclonic storms to enter off the Mediterranean Sea. These storms are confined to the northern coastal and mountain areas, where the moisture usually comes as snow. Except for the mountains, most of the Middle East receives less than 10 inches (25.4 cm) of precipitation a year, and most of the Arabian Peninsula gets even less than that. (See Figure 13.6 on page 402.)

The Arabian Desert is a continuation of the Sahara Desert of North Africa, and like the Sahara it is very dry. The desert area covers about one million square miles (2.6 million km^2) of the Arabian Peninsula. The Arabian Desert is noted for having a greater percentage of its surface covered by sand dunes than any other desert in the world; about one-third of the desert is sand. The Iranian Desert, the second major desert of the Middle East, covers about 150,000 square miles (390,000 km^2) of interior Iran. It contains some of the world's highest sand dunes and many traces of the ancient agriculturalists of historic Persia.

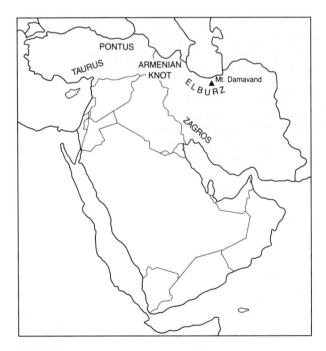

FIGURE 13.3

Locations of Important Mountain Ranges of the Middle East (see also Figure 13.4).

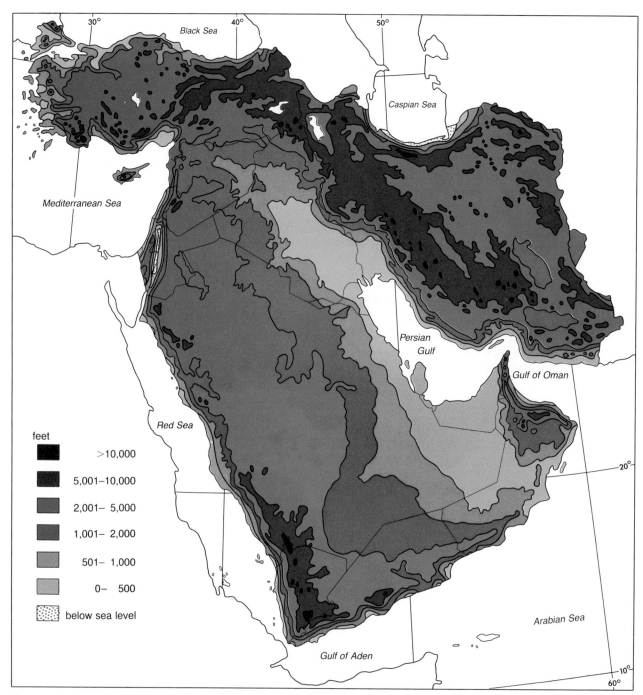

FIGURE 13.4
Land Elevations and Water Bodies of the Middle East.

siroccos A hot, steady, oppressive wind blowing from the deserts of the Middle East and North Africa across the wetter lands along the coast and sometimes continuing into Southern Europe.

Temperatures in the Middle East are similar to those of other hot desert areas of the world. Summers are long and hot, and in many places average monthly temperatures exceed 100° F (38° C) and daily highs of 120° F (49° C) are common. The high temperatures persist even into relatively high elevations, where they would be expected to be lower. In the lowland areas winters are mild, but temperatures in the mountains get quite low. The people living in the mountain areas of Turkey and Iran experience severe blizzards, and sometimes the mountain snow cover lasts until May. When the snow melts, the short rivers bulge with meltwater but soon return to their normally low amounts of flow.

The desert regions of the Middle East are subject to periodic hot, dry winds known as **siroccos.** These winds sometimes cause severe dust and sand storms, but often they just cause crops to wither from the persistence and heat of the storms. The hot winds are not only physically discomforting to humans and animals, they also cause psychological distress. If a person accused of a crime can prove that a sirocco was blowing at the time of the crime, his chances of being exonerated are good.

■ Human Geography of the Middle East

Population and Urbanization

Of the 162.6 million people living in the 16 countries of the Middle East, 136.9 million (84 percent) are in only 5 countries: Iraq, Iran, Saudi Arabia, Syria, and Turkey (see Table 13.1 for population figures). The people of the Middle East own about 2.5 million square miles (6.3 million km^2) of the earth's surface, but because the land is so dry the people are congregated in the few humid places or places where irrigation is possible. As a result, the overall population density is about 50 people per square mile. There is great variation in population density, however, both among the countries and within each country (see Figure 13.7).

Low population density figures (Table 13.1) indicate that the Middle East has large stretches of uninhabited land. This is especially true in the harsh desert countries of Saudi Arabia, Oman, and South Yemen. Nevertheless, the states of the Middle East have some of the highest population growth rates of any region in the world. The rate of natural increase for the region is 2.8 percent a

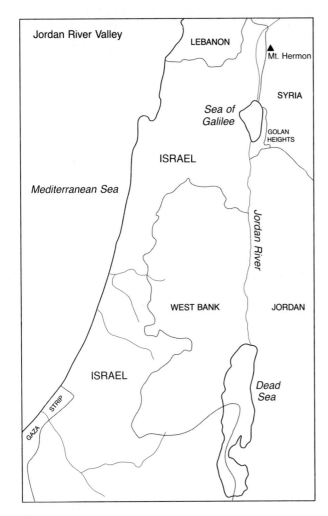

FIGURE 13.5

The Jordan River Valley. The river rises west of Mount Hermon and flows south through the Sea of Galilee to the north end of the Dead Sea. It is narrow and sluggish and not suitable for navigation. The valley is noted for its associations with Bible history.

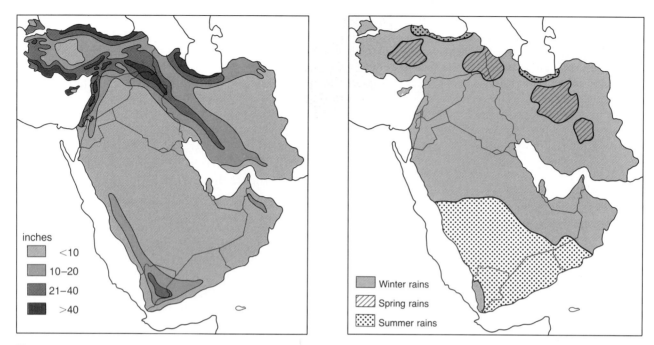

FIGURE 13.6

Precipitation Patterns of the Middle East. Annual precipitation (A) and seasonal patterns (B) are shown on these two maps.

TABLE 13.1

Population of the Middle East, Including Density and Percentage of Urban-Dwellers, 1986

COUNTRY	POPULATION (IN MILLIONS)	DENSITY (PER SQ. MILE)	% URBAN-DWELLERS
Bahrain	0.4	1,647	81
Cyprus	0.7	186	53
Iran	46.6	71	50
Iraq	16.0	92	68
Israel	4.2	526	86
Jordan	3.7	71	60
Kuwait	1.8	249	91
Lebanon	2.7	652	64
Oman	1.3	15	9
Qatar	0.3	68	10
Saudi Arabia	11.5	13	73
Syria	10.5	147	48
Turkey	52.3	168	45
United Arab Emirates	1.4	40	12
Yemen, North	6.3	122	11
Yemen, South	2.3	17	33

SOURCE: Population Reference Bureau and Central Intelligence Agency, *The World Factbook, 1986* (Washington, D.C.: Government Printing Office, 1986).

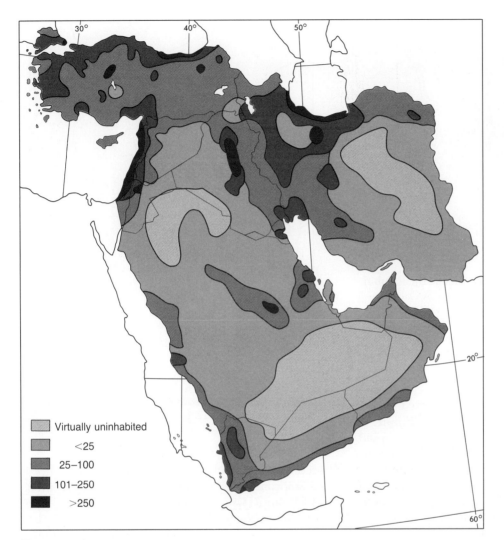

FIGURE 13.7
Population Patterns of the Middle East.

year, and it varies from a low of 1.3 percent on the island nation of Cyprus to a high of 3.8 percent in Syria. The rate of 2.8 percent is much greater than the world rate of 1.7 percent and is similar to the rate for the rapidly growing countries of Africa.

The four largest cities in the northern part of the Middle East are Istanbul (2,772,708 population) and Ankara (1,877,755) in Turkey, Tehran (5,734,199) and Mashhad (1,119,000) in Iran, and Baghdad (3,800,000) in Iraq. Istanbul, one of the unique cities of the world, is especially fascinating to Western tourists. It is situated on both sides of the Bosporus, a narrow body of water that separates Europe from Asia (see Figure 13.8). Istanbul, the leading Turkish port, has been an international city for centuries and is known for the variety of people found on its streets every day. The Bosporus connects the Black Sea and the Sea of Marmara. Although it is less than 20 miles (32 km) long, it has been a strategic passageway for shipping for hundreds of years. Ships from the Soviet Union on the Black Sea must pass through the Bosporus in order to get to the Mediterranean Sea.

Ankara is not as cosmopolitan as Istanbul, but it is the most Westernized city in Turkey. In 1923, when it was a sleepy provincial town of only

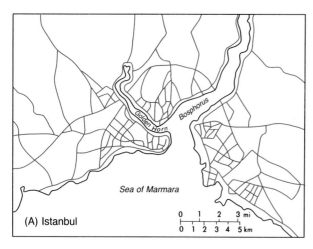

FIGURE 13.8

Street Plans of Istanbul (A) and Ankara (B). Istanbul is located adjacent to the Bosporus and Sea of Marmara, which are used by the Soviet navy to gain access to the Mediterranean Sea. Ankara, on the other hand, is located inland on the Asian side of Turkey.

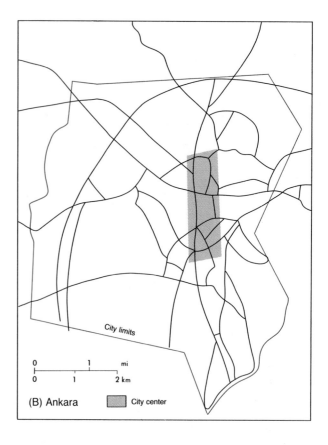

27,000 people, it was designated the capital city. The choice of Ankara over Istanbul was intended as a symbolic break with the Ottoman Empire of the past and to centralize the location of the capital. The influx of administrators and military personnel gave the town its initial spurt in population growth, but industry soon followed, bringing many more people. Today, Ankara is a modern, developing city with nearly 2 million people and is growing faster than Istanbul.

The site of present-day Baghdad was settled in ancient times, and by A.D. 634, when it was sacked by Muslim Arabs, it had become a small city. The city is located on both sides of the Tigris River and has been a commercial and administrative center for centuries. In 1921 it became the capital of Iraq, and it is now a modern city that continues to be a trade and cultural center.

Tehran was founded in the twelfth century and became the capital of Iran in 1788. The city is located at 3,810 feet above sea level on the slopes of the Elburz Mountains. The old part of the city has narrow, winding streets, but modernization has occurred in other parts. The city was the site of a conference between Roosevelt, Churchill, and Stalin in 1943, during World War II. It is most noted recently as the place where 62 hostages from the American embassy were held from November 1979 until January 1981. The Americans were caught in the middle of an internal struggle between the ruling government, which the United States supported, and conservative Muslim revolutionaries.

Mashhad (Meshed), the second city of Iran, is located in the northeastern part of the country about 475 miles (763 km) directly east of Tehran. Mashad has for centuries been an important trade center and junction point on caravan routes and highways from India to Tehran and from north to south between towns in Turkestan and the Gulf of Oman. The city has a Shiite shrine and is a place of annual pilgrimage. Today, Mashad is important strategically because of its proximity to the Soviet and Afghan borders.

The largest city on the Arabian Peninsula is Riyadh, the capital of Saudi Arabia. It was originally the site of an oasis in the middle of the

FOCUS BOX

Jerusalem

The city of Jerusalem is closely tied to Christian, Muslim, and Jewish history. It started as a small hilltop fortress known as Salem and was ruled by people known as the Jebusites. King David decided that because the location could be easily defended it would be ideal for his capital. He sent his men through a tunnel, normally used to carry water, to capture the small town. The unimportant town of Salem soon became Jerusalem, the "City of David," and eventually one of the most important religious centers in the world.

Earlier in his life, King David had defeated the Philistine giant, Goliath of Gath. The David-Goliath encounter became one of the best known and most spectacular duels related in the Bible. David, the young shepherd, dropped the huge Philistine with a well-aimed rock from his slingshot. Before the giant could recover, David decapitated him with the giant's own sword. The event provided David with widespread popularity and eventually changed his life entirely. David was summoned to the court of King Saul, the first Jewish monarch to rule over the 12 tribes of the Israelites. The gifted shepherd boy soon became a court favorite and eventually married one of King Saul's daughters. Upon the king's death, David was crowned king.

Jerusalem expanded significantly under David's son, King Solomon. Solomon brought in architects and builders from Phoenicia to build a temple, the first holy building of the Israelites. The kingdom of Solomon achieved great prosperity, and as a result many other buildings were built in Jerusalem, as well as roads and aqueducts throughout the countryside. Solomon had a thousand wives and concubines, and he built many palaces to satisfy their needs.

After David and Solomon, many epochal religious events took place in the city of Jerusalem. Jesus' last days were spent in the city, he attended what Christians call the Last Supper there, and he was crucified there. Islam's prophet, Muhammad, departed for heaven from Jerusalem, from the Dome of the Rock near the Wailing Wall. The Wailing Wall is a remnant of the Second Temple and is another Jewish shrine. Jerusalem is also the cite of the first Crusader invasion.

The modern city of Jerusalem became the capital of Israel in July 1980 and has a population of 431,000. Many of the buildings in the city are constructed from the nearby limestone, which is yellow-white with pink streaks. The indigenous stone allows Jerusalem to grow from the hills that surround it and to blend into them with one color. In the center of the city is the Israel Museum, which houses the Dead Sea Scrolls and many other priceless items of history. Jerusalem, the center of the ancient Bible lands, is one of the most famous cities in the world.

Arabian Desert. The old city was walled and contains the royal palace, a great mosque, and many smaller mosques, but today the city sprawls far beyond the ancient walls and has grown to nearly 2 million in population. It contains numerous palaces built with the deluge of oil money. Jerusalem in Israel and Mecca in Saudi Arabia, both famous religious centers, are relatively small for twentieth-century cities. Each is about 450,000 in population, but both are important tourist and pilgrimage attractions.

Oman is the least urbanized country in the Middle East (see Table 13.1). The capital and largest city of the country is Muscat, with 85,000 people. Israel is the most urbanized country, with 86 percent of the people living in cities. In spite of the large cities, however, the urbanization factor for the entire Middle East is only 55.8 percent. Nearly half the people live in rural areas, many in small farming villages and towns, and some are nomadic herders.

Language

Arabic is the official language of most of the states of the Middle East, and the majority of the people speak it, but Arabic is not the language of the people in Turkey, Iran, Israel, and Cyprus. Turkic, the official language of Turkey and Cyprus, has much of its written origin in Arabic script, so the written forms are similar. Greek and English also are spoken on Cyprus, and Greek is the second official language of the country. Hebrew is spoken in Israel, but the language is similar enough in structure to Arabic (both are Semitic) that many Israelis speak both Hebrew and Arabic. English is common in Israel too. Farsi, a remnant of the ancient Persian languages, is the language of Iran, but again the written form is Arabic. Other small, scattered groups of people, most of whom live near the Persian Gulf, speak Hindi, Baluchi, and Urdu, languages imported from India.

Religion

In any study of the Middle East, religion is an important topic because the Middle East is the source area for the world's three great monotheistic faiths. Christians, Jews, and Muslims all look on certain parts of the region as their "Holy Land." Today, most of the people of the region are Muslim. In fact, Islam predominates in every country except Israel and Cyprus, and both those countries have significant minorities of Muslims. Judaism, of course, is the faith of the Israelis, and Greek Orthodox Christians make up about 80 percent of the population of Cyprus. As late as 1975, about half the people of Lebanon were Christians, but the number has been reduced to about one-fourth the population now because of the recent political trouble there. Lebanon has become a predominately (60 percent) Muslim state. Syria, Israel, Jordan, and Iraq all have Christian minorities, but the remainder of the countries in the region are entirely Muslim. Various sects of

---

Canaan The old, native name of the part of Palestine between Jordan and the Mediterranean Sea, but sometimes used to mean vaguely all of Palestine. The region was taken over in 1200 B.C. by the Jews returning from Egypt. They called it "the land of milk and honey" and "the promised land."

---

the Muslim faith do exist, however. For example, the people in Oman belong to the Obadhi sect, those in Qatar to the Sunni sect, and the Muslims in Iran follow the Shia branch. Religious mixtures also exist, as illustrated by the Druze, who are basically Muslim but have many religious tenets that are based on Christianity.

Judaism is the oldest of the three major faiths in the Middle East. It originated among the ancient tribal herders and eventually came to predominate along the Mediterranean in an area known as **Canaan** (Moses' "land of milk and honey"). The Jewish nation was strong for the time and place, but in 46 B.C., Judaea (part of Canaan) fell under the rule of the Roman Empire. Later, during the explosive spatial diffusion of the Muslim faith in the sixth century A.D., the region came under Muslim control. About 1,000 years later, the Ottoman Empire (1516–1917) administered the region. The name "Canaan" was changed to "Palestine" and later to Israel. It was not until the defeat of the Turks (Ottoman Empire) at the end of World War I that Muslim control gave way to a British Mandate and eventually to Jewish control. During the long period of outside control of the Jewish Holy Land, Jews emigrated and settled in many other parts of the world. With the British in control, however, the Jews began to return to Palestine, and in 1948 the country of Israel was established. Today, Israel is a Jewish state, completely surrounded by Arab countries but firmly independent and in control of its own destiny. (See Figure 13.9 for the growth of the state of Israel.)

The next faith to emanate from the Middle East was Christianity. No discussion of the historical aspects of this religion is necessary, except to point out that it has had a major impact on the people of the world. All Christians, and most other literate people, are aware of the striking effects of Christianity. For example, the world's

Kaaba The sacred shrine of Islam at Mecca in Saudi Arabia, from the Arabic word *ka'bah*, meaning "square building." The building contains a small black stone (a meteorite) supposedly given to Abraham by the angel Gabriel. The Kaaba is located in the courtyard of a large mosque toward which believers turn when praying.

Allah The name believers in Islam have given to God.

Koran The sacred book of Islam, written in Arabic. The book's contents are reportedly revelations Allah made to Muhammad.

---

reckoning of time is based on the birth of Jesus of Bethlehem in Judaea. Christians are aware of the geographical place names associated with Jesus' life, such as Bethlehem, Jerusalem, the Sea of Galilee, and the Jordan River. All these and many others mentioned in the Bible are part of the Christian Holy Land and are visited today by the faithful from all over the world (see Figure 13.10 on page 410).

The third major world religion originating in the Middle East began with the birth of the founder of Islam, the prophet Muhammad, about A.D. 570 in the city of Mecca in Saudi Arabia. Mecca is located on the west of the Arabian Peninsula about 60 miles (100 km) inland from the Red Sea and about halfway between the Gulf of Aqaba and the Gulf of Aden. The location is very important to Muslims because the city is the focal point of their religion. As the birthplace of Muhammad, Mecca is also the location of the holiest shrine for Muslims. A cubed-shaped building known as the **Kaaba** houses a small, black meteorite that Muslims believe was given to Abraham by the Angel Gabriel. One of Muhammad's last acts was to make a pilgrimage to Mecca. It is now the desire of all the believers of Islam to visit and worship at the shrine. About 500,000 Muslims from more than 50 nations make a pilgrimage to Mecca every year, and all Muslims bow toward Mecca during their daily prayers.

Medina, the second holy city of Islam, is located about 220 miles (354 km) north of Mecca. Muhammad died in Medina in A.D. 632, and today his tomb is enshrined there in the Mosque of the Prophet. During the next 100 years, the Islamic faith spread rapidly outward from the Mecca-Medina source area—from Arabia westward across Northern Africa and into Europe, and eastward beyond the Indus River in India. Most of the people in this vast area today, with the exception of Europeans, remain true to **Allah** (God), his prophet Muhammad, and the five pillars of faith: belief in one God, daily prayer, the giving of alms, the fast of Ramadan, and a pilgrimage to Mecca. Muhammad's book, the **Koran,** is a guide for the devout and the basis of law in Muslim countries. It also serves as a primer for children learning to read.

Economic Geography of the Middle East
Crude-Oil Production

If crude oil production suddenly ceased, there would soon be no gasoline, grease, motor oil, jet fuel, kerosene, paraffin, benzine, naphtha, or petroleum jelly. In fact, an estimated 10,000 products come from petroleum and the petrochemical industries. Oil is probably the single most valued commodity on earth, and the Middle East has much of the world's known oil reserves (see Figure 13.11 on page 411). Seven of the countries are among the world's leaders in the production of crude oil. Oil was first discovered in Iraq, and modest production began in 1935. Three years later, substantial quantities were found in Saudi Arabia, and almost the entire country of Kuwait is now known to be located on top of a huge pool of oil. Very little production ensued, however, until after World War II. Oil was first exported from Kuwait in 1946, and the real boom began after 1950, when pipelines were built across the desert from the Persian Gulf to the eastern shores of the Mediterranean Sea, ports for oil tankers were built, and existing ports were improved.

Oil income for Iraq rose from $9.0 million a year in 1949 to $2.5 *billion* in 1975 and more than $8.0 billion in 1985. The upward spiral of production has been similar for all the countries around the Persian Gulf. Saudi Arabia is one of the top three oil producers of the world, along with the Soviet Union and the United States. The Arab

OPEC (Organization of Petroleum Exporting Countries) nations accounted for one-third of the oil produced in 1980. By 1985, however, the war between Iran and Iraq and a drop in oil prices caused OPEC production to drop to 17 percent of the world's total. In 1978, before the Iranian hostage crisis, the United States imported half a million barrels of oil a day from Iran, but that dropped to none in 1981 and is now back up to only 27,000 barrels a day. Less than 10 percent of the oil the United States imports now comes from the Arab OPEC nations, but the countries are getting wealthy from oil sales to European countries and Japan.

The wealth acquired from oil profits in the Middle East is truly fantastic. At first, much of the

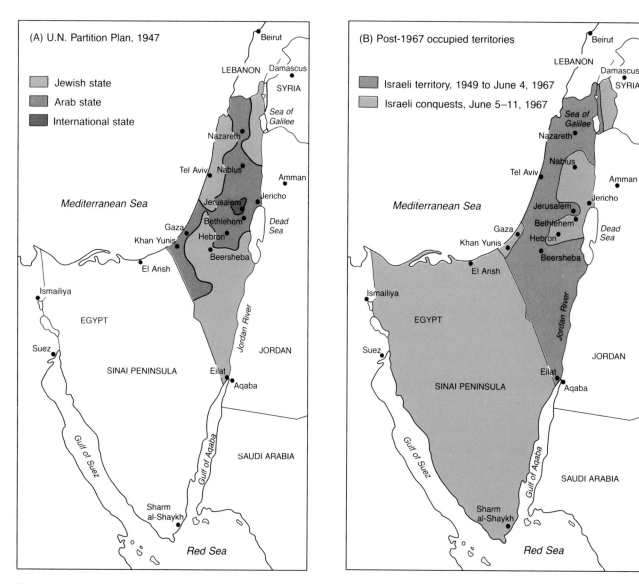

FIGURE 13.9

The Evolution of Israel. The United Nation's 1947 plan for the partition of Palestine (A), the post-1967 occupied territories (B), and Israel today (C) are shown on these three maps.

Organization of Petroleum Exporting Countries (OPEC) An organization created on November 14, 1960, intended to determine world oil prices and to advance members' interests in trade and dealings on development with the industrialized oil-consuming nations. Venezuela led the initiative for the development of the organization. Member nations today include Algeria, Ecuador, Gabon, Indonesia, Iran, Iraq, Kuwait, Libya, Nigeria, Qatar, Saudi Arabia, United Arab Emirates, and Venezuela.

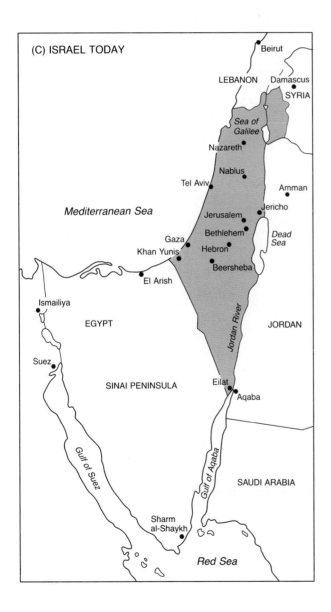

money was wasted on a privileged few. This was especially true in Saudi Arabia, where the family of the Saudi kings used the income on palaces, fleets of luxury automobiles, gambling, parties, and anything that would absorb the money and bring some pleasure. More recently, however, the deluge of money has been channeled into irrigation projects, desalinization plants, universities, streets and roads, and other projects for the benefit of the common people. The United Arab Emirates has the highest per capita gross national product of any country in the world ($23,770), and Qatar and Kuwait are not far behind. These countries extend loans to other Arab nations and pamper their people with welfare. They provide free medical care and education, levy no taxes, and guarantee social security free to all citizens.

Agriculture

After oil, agriculture is the major economic activity of the Middle East (see Figure 13.12). It is a major industry in nearly every country, even though much of the land is not fertile and rainfall is scarce. About one-third of the people in the region are engaged in agriculture, but the percentages vary from more than 90 percent in North Yemen to less than 5 percent in Bahrain and the United Arab Emirates. Other countries with high percentages of the work force in agriculture are Turkey (61 percent) and Oman (60 percent).

In the states along the Eastern Mediterranean —notably Lebanon, Israel, and Jordan—fruits and vegetables are the main agricultural products. Citrus fruits provide a major income for Israel, but melons, tomatoes, bananas, grapes, and olives also are grown. Similar crops are grown in Jordan, but only 12 percent of the land area there is arable. In 1975, nearly half the labor force in Lebanon was in agriculture, but the percentage has dropped to about 17 percent in the last 10 years as a result of internal political problems. Agriculture in Israel has increased because of extensive irrigation projects. The water comes primarily from the Sea of Galilee. Besides the fruits and vegetables, tobacco, cotton, and cereal grains also are grown in the Eastern Mediterranean region. Turkey is a leading tobacco exporter.

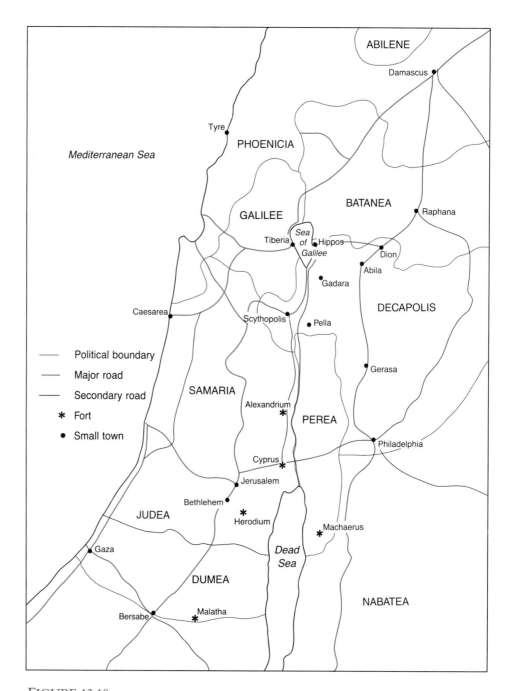

FIGURE 13.10
The Holy Land of Biblical Times. The political boundaries as of A.D. 44 are shown with the major cities, roads, and forts of the time.

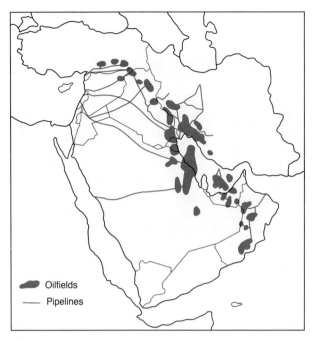

FIGURE 13.11
Oilfields and Major Pipelines of the Middle East.

Mineral Production and Industry

Minerals other than petroleum generally are not found in great abundance in the Middle East but some minerals are important to individual countries. The people of Cyprus mine copper, asbestos, gypsum, and chrome, while potash and phosphates come from Jordan. Turkey, a world leader in the production of chrome, has good deposits of zinc, iron ore, coal, manganese, borate, and antimony. Israel has a few important minerals, but such building materials as limestone and sandstone usually lead in production. Turquoise and emerald mining adds to the economy of Iran, and a few precious stones come from North Yemen. Centuries ago the Queen of Sheba sent gems to King Solomon from the region of North Yemen, and the area has produced small amounts ever since.

Based on the lack of minerals, industry is not advanced in the Middle East. Most of the countries have the basic processing plants for foodstuffs, textiles, and cement, and the oil-rich countries all have refining facilities. But heavy industry, particularly

Agriculture in Iraq, Iran, and Saudi Arabia is dominated more by wheat, barley, cotton, and millet than by fruits and vegetables. Although Iran does have a wine industry. The country also produces caviar from the sturgeon caught in the Caspian Sea. Saudi Arabia exports hides and wool from the numerous camels, donkeys, horses, and sheep raised by the nomadic herders in that country. Iraq exports tobacco grown in the Kurdish Hills, as well as wool and skins.

Aside from the nomadic herders, a few oases, and some irrigated crop land, agriculture on the Arabian Peninsula is extremely limited. In South Yemen, only one percent of the land is arable, and there is no source of water for irrigation. North Yemen, however, produces mocha coffee, sesame, barley, and cotton on the lightly watered plateau of El Jebel. Oman exports dates, limes, pomegranates, dried fish, and frankincense. Qatar, Kuwait, and the United Arab Emirates import food, paid for by their oil incomes.

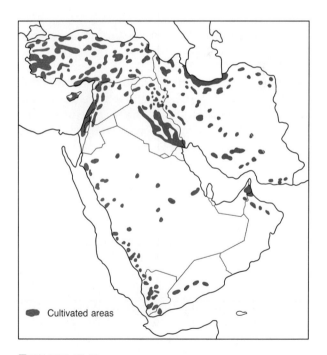

FIGURE 13.12
Cultivated Areas of the Middle East.

The Middle East

> ## FOCUS BOX
>
> ### *The Jews*
>
> The history of the Hebrews begins in Mesopotamia, where the descendants of Shem (Semites) settled after the famous Biblical flood. One of these descendants was called Abraham, and he is given the honor of having discovered the one true God, called Yahweh. For centuries after Abraham, the Hebrews were wandering tribes of desert nomads, but eventually one particular sojourn became very important to Jewish history: their movement into Egypt. The Hebrews lived a quiet life in Egypt, with their own tribal organization, a separate language, and distinct religious beliefs; they multiplied, grew wealthy, and went unmolested by the Egyptians. Then suddenly conditions changed. For some reason a sharp, anti-Asian reaction began that affected the alien Hebrew tribes. The reaction turned to oppression and eventual serfdom of the Hebrews.
>
> The most influential personality in Jewish history was Moses, who was born in Egypt during the bitterest years of oppression. The reigning Egyptian Pharaoh had carried the persecutions to the point that he had ordered all male children of Hebrew parents to be killed. The baby Moses, however, was hidden in the rushes along the Nile, where he was rescued from a watery grave by the Pharaoh's daughter. Moses, therefore, was reared in the luxury of the Pharaoh's household, but he never forgot the bondage of his people. While still only a young man, Moses saw an Egyptian beating a Hebrew serf and was compelled to kill the Egyptian, thereby sacrificing his protected position.
>
> Moses escaped the wrath of the Pharaoh by taking refuge in the desert. While in the desert, Moses had a vision of Yahweh, and from that experience gathered strength to return to Egypt to free his fellow Jews from the bondage of the Egyptians. His mission was successful, and he led the Jews out of Egypt. After years of wandering, Moses guided the Hebrews to Canaan, which they were destined to make their own. The Jews settled in Canaan (the land of milk and honey) in about 1,200 B.C. The small region eventually became a larger area known as Palestine, the southern division of which was called Judaea.
>
> Palestine thrived during the years before the birth of Christ, the Jews multiplied, and many migrated to all parts of the Mediterranean region. The Jewish state came to ruin, however, shortly after the life of Christ, when the Roman Empire extended its control over the region. Most of Palestine fell to the Romans in A.D. 68, and the Holy City of Jerusalem was conquered the next year, completing the Roman take-over. The Romans ruled the area for about 500 years, but during the sixth and seventh centuries Persian armies challenged the Roman Empire for control of the Eastern Mediterranean region. After nearly two hundred years of deadly combat, both empires were exhausted, which paved the way for the triumph of Islam.
>
> Islam emerged from the wastes of the Arabian desert to conquer the world. By 641 all the area that was once a Jewish state was conquered. The Jews were allowed to live in peace, however, as long as they paid their taxes and ground-rents. From the eighth century onward, the Jews of the East enjoyed an unprecendented period of freedom, although under Arab authority. Many Jews migrated to all parts of Europe during the centuries that followed the Arab conquests. The history of the Jews in Europe is filled with the dual tragedies of physical suffering and mental degradations, because they became unwelcome in nearly every place that they settled. In the 13th century, the Inquisition was established in Europe to search for heretics, or those who did not believe in Christianity.

Once found, the nonbelievers were punished or expelled from the country where they were living. Because they were non-Christian, the Jews were among those that were expelled.

As the doors of Christian Europe closed on the Jews, new doors opened in Turkey and Poland. Eastern Europe became the leading center of Jewish life. The powerful Ottomans had conquered Turkey, and by 1389 they ruled practically all the Balkans. The hospitality of the Turkish rulers to the Jews was a godsend as they fled to escape the persecution and bigotry of Western European countries. By 1550, Constantinople (Istanbul) boasted the largest Jewish settlement in Europe. Turkey served as a refuge for Jews from Southern Europe, especially from Spain and Portugal, and Poland became the promised land for the harried German Jews.

At the threshold of modern history, in the middle of the eighteenth century, the Jews were scattered to all parts of the world, but they probably did not total more than 3 million and formed only a small proportion of the general population. The vast majority of Jews lived in Eastern Europe, and more than a million and a half lived in Poland. Small communities thrived in the other Balkan states, and about 100 thousand lived in Asiatic Turkey. In central Europe the Jews were concentrated in the large cities of Germany and Austria, especially Berlin, Frankfurt, Hamburg, and Vienna. Small communities of Jews lived in England, France, Italy, and Holland. In America the Jewish population was quite small and scattered. A few hundred lived in the Spanish colonies, and about two thousand had migrated to the English colonies. In addition small Jewish settlements were scattered through Africa and inner Asia.

The darkest days for the Jews began during the 20th century, especially during the two great wars, World War I and World War II. The wars completely changed the Jewish situation in Europe. During World War I, millions of soldiers advanced and retreated through contested territory, especially along the eastern front where about 3.5 million Jews were concentrated. In normal times their legal and social disabilities were distressing enough, but the war brought looting, explusions, and death. One important positive feature to Jewish history occurred as a result of the war, however; the British conquered Palestine. During World War II, the new term "Genocide" had to be created to describe the Nazi leaders' criminal intent of exterminating the entire Jewish population of Europe. The victories of the Nazis put virtually all of Jewish life in Europe under their control, and the master plan was to kill off every last Jew.

On May 15, 1948 the British relinquished their mandate over Palestine, and the Jewish state of Israel was born. During the 30 years of British mandatory administration, the British offered thousands of uprooted Jews their first opportunity for freedom, thus, Jewish settlements in Palestine grew slowly, but steadily. The Israeli Declaration of Independence pledged that immigration of Jews would be welcomed from every part of the world. Under the British, immigration averaged about 18 thousand a year, but during the first year of Israeli control, immigration swelled to a quarter of a million. Eventually, the average reached 18 thousand a month, and within three years of independence, a total of 638,000 Jewish immigrants had been brought into Israel.

The influx of thousands of people into a small area of the Middle East was not without its problems. The problems were cultural, social, and religious, because the people were from very different parts of the world. The Jews from central Europe (Germany, Austria, and Czechoslovakia) were highly educated, sophisticated, and somewhat worldly in outlook. They also spoke Yiddish, a dialect of High German that uses the Hebrew alphabet. The Jews from eastern Europe (Poland, Hungary, and the Soviet Union) had been living under Communist influence, and spoke the languages of their countries. Thousands of others came from the slums and bazaars of Arab countries (Egypt, Morocco, Tunisia, and Algeria). The Jews from the Arab countries were poorly educated, sickly, and undernourished. They essentially flew from the Middle Ages into the 20th century. These Oriental Jews had to learn everything from the alphabet to elementary sanitation. Finally, there were the local Jews whose families had lived in the Middle East for thousands of years. The way these diverse people dressed, their ethical standards, their cultures, and their way of life all varied. Israel's first problems were concerned with the absorption of these immigrants into the society as a whole.

Despite the enormity of its internal problems, Israel gained strength. Its population mounted steadily, both through a high birthrate and through immigration. Each major crisis in other countries brought thousands of refugees to Israel. Yet with all the pride and exaltation of success, persistent danger continues to haunt every aspect of national life; Israel remains a tiny enclave almost completely surrounded by seven Arab states that are fanatically committed to the destruction of the Jewish nation. The Israeli government finds it necessary to assign a vastly disproportionate share of its budget to national defense. These perils notwithstanding, the Jewish nation today remains extraordinarily optimistic.

This Bedouin family in the Arabian Desert poses in front of their tent home. Note how they cover their bodies almost completely for protection from the desert sun. Photograph by Mohammed Al Wohaibi.

This vegetable market in Jordan attests to the variety of farm crops grown in the region.

steel production, is not common. Turkey is the industrial leader of the region and has the most varied production, and Turkey does produce steel. In fact, the country's steel output for 1984 was 4.3 million metric tons. Other countries of the region, such as Iraq, Iran, Saudi Arabia, and even Qatar, do have some individual steel mills but large industrial complexes headed by steel production simply do not exist in the Middle East.

Individual countries of the Middle East are noted for the production of specialty items. Diamond cutting and polishing, for example, are important in Israel, although the country produces no diamonds. The famous Persian carpets are made in Iran, glassware and brassware are produced in Syria, and fine silk comes from Turkey. The Lebanese are noted for their trading and the transshipment of goods. All these specialties produce good income for the countries involved.

Tourism has been a major industry in the Middle East, but the flow of tourists is often interrupted by local battles, terrorism, and war. Tourism, however, remains second only to citrus fruits as the major income for Israel, and it is important for Cyprus. The religious shrines draw many people to the Christian and Jewish Holy Land in Israel, and the wonderful climate and secluded beaches attract visitors to the Mediterranean island of Cyprus. Saudi Arabia does not allow tourists into the country, but religious pilgrims do produce a type of tourism income for the Saudis.

Transportation and Communications

For an area known as the "crossroads of the world," the Middle East does not have a well-developed transportation network. Aside from the

Part of a Bedouin tent made from the black wool of the sheep. All the household and personal items belonging to the Bedouins are stored in these tents. Photograph by Mohammed Al Wohaibi.

famous Berlin-to-Baghdad railroad and the routes associated with it, railroad mileage is limited. One railroad runs through Saudi Arabia from Riyadh to the Persian Gulf, another connects Tabuk with Medina. The latter railroad is an extension of the Turkish route terrorized by the British soldier T. E. Lawrence (Lawrence of Arabia) and his Bedouin raiders during World War I.

Turkey is the only country in the region with a railroad system approaching what could be called a network. Most of the major cities and ports are connected by railroads, and newly constructed routes are bringing the smaller eastern towns into the system. The railroad map of Iran resembles a large T, with Tehran located at the juncture of the two lines. The route west of the city leads to Tabriz and then on to Turkey, the route east goes to Mashhad, near Afghanistan, and the southern route winds its way through the Zagros Mountains to the Persian Gulf. Iraq's two main railroad lines run through Mesopotamia from Turkey to the Persian Gulf, and both lines go through Baghdad. Railroads also run parallel to the Eastern Mediterranean, with local networks around Beirut, Haifa, and Tel Aviv. (See Figure 13.13 for the location of roads and railroads in the Middle East.)

In the Middle East, paved roads are more common than railroads, and most of the countries are connected to the highway network. The Arabian Desert can be crossed on at least two major highways, but local drivers often take off across the desert without regard for roads. The island country of Bahrain is now connected by bridge and causeway to the Saudi Arabian mainland. Turkey has the most hard-surface road mileage in the region, followed by Iran. It is somewhat ironic that the two countries with the most rugged terrain, where road construction costs are highest, have the most extensive road systems in the region. Hard-surface roads in Qatar, Oman, and the United Arab Emirates exist only in and around the towns.

Airplane travel is common in the Middle East, and most countries boast a modern air force, commercial airlines, and one or two international airports. About 850 usable, hard-surface air strips have been constructed in the region, and Saudi Arabia has the most with 156, followed by Iran with 128; Bahrain and Qatar each have 2. Most of the countries have modern combat jet aircraft,

In the Middle East, farm products often are carried to market on two-wheeled carts drawn by donkeys. Photograph by Jim Curtis.

and many pilots were trained either in the United States or by U.S.-trained pilots. The oil-rich countries spend large sums on obtaining military aircraft, and Israel manufactures many of its own.

Radio, television, and newspapers are found in all the major cities of the Middle East, but the latter two are not common in the rural areas. Communications media are enhanced by the primary use of a single language, but newspapers are restricted because of high illiteracy rates in most countries. The overall literacy rate is only 53.6 percent of the total population of the region. It varies from 15 percent in North Yemen to 89 percent in Cyprus and Israel. With nearly half the people of the Middle East unable to read and write, newspapers are of little use. But the tattered-looking Arab leading his goat herd through the open desert is not as ill informed as his appearance might indicate. Transistor radios, small, cheap, but powerful entertainment centers, have become a boon to information transmission in Third World countries. Music from transistor radios is becoming a worldwide phenomenon.

■ Political Geography of the Middle East

The Middle East has been in various stages of political turmoil for centuries, but modern weapons and techniques have made the region especially volatile today. The United States has been directly involved in the region since the oil discoveries of the 1930s, and especially since the exploitation of the oil by American companies. Within the last decade, 62 Americans were held hostage in Iran for 14 months, 241 American servicemen died in Muslim terrorist attacks in Lebanon, and another 31 American sailors were killed by an Iraqi rocket attack on a ship in the Persian Gulf. Kidnapping of foreign nationals by Islamic militants has became common in the 1980s. American foreign policy with regard to the countries of the Middle East is riddled with complications.

Despite the economic and military support offered by the Soviet Union, and despite that country's proximity to the Middle East, none of the countries is Communist. Both Syria and South Yemen, however, are strongly socialist and lean to-

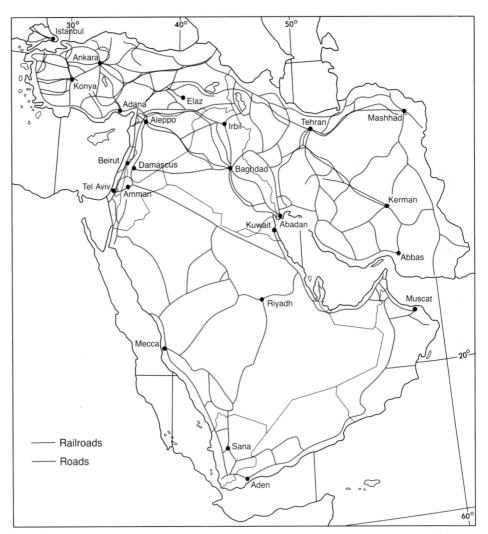

FIGURE 13.13
Roads and Railroads of the Middle East.

ward the Soviets. Communism has failed primarily because communism does not make allowances for religion and most Arabs are religious people. But in the eyes of the Arabs the West (especially Great Britain and the United States) is failing them because it was the midwife for the birth of Israel and continues to support that country.

The establishment of Israel in 1948 meant the physical loss of one fertile area in the vast deserts of the Arab World. The continuous growth of the State of Israel has resulted not only in the loss of Arab territory but also in the displacement of well over a million Arab refugees fleeing their Palestinian homeland. The resulting Arab-Israeli dispute constitutes a major threat to world peace. The United States tries to maintain friendly relations with both sides and offers military aid to Arabs and Jews alike.

Iran and Cyprus have been rocked by internal political problems, and Iran has fought a long and bitter war with Iraq over the border area between the two countries. Iran, ruled by the conservative

The Arabian Desert is often traversed without the aid of roads. Here a German-made truck has been used to bring water for the animals at a Bedouin camp. Photograph by Mohammed Al Wohaibi.

religious extremists led by the Ayatollah Khomeini since the downfall of the Shah in 1978, maintains an anti-American stance because the United States backed the Shah. Cyprus is split between Greek and Turkish sympathizers, and each group maintains its own government. Oman is ruled by a sultan, and Jordon and Saudi Arabia have kings. Because people are allowed to vote in only half the countries of the Middle East, many are not able to control their own destinies. Still, there are few Communists or Communist sympathizers in any of the countries.

The West must react impartially to the conflicts in the Middle East yet keep its own petroleum interests in mind. Military support can be dangerous, but the countries must be capable of defending themselves. Fortunately, the religion of the Arabs makes them tolerant people. They are more concerned with their domestic problems than with strategic or geopolitical problems, but they are determined to win for themselves full partnership in the modern world. The three factors of location, religion, and oil have been important historically and continue to be the major influences shaping the future of the Middle East.

This barbed wire along the border between Israel and Lebanon is a reminder of the tensions in the Middle East. Photograph by Lawrence Gibbs.

■ *Key Words*

Allah
Armenian Knot
Babylon
Canaan
Kaaba
Koran
Organization of Petroleum Exporting Countries (OPEC)
siroccos

■ *References and Readings*

Adams, Robert M. *Land Behind Baghdad: A History of Settlement on the Diyala Plains.* Chicago: University of Chicago Press, 1965.

Adler, Ron K., et al. (eds.). *Atlas of Israel.* New York: Macmillan, 1985.

Aharoni, Yohanan. *The Land of the Bible: A Historical Geography.* Translated from the Hebrew by A. F. Rainey. Philadelphia: Westminster, 1967.

Al-Farsy, Fouad. *Saudi Arabia: A Case Study in Development.* New York: Methuen, 1986.

Al-Wohaibi, Mohammed A. "Nomads in Al-Hejaz Province: A Geographic Study of Nomads near the City of Taif." Master's thesis, Oklahoma State University, 1974.

Amirahmadi, Hooshang, and Parvin, Manoucher (eds.) *Post-Revolutionary Iran.* Boulder, Colo.: Westview, 1987.

Anthony, John. *Arab States of the Lower Gulf: People, Politics, and Petroleum.* Washington, D.C.: Middle East Institute, 1975.

Asad, Talal, and Owen, Roger (eds.). *The Middle East.* London: Macmillan, 1983.

Beaumont, Peter, and McLachlan, Keith (eds.). *Agricultural Development in the Middle East.* New York: Wiley, 1986.

Beaumont, Peter, et al. *The Middle East: A Geographical Study.* New York: Wiley, 1976.

Blake, Gerald H., and Lawless, Richard I. (eds.). *The Changing Middle Eastern City.* Totowa, N.J.: Barnes & Noble, 1980.

Brice, William C. *South-West Asia: Systematic Regional Geography.* London: University of London Press, 1967.

Central Intelligence Agency, *The World Factbook, 1985.* Washington, D.C.: Government Printing Office, 1985.

Clarke, John I., and Bowen-Jones, Howard (eds.). *Change and Development in the Middle East.* New York: Methuen, 1981.

Cohen, Saul B. *The Geopolitics of Israel's Border Question.* Boulder, Colo.: Westview, 1987.

Coon, Carleton S. *Caravan: The Story of the Middle East.* New York: Holt, 1958.

Cottrell, Alvin J. (ed.). *The Persian Gulf States: A General Survey.* Baltimore: Johns Hopkins University Press, 1980.

Cressey, George B. *Crossroads: Land and Life in Southwest Asia.* Chicago: Lippincott, 1960.

Devlin, John F. *Syria: Modern State in an Ancient Land.* Boulder, Colo.: Westview, 1983.

Drysdale, Alasdair, and Blake, Gerald H. *The Middle East and North Africa: A Political Geography.* New York: Oxford University Press, 1985.

El Mallakh, R. *The Economic Development of the United Arab Emirates.* New York: St. Martin's, 1981.

Fisher, William B. *The Middle East: A Physical, Social, and Regional Geography.* Cambridge: Cambridge University Press, 1978.

Girardet, Edward R. *Afghanistan: The Soviet War.* New York: St. Martin's, 1985.

Gordon, David C. *The Republic of Lebanon: Nation in Jeopardy.* Boulder, Colo: Westview, 1983.

Gradus, Y. "The Role of Politics in Regional Inequality: The Israeli Case." *Annals of the Association of American Geographers* 73 (1983): 388–403.

Harris, George S. *Turkey: Coping with Crisis.* Boulder, Colo.: Westview, 1985.

Helms, Christian M. *Iraq: Eastern Flank of the Arab World.* Washington, D.C.: Brookings Institute, 1984.

Hidore, John, and Albokhair, Y. "Sand Encroachment in Al-Hasa Oasis, Saudi Arabia." *Geographical Review* 72 (1982): 350–356.

Hiro, Dilip. *Iran Under the Ayatollahs.* Boston: Routledge & Kegan Paul, 1985.

Lamb, David. *The Arabs: Journeys Beyond the Mirage.* New York: Random House, 1987.

Longrigg, Stephen H. *The Middle East: A Social Geography.* London: Duckworth, 1963.

———. *Oil in the Middle East: Its Discovery and Development.* New York: Oxford University Press, 1967.

Mackey, Sandra. *The Saudis: Inside the Desert Kingdom.* Boston: Houghton Mifflin, 1987.

Manners, I. R. "The Middle East." In *World Systems of Traditional Resource Management,* ed. G. A. Klee. London: Arnold, 1980.

The Middle East and North Africa. London: Europa, 1984.

Miller, E. Willard, and Miller, Ruby M. *The Middle East (Southwest Asia): A Bibliography on the Third World.* Monticello, Ill.: Vance, 1982.

Monsouri, Mehdi. "Spatial Characteristics of Population in Iran: Growth, Distribution, and Density." Doctoral dissertation, Oklahoma State University, 1981.

Mottahedeh, Roy P. *The Mantle of the Prophet: Religion and Politics in Iran.* New York: Simon & Schuster, 1985.

Naff, Thomas, and Matson, Ruth C. *Water in the Middle East: Conflict or Cooperation?* Boulder, Colo: Westview, 1984.

Peck, Malcolm C. *The United Arab Emirates: A Venture in Unity.* Boulder, Colo.: Westview, 1986.

Perlmutter, Amos. *Israel: The Partitioned State.* New York: Scribners, 1985.

Peters, Joan. *From Time Immemorial: The Origins of the Arab-Israeli Conflict over Palestine.* New York: Harper & Row, 1984.

Reifenberg, Adolf. *The Struggle Between the Desert and the Sown: Rise and Fall of Agriculture in the Levant.* Jerusalem: Jewish Agency, 1955.

St. John, Robert. *Israel.* New York: Time Inc., 1965.

Saqqaf, Abdulaziz Y. (ed.). *The Middle East City: Ancient Traditions Confront a Modern World.* New York: Paragon House, 1987.

Smith, George A. *The Historical Geography of the Holy Land, Especially in Relation to the History of Israel and of the Early Church.* London: Hodder & Stoughton, 1894.

Steward, Desmond. *The Arab World.* New York: Time Inc., 1968.

_____. *Turkey.* New York: Time Inc., 1965.

Thesiger, Wilfred. *Arabian Sands.* New York: Dutton, 1959.

_____. *The Marsh Arabs.* New York: Dutton, 1964.

Weekes, Richard V. (ed.). *Muslim Peoples: A World Ethnographic Survey.* Westport, Conn.: Greenwood, 1984.

Weinbaum, M. G. *Food, Development, and Politics in the Middle East.* Boulder, Colo.: Westview, 1982.

Wright, Robin B. *Scared Rage: The Crusade of Modern Islam.* New York: Simon & Schuster, 1985.

Chapter 14

SOUTH ASIA

■ Introduction

The continent of Asia includes the subregions of South Asia, Southeast Asia, and East Asia (see Figure 14.1). Parts of Asia already have been considered: Southwest Asia (that is, the Middle East, in Chapter 13), the Soviet Union (Chapter 7), and Japan (Chapter 8). Both the location of Asia and its diverse lifestyles are far removed from the experiences of most people of the West. The contrasts within the huge, sprawling continent almost defy the imagination. The region considered in this chapter, South Asia, is only part of that vast, diverse continent, but it contains as many strange and wonderous things as any region of the world.

One of the most fascinating aspects of Asia is the sheer number of people that live there. The region includes 14.9 percent of the land area of the world but 52.4 percent of the world's people. Remember, this does not include the Soviet Union and Japan. Precipitation amounts and landforms within the region represent some of the earth's extremes. Thus, much of the land area is uninhabitable, so the places where people do live are extremely crowded. About 50 percent of the area is arid, but the Khasi Hills of India receive more than 450 inches (1,143 cm) of precipitation annu-

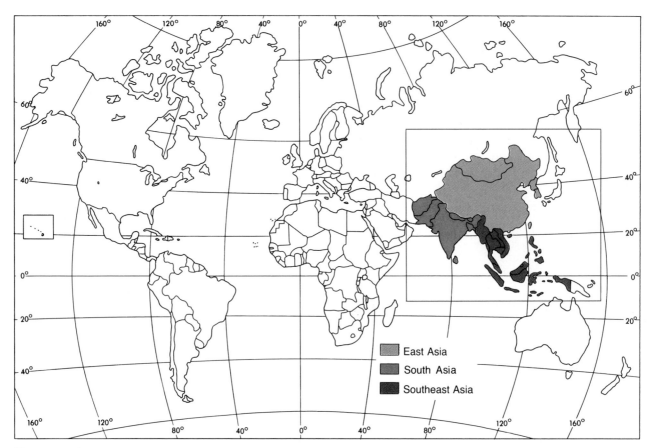

FIGURE 14.1

Three Regions of Asia. The three divisions of Asia, sometimes called Monsoon Asia, are according to the discussion in the text: South Asia, Southeast Asia, and East Asia.

ally. Many of the river valleys are low-lying and subject to periodic flooding, but the region also contains the world's tallest mountains.

Cultures within this part of Asia are as diverse as the physical setting. Nomadic herders in the desert cannot envision life in the tropical rainforest, much less the teeming, throbbing urban existence in the huge cities that dot the Asian coastlines. Asia contains a babel of languages, many races and religions, and a myriad of cultural phenomena, such as the architecture, shrines, and temples. Government and social institutions also vary among the countries. Each subregion, however, does contain a degree of internal consistency, and each has been identified historically as a separate entity. (See Figure 14.2.)

South Asia is composed of the states on and adjacent to the Indian subcontinent: Afghanistan, Bangladesh, Bhutan, India, Maldives, Nepal, Pakistan, and Sri Lanka. *East Asia* is the region sometimes referred to as the Orient: China, Hong Kong, North Korea, South Korea, Macao, Mongolia, and Taiwan. *Southeast Asia* is the group of states once under colonial dominance that occupy the peninsulas and islands on the southeastern part of the continent: Brunei, Cambodia, Indonesia, Laos, Malaysia, Myanma (formerly Burma), the Philippines, Singapore, Thailand, and Vietnam.

South Asia is about 2 million square miles (5.2 million km^2) in area, but the island countries of Maldives and Sri Lanka, as well as the mountain kingdoms of Nepal and Bhutan, are all quite small.

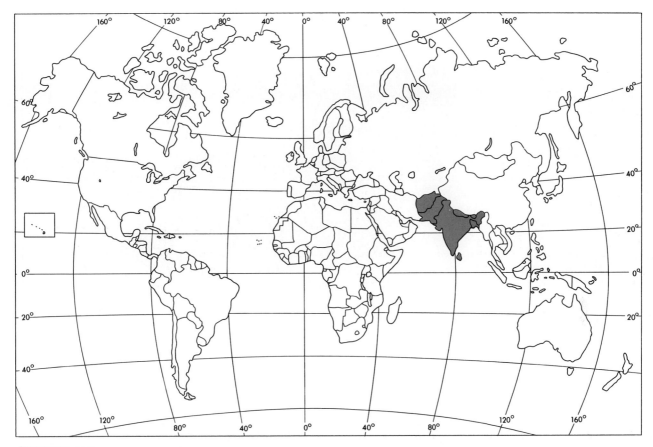

FIGURE 14.2

World Location of South Asia.

India is by far the largest country of the region, and it is the dominant economic and political force as well (see Figure 14.3). South Asia is separated from the remainder of the continent by the towering Himalayas and related highlands, especially the Hindu Kush Mountains and Khasi Hills. A few passageways breach the mountains, most notably the Khyber Pass, through which successive waves of immigrants have poured into the fertile valleys and broad plains of the Indian subcontinent.

The sedentary farmers who have lived on the plains of the Indus and Ganges rivers for 5,000 years have been culturally influenced by the repeated migrations into and through the area by Mongols from the northeast and the Caucasoids from the northwest. The most recent influence came from Western Europe, especially Great Britain.

Preceded usually by economic organizations, such as the East India Trading Company, the British colonial influence extended at one time or another over the entire region of South Asia. Great Britain did not rule completely, but the country's authority was everywhere and the region came under the sheltering umbrella of the British Empire. In some cases, British influence came about through the cooperation of the native people, but in some places it was not accepted.

Despite some conflict of culture, customs, and social institutions, British rule for the most part brought peace and mutual understanding. Not until the growth of Asian nationalism during World War II did the British realize their colonial days in South Asia were about over. They withdrew from the area without extended warfare or

FIGURE 14.3

The Countries of South Asia.

extensive bloodshed. In 1947 the last holding of the South Asian end of the "British lifeline" was relinquished. Since that year, the affairs of the region have been mostly in the hands of the people of South Asia. The major exception was the Soviet takeover of Afghanistan in 1979. The Soviets started a withdrawal, however, in 1988. Internal political bickering has occurred in India and Sri Lanka as dissident groups have vied for power, and Bangladesh fought a war with Pakistan for independence, but the majority of the region's people have enjoyed a peaceful existence.

■ Physical Geography of South Asia

The topographical relief along the northern fringe of South Asia is spectacular. Radiating outward from the Pamir Mountains (the Pamir Knot) in Western China are the great mountain ranges of Central Asia. The Hindu Kush range spreads southwest from the knot into Afghanistan, where it pushes out in many directions like the frazzled end of a rope. Eastward from the knot, the Karakoram and Zaskar ranges merge with the mighty Himalayas. Known as the "roof of the world," the

The Karakorum Mountains in the background present a stark contrast to the agriculture in the Hunza Valley of Pakistan. Photograph by Jim Curtis.

Himalayas stretch for more than 1,500 miles (2,400 km) across Central Asia. Nepal and Bhutan parallel the front (southern) edge of the range and have elevations that change from near sea level to more than 20,000 feet (6,000 m). The striking feature about the elevation change within these countries is that it occurs in less than 100 miles (160 km). Mount Everest, the world's most lofty peak, is located near the boundary between Nepal and China. Plate tectonics have forced the top of Mount Everest up to the height of 29,028 feet (8,848 m). Twelve other mountain tops within the country of Nepal exceed 25,500 feet (7,772 m) above sea level. (See Figure 14.4.)

The remainder of South Asia seems nearly flat, compared with the great wall of rock on the north. Bangladesh is exceedingly flat and low in elevation. The entire country is made up of the deltas of the Brahmaputra and Ganges rivers. India contains numerous river valleys—the Ganges Valley is the widest—but the country is also hilly and in many places quite rugged. The Deccan Plateau is the central uplifted region of India, and although it has some level surfaces on top, it is highly dissected. The flat valley of the Indus River runs through the eastern half of Pakistan, and the western half of that country is rugged mountain terrain. Afghanistan is essentially a mountain-dominated state, but the mountains give way to the high plateaus that stretch along the southern and western parts of the country. The coastal lowlands of the island of Sri Lanka surround a mountain core that rises above 8,000 feet (2,438 m). The Maldives consist of 2,000 islands grouped into 19 atolls, and these coral reefs barely rise above sea level.

The Tropic of Cancer divides South Asia into two equal parts. The precipitation pattern for the region is related to the location. The relatively warm waters of the Indian Ocean lie on the south, the great bulk of Asia is on the north. Prevailing winds blow from the water onto the land in the summer, and from the Asian land mass in the winter. In the winter, then, South Asia is dry, with no more than 5 inches (14 cm) of precipitation from November to April. When the high sun approaches the Tropic of Cancer, however, the

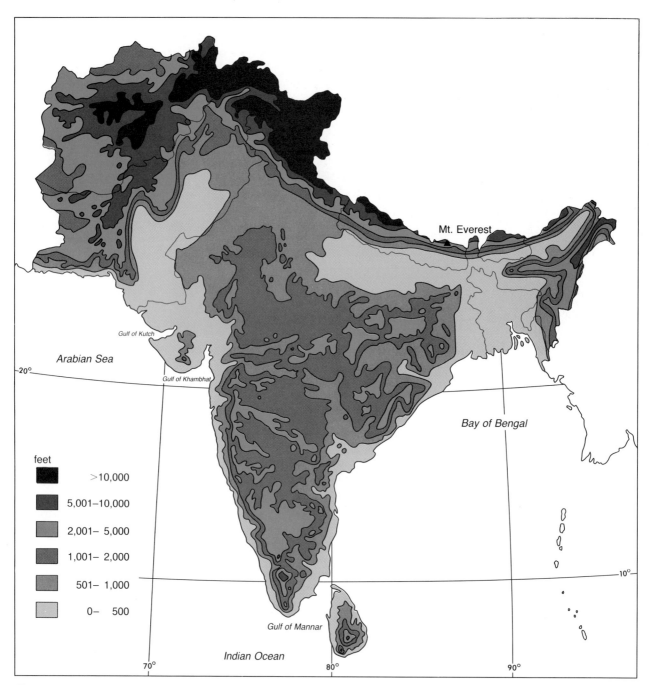

FIGURE 14.4

Land Elevations and Water Bodies of South Asia.

monsoon A seasonal wind of the Indian Ocean and South Asia blowing from the southwest from April to October and from the northeast during the rest of the year. The winds coming from the southwest carry moisture off the ocean, creating heavy rains when driven over the land.

winds begin to reverse. The dry air is replaced by the persistent warm, wet air blowing northward off the water. Only slight uplifting is required to squeeze the moisture from the air, and then the rains come. In some places it rains nearly every day during the summer. Large portions of the region receive more than 60 inches (152 cm) of rain from May to October, but along the mountain front north of Bangladesh the average is more than 400 inches (1,000 cm). Cherrapunji, in the Khasi Hills of eastern India, has been known as the wettest place in the world. In 1861, weather station attendants there recorded more than 900 inches (2,286 cm) of rainfall for an 11-month period, with 366 inches (930 cm) in the month of July.

The seasonal windshifts of South Asia that bring the wet and dry conditions are called **monsoons** (see Figure 14.5). Most of the rainfall in the lowlands, however, is associated with cyclonic storms that move westward across India from the Bay of Bengal. The summer rainy season in the lower elevations is not continuous, and clear skies may last for days or weeks. In addition, because the moisture does not penetrate all the way across India, the northwestern part of the country is always dry. Rain shadows also are pronounced, so the lee sides of highlands are dry, especially on the Deccan Plateau. The monsoon is eccentric in total amounts of rain it produces and in its beginning and ending period. Two dry years in a row will cause widespread disaster in India because so many people are closely tied to the land through subsistence agriculture.

The temperatures over South Asia are influenced by the tropical location, the winds, and the elevation. With half the region located within the

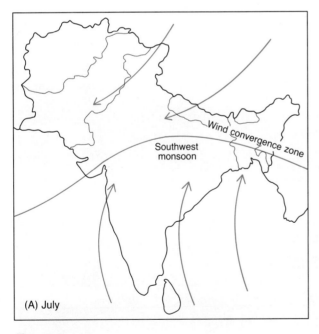
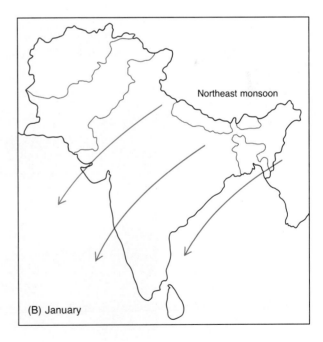

FIGURE 14.5

July (A) and January (B) Prevailing Wind Patterns of South Asia.

Tropics, generally warm temperatures would be expected. The hot season begins in March before the rainy season, when daytime temperatures often reach 100° F (37.8° C). There is a tradition that all work is suspended during midday until the heat and glare of the sun have subsided. The heat increases from south to north across India as the monsoon winds lose their effect. When the monsoon winds increase, summer temperatures are lower, but they still average over 80° F (26.7° C). The June to September averages for Bombay and Calcutta, which are located on opposite sides of India, are similar: 83° F (28.3° C). The warm temperatures, combined with the humidity, create high **sensible temperatures,** and conditions are the most unpleasant just after the daily rain. In the winter the dry, relatively warm air produces quite pleasant conditions throughout the region, although light frosts do occur in the Ganges Valley. The high plateau of Afghanistan is extremely cold in the winter, and year-round icy air is found in the mountains of the north.

South Asia is alternately lush and green or dreary and brown. Both the natural vegetation and the crops reflect the monsoon alternations of rainfall. There is also extreme variation between the scrubby, xerophytic plants in the deserts on the west of the region, and the jungle and bamboo thickets on the east. Generally, altitude is more important than latitude in determining vegetation zones. By traveling up the southern edge of the Himalayas, a distance of 100 miles (160 km), it is possible to see all the vegetation types found in a 3,000-mile (4,827 km) journey north from the southern tip of Florida to Labrador. The dense tropical forests of the lower elevations give way rapidly with elevation to pine, then oak and maple, birch and fir, mountain grasslands, and eventually bare rock.

---

sensible temperature comfort (or discomfort) index that is given in terms of temperature but is the combination of air temperature and humidity. For example, a moist, hot day is more uncomfortable than a dry, hot day, so a temperature of 100° F might be given in terms of a "sensible temperature" as 105° F.

This somewhat arid region is located on the lower slopes of the Karakorum Mountains in Pakistan in the western part of South Asia. Photograph by Jim Curtis.

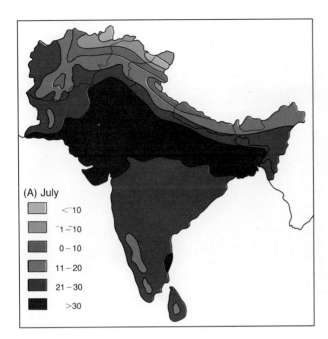
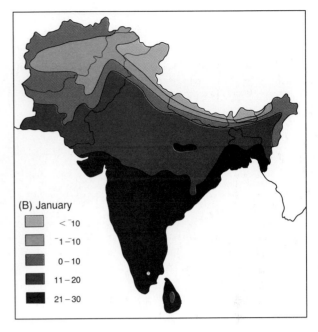

FIGURE 14.6

July (A) and January (B) Temperature Averages for South Asia.

Large areas of the South Asian forests have been cleared for agriculture, and much of the region is now treeless. By nature, however, much of the peninsula of India was a monsoon forest. The teak, banyan, palms, and bamboo have been replaced with mango trees, row crops, and grains. The pressure of people on the land is so great that all areas of possible cultivation are occupied, even the most marginal lands. Thus, the story of natural vegetation is tempered by the change in the landscape brought about by human occupation.

Human Geography of South Asia
Population

More than one billion people live in South Asia, and 785 million of them are in India. The population density of South Asia averages 678 people per square mile for the entire region, but it varies from 1,824 per square mile in Bangladesh, to 60 per square mile in Afghanistan (see Figure 14.7). Some lowland areas of Bangladesh average more than 3,000 people a square mile. The density for India is near the average for the region (606 per square mile), but it also increases in some river valleys. The Ganges Valley, from the river's delta in Bangladesh to the city of Delhi in India, is never less than 600 people a square mile for its entire width and length (see Figure 14.8 on page 433). The valley is about 150 miles (240 km) wide and more than 1,000 miles (1,600 km) long. It is the most continuously densely populated strip of land of its size in the world. Most of the people living in the Ganges Valley are dependent on the land as farmers and farm workers.

The few areas with sparse population in South Asia are found where the rigors of the physical environment set severe restrictions. One such area is the Thar (Great Indian) Desert located in northwest India along the Pakistani border. Others are the dry regions of Pakistan and Afghanistan. Besides the deserts and steppes, the mountains also restrict the number of people. Afghanistan and Bhutan have the lowest population densities of the region (60 and 78 per square mile, respectively).

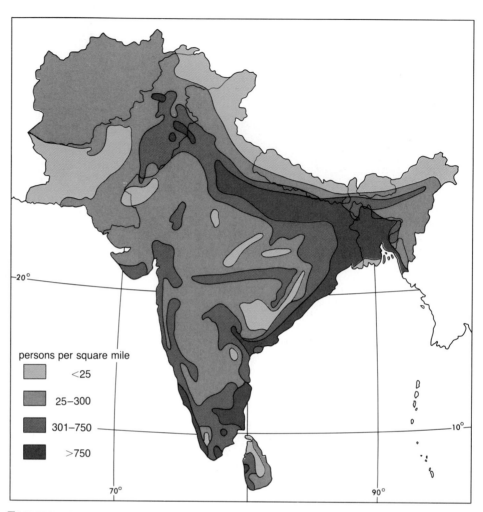

FIGURE 14.7

Population Patterns of South Asia.

The population growth rate for South Asia is 2.4 percent a year, much lower than many other parts of the Third World but higher than the world average of 1.7 percent. India has instituted programs to check the population growth of that country. The government endorses sterilization and sponsors more than 10,000 family planning centers. Millions of Indians, both male and female, have been sterilized, but the population continues to increase. The current growth rate of 2.3 percent produces 18.1 million new Indians each year. Even when no famines occur, an adequate food supply is always a problem in India. The growth rates in Bangladesh and Pakistan are greater than those for India (see Table 14.1), and pressure on land there is just as critical, especially in Bangladesh.

An enormous proportion of the population of South Asia dies at an early age. The average life expectancy for the entire region is only 52 years, and as low as 37 for Afghanistan. The life expectancy in Afghanistan is low, because the country has the highest infant mortality rate (194 per 1,000 live births) of any country of the world. Rampant diseases and malnutrition serve to make life short in the other countries of the region. Such diseases as tuberculosis, cholera, and leprosy, which are un-

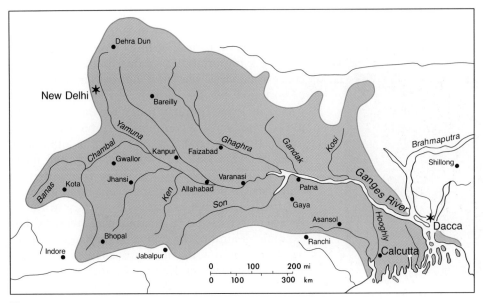

FIGURE 14.8

The Ganges River Valley. The Ganges is the sacred river of India, and the valley is one of the most heavily populated regions of the world.

Urbanization

In spite of some very large cities, South Asia is one of the least urbanized regions of the world. The average urbanization factor is 16.9 percent for the region and varies from 5 percent in the mountain countries of Bhutan and Nepal to 28 percent in Pakistan. Most of the people live in villages surrounded by the fields in which they work. It has been estimated that there are more than 600,000 villages in India. Many of these people live their entire lives without ever leaving their villages.

usual in Europe and the United States, remain common in South Asia.

TABLE 14.1

Population of South Asia, Including Density and Rate of Increase

COUNTRY	POPULATION (IN MILLIONS)	DENSITY (PER SQ. MILE)	RATE OF INCREASE (%)
Afghanistan	15.4	60	2.4
Bangladesh	104.1	1,824	2.7
Bhutan	1.4	78	2.0
India	785.0	606	2.3
Maldives	0.2	1,577	3.1
Nepal	17.4	313	2.3
Pakistan	101.9	320	2.8
Sri Lanka	16.6	645	2.0

SOURCE: Population Reference Bureau, 1987.

Not many people live in the upper elevations of the Himalayas, but these hardy Tibetans thrive mostly by herding yaks. Here a group dressed in traditional clothes pauses to throw prayer stones onto a prayer marker. Photograph by Colin Hagarty.

Besides the thousands of farm villages, India also has 10 cities with a population of more than one million each (see Figure 14.9). The largest are Calcutta and Bombay, and they are the largest cities in the region. The total number of millionaire cities in South Asia is 17, with a combined population of 55 million (see Table 14.2). Needless to say, these large cities are teeming with masses of people, and each city is important either to its local hinterland or as a port for world trade, or both.

Calcutta's origins go back to 1690, when a British jute factory was established on the banks of the Hooghly River. The growing town was captured in 1756 by the nawab of Bengal, who imprisoned the English in the famous Black Hole of Calcutta. Recaptured by the British in 1757, Calcutta was the capital of British India until 1912, when the seat of government was moved to Delhi. Today, Calcutta is a leading port city and the educational center of India.

Bombay became a town in the early fifteenth century and was part of various Hindu and Muslim dynasties. It did not become an important center until after 1534, when the Portuguese took over the area. By the time the town was ceded to the British in 1661, port activities had helped it grow to a good size. When Bombay became the headquarters for the East India Company in 1672, the city's importance as a distributing center increased rapidly. The first railroad in India was constructed by the British out of Bombay to Thana in 1853. Bombay today remains a thriving port city as well as a commercial center for western India. Bombay and Calcutta are the old, large cities of India. Before 1940, they each had more than a million people. The other large cities of India have experienced more recent rapid growth.

It is impossible to discuss every city in South Asia here, but each is fascinating in its own way. For example, Thimbu, the capital of Bhutan, is a small mountain town with cobble streets filled with yaks and Buddhist monks. Katmandu, the capital of Nepal, is a city of 150,000 nestled in a Himalayan

TABLE 14.2

Largest Cities of South Asia (1987 estimates)

COUNTRY	CITY	POPULATION (IN MILLIONS)
Afganistan	Kabul	1.400
Bangladesh	Dacca	3.500
	Chittagong	1.400
Bhutan	Thimbu	0.015
India	Calcutta	9.100
	Bombay	8.200
	Delhi	5.200
	Madras	4.300
	Bangalore	2.900
	Ahmadabad	2.500
	Kanpur	1.700
	Pune	1.700
	Hyderabad	1.500
	Nagpur	1.300
Maldives	Male	0.046
Nepal	Kathmandu	0.150
Pakistan	Karachi	5.100
	Lahore	2.900
	Faisalabad	1.000
Sri Lanka	Colombo	1.300

SOURCE: Central Intelligence Agency, *The World Factbook, 1987* (Washington, D.C.: Government Printing Office, 1987); *World Almanac,* 1987 (New York: Newspaper Enterprises Association, 1987).

FIGURE 14.9
Major Cities of South Asia.

valley. The king's palace is the showplace of Nepal, and a surprisingly large number of British and American tourists visit it each year. The city of Kabul in Afghanistan was ravaged during the Soviet takeover and has decreased in population since then. The city commands strategic routes through the mountain passes and has been a crossroads center for centuries. Famous historical leaders, such as Alexander the Great and Genghis Khan, visited Kabul during their campaigns (see Figure 14.10). Karachi is Pakistan's main seaport and the location of an important international airport. Colombo, the capital and largest city of Sri Lanka, is another important seaport. The port is a refueling station for world shipping, and the city is an important commercial center.

Religion

Religion has been a dominant force in South Asia for centuries, and it remains so today. The one with the largest following is the Hindu religion. More than 80 percent of the people of India are Hindus. Muslims, who are adherents of Islam, predominate in Afghanistan, Bangladesh, and Pakistan, and the approximately 70 million Muslims in India form a significant minority in that country. Siddhartha Gautama (Buddha), the founder of the Buddhist religion, was born in Nepal and experienced his religious enlightment at Gaya in India. His influence remains strong in both Nepal and Bhutan but has nearly disappeared in India. Buddhism is also the most common religion in Sri Lanka. Each religion dominates the lives of its followers and determines their dress, diet, friends, mates, and sometimes even their occupations. (See Figure 14.11 on page 438.)

Distinctive features of **Hinduism** include a belief in reincarnation, five principal deities, pilgrimages, and the all-prevailing caste system. The fundamental tenets of Hinduism took shape around 500 B.C. and have become fused into one philosophy. For example, reincarnation is related to the caste system in that it is held that when a person dies and is reborn that person will return to

---

Hinduism The religion and social system of the Hindus. It developed from Brahmanism and has elements from Buddhism.

---

a new station in life that depends on his or her deeds in the previous life. Once one is born a member of a certain level of caste, a person is destined to remain in that position throughout life. People who have been good can improve their position in the caste through reincarnation, but people who have been evil might come back to a lower caste or to a lowly state, such as an insect or vegetable. Government officials of modern India have attempted to modify some of the more restrictive features of the caste system, but success has been confined largely to the urban centers.

FIGURE 14.10

City Plan of Kabul. The capital city of Afghanistan has an old central part with a portion of the ancient wall surrounding it still visible. The newer, suburban area is to the left of the old central city.

FOCUS BOX

Migration from Afghanistan

One of the largest human migrations of the modern era occurred during the last decade. Some 3.5 million people left Afghanistan for neighboring Iran or Pakistan, and another 1.5 million migrated from one place to another within Afghanistan. About 3 million of these refugees from the Soviet invasion of December 1979 went to Pakistan. Most are now living in the North-West Frontier Province (NWFP), near the Afghanistan border. Although the Soviets started to withdraw in 1988, many of the refugees still live in villages established by the Pakistani government and are supported by international relief organizations. Despite the disorientation, suffering, and despair typical of refugee villages, many of the refugees are becoming assimilated into the local society. The process of assimilation is called **acculturation** by geographers.

The migration routes of the Afghan refugees follow the paths of movement that have existed in the area for centuries. The city of Kabul and the mountain passes to its north and east, especially the Khyber Pass, have been the major points of entry into South Asia. The Indo-Europeans came over the Hindu Kush (the name means "mountains of India") into the subcontinent. They were followed centuries later by the great Mogul tribes. These invaders passed through Afghanistan and left little lasting impression, as they were determined to get to the fertile plains of India. The local inhabitants melted into the mountains when the conquerors approached, and reappeared after they had departed. These periodic passages through the mountainous realm created segmentation of the country, and each partition came to be occupied by a different ethnic or tribal group. The Pashtun tribes, known as the "Afghans," eventually conquered and dominated the politically fractured region. They ruled the area

acculturation The process of becoming adapted to a new culture by one person, or the adoption by a society of the cultural traits of another society through contact between the groups.

during the 100 years prior to the Soviet invasion.

The first refugees to come to Pakistan after the Soviet invasion were mainly urban, educated, and from the merchant class. They brought with them all movable property, and, trucks, cars, buses, tractors, and animals streamed through the Khyber and adjacent passes into Pakistan. The early migrants settled in Peshawar, the major city of the region. The later migrants were more destitute and

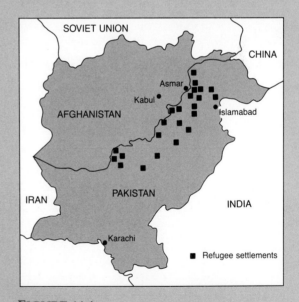

FIGURE 14.A

Locations of the Major Afghan Refugee Camps in Pakistan.

were housed in newly built villages. The refugees came from all parts of Afghanistan, but they were mainly Pashtun by ethnic origin, because the Soviet impact was most severe among the Pashtun communities.

The impact of the refugees on the landscape of the NWFP is visible everywhere. Trees have been removed for firewood, and pastures have been overgrazed. Agricultural cropping has intensified, however, not only from the need to feed more people, but also because labor is now readily available. Local commerce has benefited, and minor manufacturing has developed. The local trucking industry has been revolutionized as a result of the competition from the Pashtun drivers. Thus, despite the tremendous amount of personal suffering and the destruction of local pastures and forests, many of the refugees are better off than they were at home, and they have made a generally positive impact on the economy of the NWFP region. The refugees may never be able to return to their homeland, but many do not want to return because they can make a better living in a country with 100 million people than in a nation with one-tenth the population.

SOURCE: Nigel, J. R. Allan, "Afghanistan: The End of a Buffer State," *Focus* 36 (Fall 1986): 2–9.

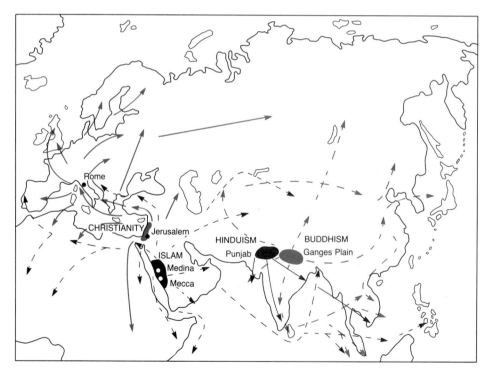

FIGURE 14.11

Source Areas and Diffusion Routes of the Major Religions of South Asia.

Buddhism A religion and philosophical system of Central and East Asia founded in India in the sixth century B.C. by Siddhartha Gautama (Buddha).

Laws now protect the "Untouchables," those who are outside the caste system and are lowest on the social scale. Certain occupations are now opening up for those who could not get such jobs under the strict caste system. Cultural heritage changes slowly, especially in the rural areas, and in India religion still determines to a large extent the lifestyles of most people.

The Islamic faith (see Chapter 13) "The Middle East") is younger than Hinduism. It was introduced to India in the Middle Ages through conquering Turkish chieftains whose bases were well established in Afghanistan. Formal inauguration of Muslim rule occurred in 1206, when Ibak, a Turkish Muslim, proclaimed himself Sultan of Delhi. After that, Islam slowly spread throughout South Asia. Present-day Muslims are of the same racial strains as Hindus and are descendants of earlier converts. As followers of the faith of Muhammad, they have escaped the social and economic burdens of the caste system.

Pakistan and Bangladesh are Muslim countries that were once one country. In 1947, East Pakistan and West Pakistan were partitioned from India, largely on religious grounds. These heavily Muslim areas wanted independence from Hindu-dominated India. After the separation, Muslims in India migrated to Pakistan, and Hindus in Pakistan left for India. This displacement of people back and forth across the new boundary line is one of the greatest mass movements of people that has ever taken place. An estimated 12 to 13 million people moved during the chaotic aftermath of partitioning. East and West Pakistan were established as one Muslim country, but the two parts were separated by 1,200 miles (1,920 km) of Indian territory and by strong racial, language, and economic differences. In 1971, animosity between the two parts of the state erupted into full-scale war. East Pakistan, with help and encouragement from India, prevailed in its separatist attempts, and the new state of Bangladesh was formed. Again, millions of migrants moved. About 10 million Bangalis left East India for the new Muslim state. Finally, in 1973, Bangladesh and Pakistani nationals who had been stranded in each other's country were allowed to return home. The episode is a classic example of the interplay between religion and politics.

The **Buddhist** religion can be traced to Siddhartha Gautama, who was born about 563 B.C One of the great spiritual leaders of the world, he came to be known as "the Enlightened One" or, simply, Buddha. Buddhism started largely as a reformation against Hinduism and is based on that religion. Buddha maintained that all life involves suffering and that the cause of suffering is desire. Once desire is controlled, suffering stops, and a state of nirvana is reached. The goal of the strict ethical code prescribed by Buddha is the attainment of nirvana. Buddhist monks dedicate their lives, through serenity, compassion, self-discipline, and meditation, to becoming enlightened and achieving nirvana. Buddhism flourished in India for nearly 1,500 years after Buddha's death, but it is now more common in other countries. Besides Bhutan, Nepal, and Sri Lanka, Buddhists also predominate in some of the mainland countries of Southeast Asia.

This crowded street in Karachi, Pakistan, is indicative of the densely populated cities of South Asia. Photograph by Jim Curtis.

A Buddhist monk prays and begs on a street in Lhasa, Tibet. Photograph by Colin Hagarty.

Language

Language differences have been a barrier to social and political unity in South Asia since the beginning of modern history. The complexity of variability from place to place, and even from village to village, is beyond any attempts to map it (see Figure 14.12 on page 442). India has 15 main languages, 12 originating from **Sanskrit,** but that is only the beginning—845 languages and dialects are spoken in the country. British India maintained 179 "official" languages. The complications of having so many languages are difficult for Americans to comprehend, but one example may help. Daily news broadcasts by radio are often given in as many as 14 different languages. When we watch a national television broadcast in the United States, where nearly everyone understands English, it is difficult to realize that people in India cannot even communicate with their neighbors in the next village.

About half the people in India speak **Hindi,** and about 10 percent speak **Urdu,** but English is

---

Sanskrit The classical Old Indic literary language that developed in India from the third century B.C. onward. It is still used in the ritual of the Northern Buddhist church.

Hindi The group of Indo-European languages associated with northern India and including Assamese, Bengali, Marathi, Punjabi, and Hindustani, among others.

Urdu A language used by the followers of Islam living in India. It developed from Hindustani but has Arabic characters.

---

the only common language spoken throughout the country. Hindi became the official language of India in 1965, but English remains an associate official language. Much government work and university instruction is conducted in English.

Other South Asian countries are only slightly more fortunate than India in their linguistic unity. Only Bhutan enjoys a high measure of unity based on one native language, Dzongkha. In Afghanistan, the languages are Pashto (Iranian), Dari Persian, and Uzbek (Turkic). Urdu and English are both official languages in Pakistan, and Bengali is both the most common language and the official language of Bangladesh. The small country of Nepal has 13 language groups, but Nepali is the official language. Sinhala (official), Tamil, and English are spoken in Sri Lanka, and Divehi, a Sinhalese dialect, is the main language of Maldives. So there are language barriers not only from village to village in India, but also from country to country throughout South Asia.

■ *Economic Geography of South Asia*
Agriculture

South Asia is primarily a region of farmers and herders. National well-being in the future will depend on whether the productivity of agriculture can be improved in each country, and especially in India. Although agriculture in India has made some progress through the introduction of high-yield seeds, chemical fertilizers, irrigation, and

> **jute** Either of two East Indian plants whose strong, glossy fibers are used in making burlap, sacks, mats, rope, and other items. (Sometimes the fiber itself is called jute.)

limited mechanization, the country still produces barely enough to feed its people. At least India does not have to import as much cereal as it did only a few years ago, but malnutrition remains a common problem.

Agricultural production in South Asia is concentrated on foodstuffs and is mostly subsistence, but some commercial crops are important in certain areas (see Figure 14.13). Rice, the staple grain, is grown over the largest area—in the coastal lowlands from Bombay on the west coast all the way around the southern tip of the country, up the east coast, and then up both the Brahmaputra Valley and the Ganges Valley. In fact, rice is grown throughout the eastern half of India anywhere sufficient rainfall permits. In the extremely wet areas, such as Bangladesh and the Assam region of eastern India, jute is grown. Both rice and jute require large amounts of water, but jute even more than rice. The **jute** plant is a bush that gets about 10 feet tall and produces a strong, glossy fiber used for making mats, rope, sacks, and burlap. Bangladesh and Assam produce most of the world's supply of jute. In the drier regions of western India, rice growth gives way to cotton. The major cotton-producing area is north of Bombay in a region surrounding the Gulf of Cambay. Cotton also is grown in scattered regions throughout South Asia where rainfall is not sufficient for rice. (see Figure 14.13 on page 443.)

Other food crops, besides rice, are important to the people of South Asia. For example, wheat is produced in the dry areas of western India, Pakistan, and Afghanistan. Millet, a type of small cereal grass, also is grown throughout South Asia. Other major crops of the region, sometimes grown in large quantities, include corn, barley, sugar cane, tea, coffee, bananas, peanuts, and olives. Sri Lanka, one of the world's major tea producers, exports tea to many countries, including both the United States and Great Britain. India and Bangladesh also export

The Ptala was the Dalai Lama's winter palace. (The Dalai Lama was the religious and political leader of Tibet.) Before the Chinese occupation of Tibet and the subsequent flight of the Lama to northern India, this palace was home to more than 10,000 Tibetan monks. Photograph by Colin Hagarty.

tea, but they both must import food grains. The Maldives have a surplus of coconuts for export but a shortage of rice. Many of the countries of South Asia trade various food commodities among themselves.

No discussion of South Asian agriculture would be complete without mention of the livestock problem in India. The Hindu religion so prominent in that country bestows particular veneration on cattle. Cattle are not actually worshiped as deities, but they are treated with respect, and there is an absolute prohibition against killing cattle, and eating their flesh would be absolutely disgraceful. Consequently, there are more cattle in India than in any other country in the world, and there is no market for them or means of getting rid of them. Millions of head of cattle (estimates range as high as 300 million) roam at will around the country. Weak, skinny, humpbacked, cows often block traffic, wander village streets, enter homes, and generally compete with humans for space and forage. The Indian government views the situation as a serious problem and has inaugurated sterilization programs for cattle. In addition, plans for

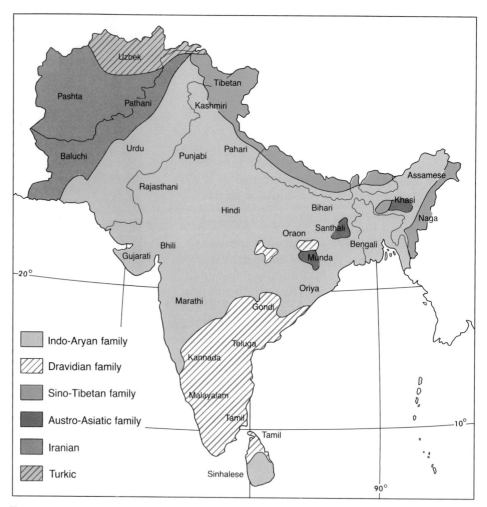

FIGURE 14.12

Language Regions of South Asia.

selective breeding aimed at the eventual development of better animals have been made on a national level.

Mineral Production and Industry

Most of South Asia is not rich in mineral resources, but India has a good supply of coal and iron ore—for example, the country produced 10 million metric tons of steel ingots in 1983. India also has manganese, mica, and bauxite deposits in such quantities that India could compete industrially with Japan and Europe if it could solve its food shortage problem (see Figure 14.14 on page 444).

More than half of all Indian investments, public and private, have gone into industrial development. The core of India's industry in recent years has been in the great river valleys outside the cities (see Figure 14.15 on page 445). Dams for irrigation, and power projects similar to the Tennessee Valley Authority in the United States, have been built. The irrigation projects are directed toward expanding agricultural output and, at the same time, power is generated for industry. India produced 144 billion kilowatt-hours (kwh) of electricity in 1984, or about 194 kwh per capita. This compares with 11,216 kwh per capita in the United States, and 4,970 kwh per capita in Great Britain.

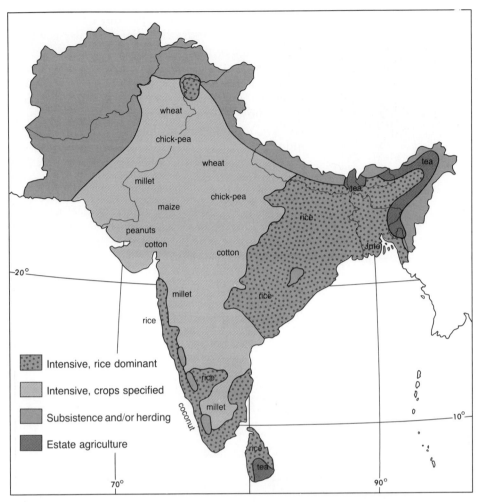

FIGURE 14.13

Agricultural Regions of South Asia.

Even though India is far from reaching its industrial potential, it is the most highly developed industrial state in South Asia. The iron and steel works located near Jamshedpur is one of the largest in Asia. Along with the heavy industry, India produces various types of light manufactured goods. Engineering goods, textiles and clothing, and tea are the country's main exports. Other important industries include cement, chemicals, and machinery, although machinery and transportation equipment are the country's leading imports. Also manufactured in India are numerous small machines, such as sewing machines, typewriters, telephones, and bicycles. Some of these commodities, such as bicycles, are produced in enough volume for home consumption and overseas trade both.

Industrialization in the other countries of South Asia has been directed toward food-processing and textiles. Manufacturing in Sri Lanka is the most varied outside India, and the most divergent from the food-fiber theme. Such wood products as plywood and paper, as well as glassware, cement, chemicals, and fertilizers, are made in Sri Lanka. Nepal has food-processing plants, and the country's major exports are rice and other food

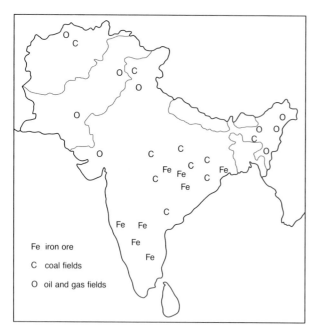

FIGURE 14.14

Coal Fields, Oil and Gas Fields, and Iron Ore Deposits of India.

products. The mountain splendor of the country holds potential for drawing many tourists as well as the production of hydroelectricity.

More than 90 percent of Afghanistan's income from exports is from agricultural products. The country's major manufacturing plants are textile mills, mostly for processing wool from the country's 4 million sheep. In Pakistan, the largest industry is the manufacture of cotton textiles, but the country also sells wool, silk, and rayon, as well as such foodstuffs as rice, wheat, sugar, oilseeds, and fish. Industry in Bangladesh is based mostly on jute, and both raw and manufactured jute accounts for most of the country's exports. The principal products of Bhutan are rice, corn, wax, cloth, and yak butter.

Transportation and Communications

Automobile traffic in most of South Asia is scarce, and very few hard-surface roads are available.

A Pakistani family poses in front of their home in Quetta. Notice their traditional attire and the mud walls of their home. Photograph by Jim Curtis.

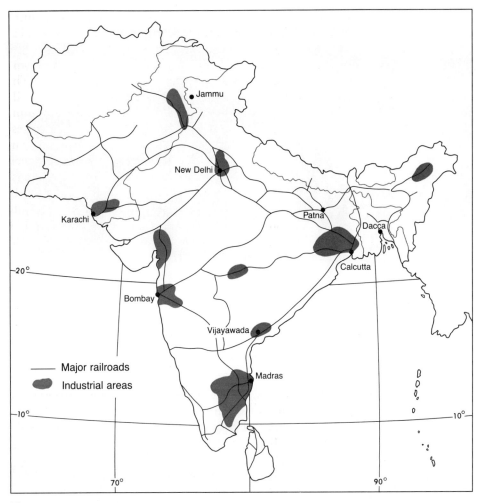

FIGURE 14.15

Railroads and Industrial Areas of India.

Local traffic moves by foot and animal carrier or by bus. Long-distance travel is usually done by railroad. The British legacy of railroads in India, Pakistan, and Bangladesh has been maintained and increased. The colonial British administration in India constructed 42,000 miles (67,578 km) of railroads (see Figure 14.15). Consequently, India now ranks second to Japan in total railroad mileage in Asia and fourth in the world. Connections by rail between major cities is adequate and often excellent. Railroads in Afghanistan are only spurs of Soviet lines, and no railroad runs to the capital city, Kabul. Bhutan has no railroads at all, and only one narrow-gauge line runs for about 39 miles (63 km) into Nepal from India.

International trade through shipping moves through the port cities of Bombay, Madras, and Calcutta in India, through Karachi in Pakistan and Colombo in Sri Lanka. The landlocked mountain countries must move their goods through other countries in order to get them to port cities. The number of local passengers moving by air increases slowly each year, but most of the air traffic is international. During World War II, Karachi was the American base for the trans-Burma connection to China, and the large airbase built there has been

Much of the passenger traffic in South Asia is carried by buses. Here a highly decorated bus stops at a station in Bahrain, Pakistan. Notice the well-stocked outdoor market. Photograph by Jim Curtis.

improved. Karachi now has the largest airport in South Asia.

The multiple language barrier severely limits communications in South Asia. In addition, many of the countries have poor telephone and telegraph services. None of the countries, for example, has more than one telephone for each 100 people, compared with 79 telephones per 100 people in the United States. In the places where it is available, India's domestic telephone service is only fair, and it is the best in the region. Television did not reach South Asia until the 1970s. India now has 17 broadcasting stations, but there are no television stations or receivers in Bhutan or Nepal. The remainder of the countries in South Asia have at least one television broadcasting station. All the countries have at least one AM radio station, but FM radio is rare.

■ Political Geography of South Asia

Politically, South Asia is a developing complex of nationalistic and regional loyalties. Although the region maintains a somewhat peaceful image, internal and external violence lie just below the surface. Since receiving independence in 1947, for example, India has seized Goa from Portugal, taken part of Kashmir from Pakistan, fought a border war with China, and destroyed the Pakistani army in the east while backing the establish-

In the cities the buses are more modern than in the rural areas, and automobiles are more abundant. This street scene is from Karachi, Pakistan.

Focus Box

Storms and Crowding in Bangladesh

The devastation of the islands of southeastern Bangladesh caused by a deadly storm on May 25, 1985, is a tragic example of the results of overcrowding at its worst. The death toll in the disaster is uncertain, but Bangladesh authorities estimated it at 10,000. The storm damaged or destroyed 200,000 homes and ruined 135,000 acres of cropland.

The islands, which consist of silt, have risen from the sea only in the last five years. Their

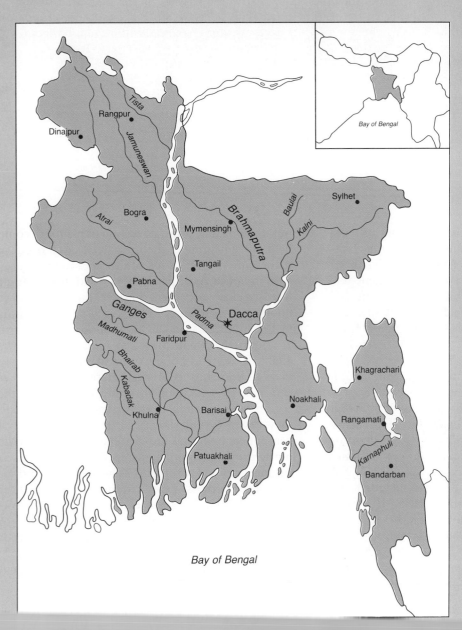

FIGURE 14.B

Bangladesh. The low ground along the coast is part of the Ganges-Brahmaputra delta and is subject to seasonal flooding. The country is bisected by the great rivers and their distributaries.

vulnerability to storms is well known, but despite a national law prohibiting settlement on an island for 10 years after it emerges, many land-starved poor people have flocked to them. Ironically, a highly sophisticated storm-warning system had been installed by the United States and the United Nations just weeks before the disaster. But being aware of the impending storm and informing the populace about it are two very different things. The only infrastructure for warning available to Bangladesh officials was radio, which proved ineffective because few of those affected could afford to own a radio.

Although the proportions of the 1985 disaster were enormous, they are dwarfed by the death counts from Bangladesh's 1970 storm and tidal wave, which left at least 300,000 dead. Clearly, if the effects of these inevitable seasonal storms are to be minimized, greater resources will have to be spent on development of more effective warning, evacuation, and fortification systems.

SOURCE: Arthur Haupt (ed.), "Crowded Islands, Lack of Warning," *Population Today* 13 (July–August 1985): 3.

---

ment of Bangladesh. Internally, Indian officials have been frustrated over Punjabi dissidents clamoring for self-government. Afghanistan has been overrun by the Soviets, and antigovernment terrorists in Sri Lanka have killed more than 100 people. Pakistan and India bicker constantly, and western Pakistan is now loaded with refugees from Afghanistan. The potential for further violence in South Asia is high.

Internally, riots, street fighting, and natural disasters, sometimes augmented by lack of appropriate political action, result in loss of life that Westerners find hard to believe. Just how many deaths in South Asia result directly from government neglect cannot be measured, but flooding alone has killed many thousands since 1947. For example, recent major flood disasters include 1,700 deaths in India and Pakistan in 1955, 10,000 in Bangladesh in 1960 and another 50,000 in 1965, 1,780 deaths in India in 1968, 300,000 deaths in Bangladesh in 1970, 15,000 in India in 1979, and 10,000 more in Bangladesh in 1985. Bangladesh, a low-lying country, receives great amounts of rain, but it seems that some government planning could help prevent some of the mass deaths that occur in that country. Every type of disaster strikes doubly hard in South Asia, partly because so many people live so close together. Besides the storms and flooding, earthquakes killed 1,500 people in India in 1950 and 2,000 in Afghanistan in 1956.

South Asia will play a significant role in the affairs of the world in the years to come. The sheer mass of humanity is of vital interest to all observers. The Soviet Union took a giant step into the region when it took over Afghanistan. Although they began to withdraw in 1988, the geopolitical consequences of that move are quite large, and the burden to Pakistan of housing Afghan refugees is an international problem. The strong religious inclination of most people in the region influences them toward a dislike of communism, but memories of the British colonial days lead them to mistrust the West. In any case, the region needs material help and technical aid from the outside world, and hard work, patience, and intelligent direction from within.

Key Words

acculturation
Buddhism
Hindi
Hinduism
jute
monsoon
Sanskrit
sensible temperature
Urdu

References and Readings

Ahmad, Kazi S. U. *A Geography of Pakistan*. Pakistan Branch: Oxford University Press, 1964.

Bayliss-Smith, Tim P., and Wanmali, Sudhir (eds.). *Understanding Green Revolution: Agrarian Change and Development Planning in South Asia*. New York: Cambridge University Press, 1984.

Baxter, Craig. *Bangladesh: A New Nation in an Old Setting*. Boulder, Colo.: Westview, 1984.

Berry, Brian J. L. *Essays on Commodity Flows and the Spatial Structure of the Indian Economy*. University of Chicago Department of Geography Research Papers, no. 111. Chicago: University of Chicago, 1966.

Boyd, Andrew. *An Atlas of World Affairs*. New York: Praeger, 1960.

Brown, Joe David. *India*. New York: Time Inc., 1964.

Burki, Shahid J. *Pakistan: A Nation in the Making*. Boulder, Colo.: Westview, 1986.

Central Intelligence Agency, *The World Factbook, 1985*. Washington, D.C.: Government Printinig Office, 1985.

Cressey, George B. *Asia's Lands and Peoples*. New York: McGraw-Hill, 1951.

Davis, Kingsley. *The Population of India and Pakistan*. New York: Russell & Russell, 1968.

DeEast, William Gordon, and Spate, Oskar H. K. (eds). *The Changing Map of Asia: A Political Geography*. 4th ed. New York: Barnes & Noble, 1961.

DeSilva, K. M. *Managing Ethnic Tensions in Multi-Ethnic Societies: Sri Lanka, 1880–1985*. Lanham, Md.: University Press of America, 1986.

Dutt, Ashok K. "Cities of South Asia." *In Cities of the World: World Regional Urban Development,* ed. Stanley D. Brunn and Jack F. Williams. New York: Harper & Row, 1983.

———. (ed.). *Contemporary Perspectives on the Medical Geography of South and Southeast Asia*. Elmsford, N.Y.: Pergamon, 1981.

Dutt, Ashok K., and Geib, Margaret. *An Atlas of South Asia*. Boulder, Colo.: Westview, 1987.

Farmer, Bertram H. *An Introduction to South Asia*. New York: Methuen, 1984.

Fernando, Tissa. *Sri Lanka: Portrait of an Island Republic*. Boulder, Colo.: Westview, 1987.

Ganguly, Sumit. *The Origins of War in South Asia: Indio-Pakistani Conflicts Since 1947*. Boulder, Colo.: Westview, 1986.

Ginsburg, Norton et al. *The Pattern of Asia*. Englewood Cliffs, N.J.: Prentice-Hall, 1960.

Gupte, Pranay. *Vengeance: India After the Assassination of Indira Gandhi*. New York: Norton, 1985.

Hardgrave, R. L., Jr. *India: Government and Politics in a Developing Nation*. New York: Harcourt Brace Jovanovich, 1980.

Johnson, Basil L. C. *Bangladesh*. Totowa, N.J.: Barnes & Noble, 1982.

———. *Development in South Asia*. New York: Viking, 1983.

Karan, Pradyumna P. *Bhutan: A Physical and Cultural Geography*. Lexington: University of Kentucky Press, 1967.

Kosinski, Leszek A., and Elahi, K. Maudood (eds.).*Population Redistribution and Development in South Asia*. Hingham, Mass.: Reidel, 1985.

Lall, Arthur S. *The Emergence of Modern India*. New York: Columbia University Press, 1981.

LaPierre, Dominque. *The City of Joy (Calcutta)*. Translated by Kathryn Spink. Garden City, N.Y.: Doubleday, 1985.

Lodrick, D. O. *Sacred Cows, Sacred Places*. Berkeley: University of California Press, 1981.

Lukacs, John R. (ed.). *The People of South Asia: The Biological Anthropology of India, Pakistan, and Nepal*. New York: Plenum, 1984.

Mamoria, C. B. *Geography of India*. Agra: Shiva Lal Agarwala, 1975.

Mitra, Sujata. "The Diffusion of Cholera in India: A Markov Chain Analysis." Master's thesis, Oklahoma State University, 1976.

Noble, Allen G., and Dutt, Ashok K. (eds.). *India: Cultural Patterns and Processes*. Boulder, Colo.: Westview, 1982.

Norton, James H. K. *The Third World: South Asia*. Guilford, Conn.: Dushkin, 1984.

O'Donnell, Charles P. *Bangladesh: Biography of a Muslim Nation*. Boulder, Colo.: Westview, 1984.

Rawson, R. R. *The Monsoon Lands of Asia*. Chicago: Aldine, 1963.

Sanders, Sol. *A Sense of Asia*. New York: Scribners, 1969.

Shabad, Theodore. *China's Changing Map: National and Regional Development, 1949–1971*. New York: Praeger, 1972.

Sopher, David E. (ed.). *An Exploration of India: Geographical Perspectives on Society and Culture*. Ithaca, N.Y.: Cornell University Press, 1980.

Spate, Oskar H. K., and Learmonth, A. T. A. *India and Pakistan: A General and Regional Geography*. New York: Barnes & Noble, 1967.

Stamp, Laurence Dudley. *Asia: A Regional and Economic Geography*. London: Methuen, 1967.

Sundaram, K. V. *Geography of Underdevelopment*. New Delhi: Concepts, 1983.

Tayyeb, Ali. *Pakistan: A Political Geography*. London: Oxford University Press, 1966.

Wint, Guy (ed.). *Asia: A Handbook*. New York: Praeger, 1966.

Chapter 15

SOUTHEAST ASIA

■ *Introduction*

Southeast Asia is a unified world region in name only. Its physical fragmentation is matched by its cultural and political diversity. The physical pattern is a result of relatively recent geologic changes that caused shallow seas to separate much of the region from the Asian mainland. A long history of migrants, explorers, and merchants entering the region imposed cultural diversity on the area. For centuries, Southeast Asia became home for waves of people from China and the sponge for religious beliefs from India. It has also been the source for many European fortunes.

During World War II, the British labeled the region "Southeast Asia" in an attempt to describe the vast water-oriented realm of Lord Mountbatten's military command. Because it was a convenient way to describe the general location of the region, it soon became a popular label worldwide. Only half the region's land area is attached to the Asian continent, however. The remaining surface area is divided among thousands of islands.

The region (see Figure 15.1) stretches east to west for 3,800 miles (6,080 km) from the western half of the island of New Guinea to the eastern border of India. Latitudinally, it lies mostly within the Tropics. Numerous bodies of water are on the periphery of the region. On the west from north to south are the Bay of Bengal, the Andaman Sea, and the Indian Ocean. The South China Sea lies to the north, penetrating into the middle of the region. The Pacific Ocean is to the east, and the Arafura Sea separates the region from Australia on the south. Many smaller seas lie between the islands, and they include the Flores, the Java, the Savu, the Banda, the Ceram, the Molucca, the Celebes, and the Sulu seas. The region also contains innumerable bays, inlets, gulfs, and straits. The most important strait is the **Strait of Malacca,** which is a major route for world shipping. Standing on high ground on the island of Singapore, one can always see at least one ocean freighter moving through the passage. As one

Strait of Malacca The channel between the southern part of the Malay Peninsula and the island of Sumatra, connecting the Indian Ocean with the South China Sea. It is about 500 miles (805 km) long and varies in width between 35 miles (56 km) and 185 miles (298 km).

freighter disappears, another comes along, for 24 hours, every day of the year.

Ten countries make up the region of Southeast Asia. They include the countries on the Asian continent—Thailand, Laos, Myanma (formerly Burma), Cambodia, and Vietnam—and the island nations (see Figure 15.2). Malaysia is partly on the

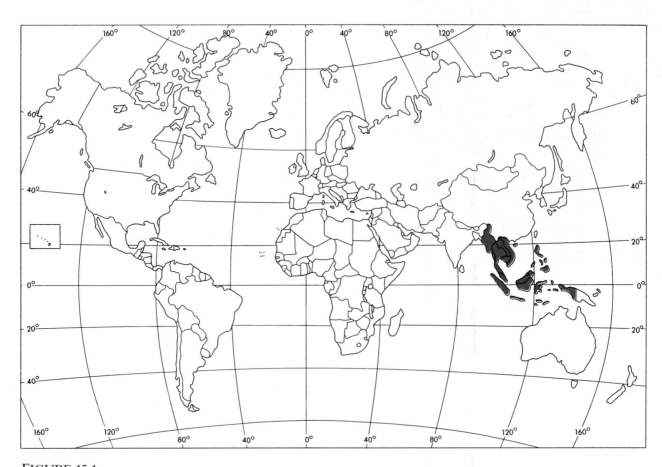

FIGURE 15.1

World Location of Southeast Asia.

FIGURE 15.2

The Countries of Southeast Asia.

mainland, where it occupies the southern portion of the Malay Peninsula, but the Malaysian territories of Sarawak and Sabah, which are also part of the Malaysian Federation, are on the north coast of the island of Borneo. The continent section is known as Western Malaysia, and the Borneo sections make up Eastern Malaysia. Singapore is the name of a city on an island by the same name. It is located at the southern tip of the Malay Peninsula, close enough to the mainland to be linked by a causeway. The tiny nation (about 225 square miles or 583 km^2) once belonged to the Malaysian Federation, but it became independent in 1965 and joined the United Nations that same year.

The Republic of Indonesia is probably the most physically fractured country in the world. Its 14,000 islands are scattered across the southern part of the region from the Indian Ocean to the

Pacific Ocean. The Philippines is a country with 7,000 islands, but most of them are located in a fairly compact area between the two largest islands of Mindanao and Luzon. Brunei, also located on the island of Borneo, is completely surrounded by Eastern Malaysia, except for frontage on the South China Sea. Although Brunei is a very small country, the people there enjoy the third highest per capita gross national product in the world (after Qatar and the United Arab Emirates). The money comes from oil production. The island of New Guinea is divided between Indonesian territory on the west and the country of Papua New Guinea on the east. The international boundary follows a longitude line, and is a remnant of colonial occupation of the Dutch on the west and Germans on the east.

Physical Geography of Southeast Asia

The essential physical features of the Southeast Asia mainland are the north-south trending mountain ranges and the several major river valleys and delta areas between the mountains. The mountain ranges become steep hills as they reach southward in rows, continually decreasing in elevation from the massive Himalayas in interior China. The large rivers of the region also originate in China and flow southward in the valleys between the mountain ranges. Most of the people on the mainland live along these great rivers, and the entire region is water oriented.

The major rivers of Myanma are the Irrawaddy and its largest tributary, the Chindwin. Rangoon and Mandalay, the largest cities in Myanma, are located on the banks of the Irraway. The Salween River flows through Myanma too, but it originates on the Plateau of Tibet and flows for about 800 miles (1,287 km) through China before entering Myanma. The delta and the flood plains of the Irrawaddy are much more extensive than those for the Salween, leaving room for the core of the country to develop. (See Figure 15.3.)

The Mekong River parallels the path of the Salween but flows in a valley to the east of the Salween. At times the two valleys are no more than 20 miles (32 km) apart. After leaving China, the Mekong is used for the boundary between Myanma and Laos, then serves as part of the boundary between Laos and Thailand, continues through the heart of Cambodia, cuts across the southern tip of Vietnam, and finally empties into the South China Sea. The Mekong, well known to Americans because of the Vietnam War, is the largest and most significant river in Southeast Asia. The capital cities of Vientiane (Laos) and Phnom Penh (Cambodia), are located along the river. Hanoi (Vietnam) is on the banks of the Red River, and Bangkok (Thailand) is divided by the Chao Phraya River. The political cores and cultural hearths of all the mainland countries, then, have developed along rivers.

The north-south orientation of the highlands and rivers of mainland Southeast Asia have kept the countries separated as effectively as the waters separate the island nations. Each country is oriented downstream of its major river and outward from the center of the region. Vietnam looks eastward toward the South China Sea, and Myanma looks westward toward the Bay of Bengal. Each country between is similarly outward oriented. The rugged physical landscape and dense vegetation have served as effective barriers to movement. Both China and India have influenced Southeast Asia, but the region has been primarily a buffer between the two giant neighbors because of its barrier status.

The physiography of the island section of Southeast Asia follows a pattern in that each island has its highland or mountain core surrounded by coastal lowlands. The pattern holds for both the large and the small islands. The mountain cores of Sumatra and Java are linear, whereas most of the others are more or less circular. Each island has a series of rivers radiating out of the mountains and flowing through the coastal lowlands. Because they are short, these are not well known rivers, but there are many of them, and at times they carry large amounts of water. The only continuous high range of mountains is on the Indonesian part of New Guinea. The Sudirman Mountains rise to their highest point (16,400 feet, or 5,000 m) at Sukarno Peak. Almost all the Indonesian islands, besides being rugged, are heavily vegetated.

Volcanic action created most of the islands, and many individual peaks reach heights of 10,000 to 12,000 feet (3,050–3,650 m). Many of the volcanoes are active today. The people on the enchanting

Krakatoa The Indonesian island volcano in the center of the Soenda Strait between Sumatra and Java, noted for its violently explosive eruptions. The eruption on August 26, 1883, was the most tremendous ever recorded and was heard as far away as Turkey and Japan. The resulting tidal wave killed about 36,000 people in western Java.

island of Bali, just east of Java (Figure 15.4), are continually plagued by volcanic eruptions. The 10,308 foot (3,142 m) high Mount Agung erupted in 1963, killing 1,500 people and burying more than a third of the farm land on the island, but the most famous volcanic eruption occurred on **Krakatoa,** an island between Java and Sumatra (see Figure 15.4). On August 27, 1883, most of the island disappeared in a massive explosion that is said to have

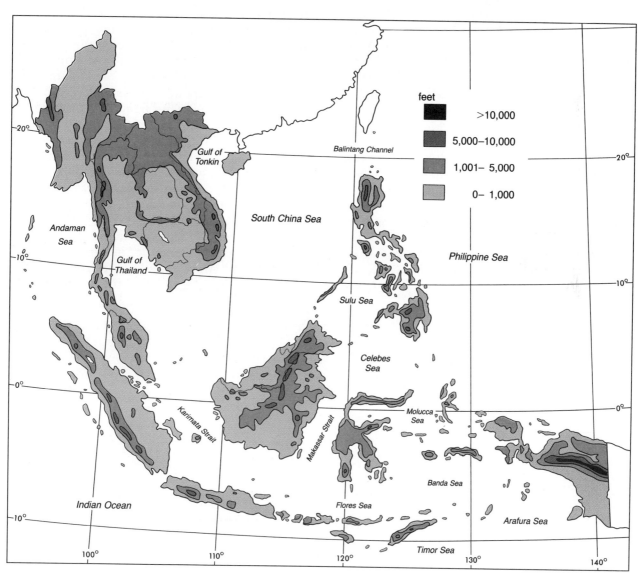

FIGURE 15.3

Land Elevations and Water Bodies of Southeast Asia.

caused the loudest noise ever heard on earth; the sound carried over 2,500 miles (4,023 km). As the island disappeared, it left a 1,000-foot (305 m) hole in the ocean water. Tidal waves resulting from the displacement of water carried onto the neighboring islands, killing more then 35,000 people. Volcanic ash and dust spewed into the atmosphere, circled the globe, and caused unusually red sunsets for months afterward. Lava eventually created another island, but in 1927 it too disappeared in yet another massive explosion. Southeast Asia is the most active volcanic region of the "Ring of Fire" that surrounds the Pacific Ocean. Of the 200 volcanic eruptions during the twentieth century, half have been in Southeast Asia.

The seas between the islands of Southeast Asia generally are quite shallow; most are 150 to 200 feet (45–60 m) deep. At the opposite extreme are the great **ocean trenches** located just to the outside of the island region. The Philippine Trench, east of the Philippines, is a 600-mile-long (965 km) canyon on the bottom of the ocean. At one point it is about 7 miles from sea level to the ocean floor. The Java Trench borders the region on the south off the coasts of Sumatra and Java, and another ocean deep has been recorded on the east of the Banda Sea.

Climates throughout Southeast Asia are remarkably consistent from place to place—

ocean trench A place in the ocean where the water is very deep, sometimes called an "ocean deep." The underwater trenches are similar to canyons on the surface of the earth. The deepest known point of any ocean is the Philippine or Mindanao Trench of the Philippine Sea in the western Pacific.

persistently hot and wet. What variation does occur is more in the precipitation than in the temperature. Near the equator and along the coastal lowlands, the temperature never varies more than one or two degrees from the 80° F (27° C) average. Farther inland and northward, temperature ranges of 9 to 10 degrees are common. Such temperature figures are for monthly averages; daily ranges are frequently greater than the ranges between July and December. A daily high of 100° F (38° C) or more may be encountered in the northern part of the region. (See Figure 15.5.)

The shallow waters of the inter-island seas are warm all year, causing constant evaporation of moisture into the air. In turn, the air is moved over the islands by the trade winds, which move back and forth across the region with the yearly movement of what is known as the "intertropical front." As the front moves south in response to the southern hemisphere summer, the northeast trade

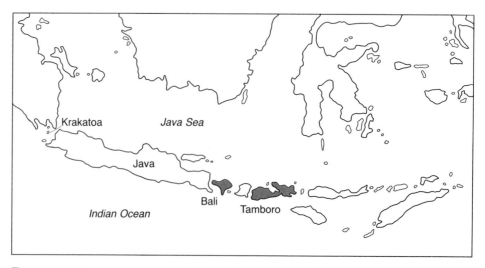

FIGURE 15.4

Part of Indonesia. The islands of Bali and Tamboro are located east of Java. The island of Krakatoa is located west of Java. Krakatoa and Tamboro are noted for violent volcanic eruptions.

typhoons Intense tropical storms of the western Pacific Ocean (called hurricanes when they are in the Atlantic Ocean) that produce wind speeds over 100 mph and torrential rains.

winds sweep across the region. When the front moves back onto the Asian mainland six months later, the southeast trade winds blow across the islands and onto the mainland. In both seasons, the islands receive copious amounts of precipitation. The uplift, caused by the mountains, cools the warm, moist air, resulting in constant condensation and the resulting rain. The mainland, however, receives its moisture primarily during the northern hemisphere summer from the southeast trade winds. Only the exposed east coast of the mainland and part of the Malay Peninsula receive winter rain.

The overall rainfall patterns may be summarized thus: (1) constantly rainy on the islands, (2) summer monsoon conditions on most of continental portion of the region, (3) both summer and winter monsoons over Vietnam and Western Malaysia. A large part of the region receives more than 80 inches (200 cm) of rain annually. It is an 80–80 climate area—always 80°F with about 80 inches of precipitation.

Between June and November, when the southeast trade winds prevail over Southeast Asia, tropical storms develop in the western Pacific (see Figure 15.6). About 100 of these storms occur each year. They are called **typhoons** and are similar to hurricanes that move over the Atlantic Ocean. These storms break off the intertropical front and are pushed by the southeast trade winds

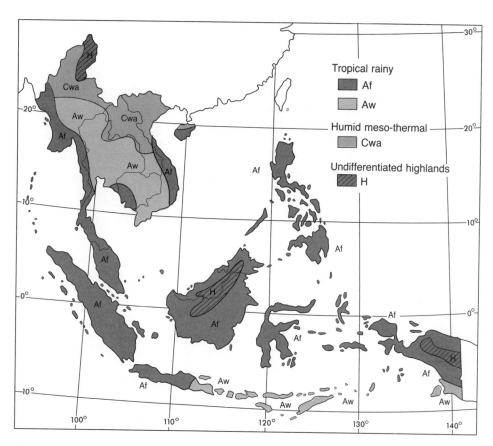

FIGURE 15.5
Climate Regions of Southeast Asia.

in a northwestwardly arc across the Philippines toward Japan (see also Chapter 8, "Japan"). They can be violent and often rake coastal areas with winds of more than 100 miles per hour. Typhoons usually dump huge amounts of rain along their paths, and the ocean churns and surges from the wind. The rain and ocean surges often cause flooding in coastal areas, sometimes causing many deaths (see Table 15.1).

The warm, humid conditions of Southeast Asia result in dense tropical forests and verdant vegetative cover in the nonforested areas. Vast regions remain uncleared for agriculture, but in some areas shifting ("slash and burn") agriculture has been carried on for centuries. The heat and moisture not only allow natural decay, mold, and fungi but also encourage cleared areas to become revegetated in a very short time. Plants as well as animals thrive in a large variety of species and subspecies. Seemingly billions of insects and parasites compete with humans for supremacy in the region. Savanna grass covers some of the drier areas that receive the summer monsoon only. These grasslands have not been utilized for herding, as in other areas of the world. Both the grasslands and the forests are insect-infested, hot and humid, and quite uncomfortable for most humans.

TABLE 15.1

Deaths in the Philippines from Typhoons, 1964–1984

DATE	NAME OF STORM	NO. KILLED
1964 June 30	Winnie	107
1968 November 18	Nina	63
1970 September 15	Georgia	300
1970 October 14	Sening	583
1970 October 15	Titang	526
1972 December 3	Theresa	169
1974 June 11	Dinah	71
1976 May 20	Olga	215
1978 October 27	Rita	400
1981 November 25	Irma	176
1984 September 2	Ike	1,363

SOURCE: Scientific Event Alert Network. Smithsonian Institute, Washington, D.C.

■ Human Geography of Southeast Asia

History

The indigenous population of Southeast Asia is made up of descendants of Malay-type people who migrated to the region approximately 4,500 years

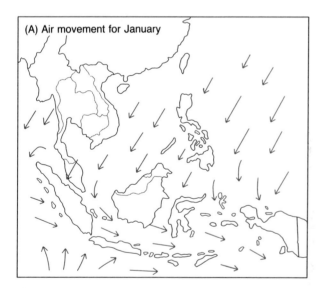
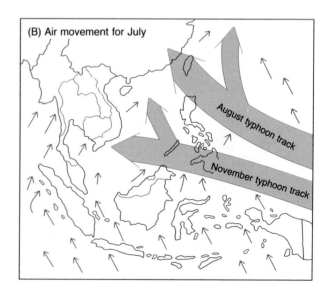

FIGURE 15.6

Wind Changes and Typhoon Tracks in Southeast Asia. The prevailing winds for January and July indicate the dramatic change from season to season. Most of the typhoons occur during the period from August to November.

Java Man A type of primitive human known from fossil remains first found in Java in 1891; also called *Pithecanthropus erectus*.

ago, but various waves of migrants had come to the area as early as 5,000 years before that. The fossilized skull of the **Java Man** (*Pithecanthropus erectus*), found along the Solo River in Java, has been dated at more than half a million years old. It was the Malays, however, who established the social and economic customs that have endured until today. They developed the technique of rice cultivation in irrigated paddy fields that allowed nomadic hunters to become sedentary farmers. The Malays also learned to raise pigs and chickens, and they domesticated the water buffalo as a beast of burden.

For centuries, the Malay people were invaded from the north. Various Chinese tribes descended on the region through the natural pathways of the great river valleys (see Figure 15.7). Some invaders were friendly migrants, others were purely military campaigners. If the invaders succeeded, they occupied the fertile river valleys and relegated the losers to the hills and mountains. The Mongoloid Viet tribes came out of China and took over some Malay lands. Later, they were displaced from some of the same areas by the Han Chinese. Then the Mongols came. Such was the pattern century after century. Today, the Chinese influence extends over almost all parts of Southeast Asia and remains strong.

During the time of the Chinese migrations from the north, Southeast Asia was visited by sailors from India. The number of these visits increased as sailing techniques improved. Seaworthy junks were built, and the pilots eventually learned to use the monsoonal shifts in wind to go back and forth between India and Southeast Asia. The Indians came in search of gold and other riches, but they influenced the region profoundly by bringing their religion—Islam. In order to gain trade favors without making demands of land, the Indian sailors married local chieftains', daughters. The wives then became propagators of their husbands' cultural habits, most notably their Muslim

The coastal lowlands of eastern Vietnam are highly utilized for farming. Notice the volcanic peak in the background. Photograph by Louis Seig.

religion. In this manner, Islam arrived in Southeast Asia. Dictated by their faith, the Indians subtly and persuasively became advisers in all facets of Malay life. In this way, Islam filtered down from the wealthy merchants to the common people and was readily accepted by the Malay. It remains the dominant faith in Indonesia today.

Europeans did not arrive in Southeast Asia until the close of the fifteenth century A.D. One item that drew them to the area was pepper. As Europe lifted itself out of the Dark Ages, it began to rebuild its cities. Urban life depends on the ability to store food, and the only known food preservatives of the time were pepper and other spices. Marco Polo, the famous world-traveling Italian, had written about the **"Spice Islands"** of the Indies nearly 50 years before serious search for them began. Another Italian, Christopher Colum-

---

Spice Islands Usually refers to the Molucca Islands, a group in the Malay archipelago between the islands of Celebes and New Guinea, noted for producing spices, especially nutmegs, mace, and cloves. The name is sometimes used to indicate all the Southeast Asia islands.

---

bus, tried to find them but failed. Finally the Portuguese, encouraged by Prince Henry and led by Vasco da Gama, found the islands where nutmeg, cloves, and especially pepper grew wild, and Lisbon became the spice capital of Europe.

Competition among the European countries for spices became brutal and often bloody, and Portuguese success in Southeast Asia lasted only a few years. The British tried to succeed them in the

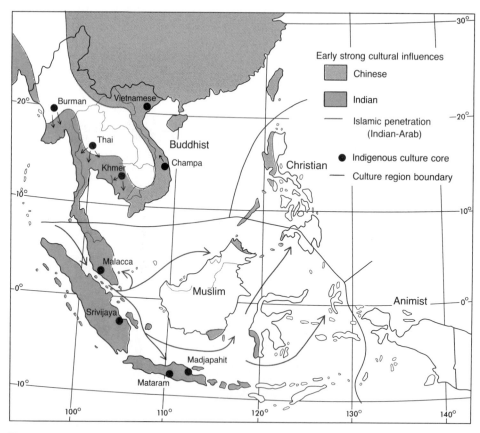

FIGURE 15.7

Cultural Cores and Historical Invasion Routes of Southeast Asia.

The pattern of these Vietnamese fields has developed over centuries of occupation. Photograph by Louis Seig.

region, but soon retired to their outposts in India and South Asia. The Dutch finally emerged as the most influential European force in the islands. The Dutch United East India Company was the vehicle for Dutch monopolization of the world spice market, and for many years millions of guilders flowed into Dutch banks from the spice trade alone. Although the East India Company went out of business in 1799, the Dutch did not relinquish their political control in the region until 1949.

While the Portuguese were going around Africa and eastward to get to the East Indies, the Spanish were proving the theory of Columbus by sailing west. When the Spanish explorer Ferdinand Magellan sailed around the world, he came on the Philippine Islands from the Pacific side and claimed the islands for Spain, setting the scene for the present-day cultural aspects of the people of the Philippines. The Portuguese and the Dutch were traders, but the Spanish were colonizers, so the Spanish made a greater impact on the culture of the islands they controlled. Islam has survived on the Portuguese and the Dutch islands, but most Filipinos are Roman Catholic.

During the nineteenth century the British returned to Southeast Asia, and France and the United States entered the region politically for the first time. In 1826, Great Britain invaded Burma (Myanma) from India, using the excuse of border problems but in reality attempting to prevent France from taking over. Around 1860, France began to assume control of Indochina (see Table 15.2), and Britain retaliated by grabbing Malaya in 1874. Twenty-four years later, Admiral George Dewey of the United States sailed into Manila Bay, wresting control of the Philippines from Spain.

The next conquest of Southeast Asia came from the Orient. In 1942, Japan swept southward and in less than one year gained control of the entire region. Even Thailand, which had never been a colony, was occupied by the Japanese. The Japanese were probably more brutal than the European rulers, because they demanded complete subservience from the people they conquered. The record indicates that the building of a Japanese railroad through Thailand cost the lives of 35,000 Burmese and Thais, as well as those of many British and American prisoners of war (part of the construction process was portrayed in the movie *Bridge on the River Kwai*). The Japanese set the scene for the demise of European control in Southeast Asia; many of the colonial empires withdrew from the area shortly after World War II. (See Figure 15.8.)

Colonialism had imposed peace of a kind on Southeast Asia, but after its demise constant fighting developed throughout the region. Nearly every country of Southeast Asia has experienced war or internal strife. After the British left, Malaya (as it was called) suffered through 12 years of civil war, an attack by neighboring Indonesia, and the eventual splintering of the country. Malaysia today is still plagued by the conflict between Malay and Chinese cultures, and there is much discrimination. Following the American withdrawal from the Philippines in 1946, **Huk Communists** attacked the government there. The Communists were officially suppressed in 1952, but internal strife, which led to martial law and a dictatorship in 1972, continued. Finally, the Philippine leader Ferdinand Marcos was forced to flee the country. The Communists still pose a major problem for the current leadership in the Philippines.

In **French Indochina,** after the Japanese were driven out, another war raged from 1946 to 1954. Local insurgents were trying to wrest control of the area from the French. After the French defeat and withdrawal, a short period of peace was followed by another war between the newly created states of North Vietnam and South

Huk Communists In the Philippines during World War II, organized guerrilla forces fought the Japanese. They were called the Hukbong Bayan Laban Sa Hapon, or People's Anti-Japanese Resistance Army, and were known as Huks. These fighters, as well as similar guerrilla forces in other parts of Southeast Asia, later became Communists and fought for national liberation, so they came to be known as Huk Communists.

French Indochina Indochina, or "Farther India," included the southeastern peninsula of Asia and all the countries from Myanma to Vietnam. For many years, the eastern part of that region, Vietnam, was controlled by France and was therefore called French Indochina.

Vietnam. The United States became involved in the war in 1964 by first sending military advisers to help the South Vietnamese. Some U.S. government officials feared that communist expansion into all of Southeast Asia would occur if North Vietnam took over South Vietnam. The fear was called the "domino effect." The war escalated, and more American troops were sent to Vietnam. A cease-fire was finally called on January 27, 1973, but the war continued until the United States was forced to withdraw in 1975. A total of 58,021 Americans died in Vietnam. Eventually, the goals of the insurgents of 1946

TABLE 15.2

Colonial Rule in Southeast Asia, 1511–1945

RULER	COUNTRY	DATES
Portugal	Spice Islands[a]	1511–1611
Spain	Philippines	1521–1898
Netherlands	Indonesia	1699–1949
Great Britain	Myanma (Burma)	1826–1948
France	Indochina[b]	1862–1955
Great Britain	Malaysia	1874–1957
United States	Philippines	1898–1946
Japan	Entire region	1942–1945

[a]Includes numerous islands mostly, of the Indonesian group.
[b]Includes Vietnam, Cambodia, and Laos.

> **boat people** When U.S. troops withdrew from Vietnam, many South Vietnamese escaped the rule of the Communist North Vietnamese by taking to the open waters of the South China Sea in small and sometimes overloaded boats. These refugees are called "boat people."

became reality as the two Vietnams united into one country, free of foreign intervention. However, thousands of people from what was South Vietnam were forced to flee their homeland, and many became **"boat people"**.

In both Laos and Cambodia, internal power struggles resulted in constant fighting between government troops and ambitious political dissidents. Spillover from the neighboring Vietnam War affected both countries too. During the 1960s, Indonesia was ruled by Sukarno, a strong-armed dictator. After extremely bloody fighting, the army seized control of that country, declared it a republic, and elected a president. Communist insurgents remain a problem in Indonesia, even though 300,000 of them were executed in 1965 after an unsuccessful coup attempt.

Population

The total land area of Southeast Asia is 1,740,289 square miles (4,507,349 km²) and is inhabited by 408 million people (see Figure 15.9). The average population density is 234 people per square mile (90 people per km²), a fairly high figure. Population pressure is much greater than the average makes it appear, however, because most of the

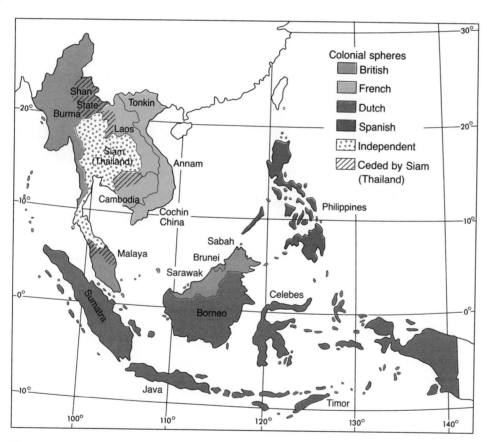

FIGURE 15.8

European Colonial Holdings in Southeast Asia.

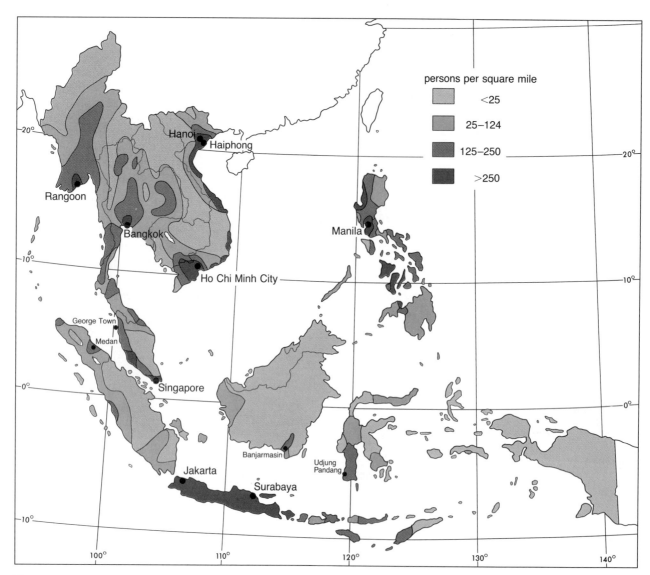

FIGURE 15.9

Population Patterns of Southeast Asia.

people live on only about 10 percent of the land. Much of the land is too rugged for human habitation. The people are concentrated in and around the urban areas, along the river plains and deltas, and along the coasts.

The islands of Singapore and Java are among the most densely populated places in the world. The population of Singapore is 2.6 million. These people live on 224 square miles (580 km^2) of land (smaller than New York City), so the density is 11,607 people per square mile (30,062 people per km^2). The population of Indonesia is 169.1 million, with an average density of 235 people per square mile (609 people per km^2). Many of the islands, including the large island of Sumatra, are sparsely populated, and some are uninhabited. Most Indonesians live on Java, where the density averages nearly 2,000 people per square mile. The Indone-

transmigration Because the island of Java in Indonesia is so crowded, the Indonesian government encourages people to migrate to some of the less densely populated islands. Free passage and free land are used as inducements. The process is called transmigration.

primate city Many Third World countries have only one large city, as opposed to industrial nations that have many. These "primate cities," usually the capitals and the economic centers, draw migrants from all parts of the countries.

sian government encourages people from Java to settle in other parts of the island country by paying for their moves and giving them free land with a home on it. This movement is called **"transmigration."** Both the Philippines and Vietnam have high population densities too (see Table 15.3).

The rate of increase for the population of Southeast Asia is 2.2 percent annually, similar to the rate for South Asia and much less than the rate for most of the countries in Africa. The population continues to increase rapidly, but the climate makes it possible for most countries to feed their people. Infant mortality rates are high (160 per 1,000 births) in the two most backward countries, Laos and Cambodia, and life expectancies are low (43 years) in those countries. Singapore has the lowest infant mortality rate in the region (9 per 1,000) and the highest life expectancy (71 years). The infant mortality rate for Singapore is actually lower than that for the United States. The other countries of the region have rates similar to those of other developing countries in the world.

In spite of the heavy concentrations of people in Southeast Asia, the urbanization factor for the entire region is only 24 percent. It varies from 100 percent in Singapore (see Figure 15.10) to 16 percent in Laos and Cambodia (see Table 15.3). In every country the capital is the largest city, and most countries in the region have only the one large, or **primate city.** The exceptions to this are in Indonesia and Vietnam. Jakarta (7.6 million) on Java is the largest city in Southeast Asia, but Surabaja (2.3 million) and Bandung (1.6 million) also are located on Java, and Medan (1.9 million) is on Sumatra. Indonesia also has many smaller cities, especially on Java. Other large cities in Southeast Asia include Bangkok (Thailand) at 4.7 million, Ho Chi Minh City (Vietnam) at 3.5 million, Rangoon (Myanma) at 2.5 million, and Hanoi (Vietnam) at 2.0 million. Phnom Penh, (Cambodia) and Vientiane (Laos) have less than a quarter of a million people each.

Table 15.3
Population Characteristics of Southeast Asia, 1985

COUNTRY	POPULATION (IN MILLIONS)	% INCREASE	DENSITY	% URBAN
Brunei	0.2	2.6	104	64
Myanma	37.7	2.0	141	24
Indonesia	169.1	2.1	235	22
Cambodia	6.4	2.3	89	16
Laos	3.7	2.3	39	16
Malaysia	15.8	2.4	121	32
Philippines	58.1	2.5	490	37
Singapore	2.6	1.1	11,607	100
Thailand	52.8	2.0	260	17
Vietnam	62.0	2.5	417	19

SOURCE: Population Reference Bureau, 1986; Central Intelligence Agency, *The World Factbook* (Washington, D.C.: Government Printing Office, 1985).

Ethnicity

The Malays are the most prominent ethnic group in Southeast Asia. Regional isolation and racial mixing have created differences among the countries of the region, but the people are basically Malay in origin. The most conspicuous ethnic minority is the Chinese.

The sizable minorities of **"overseas Chinese"** are concentrated in the urban areas of nearly every country of Southeast Asia. These are colonists from China who live in the region, and sometimes they do not even become citizens of the countries where they settle. About 25 million of these people, about 6 percent of the total population, live in the region. The percentage varies from country to country, from a high in Singapore of 77 percent to less than 1 percent in Vietnam. The economic impact of the "overseas Chinese" on the region is much greater than their numbers would indicate. Realizing their tenuous position as noncitizens, they keep their money fluid and do not invest their money locally. They are the moneylenders, bankers, importers and exporters, restauranteurs, and possibly smugglers. Essentially middlemen who control much of the money of the region, they are characterized as clannish, hardworking, intelli-

---

overseas Chinese These are Chinese nationals (citizens of China) that live in countries of Southeast Asia other than China. They are noted for their industrious work habits and their ability to obtain and wisely use money, sometimes sending large sums back to relatives in China.

---

gent, and wise with money handling. About 14 percent of Thailand's population is "overseas Chinese," but these few people control nearly 90 percent of the country's retail businesses. Despite local prejudices and discrimination, these industrious and thrifty people seem to thrive. The tiny island country of Singapore was created mostly because of discrimination against these people, so Singapore is the only country in the region where the government is in the hands of "overseas Chinese." (See Figure 15.10.)

Most of the indigenous tribal people of Southeast Asia are ethnic minorities within their own countries. These are the "hill people," various tribes of which are found in each country. Historically, the hill people and the lowland people have avoided each other, but sometimes their differ-

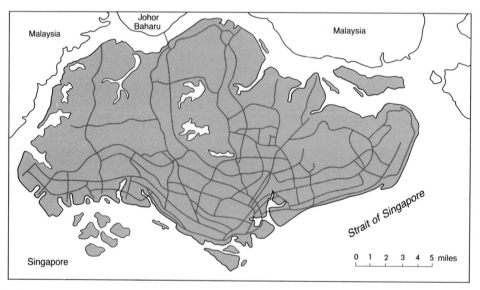

FIGURE 15.10

Singapore. The city of Singapore nearly covers the entire main island of the country. Note the causeway that connects Singapore with the Malaysian mainland.

Homes along a street in Ho Chi Minh City (Saigon). Notice the crowded street and the bicycles and mopeds that line the curb. Photograph by Louis Seig.

Sporting traditional Chinese headgear, these Vietnamese farmers, bring their products to the city in traditional Chinese baskets. Photograph by Louis Seig.

ences brought them to armed clashes. In Myanma live the Karens, the Shans, the Kachins, and the Chins. Similar tribal people live in the hills and mountains of Cambodia, Laos, Thailand, and Vietnam. Each country also contains minorities from the other countries of the region, as well as Europeans and Indians.

Religion and Language

The religions of Southeast Asia are somewhat easier to classify by area than the languages. People in groups of countries adhere to similar faiths, but each country has languages that are unique to it. Many regional dialects and sublanguages complicate the language map too. Most of the people of the region follow the two faiths imported from India: Buddhism and Islam. More than 90 percent of the people in Myanma and Thailand are Buddhists, and Buddhism is the main religion in Laos and Cambodia. Vietnam has a greater variety of religions than any of the other mainland countries, but the 20 percent of the Vietnamese who are Buddhists comprise the largest religious group.

Hinayana The Mahayana (Greater Vehicle) form of Buddhism has a profusion of relics and liturgies and is a dramatic, colorful form of the religion. It spread northward from India into Tibet, China, and Japan. The Hinayana (Lesser Vehicle) form of Buddhism is more fundamental and more subtle than the Mahayana form, and it spread eastward into Southeast Asia.

Buddhism, then, is the major religion of the continental portion of Southeast Asia. A form of the faith known as **Hinayana** (Lesser Vehicle) passed from southern India and Sri Lanka into Southeast Asia. It arrived in Myanma with the early Indian sailors and eventually diffused throughout the other mainland countries.

Islam arrived in the islands of the region much later than Buddhism, but it came in a similar manner—by way of the Indian sailors. Today, most of the people of Malaysia and Indonesia are Muslims, adherents of Islam. Islam is strongest in Indonesia, where about 90 percent of the people follow that faith. Besides the Muslims in Malaysia,

The rickshaw (*jinrikisha*) is also a traditional Chinese urban taxi, but these are lined up at a station in Ho Chi Minh City. Photograph by Louis Seig.

Tagalog The language and name of the most numerous of the native Malayan peoples of central Luzon in the Philippines.

significant numbers of people follow the Chinese religions of Confucianism and Taoism. Because of the Chinese majority, the most common religions in Singapore are Chinese in origin.

Religion in the Philippines is considerably different from the religions in the other parts of Southeast Asia. About 85 percent of the people in the Philippines are Roman Catholic, a legacy of the 377 years of Spanish rule that ended as long as 90 years ago. Also, because of the French influence, about 12 percent of the Vietnamese are Roman Catholics. Roman Catholics comprise the second largest religious group in Vietnam, after Buddhism. Other Christian minorities are found in other countries of Southeast Asia, but none make up more than 1 or 2 percent of the total population.

Each national state in Southeast Asia has sought to develop its own national tongue. Many people speak the modern counterparts of the ancient languages of their local region, but some modern languages have been introduced. Many sublanguages, dialects, and mixtures also are used. For example, about three-fourths of the people in Myanma speak Burmese, but the remainder speak 126 different local languages. Khmer is the national tongue of Cambodia, Lao and French are used in Laos, and Malay and English are common in Malaysia. Thai, derived from Sanskrit, is the language of Thailand, and Bahasa, derived from Malay, is spoken in Indonesia. The official language of the Philippines is Filipino, which came from **Tagalog,** another ancient language, but English is widely spoken in the Philippines, and some people there still speak Spanish.

Many languages overlap from one country to another, and some were introduced more recently by the colonial rulers and the Chinese. The imposed languages vary, however, because no one country had control of the entire region except Japan, and the Japanese were not there long enough to make a language impact. Mandarin Chinese and English are the only languages that are common throughout Southeast Asia. Chinese is retained as the language of the Chinese minorities of each country, and English is used in many of the urban areas and for commerce.

■ Economic Geography of Southeast Asia

Agriculture

Despite the many steep hills, mountains, and generally rugged terrain, as well as the vast tropical forests, Southeast Asia is primarily an agricultural region (see Figure 15.11). As noted above, only about one-fourth of the people of the region live in urban areas. It is a land of farmers. More than 80 percent of the people of Cambodia, Laos, Thailand, and Vietnam live in tiny villages and work as farmers or farm laborers.

The common agricultural crop grown throughout the region is rice. Rice requires a good deal of rain, so Southeast Asia is an ideal place to grow it. Rice is the chief crop of every country of

These farmers in Malaysia are harvesting rice. They use the straw for livestock bedding and for fertilizer, and the rice kernels for human food. Photograph by Peter McKinley.

Southeast Asia 471

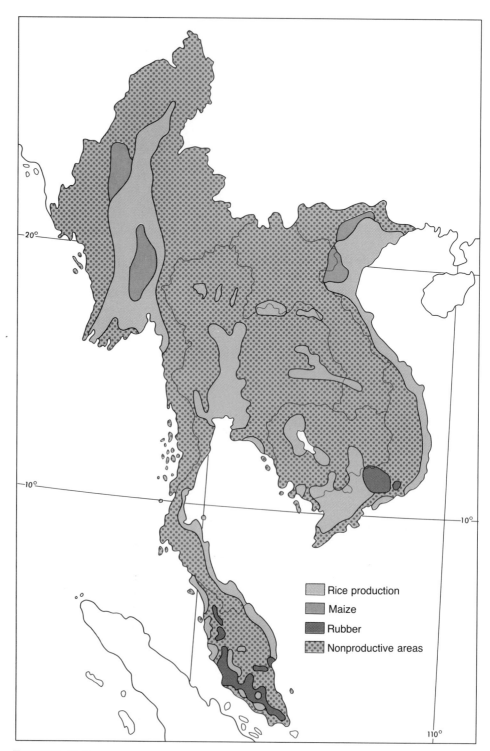

FIGURE 15.11
Agricultural Regions of Mainland Southeast Asia.

> **high-yield rice hybrids** Hybrid plants are produced by crossbreeding two plants of a different variety or species. Plant geneticists strive to grow such plants to produce a higher yield per unit area of land than either of the parent plants. This process has been successful with the rice plant.

the region, and the staple food for most of the people. In fact, so much rice is produced that for some countries it is an export commodity. Rice can account for more than half of Myanma's export income and about one-fourth of Thailand's. Since the early 1970s, rice acreage in the region has been decreasing, but output has increased steadily. The **high-yield rice hybrids** have been used to accomplish this feat. On the same amount of land, an individual farmer can produce twice as much rice as he did before the high-yield varieties were introduced. The ancient crop, well suited to the environmental conditions, has been improved by modern technology.

The second most important crop, after rice, varies greatly from country to country, but the second-ranking commodities are usually tree products of some type. Teak and other tropical hardwoods come from the forests of Myanma, Laos, Thailand, and Cambodia. Coconuts and their assorted products are exported from most of the islands. Paper production, plywood manufacturing, and other lumber-processing are common activities in Singapore, even though the country cuts no trees. Unfinished logs are imported to Singapore from nearby countries and processed, then the products are shipped to world markets. The most important tree product of the region is rubber.

It may be surprising that the rubber tree (*Hevea brasiliensis*) is not native to Southeast Asia. Originally, the trees were found scattered at random in the Amazon Basin of South America, specifically along the banks of the Jari River in Brazil. The sap of these trees is a sticky, gummy latex that was of little use until 1839, when the American Charles Goodyear, discovered the process of vulcanization. After that, thousands of uses were found for the product, and Brazil controlled the world market because it had the only source. Production was slow because the latex had to be gathered by hand from the scattered trees. The British broke Brazil's rubber monopoly by sending a young botanist to the Jari River region with the express purpose of smuggling out some rubber trees. Posing as a tourist, the botanist managed to leave the country with a stuffed alligator filled with rubber-tree seeds. The seeds were taken to the British-controlled territory of Malay in Southeast Asia. The hot-wet environment was ideal for the rubber trees, which thrived. The trees were planted in rows on large plantations, and the resulting lower production costs destroyed the Brazilian stranglehold on the world's rubber.

The imported rubber tree has had a major impact on the economy of Southeast Asia. Today, rubber accounts for about 40 percent of Malaysia's export dollars. Raw rubber is also exported from Brunei, Indonesia, Thailand, and Vietnam. Singapore imports raw rubber from its neighbors, processes it, and exports rubber products to the world.

Other agricultural products of Southeast Asia include corn (also found originally in the western hemisphere), cotton, tobacco, and pepper. Cinchona, the tree whose bark is used to make quinine, grows in most countries and has been a major crop in Indonesia, Vietnam, and the Philippines. Quinine has been used for years in the making of various medicines, but it is especially useful in the treatment of malaria. Coffee, tea, sugar, tapioca, sweet potatoes, and beans are produced in some of the region's countries. Laos has been a major supplier of opium, which comes from a type of poppy. It also is the major citrus fruit grower of Southeast Asia.

The constant heat and plentiful moisture make it possible for nearly anything to grow in Southeast Asia. Sticks pushed into the ground will soon sprout leaves, and fences made with uncured fence posts will eventually become a row of trees. In many places, especially in the flat areas, the natural tropical forests have been removed so that food products can be grown. Subsistence agriculture, through the "slash and burn" method, is used in many parts of Southeast Asia. The tie between the people and the land of the region is very strong. Despite the population density in some Southeast Asia countries, most countries are able to feed their people, and some produce a surplus.

> **FOCUS BOX**
>
> ## *The World of Rice*
>
> The origin of wet-field rice culture is obscure. Rice was domesticated in Southeast Asia thousands of years ago, but we do not know whether it started as a dry-field crop or whether it grew wild in the swamps and marshes of the region. Many wild rices are found in eastern Asia, so it is possible that both the dry-land and wet-land species developed without human intervention. Whatever the origin, wet-land agriculture evolved into a specific system in which similar methods are used from place to place.
>
> The control of water is probably the most important aspect of wet-land rice cultivation. An important feature of water control was the development of terraces. The terraces were designed not only to hold water but also to allow the farming of steep slopes. In the lowland areas the fields had to be leveled and long, so low dams for holding irrigation water were designed. Plant selection became important. Then, seedbeds were designed, and the system of transplanting from the seedbeds to the fields was developed. The puddling of soils by wading humans and animals, plowing and harrowing, and other methods of soil preparation were developed.
>
> All the complex methods used in wet-field rice cultures are products of the moist Southeast Asia region. It may seem strange that in a region that experiences copious amounts of rain, water control for irrigation became so highly developed. But rice is planted and harvested every month of the year, and the rains tend to be seasonal. In addition, rice grows best in standing water, so the water must be retained in the terrace basins rather than allowed to run off. The planned staggering of planting and harvesting throughout the year eases the problems of storage and loss in these areas of heavy rainfall. Rice is grown in small fields using a large volume of hand labor, simple tools, and a minimum of draft animals. The main draft animal, the water buffalo, is a product of Southeast Asia too.
>
> Rice is usually a subsistence crop, consumed primarily as a grain and seldom as a flour. Unlike wheat, rice has few basic differences from variety to variety. Most of the rice crop of Southeast Asia is merely hulled, a process that does not remove the inner coat of brownish-colored bran. This *brown rice,* or country rice, is high in food content because the brown coat contains most of the oils, minerals, and vitamins. Because of the oils, however, brown rice spoils quickly, so commercial rice is hulled and polished in commercial mills. The *polished rice* has a reduced food value, but it looks better and, more important keeps better than the brown rice. The polished white rice is the most popular even with the poor people, but it is a novelty for the poor.
>
> In spite of the tedious, labor-intensive work involved in growing rice and the complex methods employed in water control for the rice fields, rice probably will continue to be the basic foodstuff of most of Asia's millions of people. Asians' lives and lifestyles are so entwined with the planting and harvesting of rice that dramatic cultural changes would have to occur before rice is replaced as the major dietary staple.

Mineral Production and Industry

Myanma, Indonesia, Laos, Malaysia, and Thailand all produce tin. Malaysia supplies about one-third of the world's output and is by far the world's greatest tin producer; Thailand is the world's fourth largest producer. Tin once was used primarily to make cans for storing foodstuffs, and the phrase "tin can" became synonymous with "container." Now that aluminum and plastic have taken over in the container industry, tin has been relegated primarily to use as an alloy. It is, however, a valuable alloy for making tinfoil, solder, type metals, utensils, and even some containers.

Vast supplies of petroleum also are found in Southeast Asia (see Figure 15.12). Indonesia is one of the world's largest petroleum producers, and the United States imports nearly as much crude oil from Indonesia as it does from the Arab nations. About one-third of Indonesia's exports are petroleum products. Oil supplies nearly the entire income of Brunei and provides the tiny country with a very high standard of living. The Philippines, Thailand, and Singapore must import petroleum, however, so crude-oil production is not found in every country of the region.

Tin and oil are the major income-producing minerals of Southeast Asia, but deposits of other useful minerals also are found in the region (see Figure 15.12). Myanma, Indonesia, the Philippines, and Thailand are the most richly endowed

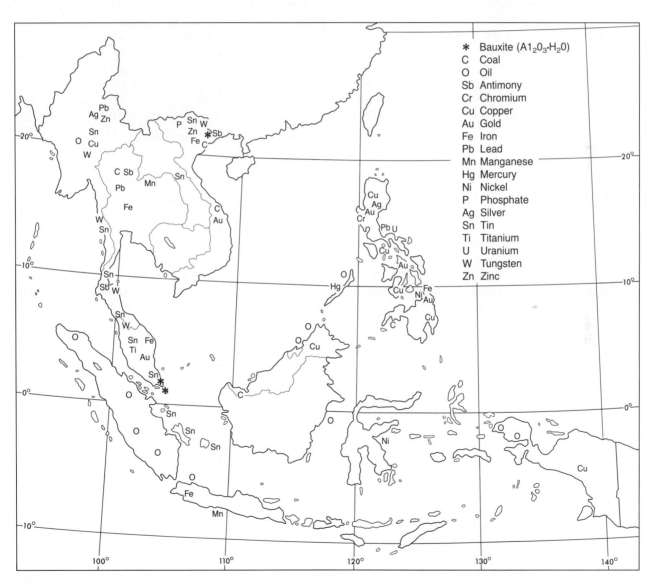

FIGURE 15.12
Mineral Deposits of Southeast Asia.

with minerals. Myanma has mines for lead, silver, tungsten, and zinc, as well as for such precious gems as rubies, sapphires, and jade. Indonesia has small supplies of coal, and bauxite, copper, manganese, nickel, and gold are produced too. The Philippines has most of the minerals that Indonesia has, as well as chromite, asbestos, and iron. Thailand produces iron ore, tungsten, and antimony. Sizable coal deposits have been found in Vietnam.

Regardless of all the mineral wealth, very little heavy industry is carried on in Southeast Asia—partly because the colonial powers did not want industry to develop and partly because both coal and iron ore have not been found within the same country. The European powers wanted to extract the resources for their own use and sell manufactured goods back to the colonies. Regional exchange of iron ore and coal requires regional agreements and planning, which have been slow in coming in Southeast Asia. The manufacturing that does exist in the region is primarily the type found in the Third World in general. Food- and textile-processing are the main activities. Such light industries are geared to meet the immediate needs of the people. Singapore is the leading processor and manufacturer of the region, and its port is one of the largest and busiest in the world. Many raw products are brought in for processing and sent out as finished or semifinished goods. In Singapore, oil is refined, electronic assembly plants thrive, and rubber, copra, and lumber are processed. Even large ships are built in Singapore.

Thailand has automobile assembly plants, but most automotive products in the region come from Japan. Thailand also makes various electrical items. Furniture, soap, fertilizers, and electronic goods are made in Malaysia. The Philippines is increasing its manufacturing output, but it has been geared toward making paper and other wood products. Today, electrical equipment and appliances are made in that country. The fish industry in the Philippines is among the largest in the world, with nearly 2 million metric tons of fish caught each year.

Another major industry in Southeast Asia is tourism. Singapore leads the region in that category, and each year the small country draws nearly 2 million tourists. Thailand has a tourist connection with European countries, especially France and Germany. Many people also visit the Philippines as tourists. The lush beauty of Southeast Asia, the exotic oriental architecture, good prices on gold and gems, and the friendly people all draw foreigners to the region, but Vietnam and the politically unsettled countries of Laos and Cambodia are rarely visited.

Much of the food in Southeast Asia goes directly from the farms to the city streets without any processing. Here a street vendor has left his offering of coconut meat and beans unattended on a sidewalk in Ho Chi Minh City. Photograph by Louis Seig.

Transportation and Communications

The rugged terrain in Southeast Asia limits overland transportation. Most of the railroad mileage is on the mainland, and the routes follow the great rivers (Figure 15.13 on page 478). Very few railroad lines connect the countries. The only exceptions are the route from Ho Chi Minh City through Vietnam and into China, the route from Bangkok down the Malay Peninsula to Singapore, and the extension of the latter into Cambodia. These routes are heavily traveled, but few people or goods actually go from one country to another. Also, the lines change gauge at each border crossing, so trains cannot continue across borders. On the islands, Java has the most rail mileage, and a coastal railroad circles the entire island. One or two short railroad lines have been built on some of the other islands, but most have none at all.

Highways do cross between the countries on the mainland, but they are few and poor (see Figure 15.13). Construction costs are very high, and because not much movement occurs there is no need to spend the money. The Burma and Ledo roads, carved out of the jungle during World War II, are typical of road construction in the region. The Burma Road winds for 700 miles (1,126 km) in order to cover an airline distance of about 250 miles (400 km) between Lashio in Myanma, and Kunming in China's Yunnan Province. The Ledo Road covers less rugged terrain but must cross many major rivers and smaller streams as it winds between Myanma and the Assam region of India. For most of Southeast Asia today, roads do not have hard surfaces, rivers are unbridged, and networks appear only around the major cities. The Indonesian portion of the island of New Guinea has no roads, and most of the islands have only meager trails unfit for automotive traffic.

Internal transportation in Southeast Asia relies heavily on inland waterways. Ships, barges, small boats, junks, sampans, and other small craft move goods and people, on the rivers and canals that interlace the populated parts of each country. Raw materials are generally moved downstream, and manufactured goods are moved upstream. Farm products are taken to market on small watercraft, and the boats themselves are often used as mobile markets. Many of the waterways are lined with homes, some on stilts, and boats are used for living quarters. This is what is meant by the term "water-oriented society."

Communications in Southeast Asia are hampered by the multitude of languages and by high illiteracy rates, especially among the poor farmers. Telephones are difficult to find in most countries, and calling long-distance may be impossible. Transistor radios are becoming popular, and new radio stations, most of which broadcast in AM only, are appearing. Every country has at least one television station (see Table 15.4), but television sets are still too expensive for most people.

Transportation and communications among the countries of Southeast Asia are meager at best.

Notice how the paved road ends at the helicopter landing pad built by the Americans. All the other roads in this Vietnamese village are dirt. Photograph by Louis Seig.

Southeast Asia

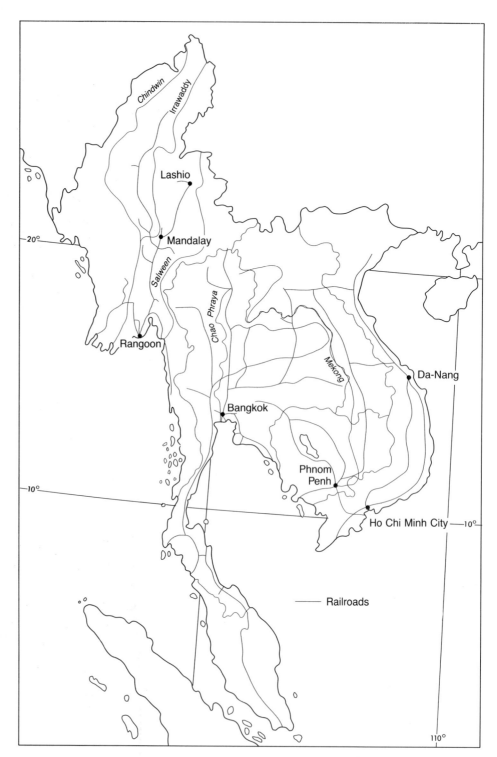

FIGURE 15.13

Roads and Railroads of Mainland Southeast Asia. The routes all parallel the mountains and rivers of the region.

TABLE 15.4

Telecommunications of Southeast Asia, 1984

COUNTRY	TV STATIONS	RADIO STATIONS	TELEPHONES PER 1,000 PEOPLE	NEWSPAPERS PER 1,000 PEOPLE
Brunei	1	6	80.000	0.1
Myanma	2	1	1.000	15.0
Indonesia	14	251	0.002	10.0
Cambodia	1	1	—[a]	0.1
Laos	1	10	—[a]	0.1
Malaysia	20	26	51.000	133.0
Philippines	33	322	13.000	38.0
Singapore	2	17	265.000	249.0
Thailand	10	170	11.000	48.0
Vietnam	6	18	0.010	8.0

[a]Less than 2,000 total telephones.
SOURCE: CIA reports; and Central Intelligence Agency, *The World Factbook* (Washington, D.C.: Government Printing Office, 1985).

The rugged landscape of the mainland countries and the open waters between the islands make the movement of people, goods, and ideas difficult. Mass communications within the countries are crippled by the large number of languages and the high illiteracy rates. The region is almost totally lacking in the unity that a common language or economic plan, such as the European Economic Community, can bring about.

Because inland waterways are the most important mode of travel in much of Southeast Asia, the villages tend to be located as close as possible to a river. Photograph by Louis Seig.

Political Geography of Southeast Asia

The small countries of Southeast Asia lie between major power blocs. The poverty of the region, combined with young and sometimes unstable governments and cultural diversity, has led to both internal and external turmoil. The people generally hold anti-Western attitudes that are residues of colonialism, and they also fear the might of the heavily populated powers near their borders to the north and west. They resent minorities—especially but not exclusively the "overseas Chinese" within their borders, and they are suspicious of all political leaders, foreign and domestic. Since independence, most of the states have been strained by internal wars, which have usually led to military-dominated governments.

Much of the political instability in the region has resulted from an insufficient base for making the governments economically viable. Except for tiny Brunei and Singapore, average per capita incomes in the region are low. Much-needed regional cooperation is nearly nonexistent. Malaysia invested wisely in its tin and rubber and has recently diversified to avoid reliance on those commodities. Much money has been put toward educating its people. The Malaysian government supports many college students being educated in the United States. Laos and Cambodia made up of subsistence farmers and hill tribes, are nearly hopelessly poor countries. Thus, in the search for political stability, poverty in a relatively rich resource area is a major obstacle. This has not escaped the notice of the Communists, and almost all the countries in the region have a strong Communist minority. The plight of Southeast Asia is of concern to the West, and the United States sends aid to most of the countries in Southeast Asia. The Philippines receives the most—about $334 million a year. People of the United States, however, are very reluctant to become too involved in the affairs of Southeast Asia after the bitter experience of the Vietnam War.

A key political problem in Southeast Asia today is the status of Cambodia and its relationship with Vietnam. Over a million people have been killed in executions and enforced hardships brought on by the Cambodian government, and thousands of refugees have flowed into Thailand to escape invasions by the Vietnamese. Neglect of the rice fields has caused famines, and many Cambodians rely on international food assistance.

Some hope for the future in Southeast Asia has been placed in the **Association of Southeast Asian Nations (ASEAN),** which was formed in 1967 to promote economic and political cooperation among countries of the region. The only countries that joined, however, are the non-Communist group, including Brunei, Indonesia, Malaysia, the Philippines, Singapore, and Thailand. The organization concentrates on problems related to agriculture, trade, transportation, communications, science, finance, and culture.

Association of Southeast Asian Nations (ASEAN) An organization formed in 1967 to promote political and economic cooperation among the non-Communist states of Southeast Asia. Members include Brunei, Indonesia, Malaysia, the Philippines, Singapore, and Thailand.

Key Words

- Association of Southeast Asian Nations (ASEAN)
- boat people
- French Indochina
- high-yield rice
- Hinayana
- Huk Communists
- Java Man
- Krakatoa
- ocean trenches
- overseas Chinese
- primate city
- Spice Islands
- Strait of Malacca
- Tagalog
- transmigration
- typhoons

References and Readings

Ablin, David A., and Hood, Marlowe (eds.). *The Cambodian Agony,* Armonk, N.Y.: Sharpe, 1987.

Armstrong, Warwick, and McGree, Terence G. *Theatres of Accumulation: Studies in Asian and Latin American Urbanization.* New York: Methuen, 1985.

Barker, Randolph, et al. *The Rice Economy of Asia.* Washington, D.C.: Resources for the Future, 1985.

Becker, Elizabeth. *When the War Was Over: The Voices of Cambodia's Revolution and Its People.* New York: Simon & Schuster, 1986.

Burley, T. M. *The Philippines: An Economic and Social Geography.* London: Bell, 1973.

Burling, Robbins. *Hill Farms and Padi Fields: Life in Mainland Southeast Asia.* Englewood Cliffs, N.J.: Prentice-Hall, 1965.

Costello, Michael A., et al. *Mobility and Employment in Urban Southeast Asia: Examples from Indonesia and the Philippines.* Boulder, Colo.: Westview, 1987.

Dobby, Ernest H. G. *Southeast Asia.* 10th ed. London: University of London Press, 1967.

Dutt, Ashok K. (ed.). *Southeast Asia: Realm of Contrasts.* Boulder, Colo.: Westview, 1985.

The Far East and Australasia, 1986. London: Europa Publications, 1986.

Firth, Raymond W. *Malay Fishermen: Their Peasant Economy.* Hamden, Conn.: Shoe String Press, 1966.

Fisher, Charles A. *Southeast Asia: A Social, Economic, and Political Geography.* New York: Dutton, 1966.

Fuchs, Roland J. (eds.). *Urbanization and Urban Policies in Pacific Asia.* Boulder, Colo.: Westview, 1987.

Hainsworth, Geoffrey B. (ed.). *Village-Level Modernization in Southeast Asia: The Political Economy of Rice and Water.* Vancouver, Can.: University of British Columbia Press, 1982.

Hansen, Gary E. (ed.). *Agricultural and Rural Development in Indonesia.* Boulder, Colo.: Westview, 1981.

Hill, Ronald D. (ed.). *South-East Asia: A Systematic Geography.* New York: Oxford University Press, 1979.

Karnow, Stanley. *Southeast Asia.* New York: Time Inc., 1967.

_____*Vietnam: A History.* New York: Viking, 1983.

Kent, George, and Valencia, Mark J. (ed.). *Marine Policy in Southeast Asia.* Berkeley, Calif.: University of California Press, 1985.

Kumar, Raj. *The Forest Resources of Malaysia: Their Economics and Development.* Singapore: Oxford University Press, 1986.

Leinbach, Thomas R., and Ulack, Richard. "Cities of Southeast Asia." In *Cities of the World: World Regional Urban Development,* ed. Stanley D. Brunn and Jack F. Williams. New York: Harper & Row, 1983.

McCloud, Donald G. *System and Process in Southeast Asia: The Evolution of a Region.* Boulder, Colo.: Westview, 1986.

McGee, T. G. *The Southeast Asian City: A Social Geography of the Primate Cities of Southeast Asia.* New York: Praeger, 1967.

Miller, E. Willard, and Miller, Ruby M. *Southeast Asia: A bibliography on the Third World.* Monticello, Ill.: Vance Bibliographies, 1982.

Milne, Robert S., and Mawzy, Diane K. *Malaysia: Tradition, Modernity, and Islam.* Boulder, Colo.: Westview, 1986.

Morgan, Joseph R., and Valencia, Mark J. (eds.). *Atlas for Marine Policy in Southeast Asian Seas.* Berkeley: University of California Press, 1983.

Pryor, Robin J. (ed.). *Migration and Development in Southeast Asia: A Demographic Perspective.* New York: Oxford University Press, 1979.

Purcell, Victor W. W. S. *The Chinese in Southeast Asia.* New York: Oxford University Press, 1965.

Robinson, Harry. *Monsoon Asia: A Geographical Survey.* New York: Praeger, 1967.

Samuels, Marwyn S. *Contest for the South China Sea.* New York: Methuen, 1982.

Sombitsuri, Virawan. "Migration to Bangkok: A Geographical Analysis," Doctoral dissertation, Oklahoma State University, 1977.

Spencer, Joseph E. *Shifting Cultivation in Southeast Asia.* Berkeley: University of California Press, 1976.

_____"Southeast Asia." *Progress in Human Geography* 6 (1982): 265–269.

Spencer, Joseph E., and Thomas, William L. *Asia, East by South: A Cultural Geography.* New York: Wiley, 1971.

Steinberg, David J. *The Philippines: A Singular and Plural Place.* Boulder, Colo.: Westview, 1982.

Stewart, I. *Indonesia: Portrait from an Archipelago.* London: Kegan Paul, 1984.

Taylor, Alice (ed.). *Focus on Southeast Asia.* New York: Praeger, 1972.

Thrift, Nigel, and Forbes, Dean K. *The Price of War: Urbanization in Vietnam, 1954–1985.* Winchester, Mass.: Allen & Unwin, 1986.

Wernstedt, Frederick L., and Spencer, Joseph E. *The Philippine Island World: A Physical, Cultural, and Regional Geography.* Berkeley: University of California Press, 1967.

Whitmore, Timothy C. *Tropical Rain Forest of the Far East.* New York: Oxford University Press, 1984.

Wilhelm, D. *Emerging Indonesia.* Totowa, N.J.: Barnes & Noble, 1980.

Williams, L. E. *The Future of the Overseas Chinese in Southeast Asia.* New York: McGraw-Hill, 1966.

Chapter 16

EAST ASIA

■ Introduction

East Asia is a world region sometimes called the **Orient** or the Far East. In terms of country size, the number of people, and political strength, China dominates the region. China comprises more than 80 percent of both the land area and the number of people in the region. The country's official name is the People's Republic of China, but it is called Communist China and sometimes Red China. Off the east coast of the mainland on the small island of Taiwan (Formosa) is Nationalist China, known officially as the Republic of China. In addition to the two Chinas, the other countries of the region are North and South Korea, Mongolia, Xianggang (the British Crown Colony of Hong Kong), and the Portuguese province of Macao. Japan also is located within the region, but that country is covered in a separate chapter (see Chapter 8). (See Figure 16.1.)

More than a quarter of the world's people live in East Asia. The total population of the region is 1.14 billion, and China alone has 1.05 billion. China has been the most populous country in the world for hundreds of years. The region contains less than 10 percent of the land area of the world, but 26 percent of the world's people. Much of the land is mountainous, desert, or otherwise unfit for farming, so the overwhelming problem is the lack of cultivable land to feed the constantly increasing numbers of people.

East Asia is bordered on the north and on much of the west by the Soviet Union. On the south are the countries of South Asia and Southeast Asia, and the Nan Hai (South China Sea). Between the islands of Japan and Taiwan and the China mainland is the Dong Hai (East China Sea). The Huang Hai (Yellow Sea) separates the Korean Peninsula from China. All these seas are extensions of the western waters of the Pacific Ocean. The **landlocked** country of Mongolia is tucked into Central Asia, between China and the Soviet Union. (See Figure 16.2.)

Physical Geography of East Asia

Landforms

The mountains of southwestern China are the highest in the world. The **Qing Zang Gaoyuan (Plateau of Tibet)** (see Figures 16.3 and 16.4) is a huge area that averages more than 15,000 feet (4,600 m) in elevation and is surrounded by mountain ranges that are over 20,000 feet (6,000 m) high. The 500,000-square-mile area was once the country of Xizang (Tibet), but in 1951 the Chinese seized political control, and two years later the province became the Xizang Autonomous Region of China. The major mountain ranges that separate the plateau from the remainder of Asia are the Himalayas on the south and the Kunlun Shan on the north. Numerous lesser-known ranges trend west to east through the plateau itself. These highlands act as a barrier to the southern winds that might otherwise bring moisture to Central Asia.

Two mountain spurs to the north of the Qing Zang Gaoyuan are the Altyn Tagh and the Nan Shan. These ranges both contain individual peaks over 20,000 feet (6,096 m) high. The Tien Shan, another high range, separates western China from the Soviet Union. The range is a distinct entity from the Qing Zang Gaoyuan. It joins the Pamir Knot on the north and runs northeastward, whereas the mountains of Xizang join the Knot on the east and run eastward.

Orient From the Latin, meaning "east," as opposed to "occident," meaning "west." Thus, the term refers to Asia, especially East Asia.

landlocked Countries that do not have a border along an open ocean, and are thus locked in by neighboring countries, are landlocked.

Qing Zang Gaoyuan (Plateau of Tibet) A broad, somewhat oval-shaped plateau in South-Central Asia that contains the highest mountains in the world.

China's Sorrow A phrase used to refer to the Huang He (Hwang Ho, or Yellow River) in China, because its periodic flooding destroys crops and kills people.

Compared with the massive mountains of southern and western China, the landforms in the remainder of the region seem insignificant. Relatively tall mountains do occur, however. The Altay Range along the western border of Mongolia rises to heights of more than 10,000 feet (3,000 m). Taiwan is dominated by Yu Shan, a peak that is 12,959 feet (3,950 m) high. Even the lowly mountains along the eastern portion of the Korean Peninsula are significant barriers locally.

Rivers

The mountains are the most spectacular physical features of East Asia, but the rivers are more important to the people. Most of the people live along the rivers and coastal lowlands. The two major rivers of China are the famous Chang Jiang (Yangtze) and Huang He (Hwang Ho or Yellow River). Both rivers begin on the Qing Zang Gaoyuan and flow eastward through the heartland of China (see Figure 16.4).

The Huang He begins in the Kunlun Shan and meanders in great loops before crossing the eastern lowland and finally emptying in the Huang Hai. The river is more than 3,000 miles (4,827 km) long. It has been called **"China's Sorrow"** because of the disastrous floods that have occurred over the centuries. The silt-laden river has gradually built its bed upward, forming levees that Chinese laborers have increased with dikes. As a result, the river now flows 25 to 30 feet (8–9 m) above its floodplain. When it does breach the levees and dikes, the river pours out of its raised bed and onto the surrounding floodplain. In 1959, some 148 million acres (60 million hectares) of farmland were damaged by floods, causing serious food shortages. Not only is cropland damaged, but countless people have died. The worst disaster occurred in 1931, when 3.7 million people died from the flooding, but 900,000 drowned in 1887 and another 200,000 in 1939. The river is being controlled much better today than when those disasters occurred. Large engineering projects have helped keep the river within its banks, but it still poses a threat to the people living near it.

The Chang Jiang also begins in the Kunlun Shan, but its source is 300 miles (483 km) farther west than the source of the Huang He. The Chang Jiang, then, is the longer of the two. It flows southeastward off the highlands of the Qing Zang Gaoyuan, turns eastward through the Chengdu Plain (Red Basin), then meanders across the East China lowland before emptying into the Dong Hai (see Figure 16.3). Many large old cities of China are located along this great river. Shanghai, the city once known as the "sin city" of the world, is built on the mud flats at the mouth of the river. Upriver from Shanghai are Suzhou (Soochow), Nanjing (Nanking), Chongquig (Chungking), and Wuhan as well as many other lesser places. For

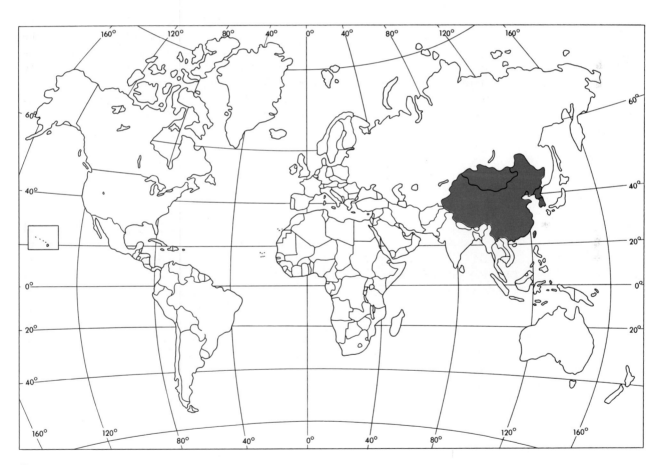

FIGURE 16.1
World Location of East Asia.

centuries, large flat-bottomed **junks** sailed partway and were pulled by hand partway from Shanghai up through the gorges and rapids of the Chang Jiang to Chongquig. Narrow walkways were chiseled into the rock walls of the gorges so that lines of **coolies** could pull the boats upstream. Because much of the course of the Chang Jiang is confined within mountain gorges, it does not usually flood, but in 1911 100,000 people did die from a devastating flood. (See cover photo.)

junk From the Spanish and Portuguese word *junco* and referring to a Chinese flat-bottomed ship with one accordian-like sail.

coolies From the Hindi word *quili,* meaning "hired servant," but used to describe an unskilled native laborer, especially in East Asia.

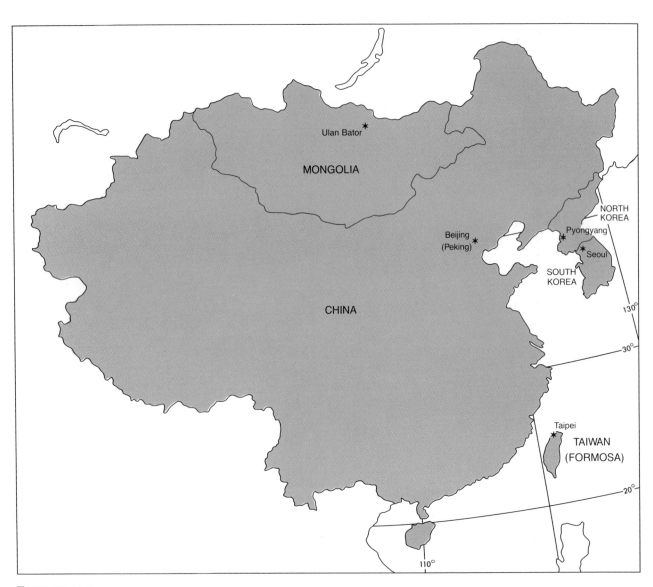

FIGURE 16.2

The Countries of East Asia.

486   Chapter 16

World Levels of INCOME

PER CAPITA INCOME is an important measure of a country's wealth, and the range is great among the countries of the world. In the **highly developed world**, the average yearly income per person is about $10,500, and in the **underdeveloped world**, the average is less than $500 per person. The actual range varies from $23,000 for Qatar, to $88 for Chad. Besides the variation in the range of incomes, the number of countries also vary among the categories. Only 31 countries have per capita incomes that allow them to be ranked in the **highly developed world**. On the other hand, there are about 50 countries in the **underdeveloped world**.

World Levels of Income

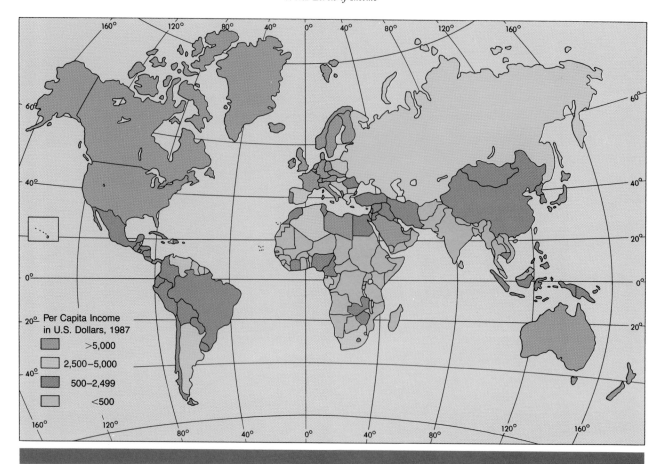

The above map contains one of the three factors used in determining the levels of economic development for the world: per capita income. Income is based on U.S. dollars, and the categories are: (1) countries with per capita income of more than $5,000, (2) countries with per capita incomes between $2,500 and $5,000, (3) countries with per capita between $500 and $2,499, and (4) countries with less than $500 income per person. It should be noted that some countries such as Libya and Saudi Arabia appear as **highly developed countries** based on their per capita incomes. The oil income produced in these countries, however, has gone to only a few wealthy people. Only recently have public works started to benefit from the oil money, so most people in the two countries still live under poverty or near poverty conditions.

World Levels of Income

(Left) A few wealthy people can afford luxuries unknown to most. Here a chauffeur helps a four-year-old boy into a limousine. This elegant display of wealth is enjoyed by the upper class of many countries, and not just those of the developed world.

(Below) These members of the Los Angeles Rolls Royce and Bentley Club enjoy a picnic under the warm California sun. The club members afford their luxuries through incomes that are among the highest in the world.

World Levels of Income

■ *(Above)* In stark contrast to the photographs on the previous page, these people in Chile barely scratch out a living. Their family income is less than $500 per year. Their tiny house is constructed of corrugated metal. Notice the rain barrel for catching rain water for use as fresh water, the outdoor wash basin, and the lady sweeping the dirt front yard.

■ *(Right)* Poverty created by extremely low incomes and in many cases no incomes is evident in this shantytown suburb of Bombay, India. As can be imagined, health conditions here are very bad, people die young of all sorts of infections and diseases, food is scarce, and life in general is quite miserable. These people are among the poorest in the world.

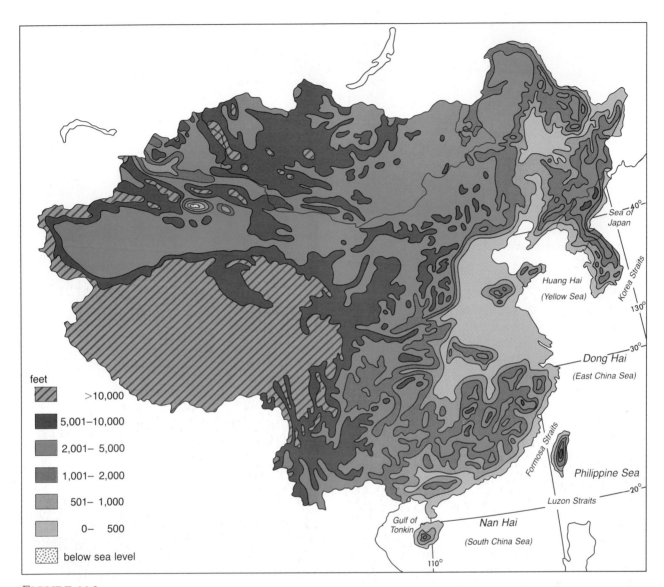

FIGURE 16.3

Land Elevations and Water Bodies of East Asia.

Two other rivers of mainland East Asia are significant for their political importance. Much of the boundary between North Korea and China follows the Yalu River. It was an important checkpoint for American flyers during the Korean War, because penetration beyond it was forbidden. The Amur River and its tributary the Ussuri River are used for most of the eastern boundary between China and the Soviet Union. A series of border clashes have occurred between the two countries along these waterways. The Tarim He (Tarim River) is not important politically, but it is interesting for another reason. It flows northeastward from the Karakoram Mountains in western China into the very dry Tarim Basin. The river actually disappears into the sands and saltbeds of that large interior drainage basin.

Except for China, the heavily populated lowlands of East Asia generally are narrow sea coasts or river valleys. The coastal lowlands of Taiwan and the

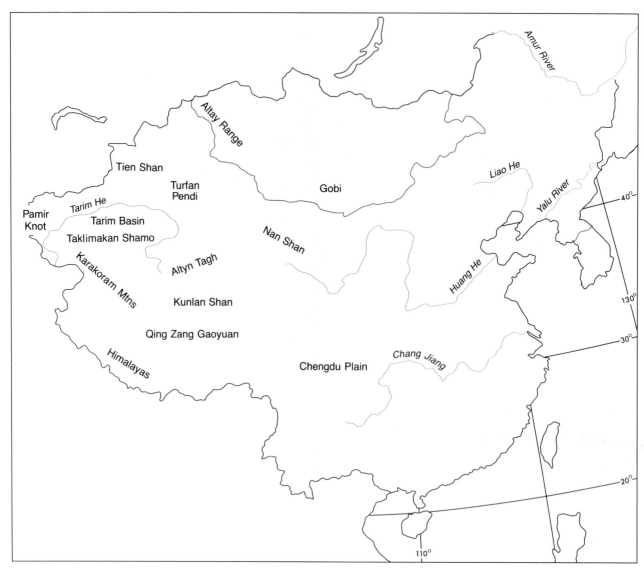

FIGURE 16.4

Physiographic Features and Rivers of East Asia.

Korean Peninsula average only about 25 miles (40 km) in width. The southeast coast of China also is indented with numerous, small isolated coastal lowlands. Only in the region between the Chang Jiang and Huang He and up the Liao He into Manchuria do the lowlands penetrate with width into the interior. These flat, low plains are heavily populated. Another area where the population density is extremely high is the Chengdu Plain. The plain is contained in a large, relatively flat and fertile mountain basin. Chongquig is the largest city of the basin, and it was the capital of China during World War II; the mountain stronghold was never captured by the Japanese. Mongolia has no lowland area, only mountains and high plains that average more than 2,000 feet (610 m) above sea level.

Part of the rugged terrain of Central China can be seen from this view from the Great Wall. Photograph by Marvin G. Bays.

Hand labor is used in construction for flood control along this river in China. Photograph by Peter McKinley.

Climate

The most typical type of climate of East Asia is the humid subtropical climate with conditions similar to those in the southeastern United States. This climate is found in all of southeastern China from the Huang He south and from the eastern edge of the Qing Zang Gaoyuan eastward to the coast (see Figure 16.5). Humid subtropical conditions also prevail in Taiwan, Macao, Hong Kong, and South Korea. This means that the summers are hot and the winters are mild with sufficient rainfall for agriculture. In many areas, two agricultural crops are produced each year. Much of the south coast of China and Taiwan have extremely wet summers. The rain is produced by the same type of winds that cause the monsoonal rains in Southeast Asia. The heavy rains do not penetrate far into the mainland of China, however, and west of the 100th meridian the humid conditions change abruptly to the dryness of the interior deserts.

Two of the great desert regions of the world are found in interior China. The Taklimakan

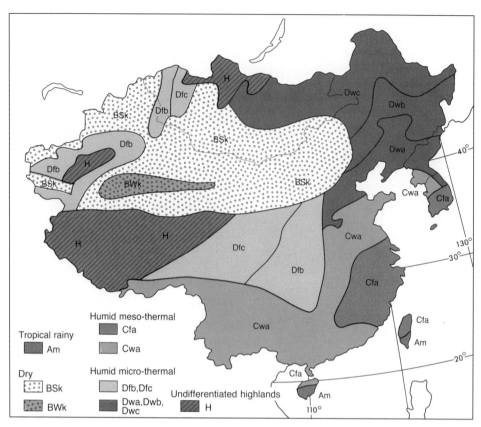

FIGURE 16.5
Climate Regions of East Asia.

Shamo (Takla Makan Desert) is located in the Tarim Basin just north of the Qing Zang Gaoyuan (see Figure 16.4) The desert covers 200,000 square miles (501,800 km²) and is surrounded on all sides except the east by the lofty mountains of western China. The desert is very dry because the mountains do not allow moisture-laden winds to come into the basin. The Chinese are building extensive irrigation ducts to water the desert from the glacial meltwater of the surrounding mountains. The region also contains a store of strategic minerals. The Chinese government has encouraged migration to the region from the overpopulated east, but many people from humid climates do not like the desert.

Eastward and slightly north of the Taklimakan Shamo, and blending into it, is the Gobi Shamo (Gobi Desert). Part of the Gobi is in Mongolia, but most of it extends into China's Inner Mongolian Autonomous Region. Compared with other deserts of the world, the Gobi is the highest in both elevation and latitude, making it also the coldest of all deserts. Between the two deserts, in northeastern Xinjiang Uygur (Sinkiang Autonomous Region), the Turfan Pendi (Turfan Depression) is one of the low spots of the world that is not covered by water. Its bottom lies 505 feet (154 m) below sea level.

Surrounding the deserts on the north and east are vast areas of semiarid climates with **steppe** grasslands. These dry, cool, short-grass regions cover nearly all of Mongolia and northern China from Manchuria to the Huang He. Both the desert and the semiarid areas are dry because of their continental interior location, where neither pre-

steppe Usually refers to one of the grassy, treeless plains of southeastern Europe and Asia, but also describes regions anywhere in the world with a semiarid climate.

Peking man An extinct species of humans (*Sinanthropus pekinensis*) determined through the 1926 discovery of skull and skeletal parts in Choukoutien, China.

vailing winds from the east coast nor cyclonic storms from the west can reach them. In the summer the southern mountains block the monsoons from coming into the interior, and in the winter the continental high pressure over Siberia blocks the cyclonic storms. Enough moisture-laden wind slips into the region to furnish rain to support the grass, but the rain fall is too unreliable for cultivated crops.

East Asia also has large areas of humid continental climates that have cold winters. The people of northeast China, notably Manchuria, and North Korea experience very cold winters. The mountains and northern lowlands receive large amounts of snow. The humid continental climate areas, however, are noted for having hot summers, even hotter than in the subtropical regions. Manchuria, located close enough to the coast to receive summer rain along with the high summer temperatures, is one of the major agricultural areas of China.

■ Human Geography of East Asia

History

Humans have been in East Asia for a very long time, and the Chinese civilization is centuries old. The skull of the **Peking man** (*Sinanthropus pekinensis*), found in northern China in 1929, has been dated at more than 360,000 years old. Modern Chinese history began with an agricultural society known as the Shang Dynasty that existed in the Huang He valley from 1700 B.C. to 1100 B.C. Nine

The land along Hong Kong Harbor is among the most densely populated in the world. Photograph by Colin Hagarty.

other dynastic periods, known as **Chinese dynasties,** followed the Shang, and each contributed to the Chinese culture. Not all the dynasties will be discussed here, but some should be mentioned. During the Chou Dynasty (1100–256 B.C.), for example, many scholars and sages were common, but the most notable was Confucius. In addition, the Great Wall of China was completed during the Chou Dynasty. (See Figure 16.6.)

The last three Chinese dynasties probably are the best known today. The Yuan Dynasty (A.D. 1279–1368) ruled China from Beijing (Peking), but the leaders were from Mongolia. The most famous Mongol emperor was Kublai Khan, a descendant of the great invader Genghis Khan. Part of the Mongol cavalry terrorized Europe, while others conquered China. The time was the zenith of Mongolian conquest, and possibly of Mongolian history. The Ming Dynasty (1368–1644) followed the Yuan, and traditional Chinese culture flourished. During this period, important novels were written and fine porcelains were produced. The porcelain pieces are now art objects that demand incredibly high prices. The last of the Chinese dynasties was the Manchu Dynasty (1644–1911). This period was significant for China because the country was again ruled by aliens. The Manchus were from Manchuria, but they absorbed the Chinese culture. The Manchus, however, are mainly responsible for China's lack of innovation and extreme introspection during the period.

Between 1911 and 1949 China's political history was one of continuous struggle. The president of the first republic, Sun Yat-sen, remained in office only 34 days. As the provincial warlords struggled for power, unity remained elusive. In the 1930s, when the Nationalist government began to solidify the unity of the country under Chiang Kai-shek, the Communist guerrillas under Mao Tse-tung began internal warfare. The Nationalist and Communist factions struggled all through the Japanese conquest of their country. During the late 1930s and early 1940s, Japan conquered all of Manchuria and all the coastal cities of China. After the Japanese were defeated, the Communist-Nationalist struggle developed into a major civil war. The Communists eventually won the war, driving Chiang Kai-shek to the island of Taiwan,

Chinese dynasties Dynasties are successions of rulers who are members of the same family. In China the dynastic period lasted from the beginning of the Chou Dynasty in 1122 B.C. until Emperor Pu Yi of the Manchu Dynasty abdicated the throne on February 12, 1912. The Chinese Empire was ruled by "barbarian" dynasties for much of its history, including the Mongol (Yuan, 1279–1368) and the Manchu (Ch'ing, 1644–1912). The Manchu Empire included not only China proper but also Tibet, Chinese Turkestan (Sinkiang), Mongolia, and Manchuria.

where the Nationalist Chinese government is located today. Mao-Tse-tung became the leader of Communist China.

The histories of the other countries in East Asia were profoundly influenced by China. Japan

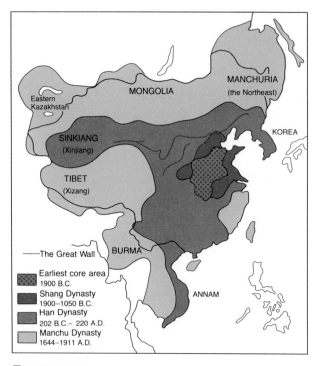

FIGURE 16.6

Evolution of China. The Chinese Empire evolved through territorial expansion from a small area near the current location of Beijing in 1900 B.C. to nearly its current size during the Manchu Dynasty, which ended in 1911.

was settled from the Asian mainland, and the Chinese culture, including the language and the Buddhist religion, was introduced to Japan from China. The Korean Peninsula also was settled from China, about 3,000 years ago. The Chinese introduced rice cultivation, religion, and a literary language. For centuries Korea was overrun by Asian armies. After the Chinese, the Tartars came, then the Mongols, followed by the Japanese and the Manchus. Finally, Japan annexed the peninsula again in 1910 and ruled it until 1945.

At the end of World War II, when the Japanese were defeated, the Korean Peninsula was divided at the 38th parallel (north latitude). The line corresponds roughly to where the American and Soviet armies met while fighting the Japanese, and it was used as the boundary for postwar administration of the peninsula. The dividing line became an international boundary when the northern and southern portions of the peninsula went their separate ways politically. The official separation occurred only after three years (1950–53) of bloody warfare between the north and the south. North Korea was supported by the Soviet Union and Communist China, while South Korea was backed by the United States and the United Nations. During the Korean War, 54,246 American servicemen died. North and South Korea have sought peaceful means for reuniting, but only an uneasy truce exists today.

This street scene in the capital city of Beijing (Peking) gives an idea of the population density in China. Bicycles are a major mode of transportation in Beijing. Photograph by Colin Hagarty.

Population

Population figures are available for most of East Asia, but totals given are sometimes only estimates. A complete Chinese census is very difficult to imagine for a country with more than a *billion* people, but one was held in 1982. China's population is four times that of the United States, and the country is only slightly larger than the United States in area. If the Chinese were evenly distributed throughout their vast territory, the density would still be 280 people per square mile. They are far from evenly distributed, however (see Figure 16.7). Large regions, such as the deserts and mountains, are very sparsely populated, and only 11 percent of the entire country is arable.

Chinese leaders understand the problems associated with having too many people and are trying to do something about it. Birth control is encouraged, and many times it is enforced through the workplace; bonuses are given to couples who have no more than one child. Young people are strongly encouraged to postpone marriage, and having only one child is becoming the patriotic thing to do. The government efforts have begun to decrease the birth rate, but the overall annual increase remains at one percent. It sounds relatively low, but one percent of 1 billion is 10 million. Thus, an average of 10 million people are added to the bulging Chinese population each year.

Except for Mongolia, China has the lowest average density of population in East Asia, but most of the Chinese people are jammed into the eastern cities, towns, and villages of the alluvial

East Asia 493

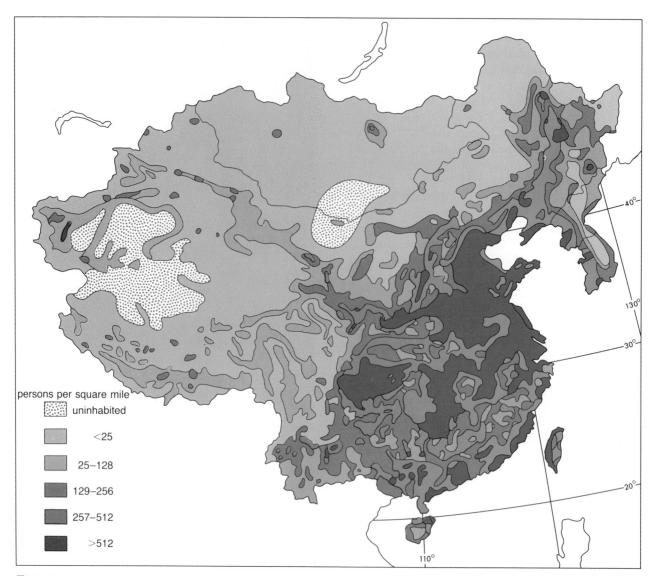

FIGURE 16.7

Population Patterns of East Asia.

plains and river valleys. The 100th meridian marks the transition of crowded China to sparsely populated China, just as it marks the transition in rainfall amounts. West of that line begin the vast areas of emptiness. The density of population for Sinkiang Uighur Autonomous Region is less than 10 people per square mile, and the density for Mongolia is about 3 people per square mile.

The islands of Taiwan and Hong Kong are very heavily populated for their area. The 14 near islands and the 64 Penghu islands of Taiwan make a total of about 14,000 square miles (36,000 km^2), and Taiwan's population is 19.6 million, giving an overall density of 1,633 per square mile. The 5.7 million people of Hong Kong are squeezed onto only 409 square miles of land, which results in an

The density of population in Hong Kong is indicated by this view of an apartment building. Photograph by Colin Hagarty.

incredible 13,936 people per square mile. The main island of Hong Kong, where the capital city of Victoria is located, is only 35.5 square miles. Most of the people live on that patch of land, and it is probably the most densely populated place on earth.

The two Koreas are similar in area, with North Korea only slightly larger. The combined populations of the two peninsular countries is 63.8 million, but South Korea has more than twice as many people as North Korea. Thus, the population density for North Korea is much less (431 per m^2) than that for South Korea (1,121 per m^2). North Korea and Mongolia have the highest rate (2.4 percent per year) of natural increase of population in the region. For the other countries, the rate is less than 2 percent a year. The countries of East Asia all have normal life expectancies (66 for the region) and infant mortality rates (45 per 1,000 live births for the region).

Urbanization

A city with a population of more than one million is a very large city by any standard, and it is difficult to conceive of a million people. An example might help. Standing at arm's length, clasping hands, one million people could form a human chain that would stretch from San Francisco to Denver. In the highly urbanized United States, only 6 cities exceed one million. By comparison, East Asia has 27 such cities. None of the countries of East Asia, however, is considered urbanized (except Japan). The urban population of East Asia is only 38 percent of the total population. The density of rural population is so great that parts of each country are continuous stretches of habitation separated only by small plots of intensively cultivated land.

Shanghai is the largest city in China, with about 12 million people. The capital, Beijing, has 8.5 million people, and Tianjin, located near the coast and about 50 miles (80 km) from Beijing, is China's third largest city, with over 7 million people. Besides the three largest cities, Guangzhou, (Canton) Shenyang, Wuhan, and Chengdu all exceed 4 million. These are only the largest cities. A dozen others either exceed a million or close to it. Other large cities of the region include

Taipei, Taiwan; Victoria, Hong Kong; and the Korean cities of Taegu, Seoul, and Pusan. (See Figure 16.8).

Religion

For centuries devotional ceremonies in East Asia were common functions of daily life. In the Communist countries, however, the religious aspects of the ceremonies have been replaced with political expressions. It is interesting to note that while both Chiang Kai-shek and Mao Tse-tung were raised as devout Buddhists, they both changed in later life. The Nationalist leader became a Methodist, the Communist leader became a Marxist. Changing religions, however, and combining the faiths of two or more religions have been traditions in China. The two indigenous Chinese religions began more as philosophies than as religions, so combining faiths may be understandable.

The major pre-Communist religions of China were Buddhism, Taoism, and Confucianism. Although Buddhism was imported from India, it shares ancestor worship with the other two religions as one of the main tenets of the faith. Thus the three religions complement each other. **Taoism** generally is regarded as the oldest of the three, as forms of the faith were practiced early in the Chou Dynasty (1110–256 B.C.). Besides ancestor worship, Taoists believe in two complementary forces known as Yin and Yang. Believers must follow Tao, or "the Way," by constantly balancing Yin and Yang. Yin is a female tenet, and Yang is a male tenet, and the two need and complement each other. Sex, therefore, is a vital part of the belief, and those who are the most active sexually are believed to live the longest. Although the religion is no longer widely practiced in China, the philosophy and its related concepts are ever present in the minds of millions of Chinese.

Confucianism developed as a faith from the philosophies of Confucius. It is not certain when Confucius lived, but his birth and death years have been estimated at 551 and 479 B.C. Confucius's philosophy was essentially an ethical system that called for moderation in all things. Many of his writings were on political matters, especially the relationship between rulers and their subjects. Confucianism is looked on as a conservative doc-

---

Taoism The Chinese word *tao* means "the way," and Taoism is the Chinese religion and philosophy based on the doctrines of Lao-tse (sixth century B.C.), who advocated simplicity and selflessness.

Confucianism The ethical teachings of Confucius introduced into the Chinese religion. The emphasis is on devotion to parents, family, and friends, ancestor worship, and maintaining law and order.

---

trine because it supports the status quo. Confucius did not think of himself as a god or a prophet, but his philosophy developed into a religion.

Buddhism came to China from India by way of Tibet and Central Asia. The Han Dynasty (202 B.C.–A.D. 220) had expanded China's borders west of the present Sinkiang-Soviet border, and that expansion established contact with the cultures of the area. Then, when the Han Dynasty collapsed, the Buddhist religion diffused into Chi-

The way this Buddhist monk is dressed makes him stand out from the crowd, but he is enjoying a tourist attraction in the city of Beijing. Photograph by Marvin G. Bays.

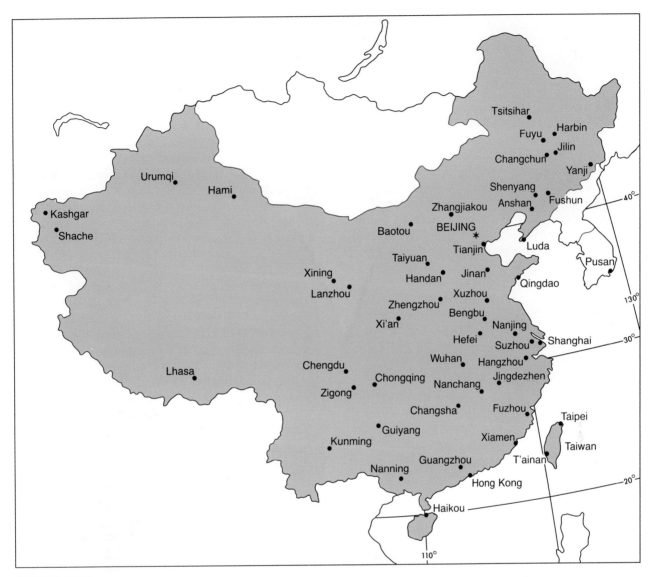

FIGURE 16.8
Major Cities of China.

na's heartland as if to fill the vacuum left by the political unrest. During the centuries that followed, Buddhism won millions of Chinese converts, who eventually outnumbered both the Taoists and the Confucianists. Buddhism appealed to all levels of Chinese society. Today, the religion is not a strong force in China but, as with China's other religions, it is a slight but constant denunciation of Communist ideology.

Islam and Christianity both arrived in China during the seventh century. Islam came from the same general area as Buddhism—the western provinces—and remains strongest there today. It has been estimated that about 20 million Muslims now live in China. Very early in the century, Roman Catholics established missions in China, and later Protestant missionaries arrived by the thousands, but Christianity never could count

> ## FOCUS BOX
>
> ### Chinese Place Names and the Pin-yin System
>
> The Chinese government adopted the Pin-yin system of spelling place names in 1958 in order to establish a common pronunciation for the entire country. The conventional English spellings were used by other countries of the world, however, until 1979 when all press reports started to appear from China in Pin-yin. Conversion to a new system is a slow process, which can be understood by Americans trying to convert from the English distance and weight measures to the metric system. During the 1980s, however, most geographic material relating to China started to appear in Pin-yin. You will have noticed that in this book, the Pin-yin spelling is given first, followed by the conventional spelling in parentheses.
>
> The most common variant of the Chinese language is Mandarin, which is spoken in the region from Beijing northward. The Pin-yin system of spelling place names is based on the pronunciation of Mandarin Chinese. The Pin-yin word-sounds in English are made by using the letter sounds from English. By pronouncing Pin-yin words as they appear, then, an English-speaking person can closely approximate the way they sound in Mandarin.
>
> Some Pin-yin words will become familiar faster than others. For example, "ho" as in the conventional spelling of Hwang Ho, has become "he" in Pin-yin, and the word means "river." So, Hwang Ho is now Huang He, or Yellow River. Furthermore, the Tarim He is the Tarim River, the Wai He is the Wai River, and so forth. The Yellow Sea is now the Huang Hai, because "Huang" means "Yellow" (from above) and "Hai" means "sea." So, the Dong Hai and Nan Hai also are seas. Another change that should be easy to remember is "shamo" for "desert"; thus, the Gobi Desert in Pin-yin is the Gobi Shamo. It should be noted that some words such as Gobi and Shanghai are the same in the conventional and Pin-yin forms. The following list contains some of the old and new spellings:

more than 5 or 6 million Chinese converts. About 75 percent of the Christians in China were Roman Catholics.

Communism has been a major ideological force in China only since 1949, and a minor force since the mid-1920s. Compared with the length of time that Taoism, Confucianism, Buddhism, Islam, or even Christianity have been in China, it has not been there very long. It is undeniable, however, that communism reigns supreme as the major ideology of China today.

Religion is not significant in the Communist country of North Korea, but South Koreans follow multiple faiths. The religions of China moved on to dominate Korea and Taiwan, and Christianity has made strong inroads in both countries. Most of the people of Mongolia traditionally were Buddhists, but now their ideologies are formed around communism.

Language

The most common spoken language in China is Mandarin Chinese, which is the national tongue, but unfortunately only about half the people speak Mandarin. The remainder speak a variety of other languages and dialects. There are so many secondary languages that they have not been accurately counted. Amazingly, however, all literate Chinese read and write the same script. The written language is not easy to learn, because it contains thousands of symbols. No alphabet for forming words exists, but each word is a type of picto-

CITIES		PROVINCES	
Conventional	Pin-yin	Conventional	Pin-yin
Canton	Guangzhou	Anhwei	Anhui
Chungking	Chongqing	Chekiang	Zhejiang
Harbin	Haerhpin	Hopeh	Hebei
Hong Kong	Xianggang	Inner Mongolia	Nei Monggol
Kowloon	Jiulong	Kansu	Gansu
Kunming City	Kunming Shi	Kwangsi	Guangxi Zhuangzu
Mukden	Shenyang	Ningsia Hui	Ningxia Huizu
Nanking	Nanjing	Shansi	Shanxi
Peking	Beijing	Shensi	Shaanxi
Shanghai	Shanghai	Singkiang	Xinjiang Uygur
Taipei	Taibei	Szechwan	Sichuan
Tientsin	Tianjin	Tibet	Xizang

RIVERS		UPLANDS	
Conventional	Pin-yin	Conventional	Pin-yin
Hwang Ho	Huang He	Altai Mountains	Altay Shan
Hsun Chiang	Xi Jiang	Astin Tagh	Altyn Tagh
Tarim River	Tarim He	Greater Khingan Mts.	Hinggan Ling
Sung Jua	Songjua Jiang	Kunlun Mountains	Kun Lun Shan
Wei Ho	Wai He	Plateau of Tibet	Qing Zang Gaoyuan
Yangtze Kiang	Chang Jiang	Tyan Shan	Tien Shan

graph. The symbols have evolved since the eighteenth century B.C.

All spoken Chinese languages are tonal, which means that the tone of voice is important to talking and understanding. Words have different meanings when different tones are used. Speakers of the various dialects cannot understand each other, so regional language differences create problems in communications. Language boundaries often exist between villages. Besides Mandarin Chinese and its various dialects, there are areas in western China where the majority of people speak Tibetan, Turkic, or Mongolic tongues.

Khalkha Mongol is the major language of Mongolia, but many people speak Russian and/or Chinese. The literacy rate is nearly 100 percent in Mongolia, at least among the younger generations. Even the children of the nomadic herders are sent away to schools to be educated. Ulan Bator, the capital of Mongolia, has eight institutions of higher education, including a university and a medical college.

The people of Taiwan, Hong Kong, and Macao are Chinese, so Mandarin is their main language, but a variety of dialects are spoken in each place. Korean, however, is a separate tongue, and the

The Forbidden City, in Beijing, is the seat of the Communist government in China (note the picture of Mao on the wall). In ancient times it was a royal palace. Photograph by Colin Hagarty.

The Mongolian ancestry of these Chinese women is evident. Note the Western-style clothing, including jeans, on these tourists visiting the Great Wall. Photograph by Marvin G. Bays.

A statue of Confucius remains as a reminder of China's past glory and as a shrine to the followers of Confucianism. This statue stands in a small park in Beijing. Photograph by Marvin G. Bays.

people of both North Korea and South Korea speak the same language. English is taught in many of the high schools of South Korea. More than 90 percent of the people of these countries can read and write, whereas the rate in China is about 75 percent.

Economic Geography of East Asia

Agriculture

A very large portion of the people of East Asia are farmers. About 62 percent of the population is engaged in agriculture. This includes the herders of Mongolia who care for their livestock. The percentage of farmers is somewhat surprising when the number of very large cities is considered, but it is a reflection of the huge number of people in the region and of the need for agriculture.

The bulk of China's agriculture is concentrated in the wetter, eastern third of the country. There are four distinct production regions: (1) the northeast, (2) the Huang He Plain, (3) the Chang Jiang drainage region, and (4) South China. The northeast, mostly Manchuria, is the area with the cold winter climate much like the upper midwestern United States. About 150,000 square miles (376,350 km^2) of gently rolling plains are used for once-a-year crops, such as soybeans, wheat, and sugar beets. Soybeans are an important protein supplement to the meat-limited Chinese diet, so they are used for human food. The area has become a model of Communist agricultural techniques; large state farms are cultivated with huge machines.

The Huang He Plain, essentially a large alluvial plain created by the Huang River, is the heartland of Chinese civilization. This lowland area has been under intense cultivation for centuries. The major crops produced are wheat, barley, corn, millet, and cotton. The region also produces most of China's apples, and hogs are found nearly everywhere. Much of the agriculture is subsistence-level, with farmers producing only enough to feed their families on tiny plots of land. The Communist government's attempts to create large collective farms in the region have been unsuccessful. Peasant farmers do not use machinery because they cannot afford to buy the equipment, and tractors cannot be used on such small plots of land anyway. Until recently they did not use commercial fertilizers, and the ancient custom of using human waste (night soil) for fertilizer continues.

To the west of the Huang He Plain lie the loess hills of Northern China. This region of windblown soil has been dissected by thousands of gullies, but the flat areas between the miniature canyons are farmed intensively. Many of the people lived in caves dug into the sides of the gullies, and in 1556 some 830,000 people died when an earthquake buried them in their homes. Today, the area is much the same as in pre-Communist days, because large collective farms are difficult to lay out in such topography.

FOCUS BOX

Contrasting Koreas

South Korea is enjoying a period of spectacular economic growth, which has averaged about 8 percent a year over the last 20 years. As of the middle of 1987, the economic growth rate was surging at 15.7 percent (compared with about 4.8 percent for the United States and 1.2 percent for Japan). South Korea has progressed from poverty to prosperity in little more than a generation. The literacy rate of 98 percent is one of the world's highest, and a third of South Korea's high school graduates go to college. Per capita income rose from $105 in 1965 to $2,300 in 1987. The economy's boom has been fed by a burst of exports. For example, the Hyundai Excel automobile, introduced into the United States in 1986, sold 168,000 units to become the most successful new car import in U.S. automotive history. Part of the economic success, however, has been based on extremely low wages for laborers, who put in extremely long hours at work. These conditions, combined with an antigovernment sentiment, have led to disturbances and demonstrations. Many Koreans, led by thousands of university students, have rallied against the military-dominated government and for labor unions and a democratic society. Meanwhile, the economy continues to boom.

North of the 38th parallel is another Korea. In every sense, North Korea is a contrast to the turbulent and economically dynamic South. Communist North Korea hides behind miles of barbed wire and mine fields and sometimes threatens life in the South. Spartan, plodding, and regimented, North Koreans seem to act with one corporate mind. The world's sixth largest fighting force of 885,000 troops helps government leaders maintain the stability. Like daily life in North Korea, the economy there is also stable. It is not grow-

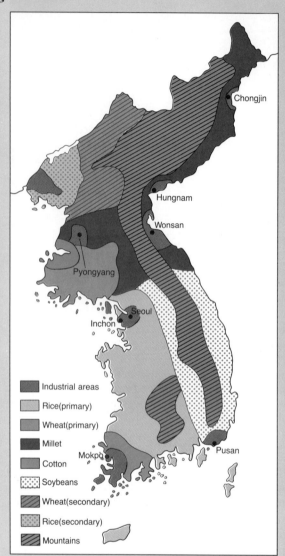

FIGURE 16.A

Agricultural Regions of North Korea and South Korea.

> ing, but the people are well fed, clothed, and housed. The regimented agriculture is often bountiful, but city streets are deserted. Bicycles and cars are reserved for official use only, and any conversations that do occur on the streets take place in murmurs. Television broadcasts are dominated by three to six hours of political programs explaining the Communist leaders' policies. Students study a blend of Marxism and Korean nationalism that stresses economic self-sufficiency and abiding loyalty to the leaders. North Korea has been a model Communist state for 39 years, but life there is dull and moves at a uniform pace. North Korea is a place frozen in time, in ideology, and in its prospects.

The Chang Jiang drainage basin, the third major farming region of China, is the rice-producing area. Rice is the major crop along the river from the Szechwan Basin to Shanghai. Other crops include sweet potatoes, barley, millet, sugar cane, cotton, and peas. The region has also been noted for the production of silk and tea. Mulberry trees for feeding silkworms are still common, although the silk industry has declined. Citrus fruits are grown along the lowland portion of the Chang Jiang, and tea bushes cover the steep banks of the upper reaches of the river. The main agricultural crop of China is rice, however, and much of it is grown along the Chang Jiang (see Figure 16.9).

South China is the poorest of the four major agricultural regions. The plain surrounding Canton is not large, and the rolling hills give way quickly to nonarable mountains. A number of these small plains exist, however, all along the coast from Hong Kong to Shanghai. The semitropical climate sometimes makes it possible to grow two crops each year, but the soils are thin and not as fertile as the river basins to the north. The typical Chinese crops grown in this region are tea, rice, and sweet potatoes, but other grains and citrus fruits are common too.

In spite of the vast areas of agriculture and the millions of farmers, China has been forced to import large quantities of grain each year. Grain imports vary between 10 million and 20 million metric tons annually, most of which is wheat and rice. The wheat comes from Argentina, Mexico, Canada, and Australia, while the rice comes from Southeast Asia. Traditionally, the wheat eaters of China have been the people in the Huang He region and northward, while the rice and fish eaters are the people in the Chang Jiang Valley and southward. These people have tended to look down on each other for their eating habits.

Rice is also the major crop of Taiwan and the Korean Peninsula. Taiwan is warmer and receives more rain than Korea, so jute and sugar cane are important secondary crops. In Korea, other crops include wheat, barley, tobacco, and beans. Agriculture in Mongolia is dependent on irrigation, and few row crops are produced. For centuries, the dry, grassy plateaus have been home for nomadic herders. The Communist regime has tried to get people to settle on farms but has been largely unsuccessful. The herders do belong to cooperatives for selling the wool, meat, hides, and other animal products, but most still live in yurts and move with their herds. The people are tough and have very few amenities. Like other nomads, their lives evolve around their animals. One of their favorite drinks is fermented mare's milk.

Industry

Textiles and food-processing have been the primary industries of East Asia, satisfying the imme-

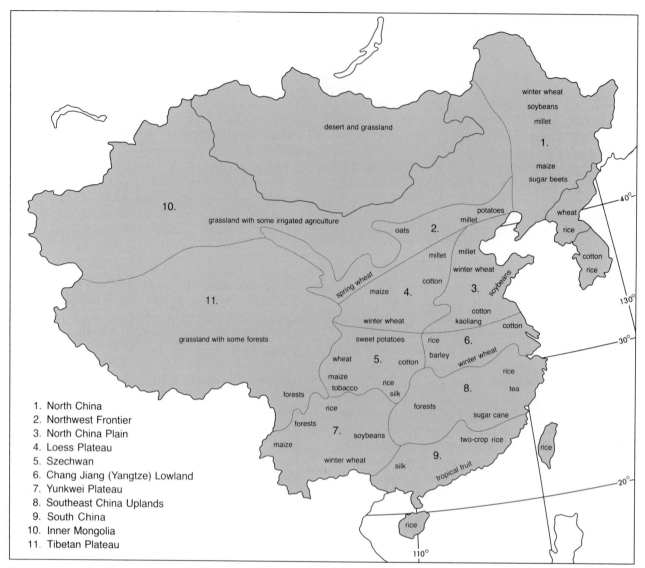

FIGURE 16.9

Agricultural Regions of China.

diate needs of the people for food and clothes. Recently, however, heavy industry has been introduced and has become much more sophisticated. Taiwan and South Korea now make electronic equipment for trade with the United States, and China produces most of the modern items, although on a limited basis.

Manchuria has been China's heavy industry region. Iron ore and coal are found there, and the area produces more than one-third of China's steel (see Figure 16.10). Other manufactured items from the region include numerous steel products, such as machine tools, ball bearings, and railroad stock. The industry in Manchuria was started by the Japanese when they held control from 1931 to 1945, but China has poured immense amounts of human energy into the struggle to increase industrial production. It is the official government goal

> **FOCUS BOX**
>
> ### *China's Fertility Debate*
>
> Despite great efforts, China's promotion of minimal reproduction has not resulted in one-child couples exclusively. Findings from a large-scale sample survey in Shanghai, Hebei, and Shanxi in April 1985 documented that a significant number of households had a second child. The plan that requires all or the majority of families to have only one child is difficult for most of the people to accept. Reservations about the one-child limit are numerous, even from a demographic standpoint. One official stated, for example, that it would cause the population to age too fast and that too many people would be without children at their side in old age. In 30 years, the policy would create an abnormal population pyramid, rendering it necessary for one young couple to support four parents and one child.
>
> An alternative plan, proposed in 1985 by Liang Zhontang, seems to be more workable. It is based on the two principal ingredients of *xi* and *shao*—longer spacing and fewer children. Specifically, for the next several decades each couple would be allowed two children, provided that the interval between the two births would be strictly kept at 8 to 10 years. The plan also calls for 30 percent of the couples to have only one child. But that provision would not take effect until 1990 and would be realized more easily through education and propaganda rather than by strong administrative measures. After a lapse of 10 years, with survival of the first child, the family routine would crystallize around one child, and a proportion of the couples would voluntarily refrain from having a second child. It has been calculated that implementation of this plan would yield a stationary population of 1.11 billion by the year 2020, but based on China's current population the figure is obviously wrong. Many Chinese couples would like a second child and would welcome the plan.
>
> SOURCE: H. Yuan Tien, "As China's One-Two Fertility Debate Turns," *Population Today* 14 (April 1986): 6–9.

for China to become a major industrial power, and the northeast is looked on as the leader in fulfilling that goal.

The Huang He and Chang Jiang valleys produce most of China's textiles. The textile industry utilizes the cotton grown in the region, as well as wool brought in from the western provinces. Shanghai, where the ancient silk industry is still located, is the country's largest textile center. Wheat and rice milling are common throughout the lowland areas of eastern China; other small industries include tanning of leather, butchering, and cement manufacturing.

The Chinese have hopes that western China will become an industrial region. Many minerals have been discovered in the mountains of the western provinces, and many more veins of ore probably will be found (see Figure 16.10). Large deposits of oil, coal, iron ore, lead, sulfur, and even gold have been found. Iron and steel plants, oil refineries, and some manufacturing plants for consumer goods have been built, especially along the railroad between Sinkiang and eastern China. China's first atomic device was exploded in Sinkiang in 1964, so the remote area, rich in natural resources, is becoming China's research area as well as a new industrial region.

On the Korean Peninsula mineral resources are adequate, but heavy industry lags far behind that of China. Trade between North and South Korea

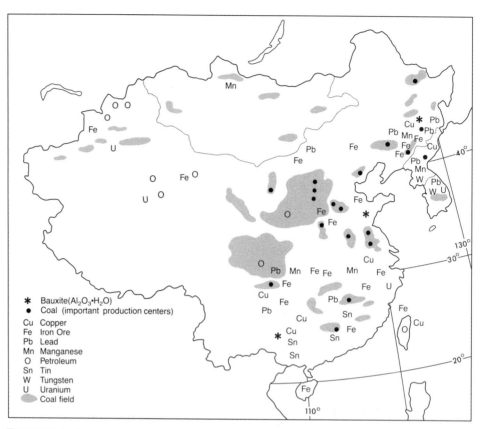

FIGURE 16.10

Mineral Deposits and Industrial Areas of China.

These industrial workers take a break and play the game of mah-jongg on the street in front of their factory in Shanghai. Photograph by Colin Hagarty.

could help cut the cost of many industries, but it does not occur. Coal, tungsten, graphite, and magnesite are found in large quantities in both the north and the south. North Korea also produces lead, zinc, iron ore, gold, and phosphate. Industries in the north are nationalized, and output details are kept secret. South Korea has heavy industry, but it could use some more iron ore for higher production. One major heavy industry of the country is shipbuilding. The South Koreans also make electronic items, appliances, and other consumer goods, many for export, especially to the United States. Cement and fertilizer plants are common, and South Korea has a few oil refineries.

Minerals on Taiwan include coal, copper, gold, and silver, but industries predominantly manufacture light consumer goods. Small electronic items, such as transistor radios, and such goods as clothing and food products are the most common commodities produced. Mongolia has large coal deposits, and some of the coal is used to produce electricity. Mongolia also has tungsten, copper, tin, and some gold. Industry in Hong Kong has the most productive, booming air about it than any region of East Asia (outside Japan). Hong Kong harbor, one of the world's largest transshipment ports, is extremely busy; goods come from and go all over the world. Shipbuilding and ship servicing are major industries for the country, but the 5,000 or more factories in Hong Kong also produce steel, electronic items, textiles, cement, and foodstuffs. Tourism is also a major industry of Hong Kong, and the tiny country leads the region in the number of tourists it attracts each year.

Transportation and Communications

The lives of many East Asians are water-oriented. No place in Hong Kong, Macao, Taiwan, and the Koreas is very far from ocean waters. The bulk of the population of China also lives near the coast or along the rivers and vast system of canals. All the nations, except Mongolia, are shipbuilders, and most of the people enjoy eating fish. Transportation by water, then, is normal and common.

Most of the million or so villages in China are not connected by roads, and all-weather highways run only between major cities (see Figure 16.11). Immediately after World War II, China had only about 100,000 miles of roads. The Communist government tripled that mileage, but rural transportation still depends on primitive trails or on the waterways. Most of the major cities in China are connected by railroad lines, but the most extensive network is in Manchuria (see Figure 16.12). Many of the original railroad beds in the northeast were laid by the Japanese during their occupation. The railroads of China carry large volumes of passengers and goods.

The inland waterways of China are used heavily for the movement of goods. Various types of small

The streets of Beijing are clogged with bicycle traffic during rush-hours. Photograph by Marvin G. Bays.

East Asia 507

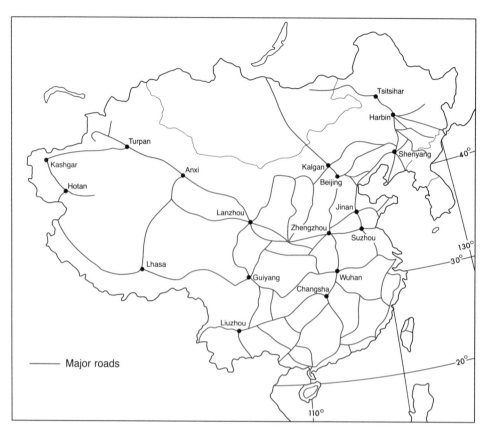

FIGURE 16.11
Major Roads of China.

crafts navigate from place to place on the rivers, and barges of various sizes are used on the canals. The Yun Ho (Grand) Canal connects the Huang He and Chang Jiang rivers and runs northward to Beijing. It is the main inland waterway of China. Besides the complex network of canals that connect with the rivers, eastern China also has a number of lakes that are part of the inland waterway system.

Automobiles are not common in China, but the volume of truck traffic has increased in recent years. The city streets are choked with thousands of bicycles, especially during the prime commuting times. Heavily laden bicycles also travel the rural roads, as farmers carry goods to market on them. Bicycles sometimes pull intricate racks for carrying live chickens and ducks. Foot traffic is heavy in China, and people garbed in clothing that is quite similar are seen walking nearly everywhere.

Western-style clothes are creeping into use, however, and a metamorphosis of color is occurring.

Automobiles are scarce on the Korean Peninsula too, especially outside the cities. Railroads link the major cities, but the rail network is disconnected at the boundary between the north and the south. North and South Korea have similar total railroad mileage, and both have primitive rural transportation facilities. The roads in Taiwan are good, hard-surface tracks but limited in mileage. A major highway, paralleled by a railroad, circles the entire island. One rail line runs through Ulan Bator, the largest city of Mongolia. The railroad is part of the international line from Irkutsk in the Soviet Union to Beijing, China. Most of the small towns of Mongolia are connected to the capital with passable motor roads. Mongolians love horses and use them whenever

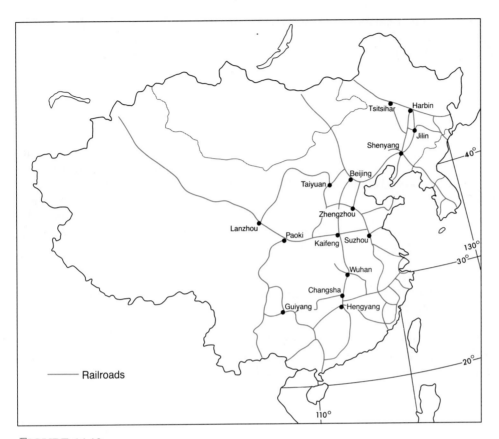

FIGURE 16.12

Railroads of China.

possible. Horse caravans, and sometimes even camels, are used to haul goods to market.

Very few people travel internally by airplane in East Asia. Most air traffic is either military or for foreign visitors, such as business people and tourists. Nevertheless, many of the countries support a modern air force. Jet planes are purchased from either the Soviet Union or the United States, depending on the political climate. North Korea has nearly 1,000 Russian-built MIG fighters, as well as 100 bombers and 200 jet trainers. South Korea counters the north with numerous squadrons of American-built F-5s and F-86Fs. China has 322 usable airfields, but only about half have hard-surface runways, and only two are longer than 4,000 feet (1,219 m). South Korea, on the other hand, has 68 permanent-surface runways, with a dozen over 4,000 feet in length.

Telecommunications in China is fair to good, but domestic and international services exist primarily for official purposes. An unevenly distributed telephone system serves principal cities and industrial regions. China has about 1 telephone for every 200 people, but in the cities the figure is about 5 telephones per 100 people. Domestic radio broadcasts reach about 65 percent of the people from 122 main AM stations, and broadcasts from 52 television stations reach about 60 percent of the population. A television set has become one of the items the Chinese prize most because it gives them a glimpse of the world beyond their own village. Many villages have only one television receiver, which all the people in the village watch.

Communications in the small countries of East Asia are much easier than in China. In those countries, most of the people speak the same language, and the area to cover is much smaller. South Korea has 121 telephones per 100 people, 256 television stations, and 79 AM and 46 FM radio stations. Compared with China, telecommunications in South Korea are quite modern. For North Korea, not much information is available, but that country's communications networks seem to be similar to those in China. Televisions, radios, and telephones are available in abundance on Taiwan and in Hong Kong. Mongolia has about 65,000 television sets and 180,000 radios for its population of 1.9 million.

■ Political Geography of East Asia

The political history of East Asia is filled with violence and upheaval. Communism seems to have brought order to much of the region, at least on the surface. Life expectancy figures are generally high for the region (66 years), which indicates that many of the centuries-old disease problems have been conquered. Education is becoming more important each year as the countries move into the electronic age. Poverty remains common, but the crunching dietary deficiencies of the past are being eased, so political stability may have a chance to endure.

Aside from border skirmishes with the Soviet Union, India, and Vietnam, China has had few external political problems since World War II. The country's Communist leaders are aggressive, however, and seem to seek ideological leadership of the Third World. For example, even though there has been a razor-thin line between starvation and full stomachs at home, China sent military and economic aid to many other countries. China sends supplies to countries in Africa and South America, as well as to allies in Asia.

Internally, China has many minority groups that the Communist leaders must appease. Western China has Tibetans and Mongols, and the heartland is filled with peasants who seem to have a natural tendency toward capitalism. Religion also is an ideological burden for the Communist leaders. If food shortages are averted, however, internal control should be relatively easy for the Communists.

Classic Chinese architecture lends beauty to part of the capital city of Beijing. Photograph by Marvin G. Bays.

The countries in the remainder of East Asia each have their own political problems. The Nationalist Chinese leaders on Taiwan still have dreams of regaining political control of the mainland and nightmares about being invaded from there. North Korea keeps pressure on South Korea and could break across the cease-fire line at any time. Mongolia, the buffer state between the giant Communist powers, has external fears from all directions. The country became independent from China with Soviet aid, and its internal political organization is identical to that of the Soviet Union. Russians do not occupy the country, but it is clear that Mongolia would not move on its own without approval from Moscow.

The region of East Asia has been a subtle force in world affairs for many years, and that force is increasing as its countries begin to develop. Led by Japan's production, the region is rapidly becoming a world economic force, but China looms as a huge potential market too. American companies, such as Coca-Cola have only recently been allowed to tap that market.

Key Words

China's Sorrow
Chinese dynasties
Confucianism
coolies
junk
landlocked
Orient
Peking man
Qing Zang Gaoyuan
  (Plateau of Tibet)
steppe
Taoism

References and Readings

Barnet, A. Doak, and Clough, Ralph N. (eds.). *Modernizing China: Post-Mao Reform and Development.* Boulder, Colo.: Westview, 1985.

Baum, Richard D. (ed.). *China's Four Modernizations: The New Technological Revolution.* Boulder, Colo.: Westview, 1980.

Benton, G. *The Hong Kong Crisis.* Dover, N.H.: Pluto, 1980.

Boyd, Andrew. *An Atlas of World Affairs.* New York: Praeger, 1960.

Buchanan, Keith, et al. *China: The Land and People.* New York: Crown, 1981.

Bunge, Frederica (ed.). *North Korea: A Country Study.* Washington, D.C.: Government Printing Office, 1981.

_____. *South Korea: A Country Study.* Washington, D.C.: Government Printing Office, 1982.

Central Intelligence Agency. *The World Factbook.* Washington, D.C.: Government Printing Office, 1985.

Chang, Kuei-sheng. "The Changing Railroad Pattern in Mainland China." *Geographical Review* 51 (1961): 534–548.

Chang, S-d. "Modernization and China's Urban Development." *Annals of the Association of American Geographers* 71 (1981):202–219.

Chiu, T. N., and So, C. L. (eds.). *A Geography of Hong Kong.* London: Oxford University Press, 1983.

Cressey, George B. *Asia's Lands and Peoples.* New York: McGraw-Hill, 1951.

_____. *Land of the 500 Million.* New York: McGraw-Hill, 1955.

Dawson, Owen L. *Communist China's Agriculture.* New York: Praeger, 1970.

Decrespigny, Rafe. *China: The Land and Its People.* New York: St. Martin's, 1974.

Ding, C. "The Economic Development of China." *Scientific American* 243 (1980): 152–165.

East, William Gordon, and Spate, Oskar H. K. (eds.). *The Changing Map of Asia: A Political Geography.* 4th ed. New York: Barnes & Noble, 1961.

Eckstein, Alexander (ed.). *Economic Trends in Communist China.* Chicago: Aldine, 1968.

Fessler, Loren. *China.* New York: Time Inc., 1963.

Geelan, P. J. M., and Twitchett, D. C. (eds.). *The Times Atlas of China.* 2d ed. New York: Van Nostrand Reinhold, 1984.

Ginsburg, Norton S. et al. *The Pattern of Asia.* Englewood Cliffs, N. J.: Prentice-Hall, 1960.

Ginsburg, Norton S., and Lalor, Bernard A. (eds.). *China: The 80s Era.* Boulder, Colo.: Westview, 1984.

Goldstein, Steven M., and Sears, Kathrin (eds.). *The People's Republic of China: A Basic Handbook.* Croton-on-Hudson, N.Y.: Council on Public Affairs, 1984.

Herman, Theodore. *The Geography of China.* New York: State Education Department, 1967.

Hofheinz, Roy, and Calder, K. *The East Asia Edge.* New York: Basic Books, 1982.

Hook, Brian (ed.). *The Cambridge Encyclopedia of China.* New York: Cambridge University Press, 1982.

Hsieh, Chiao-min. *Taiwan-Ilha Formosa: A Geography in Perspective.* Washington, D.C.: Butterworth, 1964.

Jackson, William A. Douglas. *The Russo-Chinese Borderlands: Zone of Peaceful Contact or Potential Conflict?* Princeton: Van Nostrand, 1968.

Jao, Y. C., and Leung, Chi-keung (eds.). *China's Special Economic Zones: Policies, Problems, and Prospects.* New York: Oxford University Press, 1986.

Kelly, Ian. *Hong Kong: A Political-Geographical Analysis.* Honolulu: Hawaii University Press, 1987.

Kirby, Richard J. R. *Urbanization in China: Town and Country in a Developing Economy, 1949–2000 A.D.* New York: Columbia University Press, 1985.

Knapp, Ronald G. (ed.). *China's Island Frontier: Studies in the Historical Geography of Taiwan.* Honolulu: Hawaii University Press, 1980.

Kolb, Albert. *East Asia: China, Japan, and Korea.* London: Methuen, 1977.

Lattimore, Owen. *Inner Asian Frontiers of China.* Boston: Beacon, 1962.

Leeming, Frank. *Rural China Today.* White Plains, N.Y.: Longman, 1985.

Ma, Lawrence J. C., and Hanten, Edward W. (eds.). *Urban Development in Modern China.* Boulder, Colo.: Westview, 1981.

Ma. Lawrence J. C., and Noble, Allen G. (eds.). *Chinese and American Perspectives on the Environment.* New York: Methuen, 1981.

Mallory, Walter H. *China: Land of Famine.* AGS Special Publication 6. New York: American Geographical Society, 1926.

Miller, E. Willard, and Miller, Ruby M. *The Far East: A Bibliography on the Third World.* Monticello, Ill.: Vance Bibliographies, 1982.

Murphy, Rhoads. *The Fading of the Maoist Vision: City and Country in Chinese Development.* New York: Methuen, 1980.

Ness, Gayl D., and Ando, Hirofumi. *The Land Is Shrinking: Population Planning in Asia.* Baltimore: Johns Hopkins University Press, 1984.

Pannell, Clifton W. (ed.). *East Asia: Geographical and Historical Approaches to Foreign Area Studies.* Dubuque, Iowa: Kendall/Hunt, 1983.

Pannell, Clifton W., and Ma, Lawrence J. C. *China: The Geography of Development and Modernization.* New York: Halsted, 1983.

Perkins, Dwight H. *Agricultural Development in China, 1368–1968.* Chicago: Aldine, 1969.

Sanders, Sol. *A Sense of Asia.* New York: Scribners, 1969.

Shabad, Theodore. *China's Changing Map: National and Regional Development, 1949–1971.* New York: Praeger, 1972.

———. *China's Changing Map: A Political and Economic Geography of the Chinese People's Republic.* New York: Praeger, 1956.

Sit, Victor F. S. (ed.). *Chinese Cities: The Growth of the Metropolis Since 1949.* New York: Oxford University Press, 1985.

Smil, Vaclav. *The Bad Earth: Environmental Degradation in China.* Armonk, N.Y.: Sharpe, 1984.

Smith, Michael, et al. *Asia's New Industrial World.* New York: Methuen, 1985.

Songqiao, Zhao. *Physical Geography of China.* New York: Wiley, 1986.

Stamp, Laurence Dudley. *Asia: A Regional and Economic Geography.* London: Methuen, 1967.

Tawny, Richard H. *Land and Labour in China.* New York: Octagon, 1964.

Thorp, James. *Geography of the Soils of China.* Nanking: National Geological Survey of China, 1936.

Tregear, Thomas R. A. *China: A Geographical Survey.* New York: Wiley, 1980.

———. *A Geography of China.* Chicago: Aldine, 1965.

U.S.S.R. Academy of Sciences. *The Physical Geography of China.* New York: Praeger, 1969.

Whyte, Martin King, and Parish, William L. *Urban Life in Contemporary China.* Chicago: University of Chicago Press, 1984.

Wint, Guy (ed.). *Asia: A Handbook.* New York: Praeger, 1966.

Young, Graham (ed.). *China: Dilemmas of Modernization.* Beckenham, Eng.: Croom Helm, 1985.

Yu-ti, Jen. *A Concise Geography of China.* Peking: Foreign Languages Press, 1964.

Chapter 17

THE PACIFIC ISLANDS

■ Introduction

The Pacific Ocean covers more than one-third of the earth's surface. It is the largest and deepest ocean, and the largest single earth feature. As with other regions, the geography of the Pacific Islands region is concerned with people, landforms, climates, and water. In this case, however, the water area is huge, while the land and people are relatively scarce.

Distance is an important factor in the geography of the Pacific. Distance has affected human activities from early migrations to present-day transportation of goods and has limited the spread of plants and animals. It is halfway around the world, about 12,500 miles (20,113 km), from Panama to the Malay Peninsula. From north to south, the ocean extends 9,300 miles (14,964 km), from the Bering Strait to the Antarctic Circle. The average depth of the Pacific Ocean is 14,000 feet (4,267 m) with an extreme depth of 35,400 feet (10,790 m) between Guam and Mindanao.

The Pacific Ocean is dotted with island groups, some of the islands are large, but most are small. The islands have been divided according to ethnicity of the native people into the following regions Indonesia, Micronesia, Melanesia, and

Polynesia. Indonesia was discussed as part of Southeast Asia (Chapter 15). (See Figure 17.1.)

Physical Geography of the Pacific Islands

Landforms

Except near the shores of the Americas, the eastern Pacific is nearly devoid of islands. A broad submarine platform, called the Albatross Plateau, is located west of South America. The volcanic islands rising from the platform include Easter Island, Sala y Gomez, the Juan Fernández group, the Galápagos Islands, and the Cocos islands. These islands are located above fracture lines running along the edge of the Albatross Plateau.

The western Pacific contains thousands of islands, generally located along arcs. These islands also rise from submarine platforms but, unlike the Albatross Plateau, they were created from outpourings of lava from undersea rifts. Many islands are the summits of mountain ranges that are mostly under water. The nonvisible ranges can be visualized by the location of the island arcs. Some volcanic islands have been built up from extreme ocean depths, making the individual peaks taller, from sea bottom to summit, than Mount Everest is above sea level. Both Mauna Kea (13,784 feet, or 4,201 m) and Mauna Loa (13,679 feet, or 4,169 m) on the island of Hawaii are such mountains. The peaks rise from an 18,000-foot (5,486 m) deep ocean floor, making them more than 31,000 feet (9,449 m) tall (see Figure 17.2). In the tropical Pacific the islands are fringed with coral. Thou-

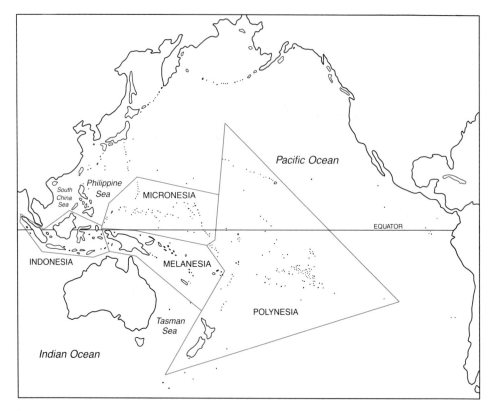

FIGURE 17.1

Ethnic Regions of the Pacific Islands. The boundaries of the ethnic regions of the Pacific Islands form an arrow that points toward South America.

FOCUS BOX

The Island World

Islands hold a mystique that attracts both ordinary people and scientists. Some of the allure undoubtedly derives from the simple fact that an island is a world unto itself. Because they have been isolated in time and space, islands are unique looking-glasses into the biological and cultural past, the story told by descendants of the few successful wayfarers who established a foothold. Much that we know about cultural, geological, and ecological change has been learned from islands, where processes are more evident and each inhabitant plays a more critical role in the entire system.

Islands change slowly because they are protected by difficulty of access. Those who colonized places like Hawaii were hardy seafarers, and many plants and animals who successfully crossed the wide expanse of water did so merely by chance. Migrants were thus few and far between. The long intervals that separated invasions allowed for gradual adjustment and relative stability. But insularity and vulnerability go together; in their isolated world, neither people nor other organisms acquired immunity to disease or the ability to resist the larger societies and stronger economies of the mainlands.

In the absence of more cosmopolitan, aggressive continental groups, island organisms and societies developed unique characteristics to utilize the environment. More than in any other location, one finds rare species, traits such as flightlessness in birds, and specialized cultural adaptations on islands. The more remote the island, the more pristine and unusual its characteristics.

Today, distance and hardship are no longer deterrents. As old methods of travel and communication become less important, the links with metropolitan areas become constant and irreversible. Islands are currently being transformed at an unprecedented rate. Tourism, in particular, is a major force. Attracted by natural beauty and isolation, tourists have traditionally been a transient element, but today they are a steady stream and their effects are lasting. With little else to sustain their economies, small island nations have been enticed to the ostensible benefits of the mass tourism industry. Yet modern tourism is little more than neocolonialism. As residents leave their farms, fishing, and other activities to take jobs in tourist facilities, the local economy becomes dependent on a single source of income. The real benefits accrue mostly to the already well-off and to expatriates, while the locals remain poor.

The world's coastlines are becoming homogenized by a multinational business. Strip development places rows of concrete monoliths against the sea and pushes land prices out of reach for all but the most wealthy. On many islands, two virtually incompatible and mutually exclusive worlds are found. The coastal fringe is a bastion of foreign values, foreign money, and luxury. In the interior, however, poverty, isolation, and declining hopes reign. Because they come to judge their own society and environment by the values of visitors, island residents retain little pride in their rapidly disappearing culture and remote island world. No longer a world unto itself, today's tropical island risks being totally dependent on the vagaries of the tourist dollar and losing the unique attributes that once made it a special place.

SOURCE: Janet Crane, "The Fragile Island World," *Focus* 35 (July 1985): 1.

sands of these coral islands and reefs, interspersed with volcanic peaks, are located in the western Pacific. Sometimes individual islands are separated from any neighbors by hundreds of miles.

The islands of the western Pacific are separated geologically from those of the mid-Pacific and eastern Pacific by a boundary called the **andesite line.** The volcanoes of the western Pacific extrude material called andesite and other sialic rocks, while those to the east emit basalt. The andesite line represents the outer limits of the Asian continental mass that once extended into the Pacific. It is the boundary between Asia-Australia and the Pacific basin. (See Figure 17.3.)

A zone of weakness in the earth's crust runs along the eastern, northern, and western edges of the Pacific Basin. Within this zone, frequent earthquakes and thousands of volcanoes make the land regions unstable. The nearly 300 active volcanoes along this horseshoe-shaped zone have been called the "Ring of Fire." Fractures of the earth's crust determine the sites of the volcanoes and uplifted mountains. Mountain ridges run in arcs along the edges of the basin, and volcanoes develop at appropriate vents where the crust is weakest. The magma emerges either quietly as lava-flows or explosively as pumice and ash. Frequently, ocean deeps parallel the uplifted mountain ranges.

As soon as an island appears above the sea, it is attacked by the agents of erosion. Strong waves, beating rain, and running water all work to wear

andesite line The andesite line is the geologic boundary in the Pacific Ocean between the Asian-Australian continents and the Pacific Ocean basin to the east. The word "andesite" comes from the Andes Mountains and refers to a small-grained, dark-gray rock of volcanic origin.

coral island Any island that exists because of coral buildup, although the island may have originally been created by volcanic activity. Coral is a hard substance made of the skeletons of small sea animals called polyps that live in tropical seas.

coral reef A reef is a line or ridge of rock (or sand) lying offshore but near the surface of the water. In tropical areas, coral growth creates such ridges.

down the island. Wave erosion, through undercutting along the shore, creates high cliffs out of what were once gentle slopes, and the force of falling raindrops over thousands of years cuts and carves the bedrock. The rapidity of erosion depends on the character of the rock, the amount of soil that has developed, the vegetation cover, and of course the amount of precipitation. New lava and cinder beds do not allow rapid erosion because the water is absorbed. Once soil forms, however, the water begins to flow, causing erosion by running water.

Coral islands are typically ring-shaped with a central lagoon (see Figure 17.4.) **Coral reefs** are

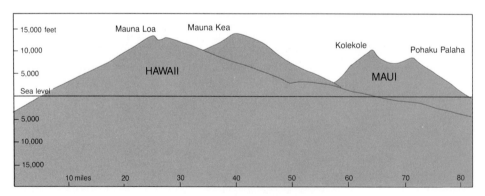

FIGURE 17.2

Mountains of the Hawaiian Islands. When measured from their base at the bottom of the Pacific Ocean to the tops, the volcanic peaks that form the Hawaiian Islands are taller than the Himalaya Mountains in Asia.

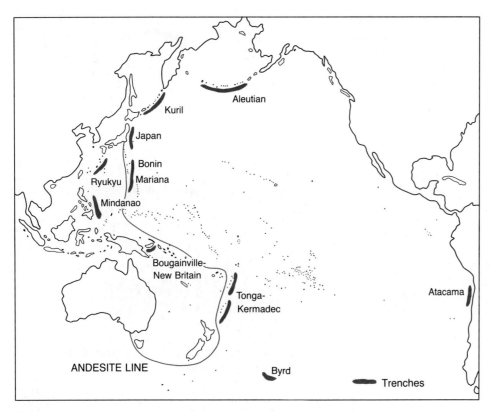

FIGURE 17.3

Geology of the Pacific. The andesite (sial) line separates Asia from the Americas, and the ocean trenches separate the various geologic structures that make up the Pacific Basin.

made by different types of lime-secreting organisms that attach themselves together, forming limestone beds. Only the outer edge of a reef is alive and active. The inner side contains the remains of the lime-secreting animals that have been consolidated into rock. The most important coral-forming animal is a polyp, but various forms of hydroids can make coral. Coral reefs are called fringing when they form a platform near the shoreline. What is known as a *barrier reef* forms at some distance from the coast, creating a large lagoon between the reef and the shore. Such lagoons vary from a few feet to several miles wide.

Atolls are islands formed when a fringing reef develops around a volcanic peak. Eventually, through erosion, the island wears away, leaving only the coral reef. While atolls vary greatly in size, they are all low in elevation and have a narrow width of land and a central lagoon. Kwajalein, in the Marshall Islands, is the world's largest atoll; it is about 90 miles (145 km) long and 20 miles (32 km) wide. Atolls are small, but rarely smaller than one mile (1.6 km) in diameter. They also are rarely circular, or perfectly donut-shaped; instead they are oblong.

Submarine earthquakes sometimes cause powerful seismic waves known as *tsunamis*. These earthquake waves have been measured as high as 100 feet (31 m). One of the most devastating such waves occurred in 1883, when the volcano Krakatoa erupted (see Chapter 15). Tsunamis affect the entire Pacific and have ravaged the coasts of Japan, Hawaii, and Chile, drowned thousands of people, and destroyed much property.

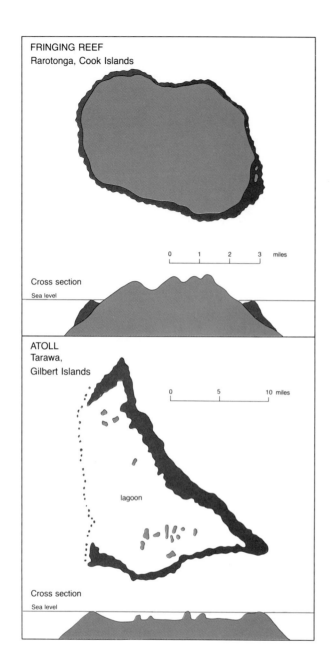

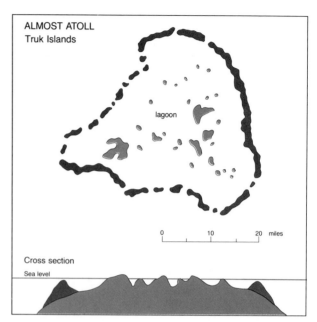

FIGURE 17.4

Formation of Atolls. Coral atolls develop from the combination of vulcanism and the growth of polyps at the water's edge. Charles Darwin developed the theory on how atolls evolve.

Climate

The climates of the Pacific Ocean area are created by the prevailing-wind belts of the region. The *doldrums,* or belt of equatorial calms, is located near the equator (see Figure 17.5). Extending away from the doldrums on either side are the trade winds, the horse latitudes, and the prevailing westerly winds, known as the "prevailing westerlies." The doldrums correspond to the heat equator, the area that receives the most direct sun and therefore the most heat. The heat causes the air to expand and rise, creating a low-pressure belt, where surface winds are light and variable. As the most direct rays of the sun move into the northern hemisphere, bringing summer, the doldrums follow. Then, as summer comes to the southern hemisphere, the doldrums move into that area. This movement of the doldrums causes the entire wind system to shift from season to season.

The spinning of the earth causes the prevailing winds to be deflected from their otherwise normal course. The deflection in the northern hemisphere is to the right, so the trade winds coming toward the doldrums from the north are deflected to become northeast winds—the "northeast trades." In the southern hemisphere the deflection is to the left, and thus the winds coming to the doldrums from the south are deflected to become the "southeast trades." Because the trade winds blow over warm, tropical water that is easily evapo-

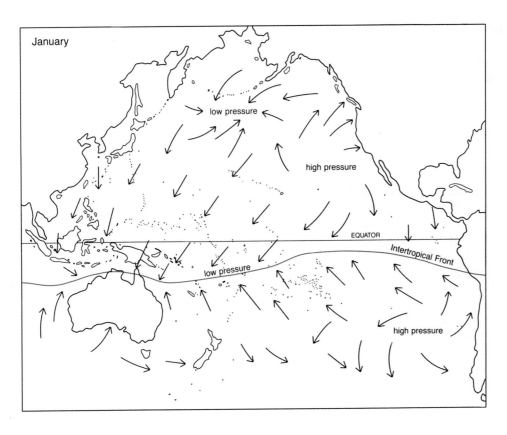

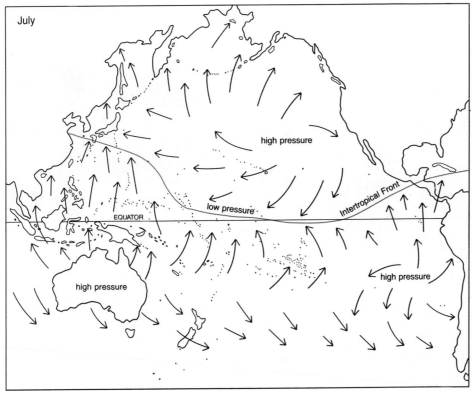

FIGURE 17.5

Prevailing Winds of the Pacific Region. The prevailing-wind pattern changes from January to July with the shift in the world wind system. The location of the intertropical front also changes from season to season.

The Pacific Islands 521

rated, they are moisture-laden. Any barrier to their movement causes uplifting of the air, which in turn causes condensation of the moisture, resulting in rainfall. Consequently, the thousands of islands in the tropical part of the Pacific Ocean usually have a windward rainy side and a drier leeward side. As the change in seasons creates a shift in the winds, the windward and leeward sides of the islands change.

The trade winds are not constant in their makeup. They vary in temperature, density, humidity, and velocity. As the northern and southern trade winds meet at the doldrums, the cooler, denser portion underrides the warmer, lighter part. The resultant cooling of the rising lighter air causes condensation, which results in precipitation. The zone of rainfall is called the **intertropical front.** Frequently, low-pressure cyclonic storms develop along the front and cause heavy precipitation. The most severe of these tropical cyclones are the typhoons (see chapters 8 and 15), and the individual storms travel north and south of the equator for many miles. About 150 tropical cyclones occur each year, but only a few develop into typhoons. The winds in a cyclone spiral around the center and vary in velocity from 15 to 150 miles per hour. The storm is designated a typhoon when the wind speed exceeds 75 miles per hour.

Rainfall in the tropical Pacific islands is highly variable. The maximum amounts occur on the windward slopes of the high mountains. The leeward sides of mountain island and the low-lying atolls receive the least amounts. Because of the seasonal shifting of the intertropical front, the islands in the northern hemisphere get most of their rain from July to October, and the islands in

---

intertropical front The convergence zone between two air masses is called a front, and in the tropical areas the northeast trade winds meet or converge with the southeast trade winds; thus, the zone where they meet is called the "intertropical front." The area also is called the "doldrums" and/or the "intertropical convergence zone."

Vulcanism created these jagged peaks on one of the Society Islands. The tallest peak on Tahiti (Mount Orohena) is near the center of the island and reaches up to 7,339 feet (2237 m). Photograph by Liz McKinley.

Wallace's Line The line or boundary between the Asian mainland and the islands of Southeast Asia that designates the separation between plant and animal species found in the two areas. Named after Alfred Russel Wallace (1823–1913), a British naturalist who first suggested the differences.

the southern hemisphere get their rain from November to April, or during the summertime for each area.

The Pacific Ocean largely controls the temperatures reached on the isolated islands. All islands have maritime conditions with small annual and daily ranges of temperature. The circulation of ocean water modifies the temperatures found along the coasts. In addition, sea water temperature drops about half a degree (F) for each degree of latitude away from the equator. Thus, the temperatures on the islands vary according to the circulation of the water and the distance from the equator. Most tropical Pacific islands have a uniform temperature that varies between 70° and 80° F, with daily ranges greater than the yearly range.

Vegetation

Alfred Russel Wallace, a nineteenth-century, British naturalist, drew a figurative boundary known as **Wallace's Line** between the islands of Southeast Asia to separate Asiatic flora and fauna from that of Australia (see Figure 17.6). Although the line is actually a transition zone, a definite change does occur. Tigers, squirrels, and other mammals found north of the zone do not live south of it, and the marsupials found in Australia do not exist north of the zone. Land birds and numerous types of vegetation also differ greatly on opposite sides of the zone. It is assumed that during the glacial periods the sea level was lowered, creating a land bridge between Asia and Australia and allowing the migration of plants and animals from the Asian mainland to Australia. When the glaciers melted, however, the sea level rose again, isolating the migrants. The isolation allowed different forms of life to evolve.

Zones similar to Wallace's Line have been found in successive waves eastward from Asia-Australia into the Pacific region. From west to east throughout the islands of the Pacific Basin, the types of plants and animals change. Plants and animals on oceanic islands are subjected to varied

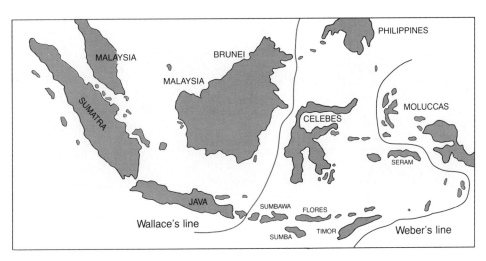

FIGURE 17.6

Wallace's Line. Other zoologists have questioned the location of Wallace's Line, and M. Weber suggested that it should be located farther east (see Weber's line on the map).

environments. The longer the elapsed time since their introduction on an island, the more the local conditions bring about modifications. The general rule for the Pacific Islands is that there are few families of flora and fauna but many species within the families that do exist. Bats are the only mammals that came to the islands without the help of humans, but more than 100,000 species of insects have been recorded. Mosquitoes, burrowing mites, stinging flies, hornets, poisonous centipedes, and scorpions make humans miserable on some islands but are completely absent on others.

The most common vegetation type found on the Pacific islands is the **strandline,** or seashore association of plants (see Figure 17.7). The band of trees, especially palms, shrubs, vines, and herbs found inland from the zone of breaking waves, make up the strandline. Many of these plants bear seeds or fruits that are capable of floating in saltwater without losing the ability to germinate, so they can migrate from island to island. Other seeds are carried by the wind and by birds. The lushness and thickness of the strandline zone is related to the amount of rainfall received on the islands. On the wetter islands or the windward side of other islands, thick jungle conditions occur, while in the drier places only a few palm trees mark the strandline.

strandline Literally means the shoreline, but often used to designate the area where particular types of vegetation grow on Pacific islands.

Vegetation changes with elevation on the high islands. Rainforests cover the valley bottoms and the low coastal plains, but they give way to grassy uplands and eventually to rocky peaks. Some islands have endemic plants that have evolved under special conditions and have not spread from their place of origin. For example, about 80 percent of the 2,500 species of flowering plants on New Caledonia are not found anywhere else on earth. Much of the virgin rainforest on many islands has been cleared for agriculture, and the secondary growth is of jungle. Repeated burning of the forests also has created savannas.

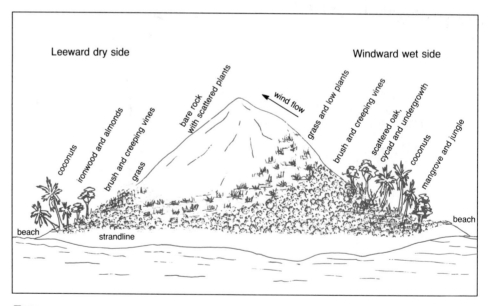

FIGURE 17.7

Typical Pacific Island. The schematic of the profile of a typical Pacific island indicates that vegetation regions differ from one side of islands to the other, depending on which direction the prevailing wind blows.

taro From Tahitian, refers to a tropical plant (or its root) of the arum family, identified by their flowers, fleshy spikes, and hoodlike leaves. The tuberous root is edible and contains much starch.

Human Geography of the Pacific Islands

History

The original people of the Pacific Islands came from the large islands off the mainland of Southeast Asia. As their oceangoing skills developed, they moved eastward throughout the Pacific Basin, using the islands as stepping-stones. Moving eastward, they found the islands to be smaller and more scattered. Consequently, many island people developed in isolation, creating their own languages and religions. The settlers that came in successive waves also varied in physical type and culture.

The first people to enter the Pacific Islands region were small, dark, wavy-haired Australian aborigines. Their migratory wave started about 50,000 years ago and spread from Indonesia and Australia to the Philippines and possibly as far north as Taiwan. A second wave of Negroid people came after the aborigines. These people, larger and taller than the aborigines, moved as far east as New Caledonia, but not northward into the Philippines. A third Negroid group migrated into the western Pacific islands. These people had fuzzy hair and coarse features, were of medium height, and were the most technologically advanced of the early Negroid groups. They built sophisticated canoes, fished, grew **taro,** ate coconuts, raised pigs and chickens, and also may have been cannibals. This third group is sometimes referred to as the "higher Melanesians" because of their more advanced culture.

About 4,000 to 5,000 B.C., the first non-Negroid migrants found their way to the Pacific Islands. They are classified as "early Indonesian" or Malays of the New Stone Age. They built

Inland and at a higher elevation than the strandline, the palm trees are larger and the undergrowth is thicker. Here the path for an island road has been carved out of the tropical rainforest. Photograph by Peter McKinley.

Many Pacific Island people maintain ties with the past through objects they use daily. The use of outrigger canoes is one very old tradition. Photograph by Liz McKinley.

Polynesians A group of South Pacific island people, light brown in color and having similar languages and cultures, including the Hawaiians, the Tahitians, the Samoans, and the Maoris.

---

thatched-roofed wood-frame houses, used stone axes and hammers, and grew millet and yams on forest clearings. About 12 percent of the current population of the Philippines are descendants of these people.

People from the Asian mainland began to migrate eastward around 1,500 B.C. They introduced upland rice and new varieties of yams to the region. Then, by 800 B.C., the terrace-builders and growers of wetland rice came from Indonesia. These people of the Bronze Age, who, had developed copper mining and smelting and bronze forging for making tools and jewelry, were the last of the prehistoric migrants into the western Pacific and continued to move eastward for many years. About one-third of the population of the Philippines are descendants of these Malay-type people who came during the last 2,000 years. Besides the Malays, elements of Hindu-Indians, Arabs, Persians, and Chinese, as well as Europeans and Americans, also found their way to the western Pacific.

The first wave of **Polynesians** moved through the Carolines and the Marshalls about the time of the birth of Christ. From the Marshalls, one line migrated directly northeast to Hawaii, the main stream followed the Gilberts to the Society Islands, and another stream turned south toward Samoa.

The major dispersals occurred between A.D. 1,000 and 1,200 but may have taken as long as 500 years. Polynesians may have lived in the western Pacific for as long as 1,000 years before the major dispersals occurred.

The people of the various migratory waves of Micronesians and Polynesians in the western, central, and southern Pacific undoubtedly discovered almost every bit of land in the tropical Pacific long before European explorers arrived. Movement among the islands for trade and warfare was common, and the occupants of dugout canoes knew the general location and geography of large numbers of islands. The record of their movements in small crafts on the open oceans is quite impressive. (See Figure 17.8.)

Population

Demographic and land area statistics for the Pacific Islands (Table 17.1) are mostly estimates. Accurate data exist for Hawaii and other American territories, but for most of the other islands regular censuses are not taken. The land area is suspect because it is very difficult to determine the aggregate area of hundreds of small patches of land in the middle of a huge ocean. Air photographs are used to calculate statistics, but the scale is so small that the bits of land are barely visible. Some land areas exist only at low tide, so photographs taken at that time would show more land than there actually is. The figures in Table 17.1 then, should be considered estimates.

The 1985 estimated population of the Pacific Islands was slightly over 6 million, and the total land area of the region is 226,377 square miles (586,316 km^2). The average population density, then, is 27 people per square mile (10 people per km^2). But population distributions and densities vary considerably throughout the islands. The density varies from a low of less than 2 people per

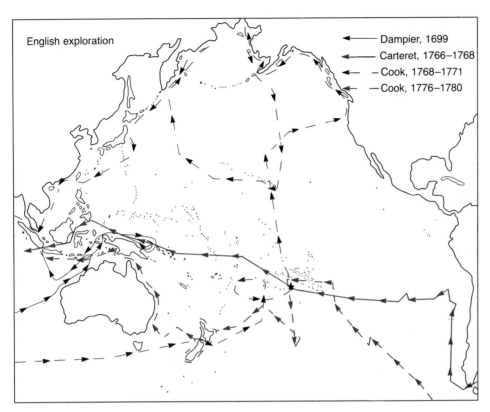

FIGURE 17.8

Exploration of the Pacific Region. The lines show the routes the early explorers of the Pacific Ocean region followed.

square mile (0.8 people per km^2) on the Aleutian Islands, to a high of 981 per square mile (2,540 km^2) on Nauru. About 70 percent of the people of the region live in the two territories of Papua New Guinea and Hawaii.

Numerous factors are responsible for the population discrepancies among the islands: how much land is arable, how fertile the soil is, the climatic conditions, the economic conditions, the type of government, local health conditions, and so on. The most important factor is the amount of arable land, and only small portions of most Pacific volcanic islands are arable. The lack of suitable agricultural land is overcome somewhat by access to the sea. Land-derived foods are supplemented with marine foods. Soil development is often nil, but there seems to be no difference between the fertility of soils derived from a lava base or those that have a limestone base. The major climatic condition that affects agriculture is the extreme variability in rainfall.

Like the population densities, the rates of population increase on the Pacific islands also vary from island to island. For the most part, the rates are very high, compared with the more developed world and the world as a whole. The overall rate of population increase for the region is 3.27 percent a year, compared with the world rate of 1.7 percent. The rate varies from 3.7 percent on the Solomon Islands to 2.0 percent on New Caledonia. These rapid growth rates are very critical in this region, because the population-support capacities of the islands are quite limited. The problems faced by the islanders today—for example, concerning health and medical services, education, communications, transportation, and limited sources of food and income—will be magnified with any further population increases.

TABLE 17.1

Demographic and Area Statistics for the Pacific Islands

ISLANDS	NO. OF ISLANDS	AREA (SQ. MI.)	POPULATION
Aleutian	150	6,821.0	8,521
American Samoa	7	76.1	33,920
Cook	15	90.3	17,227
Fiji	400	7,095.0	588,068
French Polynesia	110	1,622.0	166,753
Guam	1	209.0	111,000
Hawaii	132	6,450.0	1,038,700
Kiribati	33	277.0	56,213
Nauru	1	8.2	8,042
New Caledonia	1[a]	7,376.0	145,368
Niue	1	100.0	2,900
Norfolk	1	13.3	2,175
Northern Marianas	16	182.0	19,635
Papua New Guinea	1[a]	178,704.0	3,328,700
Pitcairn	4	1.8	65
Solomon	21	10,639.0	244,000
Tokelau	3	3.9	1,595
Tonga	172	289.0	96,448
Tuvalu	9	10.0	8,229
U.S. Trust Territory	2,125	502.0	116,974
Vanuatu	80	4,706.0	127,800
Wake	3	3.0	305
Wallis & Futuna	25	106.0	12,391
Western Samoa	9	1,093.0	156,349
Total	3,320	226,377.6	6,291,378

[a]Consists of one main island plus numerous islets.

SOURCES: Couper, *Times Atlas,* 1983; *The Far East* (London: Europa, 1984); *Demographic Yearbook, 1987.* (New York: United Nations Publishing, 1988).

Ethnicity

To varying degrees, modern Melanesians, Polynesians, and Micronesians are the outcome of the mixing of the early Australoid and Mongoloid stocks. The Melanesians, who inhabit the island chains from the Solomon Islands eastward to Fiji, are basically Australoid. The Fijians are an intermediate group, but the Polynesians tend to have the Mongoloid characteristics and the Micronesians have the strongest Mongoloid features of the island people. However, the pattern is more complex than such a simple description of the ethnic background of the island people can reveal.

Languages

More than 1,000 different languages are spoken in the Pacific Islands, but about 70 percent of these are found on Papua New Guinea and the Solomon Islands. The languages belong to two groups. Micronesia and Polynesia share linguistic kinship with the Indonesian region, and the linguistic family is the Malay-Polynesian family. Differences among the languages within this family are comparable to differences among those of the Romanic languages. Communications sometimes are possible, but, they are always difficult. The diversity among the Polynesian groups is not as great as the

copra The dried meat of coconuts used for making coconut oil.

---

diversity in Micronesian languages, where sharp differences occur. The Micronesian languages are found in Papua New Guinea and in scattered pockets of Indonesia. The Polynesian languages are spoken in coastal Papua New Guinea, most of the islands of Melanesia, and all of Polynesia and Micronesia, as well as in parts of Indonesia, the Philippines, Southeast Asia, and even Madagascar.

Religion

Most of the people of the Pacific Islands are Christians. On many islands, the percentage of Christians is as high as 100 percent. Many Christian churches are represented, but the Anglican church from England and the Roman Catholic church are the most common. The people of the islands of the U.S. Trust Territory, Guam, Kiribati, and Wallis and Futuna are nearly all Roman Catholic, while those on Cook, Tuvalu, Western Samoa, and Nauru are mostly members of the Anglican church. The people on the island of Atafu in the Tokelau Islands are all Anglican, while their neighbors on Nukunono are all Roman Catholic. Methodists dominate on Tonga, and all the people on Pitcairn are Seventh-Day Adventists. Many of the islands also have a Mormon minority. The one exception to the Christian dominance in the Pacific Islands is on Papua New Guinea, where traditional beliefs, including magic and sorcery, still prevail. In spite of the adoption of Christianity, the indigenous peoples of Papua New Guinea are mainly *pantheistic*—that is, they equate God with the forces and laws of nature.

■ Economic Geography of the Pacific Islands

Agriculture

On many of the islands of the Pacific, agriculture is entirely subsistence farming. The thin soils and lack of rainfall on the low islands restrict the growing of most crops other than coconut palms. Thus, coconuts, coconut oil, and copra are the major exports of most of the islands. **Copra** is the dried meat of coconuts, and it produces an oil. On the high islands, where deep volcanic soils and adequate moisture allow crop production, all the tropical food items are produced. These include sugar cane, bananas, cassava, yams, sweet potatoes, passion fruit, pineapples, taro, and other vegetables and fruits. Agriculture varies from strictly subsistence farming on Pitcairn, the Solomons, and Vanuatu to highly organized plantation-growing of sugar cane and pineapples on Hawaii.

The people on most of the islands raise some livestock, the most common being chickens, pigs, and goats. Some cattle graze on the higher grassy plateaus, but generally beef is not a common item in the diet. Some of the islands produce enough livestock commodities for export. Guam, for example, exports eggs and pork, as well as fruits and vegetables (see Figure 17.9). Most of the food commodity trade is among the islands themselves. Bananas and taro from one island are traded for eggs and pork from another island. All the islands produce coconuts, however, so they are all produced for shipment to the First World countries. Trade from the islands is difficult because they are so scattered, with such long distances between them. Collecting a large enough amount of commodities for export is expensive and time-consuming.

Mineral Production and Industry

Few commercially valuable deposits of any minerals have been found on the islands of the Pacific, and only seven of the island groups have any mineral production at all. One exception is New Caledonia, which has the world's largest known deposit of nickel ore. The entire economy of that island is closely tied to the mining and processing of nickel. New Caledonia also has cobalt and chromium. Vanuatu produces some manganese; a little copper, silver, and gold are mined on Papua New Guinea; Fiji too has some silver and gold; and Nauru produces some phosphate. Kiribati closed its phosphate mine in 1979, and other than pumice

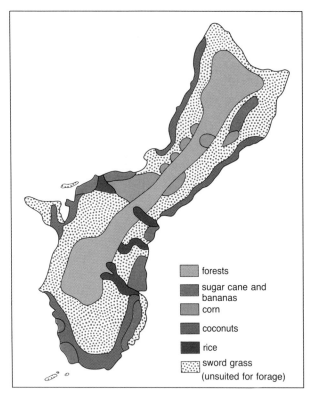

FIGURE 17.9

Agriculture on Guam. Agriculture on the Pacific Islands is limited to small patches of land, such as these shown for Guam.

Pacific Islands are too small or too barren to support facilities for tourists.

Transportation and Communications

Railroads are almost entirely lacking on the Pacific Islands, but most islands have decent roads. Road mileage is short, however, because the islands are small. Western Samoa, for example, has 229 miles (369 km) of main roads and about three times that amount of unsurfaced plantation roads. But Wallis and Futuna Islands have only one route, which circles each main island. Road construction on the high islands is very expensive because of the steep slopes and rugged terrain. Most of the roads lead from the plantations to the ports or circle the islands.

Every island has a port, but the facilities vary greatly. All goods for export and import are transported by ship. On some of the islands, the loading and unloading of freight from ships are ongoing activities, while on other islands a cargo ship may stop only once a month. Many people still travel by boat, and in places where there are no

from American Samoa, no other minerals are produced on the Pacific Islands.

Like other regions where iron ore and coal are lacking, the Pacific Islands lack industry. None of the islands has any heavy industry, but there are some production plants for making cement, cigarettes, and beer and for canning fish. Most of the larger islands have plants for producing electricity, and plants for producing palm oil are common on almost every island. One industry that is fairly common throughout the region is tourism. Hawaii, for example, gets more income from tourists than from its well-developed agriculture. South Seas tropical islands, with their sandy beaches and palm trees, have been tourist attractions for many years. The words "island paradise" have been applied to many of these islands. But many of the

Many of the small Pacific islands have room for only short runways, so large commercial jets must land on the larger islands. Tourists and mail reach the small islands via small aircraft. Photograph by Liz McKinley.

This inter-island ship is used mainly to transport tourists, but it also hauls some cargo between the islands of the South Pacific. Photograph by Liz McKinley.

airstrips, boats are the only means of access. All the larger islands have commercial airports, which are important for the tourist trade but not for local travel.

Communications among the islands are restricted because of the language differences, but communications facilities for use on particular islands are fairly good, depending on the level of development and income (see Table 17.2). Radios are common nearly everywhere, but television reception is spotty. The people of American Samoa and Western Samoa receive television programs from the United States. The station in Pago Pago is linked with U.S. television networks. Many islands, however, receive no television broadcasts at all.

Communications with the world outside the Pacific Islands has been poor, but the increased use of satellite transmissions has improved them. The shortwave radio is the most important device for local island-to-island communication, but not many people own such radios. The local people rarely need to communicate with people outside their island world.

Political Geography of the Pacific Islands

The political geography of the Pacific Islands is as complex and fractured as the physical geography. A political organization of the islands should include the dependent and independent states both. The dependent states governed wholly or partially by colonial administrations are the Pitcairn Islands (United Kingdom); Wallis and Futuna (France); and Hawaii, a state of the United States. Other dependent states are internally self-governing and include Cook Island, Niue, and Tokelau (New Zealand); Norfolk Island (Australia); French Polynesia and New Caledonia (France); and the Caroline, Marshall, and Northern Mariana islands, Guam, and American Samoa (United States). Former dependencies that have achieved full political independence are shown in Table 17.3.

When they gained political independence, Fiji, Papua New Guinea, the Solomon Islands, and Vanuatu joined the United Nations. The other countries rely on their former rulers to solve their international problems.

TABLE 17.2

Newspapers, Radio, and TV Stations in the Pacific Islands, 1986

ISLAND	DAILY NEWSPAPERS	RADIO STATIONS	TV STATIONS
American Samoa	3	1	1
Cook	3	1	0
Fiji	14	2	0
French Polynesia	3	1	1
Guam	4	4	3
Hawaii	3	38	15
Kiribati	3	1	0
Nauru	1	1	0
New Caledonia	5	2	1
Niue	1	1	0
Norfolk	2	1	0
Northern Marianas	3	4	1
Papua New Guinea	10	1	1
Pitcairn	1	0	0
Solomon	2	1	0
Tokelau	0	0	0
Tonga	3	1	0
Tuvalu	1	1	0
U.S. Trust Territory	6	6	4
Vanuatu	2	1	0
Wake	0	0	0
Wallis & Futuna	0	1	0
Western Samoa	4	1	1

SOURCE: *The Far East and Australasia, 1986* (London: Europa, 1987).

The most important regional organization is the **South Pacific Forum (SPF),** made up of the independent and self-governing islands of the region. The **South Pacific Commission (SPC)** includes the nonindependent countries and the former colonial powers. Representatives of the countries comprising each organization meet once a year to discuss regional problems. Usually the meetings center on trade agreements, but initiatives to create a nuclear–free Pacific zone and other political topics are sometimes discussed.

One important political aspect of some of the Pacific islands is their strategic location. For example, Guam remains an important naval and air base for the United States. Usually about 10,000 American servicemen and women are stationed on the island. Pago Pago in the American Samoan islands is a major mid-Pacific stopover for large passenger aircraft. And Hawaii's location has been strategically significant for the United States for many years.

South Pacific Forum (SPF) A nonmilitary regional organization of the independent, self-governing islands of the South Pacific, designed to promote economic activity and trade among the islands.

South Pacific Commission (SPC) A nonmilitary regional organization of the nonindependent island countries of the South Pacific, designed to confront all regional problems on a group basis.

TABLE 17.3

Former Dependencies of the Pacific Islands with Political Independence

COUNTRY	FORMER RULER	YEAR OF INDEPENDENCE
Fiji	United Kingdom	1970
Kiribati	United Kingdom	1979
Nauru	Australia	1968
Papua New Guinea	Australia	1975
Solomon Islands	United Kingdom	1978
Tonga	United Kingdom	1970
Tuvalu	United Kingdom	1978
Vanuatu	United Kingdom & France	1980
Western Samoa	New Zealand	1962

Key Words

andesite line
copra
coral islands
coral reef
intertropical front
Polynesians
South Pacific Commission (SPC)
South Pacific Forum (SPF)
strandline
taro
Wallace's Line

References and Readings

Atlas of Hawaii. Honolulu: Hawaii University Press, 1983.

Attenborough, David. *People or Paradise.* New York: Harper, 1961.

Beazley, Mitchell (ed.). *The Rand McNally Atlas of the Oceans.* New York: Rand McNally, 1977.

Benjamin, Roger, and Kudrle, Robert T. (eds.). *The Industrial Future of the Pacific Basin.* Boulder, Colo.: Westview, 1984.

Brookfield, Harold C., and Hart, Doreen. *Melanesia: A Geographical Interpretation of an Island World.* New York: Barnes & Noble, 1971.

Brown, DeSoto. *Hawaii Recalls: Nostalgic Images of the Hawaiian Islands, 1910–1950.* Boston: Routledge & Kegan Paul, 1986.

Brown, Paula. *Highland Peoples of New Guinea.* New York: Cambridge University Press, 1978.

Chapman, Murray, and Prothero, R. Mansell (eds.). *Circulation in Population Movement: Substance and Concept from the Melanesian Case.* Boston: Routledge & Kegan Paul, 1985.

Couper, Alastair D. (ed.). *The Times Atlas of the Oceans.* New York: Van Nostrand Rinehold, 1983.

Craig, Robert D., and King, Frank P. (ed.). *Historical Dictionary of Oceania.* Westport, Conn.: Greenwood, 1981.

Cumberland, Kenneth B. *Southwest Pacific.* New York: McGraw-Hill, 1956.

Delaplane, Stanton. *Pacific Pathways.* New York: McGraw-Hill, 1963.

The Far East and Australia. London: Europa, 1984.

Freeman, Otis W. (ed.). *Geography of the Pacific.* New York: Wiley, 1951.

Friis, Herman R. *The Pacific Basin: A History of Its Geographical Exploration.* New York: American Geographical Society, 1967.

Garbell, Maurice A. *Tropical and Equatorial Meteorology.* New York: Pitman, 1947.

Gopalakhrishnan, Chennat (ed.). *The Emerging Marine Economy of the Pacific.* Boston: Butterworth, 1984.

Grossman, Lawrence S. *Peasants, Subsistence Ecology, and Development in the Highlands of Papua New Guinea.* Princeton: Princeton University Press, 1984.

Hawthorne, David. *Islands of the Pacific.* New York: Putnam, 1943.

Howells, William W. *The Pacific Islanders.* New York: Scribners, 1973.

King, David, and Ranck, Stephen (eds.). *Atlas of Papua New Guinea: A Nation in Transition*. Bathurst, Aust.: Brown, 1981.

Levinson, M., Ward, R., and Webb, J. *The Settlement of Polynesia*. Minneapolis: University of Minnesota Press, 1977.

Miles, J. A. R. (ed.). *Public Health Progress in the Pacific*. Hingham, Mass.: Reidel, 1984.

Morgan, Joseph R. (ed.). *Hawaii*. Boulder, Colo.: Westview, 1983.

Nairn, A. E. M. *The Ocean Basins and Margins*. New York: Plenum, 1973.

Oliver, Douglas L. *The Pacific Islands*. Cambridge, Mass.: Harvard University Press, 1951.

Osborn, Fairfield. *The Pacific World*. New York: Norton, 1944.

Pataki-Schweizer, K. J. *A New Guinea Landscape: Community, Space, and Time in the Eastern Highlands*. Seattle: University of Washington Press, 1980.

Sahlins, Marshall D. *Islands of History*. Chicago: University of Chicago Press, 1985.

Shadbolt, Maurice. *Isles of the South Pacific*. Washington, D.C.: National Geographic Society, 1968.

Spate, O. H. K. *The Pacific Since Magellan: Monopolis and Freebooters*. Minneapolis: University of Minnesota Press, 1983.

———. *The Spanish Lake*. Minneapolis: University of Minnesota Press, 1979.

Vayda, Andrew P. (ed.). *Peoples and Cultures of the Pacific*. New York: National History Press, 1968.

Ward, R. Gerard (ed.). *Man in the Pacific Islands: Essays on Geographical Change in the Pacific*. New York: Oxford University Press, 1972.

Wiens, Herold J. *Pacific Island Bastions of the United States*. Princeton: Van Nostrand, 1962.

Wood, Gordon. *The Pacific Basin: A Human and Economic Geography*. Melbourne: Oxford University Press, 1950.

Wurm, Stephen A., and Hittori, S. (eds.). *Language Atlas of the Pacific Area*. Canberra: Australian Academy of the Humanities, 1982.

Yawata, I., and Sinoto, Y. H. *Prehistoric Culture in Oceania*. Honolulu: Bishop Museum Press, 1968.

Chapter 18

THE POLAR REGIONS

Introduction

The polar regions of the world form a distinct environment, one that is largely uninhabited by humans. Settlements do exist in the polar regions, but for the most part they are temporary research stations and petroleum-related bases supported by food and necessities from the temperate world. The permanent settlements include a few native fishing villages and the Soviet city of Murmansk. Thus, the polar areas of the earth are essentially devoid of human occupancy, but in our attempt to understand the entire world we cannot neglect any part, whether it is occupied by humans or not.

Boundaries of the polar regions are not distinct lines, because in some places polar conditions go beyond the normal boundaries. For the purposes here, however, the polar regions will be identified as most of the land and sea areas within the Arctic Circle and the Antarctic Circle (see Figure 18.1). These special parallels are located at 66.5 degrees north latitude, and 66.5 degrees south latitude. The regions include about 3 million square miles (7.8 million km^2) of land in the northern hemisphere, about one-fourth of which is permanently covered with ice. The Arctic Ocean is just over 5 million square miles (12.9 million km^2) in area, and the north polar seas cover another 5.7 million square miles (14.8 million km²). The Antarctic continent contains about 5.4 million square miles (14 million km²) of territory, but icy conditions extend for many miles beyond the continent.

In the north, the Arctic Ocean is centered around the pole itself, but it is covered by ice. In the south, the Antarctic continent, centered on the South Pole, is covered by a huge **ice sheet**. Extremely cold climates occur at both places. In the northern hemisphere, lands adjacent to the Arctic Ocean belong to Alaska, Canada, Greenland, Norway, and the Soviet Union. In the Antarctic region, various territorial claims have been made, but no official world agreement has sanctioned the claims. The countries involved include Argentina, Australia, Chile, France, New Zealand, Norway, and the United Kingdom. One large pie-shaped wedge remains unclaimed. The United States maintains a weather and research station at the South Pole.

Certain polar characteristics exist that set these regions apart from other areas of the earth. The distinct location at high latitudes is the most obvious, and these locations produce the cold conditions. The angle of the summer sun is always low, so its rays must penetrate large amounts of atmosphere before striking the earth. This movement through the atmosphere causes absorption and diffusion of the rays, which reduces their effective heating. During the winter months the sun disappears entirely, so very little heating occurs. The location, thus, produces short, cold summers and long, extremely cold winters.

Because both polar regions are out of the paths of the mid-latitude cyclonic storms, they receive

---

ice sheet A thick layer of ice covering a large area for a long time, such as during the Ice Age. Sometimes called "ice caps" or "continental glaciers" (as opposed to mountain glaciers), ice sheets are found today only on Greenland and on the Antarctic continent.

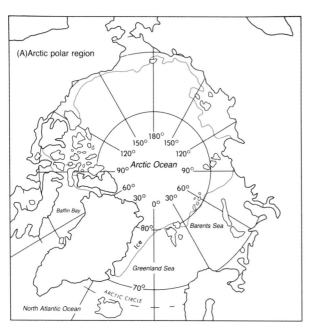 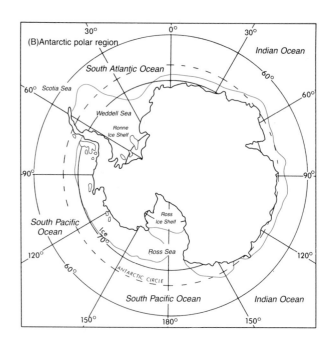

FIGURE 18.1

Polar Projections of the Arctic and the Antarctic. The two views are of the top (A) and bottom (B) of the world, with the North Pole and the South Pole in the centers of the maps. All directions from the centers outward to the edges are south from the North Pole and north from the South Pole.

windchill Both low temperature and wind cause heat loss from body surfaces, and the greater the velocity of the wind, the lower the temperature feels. The decrease in temperature continues up to wind speeds of about 45 mph; after that, there is no additional chilling effect.

mosses Very small green plants that grow in velvety clusters on rocks, trees, and moist ground.

lichens A large group of mosslike plants consisting of algae and fungi growing close together in patches on rocks and tree trunks.

tundra The vast, nearly level treeless plains of the Arctic regions.

fast ice Ice that develops in polar oceans by growing outward from the land as the water freezes, creating an ice shelf off the land.

pack ice Ice that has been broken into chunks and refrozen to create a jumbled mass over polar waters, especially in the Arctic region. The ice pack also drifts with the oceanic circulation, creating further breakage and refreezing.

polar ice The stationary ice in the Arctic Ocean around which the pack ice rotates. Polar ice is about 10 feet (3 m) thick and is located directly over the North Pole.

---

low amounts of precipitation. The cold conditions produce very little evaporation, however, and what moisture does come tends to remain. Usually the moisture is frozen and remains frozen. The polar ice contains moisture that fell as snow thousands of years ago. Precipitation amounts are difficult to determine, though, because it is difficult to differentiate between newly falling snow and snow that is blowing. Strong winds are also common to both polar regions. High pressure builds up over the poles, causing surface winds to blow constantly outward from the poles. These winds create blizzards without any snow actually falling from the atmosphere. The winds also create a high **windchill** that makes temperatures plunge well below what they would be with calm air.

The polar regions are similar also in the absence of most types of vegetation, especially trees. Microscopic soil fungi and algae, as well as **mosses** and **lichens,** are the most common living plants. More than 500 species of lichens have been found in the Antarctic, and about 70 species of mosses, but none of these grows within a 500-mile radius of the pole itself. Vegetation in the north polar region includes the mosses and lichens but also the grassland **tundra** and heath tundra. The heath tundra vegetation is mostly dwarf shrubs and birch trees, creeping willows, and some berry bushes. Agriculture occurs only on the fringes of the northern hemisphere's polar region, and there is none in the southern hemisphere.

■ Physical Geography of the Polar Regions
Landforms

The earth is shaped as if some huge thumb pushed in a dent at the north pole that popped land out at the south pole. The result is that the north polar region is a large basin and that the south polar region is the elevated Antarctic continent. The basin at the north is filled with the waters of the Arctic Ocean, and the land south of the basin (from 55° N to 80° N) slopes toward the pole. By contrast, the Antarctic continent is the highest of all the continents, averaging 6,000 feet in elevation. Most of the Arctic Ocean is frozen year-round, and the Antarctic ice sheet covers nearly the entire continent. One moving part of the Antarctic ice sheet forms the world's largest glacier, the Lambert. Ice can be considered a type of landform in the polar areas.

The north polar seas have three types of sea ice. The **fast ice** grows outward from the shoreline and in the winter joins the **pack ice,** the drifting layer of ice that is constantly on the move. It tends to circle the North Pole, moving from east to west. The **polar ice,** which occupies the central part of the Arctic Ocean, is stationary and about 10 feet thick. It does not freeze any deeper into the water, because its own thickness insulates the water from the cold air temperatures. However, rafting of broken pieces one upon another causes thick ice ridges that are common in both the pack ice and the polar ice and make travel over the surface very difficult.

Along the margins of the Arctic Ocean, the landforms are extensions of the physiographic provinces farther south—the mountains of Alaska, the Canadian Shield, most of Greenland, the mountains of Scandinavia, and the mountains and plains of the Soviet Union. The Arctic portion of Alaska can be divided into two main regions: the North (or Arctic) Slope and the Brooks mountain range. The North Slope is a great syncline that slopes down northward from the Brooks Range to Point Barrow. The underlying rocks are 20,000 feet (6,096 m) thick and made up of sandstones, shales, coal, and oil shale. This is the region overlying the huge Alaskan oil reserves. About half the structure is under the waters of the Arctic Ocean. The North Slope is a barren tundra, with permanently frozen ground (permafrost) and patterned land from the work of **solifluction**. The Brooks Range is a rugged mountain region that rises up to heights of 9,000 feet (2,743 m). The mountains are jumbled masses of barren rock covered extensively by snow and glaciers. The Brooks Range drops abruptly to the north to a 2,500-foot-high (762 m) plateau that is about 80 miles (129 km) wide. The plateau is eroded into rolling foothills, which in turn drop to the North Slope. (See Figure 18.2).

solifluction Any mass movement of earth materials (including soil and rock) that is caused by alternate freezing and thawing.

esker A narrow, sometimes winding ridge of sand and gravel deposited by a stream flowing under glacial ice, then exposed after the glacier melts.

drumlin An oval-shaped hill of sand and gravel deposited under glacial ice and then exposed when the glacier melts.

The Canadian Arctic is the largest bloc of territory belonging to one country located north of the tree line. It consists of two main areas: the mainland portion and the island portion. The surface configuration varies from low alluvial flats to rugged mountains 8,000 feet (2,438 m) high. Across the Canadian Arctic as a whole, however, the landscape is dominated by relatively flat ground covered with tundra vegetation. The mountains occur in the eastern part of the region, on Baffin and Ellesmere islands. Most of the region was heavily glaciated, so **eskers** and **drumlins** are common. (See Figure 18.2.)

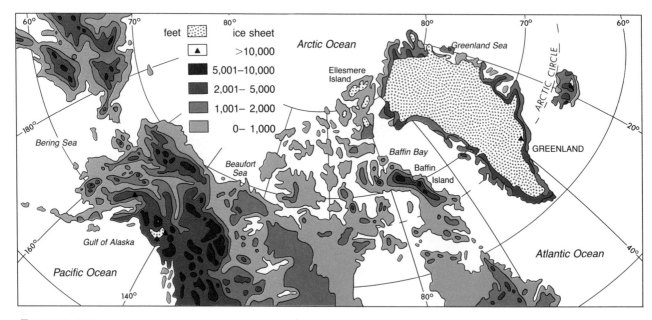

FIGURE 18.2

Land Elevations and Water Bodies of the Polar Region of North America.

Greenland is the world's largest island and has some other unique features. Most of the island (85 percent) is covered with an immense ice sheet. The ice does not extend out onto the sea, however, as it does in the Antarctic. Most of the coastal areas have bare rocks showing, but the land is deeply dissected by long fjords. Greenland is somewhat saucer-shaped. The mass of ice fills a central depression that is surrounded by a coastal mountain fringe. Near the center of the island, the ice is about 10,000 feet (3,048 m) thick, but it decreases in thickness toward the coasts. The ice sheet holds a tremendous amount of water.

Soviet Arctic territory is nearly as large as that of Canada. It consists of a tundra strip that extends along the entire northern coast and is crossed by many rivers flowing from south to north. A tip of the Ural Mountains extends into the Soviet Arctic, and the Putorana and Byrranga ranges interrupt the otherwise low-lying tundra. Soviet territory also consists of numerous islands located offshore to the north of the tundra.

Norway's Arctic islands go by the general name of Svalbard, a medieval name dating back to the year 1194, when the islands were discovered (see Figure 18.3). The islands consist of two separate groups with two main islands and numerous smaller islands. About half the area of Svalbard is ice-covered, and the islands are mountainous and have deeply indented coastlines caused by fjords. Svalbard has been important to Norway for its mineral industry, and especially coal. Coal seams several yards thick have been found, and mining has been carried on since 1904.

The Antarctic continent has been isolated at the bottom of the earth for the last 25 million years. It covers about 10 percent of the world's land surface, but the surface area changes with the seasonal expansion and retreat of the ice shelves. The continent once was adjacent to the landmasses of South America, Africa, and Australia, and exposed coal seams match those found on the other continents. The coal indicates that Antarctica once shared the warm and equable climates found

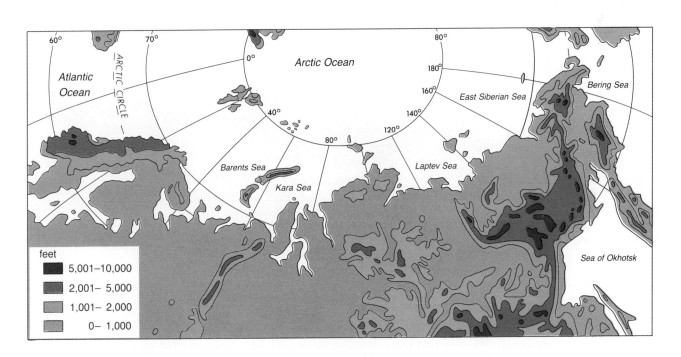

FIGURE 18.3
Land Elevations and Water Bodies of the Polar Region of Europe and Asia.

on the other continents. Today, however, the coastal landscape is spiny and mountainous, with no coastal plains and steep ice-cliff shores. Most of the inland area is a vast white wasteland interrupted occasionally by mountain peaks.

The land in the Antarctic is dominated by the huge ice sheet that covers about 98 percent of the continent. The ice sheet was formed from the compacted accumulation of about 100,000 years of snow. The average thickness of the ice is 5,250 feet (1,600 m), and it contains 70 percent of the world's fresh-water reserves. The sheer weight of the ice has warped the continent downward so that much of the land of Antarctica lies below sea level. If the ice melted, the world's oceans would rise by as much as 295 feet (90 m) and the water would inundate the homes of half the world's people. With the removal of the weight of the ice, Antarctica would rise in a process called **isostasy,** so that

isostasy Refers to the universal equilibrium in the earth's crust due to the gravitational yielding of the magma beneath the earth's surface.

pole of inaccessibility A phrase used to describe the south polar region because for many years it was inaccessible to the early explorers. Not only are the temperatures very low at the South Pole, but the elevation is very high (13,000 feet, or 3,962 m), making the lack of oxygen a major problem for explorers.

much of the land would be above sea level again. (See Figure 18.4.)

The Antarctic ice sheet forms a high dome over the center of the continent that is 13,000 feet (3,962 m) in elevation. The top of the dome is called the **pole of inaccessibility.** The ice is constantly moving outward from the center of the

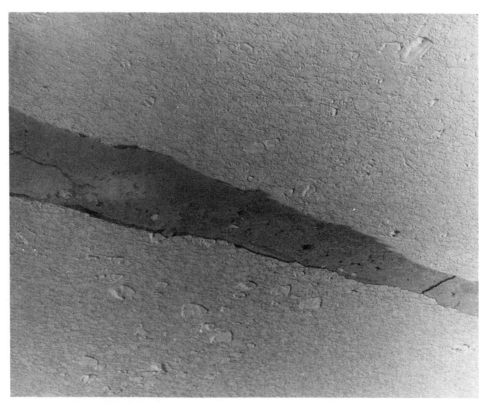

The very cold polar air causes the ocean waters to freeze. The ice pack shown here is located in the Greenland Sea very near the Arctic Ocean. Notice how the lead (crack) has refrozen. An official U.S. Navy photograph, courtesy of the Naval Polar Oceanography Center.

nunataks Individual polar region mountain peaks that protrude above glacial ice and remain dry from the ice and snow. They look like islands in a sea of ice.

dome, seeking lower levels toward the coastlines, and as it reaches the coasts it pushes right out into the open sea, still maintaining glacial form. The front face of the ice forms a barrier cliff 150 feet (46 m) high and goes down to 850 feet (259 m) below the surface of the water. In the Ross Sea region, the barrier extends for 500 miles (805 km) off the coast, and the front is 400 miles (644 km) wide. Eventually, parts of the ice break off, becoming flat-topped icebergs. These huge chunks of ice float out into the oceans for many miles before finally melting. The largest iceberg ever recorded measured 208 miles (335 km) by 60 miles (97 km), but most are less than 25 miles (40 km) in diameter.

Exposed rock of the Antarctic region is found mostly where mountain peaks jut up through the ice sheet. Many isolated peaks, called **nunataks**,

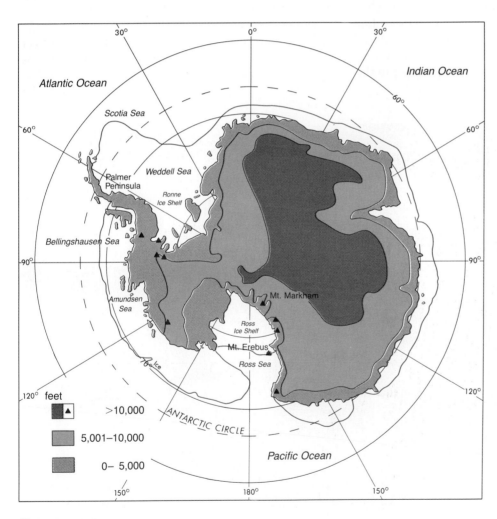

FIGURE 18.4

Land and Ice Elevations of the Antarctic. Most of the landforms of Antarctica are covered with glacial ice, but some mountains extend upward through the ice. Most elevations shown on this map are for the heights of the ice sheet.

The Polar Regions 543

loom above the ice cover along western Antarctica, but some dry valleys are found in southern Victoria Land. They are called "dry" because they lack snow and ice. A region adjacent to the Ross Sea from Victoria Land to the Queen Maud Mountains contains the most mountain tops not covered by the ice sheet. Numerous peaks of this region exceed 10,000 feet (3,048 m) in elevation, and Mount Markham is the tallest at 14,272 feet (4,350 m). Other areas not covered by the ice sheet are parts of the Palmer Peninsula that point toward the Shetland Islands and the tip of South America. This peninsula is the only part of the Antarctic continent that extends north of the Antarctic Circle. Other coastal mountains also lie exposed, but no rock landforms can be seen for hundreds of miles from the center of the pole of inaccessibility. The Antarctic's tallest peak is Vinson Massif in the Ellsworth Mountains, located near the Ronne Ice Shelf. The top of the mountain is 16,800 feet (5,121 m) above sea level. The 12,280-foot-high (3,743 m) Mount Erebus, in Victoria Land near McMurdo Sound, is the continent's only live volcano.

Climate

The polar air is not only dry, cold, and stable, it is also relatively free from impurities. No desert dust, forest-fire smoke, or industrial wastes pollute the air. Visibility extends for many miles, and it is difficult for people accustomed to the haze of temperate-latitude air to judge distances in the polar regions. Stable air means lack of turbidity that causes storms, not lack of wind. Surface winds, unrelated to storms, blow stronger in the

Nunataks in the background stick out of the top of the Antarctic glacier located near McMurdo Sound. The ice in the sound has large cracks, called leads. The huge pieces that break off usually drift out into the Ross Sea during the Austral summer. An official U.S. Navy photograph, courtesy of the Naval Polar Oceanography Center.

> **zooplankton** Zooplankton, consisting principally of algae and diatoms, is to the life of the sea what pastures are to the life of the land. Because there are smaller amounts of denitrifying bacteria in colder water, it is more abundant in cold water than in warm water.

polar region than anywhere else on earth. Gusts of more than 200 miles an hour are common. The stable air is sluggish, however, which means that air masses move much slower than the warmer air masses of the lower latitudes.

The polar winds are deflected by the earth's turning movement so that they blow from the east. They are called "polar easterlies." They meet the prevailing westerly winds in the areas adjacent to the polar regions. The cold, polar air meets the warm, temperate air in frontal conditions that create storms. These storms usually stay within the zone of the prevailing westerly winds and greatly affect the weather of mid-latitude locations, but sometimes they sweep over the polar regions, causing blizzards. The generalized wind patterns for both polar regions, then, are (1) calm air over the poles themselves, (2) constantly roaring surface winds outward from the calm centers, and (3) windy storms along the fringes of the polar regions.

The temperature of the polar air is colder than those who have not experienced it can imagine. In such very cold conditions, steel shatters and oil turns to the consistency of rubber. The coldest temperature ever recorded on earth was at a Soviet weather station in the Antarctic. The temperature for August 24, 1960 was $-126.9°$ F ($-88.3°$ C). The Vostok station is 2,300 feet (901 m) higher than the South Pole, so it usually records lower temperatures. At the U.S. South Pole station, however, the temperature often drops to 100° F below zero (-55.6° C).

The north polar region is much milder than the south polar region because the Arctic Ocean is a moderating influence on the temperature. Thus, the coldest place in the northern hemisphere is 1,500 miles (2,414 km) south of the North Pole, just north of the Arctic Circle, at the Siberian town of Verkhoyansk. Winter temperatures at Verkhoyansk have reached 90°F below zero, but the summer temperatures average above 60°F (15.5°C). The brief, warm summer seasons along the edges of the Arctic are never experienced in the Antarctic. Temperatures near the North Pole itself get down to 60° F below, but rarely colder. Summer temperatures at the pole are never above freezing.

The annual average precipitation in the polar regions is less than 5 inches, which makes the polar regions as dry as the Sahara Desert. The great blizzards are mainly loose snow being blown by the polar winds from one place to another. The Antarctic coasts and the land fringes of the north polar region are much more humid than the poles themselves. Measurable amounts of rain are recorded during the summers, but only snow in the winters. The reason for the sparse precipitation is the inability of moisture-laden storms to penetrate the high-pressure domes that build up over the poles. Upper air descends over the poles and spreads outward as surface air movement (see Figure 18.5). The descending air and cold temperatures create the high pressure. High pressure in any part of the world will create dry conditions, but such conditions are usually temporary. In the polar regions, however, the constant high pressure keeps the precipitation amounts low all year long.

Vegetation and Wildlife

Besides the mosses and lichens, the south polar region has two flowering plant species, found on the far northern tip of the Palmer Peninsula. No higher plants, such as trees, shrubs, grasses, or herbs, are found in Antarctica. This means that no land animals can exist there either, and no indigenous vertebrates are found. The southern oceans do support a very rich marine life though. A wealth of microscopic plankton and algae are nourished by the nutrient-rich waters. These small organisms are eaten by the **zooplankton,** which in turn are food for the fish, penguins, seals, whales, and sea birds. But these large populations all rely on the sea for their food, so the interior of the continent is devoid of life.

The Arctic region, with more land along the fringes, has more plants than the Antarctic. The short growing season restricts the development of

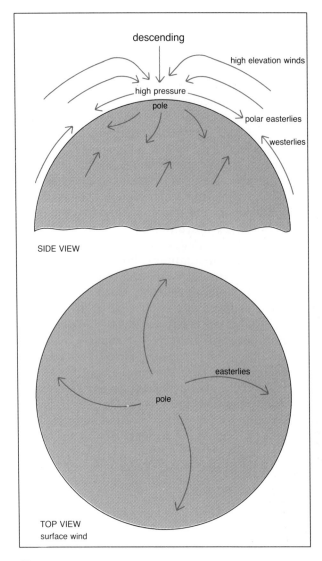

FIGURE 18.5

Atmospheric Circulation. The general circulation of the atmosphere causes air to descend over the polar regions. This creates high barometric pressure and dry air at the poles.

dwarf vegetation Extremely cold conditions restrict plant growth. Thus, vegetation in the polar regions is stunted, creating abnormally small (dwarfed) versions of otherwise large plants.

fell fields Open areas in the Arctic where the plant association is primarily grass and small shrubs found in scattered clumps.

concentrations in the thin soils, permafrost, and scanty precipitation. As a result, most species are **dwarfs** of those found in the temperate regions, and their annual growth rates are small. An Arctic willow may have 400 annual rings crowded into a one-inch-diameter trunk.

Arctic vegetation can be divided into three main types of plant communities. The **fell fields** are the open areas where plants are scattered over the surface but never cover the ground completely. The moss and lichen, as well as some flowers, such as the Arctic poppy, are included in this community. The tundra communities have continuous

Vegetation is also missing in the Antarctic, as this view of Bullfinch Bay shows. Here the ice in the bay has broken into individual ice flues in various polygon shapes. An official U.S. Navy photograph, courtesy of the Naval Polar Oceanography Center.

annuals, however, so all plants are perennials. They spring to life briefly each summer and remain dormant for 10 to 11 months. Many Arctic flowers create a burst of color for the otherwise drab tundra landscape, if only for brief periods. About 900 species of flowering plants have been identified in the tundra. Vegetation in the Arctic region must overcome the windchill, low nitrogen

A lone pine tree grows in a protected valley along the coast of Alaska, but most of the vegetation there consists of bunch grass and hardy shrubs. Photograph by Todd Reutlinger.

vegetation cover. Here a carpet of mosses and lichens covers the ground between the dwarf shrubs. In the damper areas, the tundra grasses predominate. The shore communities include the tidal flats covered by salt-tolerant grasses. There are few species of plants, but sometimes large numbers of individuals within species are found. A similar condition exists for Arctic land animals.

The best-known animal of the Arctic region is the polar bear. Polar bears feed on fish, young seals and walruses, dead whales, seaweed, berries, and rodents. They have been seen near the North Pole itself and as far out to sea as 25 miles (40 km) so they roam the entire north polar region and the surrounding waters. But they live only where three things are found together: cold water, sea food, and pack ice. If any one of the three is missing, so are the bears. Many live their entire lives without ever going on shore. The Arctic fox follows the polar bears and eats the leftover scraps of meat. The bears and the foxes are the only carnivorous land animals that roam into the far north.

Fast ice and rafted ridges of the Arctic Ocean can be seen in the background, and the small animals on one piece of floating ice are baby seals. Photograph by Todd Reutlinger.

The Polar Regions

Focus Box

Ozone Depletion

The earth's atmosphere is a blanket of gases composed of nitrogen (78 percent), oxygen (21 percent), argon (1 percent), and lesser amounts of carbon dioxide, hydrogen, neon, helium, krypton, and xenon. Ozone is a form of oxygen sometimes called "blue gas." A silent electrical discharge in the air, changes oxygen into ozone. The ozone then collects into zones of the atmosphere. The height of the ozone layer varies from about 12 miles to 21 miles above the earth, but traces have been found as low as 6 miles and as high as 35.

The ozone layers are critical to life on earth because they screen out most of the ultraviolet or visible light rays from the sun. In large quantities, these shortwave rays can be dangerous, causing skin cancer, cataracts, immune deficiencies, and the breakup of such important biological molecules as DNA. Ultraviolet rays also harm crops and aquatic ecosystems.

In 1985 at the south polar region, an alarming discovery was made about ozone. The British Antarctic Survey reported that between 1977 and 1984 springtime amounts of ozone in the atmosphere over Halley Bay, Antarctica, had decreased by more than 40 percent. Other groups of scientists soon not only confirmed the British discovery but also found that region of ozone depletion was actually wider than the continent and extended upward from 7 to 15 miles in altitude. In essence, they found an ozone "hole" in the polar atmosphere.

The ozone depletion discovery is very disturbing because it suggests that the entire global stratospheric layer of ozone may be eroding away. Because the problem is potentially so serious, many scientists are racing to discover the cause of the ozone hole. The investigators seek to determine not only why the hole forms but also whether it has global implications or if it will remain in the Antarctic.

So far, two theories on the cause of ozone depletion have been developed. One states that pollutants are to blame, and the theory actually predates the depletion discovery. Some scientists predicted in 1971 that the

---

Other Arctic mammals include caribou, reindeer, hare, musk ox, wolf, and lemming, all of which have one of two mechanisms to preserve body heat through insulation—a layer of fat or a layer of special fur—and some animals have both. The fat layer is tissue that is heavily impregnated with oil, and the fur layer contains long *guard hairs* and short underwool. The guard hairs of the musk ox are 2 feet (0.61 m) long and trail nearly to the ground. The hair of the caribou is thicker at the tip than at the base and contains air cells within the hair stem. Just as birds ruffle their feathers, these mammals can fluff out their fur to trap air and gain warmth.

The sea animals of the Antarctic region all have relatives in the north, except the penguins. Black and white flightless birds, penguins are found only in the southern hemisphere, and most of them are in the Antarctic. Only 5 of the 47 species of seals, however, live in the southern waters. These carnivorous fin-footed mammals live off fish and other seals. Three families of seals live in the Arctic: the eared seal (including sea lions and fur seals), the walruses, and the earless seals. The earless seals are sometimes called true seals or hair seals, and include all the types found in the Antarctic. These mammals are protected from the cold by layers of fat and by special fur. The fur beneath the coarse guard hairs of the fur seal is valued for making coats for humans. Seals once were hunted nearly to extinction, but they have made a recovery since an international law imposed a heavy penalty on anyone that killed them.

ozone layers would be damaged by the emission of supersonic airplanes, and then in 1974 scientists discovered that the use of chlorofluorocarbons does in fact damage the ozone layer. Chlorofluorocarbons are widely used in coolants for refrigerators and air conditioners, for propellants for aerosol sprays, for making foam, and for cleansers. The use of chlorofluorocarbons in hair sprays and deodorants was banned in the United States in 1978, but other uses continue.

The second theory developed to explain the hole in the Antarctic ozone layer states that natural processes, such as shifts in the atmospheric dynamics, may cause the ozone redistribution. According to this theory, the ozone is just moved from one place to another, not destroyed. Most of the ozone is made in the air above the Tropics and carried to the polar regions by the global air movements. Stratospheric air tends to circulate from high altitudes in the Tropics toward lower altitudes in the polar regions. The air circulation creates a concentration of ozone near the polar regions that enters and leaves the polar areas seasonally. The seasonal shifts and changes in circulation are related to the seasonal shifts in the global wind patterns. Thus, ozone depletion in the Antarctic may be likened to the dry season in monsoon areas.

It is possible that both pollutants and global air circulation are working to cause the alarming decrease in ozone in the south polar atmosphere, but there may be some yet undiscovered cause. Scientists are working to design new ozone-measuring devices for ground, air, and satellite laboratories. Detailed information is being gathered, and the analysis continues. The answers could be vital to all life on earth.

SOURCE: For further information, see Richard S. Stolarski, "The Antarctic Ozone Hole," *Scientific American* 258 (January 1988): 30–36.

Human Geography of the Polar Regions

When the Antarctic continent was first reached by explorers in the nineteenth century, no native peoples were found. The combination of the massive ice sheet, the lack of vegetation, and the vastness of the cold, surrounding ocean make the continent not only inhospitable but also difficult to reach. Currently, however, many countries have permanent research stations in the Antarctic (see Table 18.1 and Figure 18.6).

Of the 34 stations in the Antarctic, 14 are located on Palmer Peninsula and the nearby offshore islands. The largest multipurpose research and logistics center is the American-operated station at McMurdo. In the summer, the station's

TABLE 18.1

Countries with Permanent Research Stations in the Antarctic, 1983

COUNTRY	NO. OF STATIONS
Argentina	8
Australia	3
Chile	4
France	1
New Zealand	2
Poland	1
South Africa	1
United Kingdom	3
United States	4
Soviet Union	7
Total	34

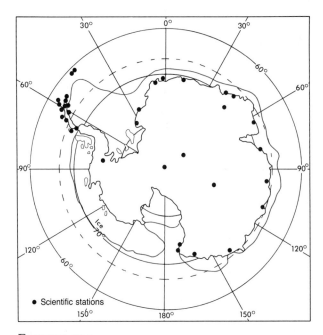

FIGURE 18.6

Scientific Stations in Antarctica. During the mid-1980s some 34 scientific stations were in operation in the Antarctic region. Most of the stations are located along the coast of the continent.

population peaks at around 700, and in addition there is a large shipboard naval contingent. The station has its own chapel, theater, and radio station, plus well-equipped laboratories. The Amundsen-Scott station operated by the United States is located at the South Pole, but the shifting of the ice sheet causes the station to keep moving away from the exact site of the pole.

The seven Soviet bases effectively surround the Antarctic, but Soviet operations have been less active than those of the Americans. The largest Soviet station is Molodezhnaya in eastern Antarctica. About 260 Soviet scientists man the station. The Argentinians have the greatest number of permanent bases in the Antarctic and seem determined to colonize the territory. Emilio de Palma, the first child born in the Antarctic, was an Argentinian. He was born at the Hope Bay Colony in 1978.

The Arctic region, unlike the Antarctic, has had a sprinkling of native people for thousands of years. These people are diverse in racial origin, culture, and language, but they are similar in their ability to exist in the extremely harsh conditions. The basis for their subsistence-level existence has been fishing, hunting sea mammals, and herding. All the groups fish, but hunting and herding are alternate activities. Generally, the groups are divided into those found in the east—the Lapps, the Finno-Ugrians, the Tungus-Manchu, and the Yakut-Turkic—and the Eskimos of the west. (See Figure 18.7).

The Eskimos, the last of the Mongoloid peoples to reach North America, have occupied the Arctic region for about 5,000 years. Today, about 50,000 Eskimos survive, living in a thinly occupied line from the Bering Strait to the western coast of Greenland. Those along the coast live off fishing and hunting sea mammals, and those on the tundra live mostly off the herds of caribou. The Lapps live in northern Scandinavia, but they are not related to the Mongoloids or the modern European

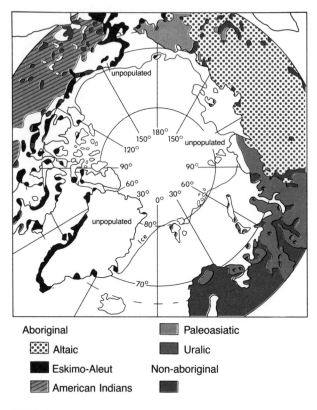

FIGURE 18.7

The native people of the Artic include those from the groups: (1) Altaic, (2) American Indian, (3) Eskimo-Aleut, (4) Paleoasiatic, and (5) Uralic.

Oil from the numerous wells on the North Slope of Alaska is collected at this Prudhoe Bay Gathering Center before it is piped south. Photograph by Todd Reutlinger.

The Alaskan pipeline begins here at mile marker zero (small sign) and winds its way southward more than 700 miles (1,126 km) across Alaska to the port of Valdez. Photograph by Todd Reutlinger.

races. They are descendants of the Komsa Stone Age people who inhabited Europe before the modern racial stocks evolved. Today, about 30,000 Lapps roam their frigid homeland, herding reindeer and hunting small game. Some Lapps are fishermen, and some maintain farms along the rocky Arctic coasts.

The native groups of the Arctic other than the Eskimos and the Lapps are Siberian races that have representatives farther south in the forest fringe of the region. Many are reindeer herders who move seasonally onto the tundra but then go back to the taiga for winter. Only a few live permanently on the coast.

Besides the native people, however, approximately 5 million Soviet citizens live north of the Arctic Circle. Four Soviet cities exceed 50,000 in population. The largest is Murmansk, with nearly half a million people (see Figure 18.8). The city is a seaport and center for Soviet mineral exploration, as well as a military garrison. The city of Norilsk is growing faster than Murmansk, however, and today it has about 250,000 people. It has been a boom town since the discovery of gold, copper, nickel, cobalt, and coal in the region nearby. Because so much of the land and resources of the U.S.S.R. lies so far north, the Soviets have been exploring and settling their northlands since World War II.

Economic Geography of the Polar Regions

So far, the only export from the Antarctic region has been the annual harvest of scientific data. Antarctica is of intense interest to scientists because of its effect on world climates. The interactions of world wind-systems and ocean currents with the atmosphere and water of the Antarctic play a major role in governing the world's weather. A circulation of heat flows from the Tropics to the poles, and colder air moves back again. The flow of cold water from the polar regions penetrates the tropical oceans and circulates back as warm water. These exchanges of heat and cold have dramatic effects on world climates. For example, an offshore cold ocean current creates coastal deserts that would otherwise be more humid. The west coasts of all continents have offshore cold water that has circulated from the polar regions, regardless of whether a continent is in the northern hemisphere or the southern. The mid-latitude cyclonic storms originate in the fronts between the cold polar air and the prevailing westerly winds, and without these storms much of the mid-latitude region would be much drier than it is now. Study of the polar regions helps scientists understand how the global ocean and atmospheric circulation systems work.

Probably the first Arctic resource that comes to mind is the huge Alaskan oil reserve (see chapter 5), but many other minerals have been discovered in the Arctic (see Figure 18.9). Both the North American and the Soviet Arctic regions have been mined for copper, gold, silver, mercury,

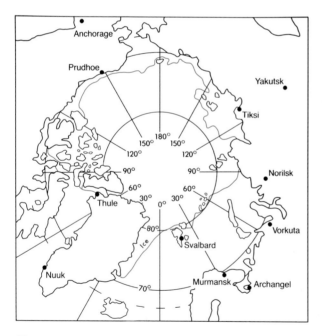

FIGURE 18.8

Very few people live north of the Arctic Circle, but some cities, permanent bases, and a few small towns do exist. Not shown on this map are the numerous small villages occupied by the native people.

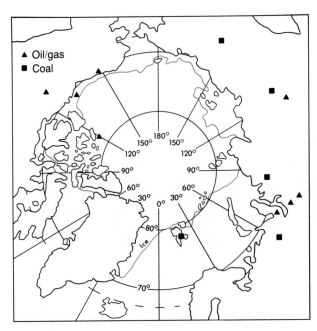

FIGURE 18.9

Mineral Deposits of the Arctic. The minerals now being produced in the Arctic region are primarily coal and petroleum.

coal, nickel, lead, zinc, and uranium, in addition to natural gas and crude petroleum. The Svalbard Islands of Norway have provided that country with minerals, and the large Kiruna iron ore mines of Sweden are located north of the Arctic Circle. The Russians are well on their way to turning parts of northern Siberia into industrial regions based on the Arctic mineral wealth. Population growth in the North American Arctic has not been as rapid as in the Soviet Union, but over the last 20 years exploitation of minerals has increased steadily. The economic future of the Arctic will be based on mineral exploitation.

Political Geography of the Polar Regions

In 1908, the British made the first formal claims of national sovereignty in the Antarctic. They laid claim to an Antarctic sector that included South Georgia, the South Orkney Islands and the Shetland Islands, the South Sandwich Islands, and the Palmer Peninsula and made these territories dependencies of the Falkland Islands. The claims had their origin in the rise of whaling operations in the region, and British magistrates became permanent residents of South Georgia and Deception Island. The British continued to make larger claims on all sides of the Antarctic continent. Later, much of the territory claimed by the British was endowed to Australia and New Zealand.

Most of the other countries' claims to the Antarctic came during the 1930s and 1940s. It was the British, however, who started the concept of using pie-shaped wedges as territorial claims. These unusual political boundaries remain in use today (see Figure 18.10), although they are largely ignored by current researchers and explorers. The current political status of the Antarctic is based on the Antarctic Treaty. This American proposal to the members of the 1957–58 International Geophysical Year (IGY) meetings declared the Antarctic to be an international laboratory for scientific research to be used only for peaceful purposes. Adopted in 1959, the proposal went into force in 1961 for 30 years, at which time the terms can be renegotiated. The treaty forbids any military activity in the Antarctic and any more claims of sovereignty.

Administration of the northern polar region is more clear-cut than that in the south. Alaska became the 49th state of the United States in 1959. The political circumstances of that area are unambiguous, as all the land territory is within the framework of the United States, but claims to the Arctic Ocean have been disputed. The Canadian Arctic territories are similar in status to Alaska. They are part of Canada and make up parts of the provinces of Quebec, Newfoundland, and the Northwest Territories. These provinces are administered directly by the Canadian government in Ottawa.

The people of Greenland achieved full status as part of Denmark in 1953. Two members of the Danish parliament are from Greenland, and Greenlanders have the same voting rights as the homeland Danes. The Svalbard group of islands is officially part of Norway, but a treaty defines some

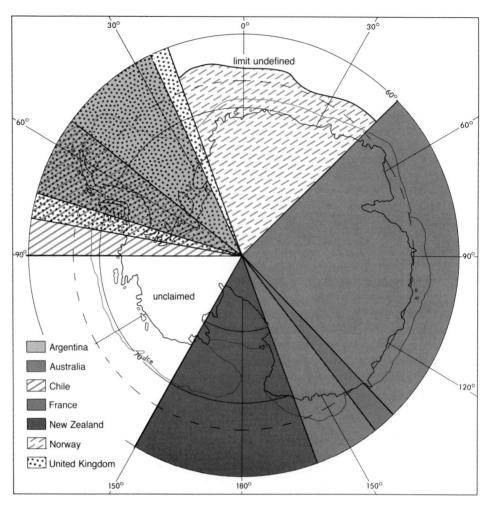

FIGURE 18.10

Political Divisions of Antarctica. The pie-shaped wedges shown on this map are the political claim areas, but none of these has been officially sanctioned.

differences between the administration of that territory and the administration of the remainder of Norway. For example, the Svalbard is not a county of Norway, and Norwegian laws do not apply there unless specifically stated. As a result, the Norwegian miners pay lower income taxes than other Norwegians, and the mineral export duties are less for the Svalbard than for the remainder of Norway.

Administration of the Soviet Arctic is under the strict control of the Soviet Union. The mainland Arctic fringe and offshore islands are part of the territory of the Soviet Union. The Soviets, however, also claim all the territory in a sector north of their mainland to the North Pole. The Arctic region is a very sensitive area for the Soviets, and the mainland bristles with aircraft- and missile-detection devices. The Arctic fringe is becoming the new home for many Soviet scientists and settlers and is one of the fastest-growing regions in the Soviet Union.

The riches of the polar regions may be greater than anyone expects, and political possessiveness is probably merited. However, the climate and the transportation problems will stand in the way of full development for many years to come.

Key Words

drumlin	nunataks
dwarf vegetation	pack ice
esker	polar ice
fast ice	pole of inaccessibility
fell fields	solifluction
ice sheet	tundra
isostasy	windchill
lichen	zooplankton
moss	

References and Readings

Adams, John Q. "Settlements of the Northwestern Canadian Arctic." *Geographical Review* 27 (1941): 112–126.

Baird, Patrick D. *The Polar World.* New York: Wiley, 1964.

Barnes, James N. *Let's Save Antarctica.* Richmond, Aust.: Greenhouse, 1982.

Bentley, C. R. "Glacial and Subglacial Topography of West Antarctica." *Journal of Glaciology* 3, no. 29 (1961): 882–911.

Bertram, G. C. L. *Arctic and Antarctic: A Prospect of the Polar Regions.* Cambridge: Heffer & Sons, 1958.

Boyd, Andrew. *An Atlas of World Affairs.* New York: Praeger, 1960.

Brewster, Barney. *Antarctica: Wilderness at Risk.* San Francisco: Friends of the Earth, 1982.

Brown, R. N. R. "Plant Life in the Antarctica." *Discovery* 4, no. 42 (1923): 149–153.

Bruemmer, Fred. *The Arctic World.* San Francisco: Sierra Club Books, 1985.

Cameron, Ian. *Antarctica: The Last Continent.* Boston: Little, Brown, 1974.

Deacon, George E. R. *The Antarctic Circumpolar Ocean.* New York: Cambridge University Press, 1984.

Debenham, Frank. *Antarctica: The Story of a Continent.* New York: Macmillan, 1961.

Hatherton, Trevor (ed.). *Antarctica.* New York: Praeger, 1965.

Kimble, G. H. T., and Good, D. (eds.). *Geography of the Northlands.* New York: Wiley, 1956.

Lamb, H. H., and Johnson, A. I. "Climatic Variation and Observed Changes in General Circulation." *Geografiska Annalar* 41, nos. 2 and 3 (1959): 94–134.

Ley, Willy. *The Poles.* New York: Time Inc., 1962.

Liversidge, D. *White Horizon.* London: Oldhams, 1951.

Lovering, J., and Prescott, J. R. V. *Last of Lands . . . Antarctica.* Melbourne, Aust.: Melbourne University Press, 1979.

MacDonald, R. St. J. (ed.). *The Arctic Frontier.* Toronto: University of Toronto Press, 1966.

Mills Hugh. *Kingdom of the Ice Bear: A Portrait of the Arctic.* Austin: University of Texas Press, 1985.

Muller, Fritz B. *The Living Arctic.* New York: Methuen, 1981.

Nordenskjold, Otto. *The Geography of the North Polar Regions.* New York: American Geographical Society, 1928.

Orvig, S. *Climates of the Polar Regions.* New York: Elsevier, 1970.

Payne, Donald G. *Antarctica: The Last Continent.* Boston: Little, Brown, 1974.

Pederson, Alwin. *Polar Animals.* New York: Taplinger, 1966.

Polunin, G. E. *Introduction to Plant Geography.* London: Longmans, Green, 1960.

Porsild, A. E. "Plant Life in the Arctic." *Canadian Geographical Review* 42 (1951): 122–145.

Porter, Eliot. *Antarctica.* New York: Dutton, 1978.

Price, R. J., and Sugden, D. E. *Polar Geomorphology.* London: Institute of British Geographers, 1972.

Priestley, Sir Raymond; Adie, Raymond J.; and Robin, G. D. Q. (eds.). *Antarctic Research.* London: Butterworths, 1964.

Sater, John E. *The Arctic Basin.* Centreville, Md.: Tidewater, 1963.

Smiley, Terah L., and Zumberge, James H. *Polar Deserts and Modern Man.* Tucson: University of Arizona Press, 1974.

Stolarski, Richard C. "The Antarctic Ozone Hole." *Scientific American* 258 (January 1988): 30–36.

Sugden, David E. *Arctic and Antarctic: A Modern Geographical Synthesis.* Totowa, N.J.: Barnes & Noble, 1982.

Tedrow, J. C. F. *Soils of the Polar Landscape.* New Brunswick, N.J.: Rutgers University Press, 1977.

Zegarelli, Philip E. "Antarctica." *Focus,* September–October 1978, pp. 1–9.

Chapter 19

WORLD PROBLEMS AND PROSPECTS

■ Introduction

Many world problems can be discussed better on a worldwide basis than on a regional basis. In this book the world has been divided into 14 regions, but in reality the world is one large system. Things that affect one part reverberate throughout the entire system. Loss of a rice crop in Southeast Asia does not directly affect the average American urban-dweller, but it does affect the lives of the local people and eventually may affect the world economy and then indirectly Americans. A more direct causal link exists between Americans and others when Arab oil prices are reduced. Within weeks, gasoline prices in the United States go down, exploratory drilling slows, and an overall economic slump occurs in the oil-producing regions of the United States. These examples from economic geography have their counterparts in other areas of geography.

The focus of this chapter is on world problems that are related to the environment (physical geography), to population (human geography), to the economy (economic geography), and to politics (political geography). These topics make up the major parts of geography. The problems confront the citizens of the entire world, not just the people of the region in which they occur. It will take world planning to solve these problems, and world cooperation is just beginning to occur.

Environmental Problems

Air Pollution

The quality of the environment is reduced by the pollution of the world's air. The countries of the Developed World are the worst offenders, but **air pollution** occurs to some extent wherever people live (see Figure 19.1).

One concern has to do with the burning of fossil fuels. In the United States, the use of coal for energy is increasing, and it has remained steady in

---

air pollution Contamination of the air by vehicle exhaust fumes, factory smoke, and other human activities that make the air impure or dirty.

acid rain Precipitation (both rain and snow) that carries dioxides, such as those from sulfur and nitrogen, that have been emitted into the atmosphere from the burning of fossil fuels and that are dangerous to plants and animals, including humans.

---

other industrial nations. As a result, however, there has been an increase in sulfur dioxide and nitrogen dioxide emissions into the atmosphere. These dioxides are known to cause acid rain and sulfate haze. **Acid rain** pollutes rivers and lakes and can become

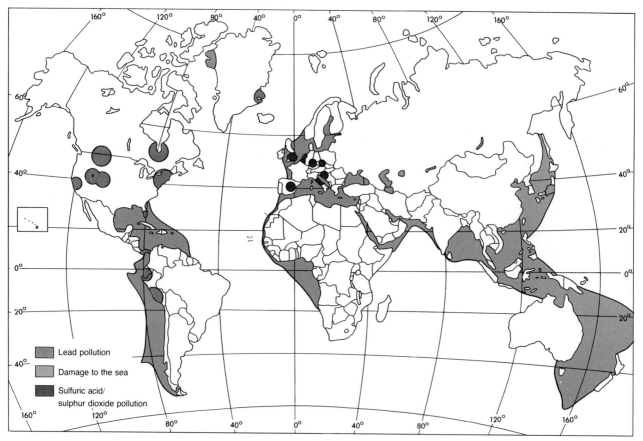

FIGURE 19.1

World Pollution. Air and water pollution are worldwide problems, and some of the worst polluted areas are indicated on this map.

> **greenhouse effect** An atmospheric condition in which the sun's heat energy is trapped by layers of carbon dioxide in the atmosphere, similar to the way heat energy is trapped inside a greenhouse by the panels of glass.
>
> **desertification** Literally, making a desert out of land that was not a desert. This occurs through human misuse of the land, such as allowing animals to overgraze the grasslands adjacent to the deserts, and through atmospheric heat balance changes, like the changes caused by the "greenhouse effect."
>
> **water pollution** Contamination of the earth's water, especially the rivers and oceans near port cities, with garbage, factory waste, and other refuse.

a soil problem. Sulfur dioxide also becomes sulphuric acid and, combined with water vapor, the mixture becomes a prime cause of chronic lung disease. Lead from these emissions can permanently damage the human nervous system and is particularly dangerous to children. The urbanized areas of the world, with all their cars, trucks, and power plants, produce carbon monoxide, a colorless, odorless gas that is highly poisonous.

Another world concern is the destruction of the ozone layer of the atmosphere. The ozonosphere lies approximately 25 miles above the surface of the earth and contains a faintly blue form of oxygen that is known for its purity. When this layer is reduced or polluted, the air we breathe is less pure and less effective in filtering out the harmful ultraviolet sun rays, which are destructive to life on the earth (see the Focus Box on Ozone Depletion in chapter 18). Air pollution is neither local nor isolated, and it occurs throughout the urbanized world. Anything spewed into the atmosphere in one part of the world affects all the people of the world.

The Greenhouse Effect

The burning of fossil fuels also creates carbon dioxide. Controlling the production of carbon dioxide is a serious world problem. The level of carbon dioxide in the atmosphere has steadily increased ever since the beginning of the industrial revolution in the nineteenth century. Too much carbon dioxide in the atmosphere can change the world's climates by allowing the lower layers of the atmosphere to heat up more than they would with normal amounts of the gas. This occurs because the gas allows the sun's rays to penetrate through it, but when the rays are converted to infrared by reflection from the earth's surface, the carbon dioxide traps the heat energy.

The process is similar to the process that takes place in a greenhouse, so it is called the **greenhouse effect** (see Figure 19.2). As a result, the air layers near the surface of the earth have been getting steadily warmer. Climatologists are attempting to model how this effect might shift global air masses and how that shift would in turn affect crop production in the great mid-latitude grain belts. Along with human misuse of the land, the greenhouse effect has resulted in an increase in the size of the desert regions of the world. The Sahara, for example, has steadily marched southward for the last half-century. More than 250,000 square miles (647,500 km^2) of territory along the southern edge of the Sahara have been changed from productive land to desert. About 100 million people are directly affected by this **desertification** process, not only in the Sahara but in many other parts of the world as well.

Besides the decrease in productivity associated with desertification, an even worse disaster could occur. The great ice caps of Greenland and the Antarctic could melt if the world's climates continue to warm. Melting of the ice caps would raise the ocean level to a point where the homes of half the world's people would be inundated (see Chapter 18).

Water Pollution

Water quality continues to decline in spite of some countries' efforts to control **water pollution**. The worst offender is industrial wastewater containing contaminants left from industrial processes, but non-point sources in the United States (such as farm land and roadways) are the most difficult to control. Many of America's rivers and lakes contain hundreds of contaminants. A recent survey found that samples of ground water contained 300

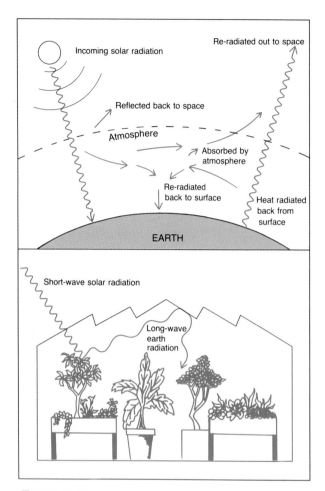

FIGURE 19.2

The Greenhouse Effect. The greenhouse effect works in the atmosphere but is now considered to be less effective for warming greenhouses than originally thought. Radiation from the sun passes through the atmosphere as shortwave energy (ultraviolet light). When it strikes the earth's surface it is converted to longwave energy (infrared light), which is trapped by the water vapor and carbon dioxide in the troposphere. The increased levels of carbon dioxide caused by the burning of fossil fuels results in increased warming of the lower levels of the atmosphere. Real greenhouses trap heat mainly because the glass roofs let sunlight in but inhibit convection (vertical air movement), which stops the warm air from rising and escaping.

Environmental Protection Agency (EPA) An independent agency of the U.S. government charged with overseeing the protection of the environment. It sets standards, partrols abusers, and fines abusers.

contaminants, when the **Environmental Protection Agency (EPA)** standard is 22 or less. More than half the people in the United States drink ground water, and it is also used extensively for irrigation of cropland.

Probably the most polluted body of water in the world is the Mediterranean Sea. In addition to all the industrial wastes, it contains tar, sewage, and chemical fertilizers. The sea water does not circulate out of the basin into the world's oceans, so the collection of pollutants does not diffuse, but keeps building up. Control is difficult because many countries border the Mediterranean. The European countries, especially the Common Market countries, are attempting to control their industries, but the countries of North Africa continue to pollute. Many other areas of the world's waterways are being polluted, and the contaminants are hazardous not only to marine life but also to the people who eat the fish.

Destruction of Farm Land

Farmers worldwide generally put financial survival ahead of soil conservation, so in many places the land continues to deteriorate. In the United States, approximately 3 billion tons of soil each year are being stripped from the nation's farm lands. Such losses cause significant reductions in crop yields and add to the pollution of lakes and rivers. Many of the productive areas of other parts of the world experience similar losses of topsoil.

The continued irrigation of fields in southern California and Arizona has caused a salt buildup in the soil, to the point where the land is useless for crop production. This is not a localized problem, but one that affects thousands of acres. Fields ruined in this manner must lie fallow for years, perhaps as many as 50 years.

Destruction of the natural vegetation of an area in order to create new farm land also creates

hazardous waste Any of the 403 toxic chemicals that the EPA has identified as harmful to humans, animals, plants, or other parts of the environment and that have been stored or dumped and are of no further use.

problems. Not only does destruction of the tropical rainforest reduce the number of trees that produce oxygen, but continued burning can reduce the rainforest to savanna and eventually to a desert. Left alone after "slash and burn" agriculture has depleted the tropical soils, the rainforest needs 100 years to return to its natural state. Destruction of the land, then, is occurring both in the Developed World and in the underdeveloped world.

Hazardous-waste sites ruin the land adjacent to the sites. While this may seem to be a minor problem, the EPA has identified 20,766 such sites in the United States (see Figure 19.3). Not only is the local land ruined, but for miles around each site the ground water is poisoned. The EPA has identified 403 toxic chemicals that pose an immediate threat to life and health, and many of these are stored or dumped in the hazardous-waste sites. **Hazardous wastes** include such things as acids, cyanides, chemical warfare agents, industrial gases, pesticides, pharmaceuticals, solvents, and miscellaneous compounds.

Depletion of Resources

The world's mineral reserves continue to be depleted, some at alarmingly fast rates. Mineral resources are nonrenewable resources because they are not replaced. Once mined from the ground and used for energy or to make metal objects, the mineral resources are gone. Much effort has been devoted to recycling aluminum containers and scrap iron, but recycling cannot keep pace with world use. Burning of the fossil fuels not only depletes them forever but also contaminates the

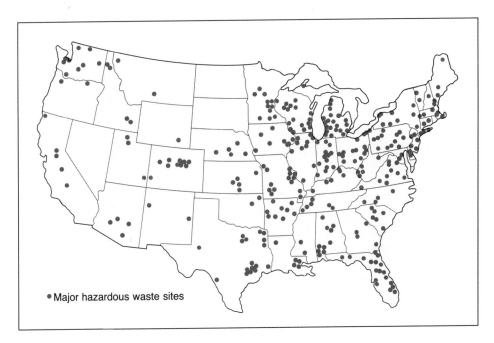

FIGURE 19.3

Major Hazardous Waste Sites in the United States. Hazardous waste dumps tend to be located in the industrial areas because about 93 percent of the waste produced comes from the chemical, petroleum, and metal-related industries.

World Problems and Prospects

atmosphere. The consumption of nonfuel minerals occurs mainly in the Developed World, but more and more products are being shipped to the underdeveloped areas.

The distribution of the world's mineral resources is extremely uneven. Only five countries have abundant and diverse minerals that are considered to have economic and strategic importance: Australia, Canada, South Africa, the United States, and the Soviet Union. The United States lacks such things as antimony, bauxite, chrome, diamonds, manganese, nickel, and tin. In spite of the lack of strategic minerals that must be imported, however, the United States is considered to be a "mineral power." The Soviet Union, on the other hand, has a good supply of all vital minerals except tin and uranium. Many minerals are found in only a few places. For example, some 99 percent of the world's chrome comes from South Africa and Zimbabwe. South Africa and the Soviet Union have 90 percent of the world's reserves of manganese and platinum. Some minerals, however, are distributed widely—for example, silver and zinc are mined in more than 50 countries. The relative scarcity of the world's vital minerals, and the disparity in their distributions, adds to the world's problems.

■ Population Problems

Rate of Increase

The most rapid expansion ever of world population size is now occurring. In mid-1987 the world's population surpassed 5 billion—an increase of 3 billion since 1950—and another 3 billion increase is expected in the next 30 years. At the current rate of increase, the world's population will double to 10 billion in 41 years. The pace of growth is about 80 million people a year. These facts, which have been well documented, represent what has been called the **population explosion.** But rapid growth is not occurring at an equal pace all over the world. (See Figure 19.4).

The industrialized nations, which are part of the Developed World, have the lowest population growth rates. These wealthy nations enjoy high life expectancies (approaching 80 years in some

---

population explosion The very rapid rate of growth of the world's population during the last 50 years.

zero population growth (ZPG) Occurs when a country's rate of natural increase is zero percent. It can be achieved if no family has more than two children and if immigration does not exceed outmigration.

---

countries) and have fertility rates of less than two children per couple. When couples average about two children each, populations cease growing; each generation of couples just replaces itself, and **zero population growth** (ZPG) is achieved. The low fertility rates and high life expectancy rates produce an "old" population, which would be expected to maintain ZPG.

On the other end of the scale are the poor developing countries, which, although high fertility rates prevail, feature economies based on subsistence agriculture, poor health conditions, life expectancies below 50, and high infant mortality rates. In some of these countries, women average seven or more children each, but one child in five dies before the age of one. The populations of these countries are young, with nearly half the people less than 15 years old. These are the places where the population explosion is now occurring. In some of the developing countries, national policies have resulted in a lowering of the birth rates, but in others either no national policy exists or the high birth rates are ignored. It is assumed that economic development will bring a desire for smaller families, but development is taking place only very slowly, leaving the countries with an unknown potential for population growth.

Population Movement

Human migrations of large numbers of people can be classified into three general categories: the rural to urban movement, movement for employment, and refugees. The rural to urban movement started in the developed nations after the industrial revolution but picked up steam after World War II. That is when First World cities really

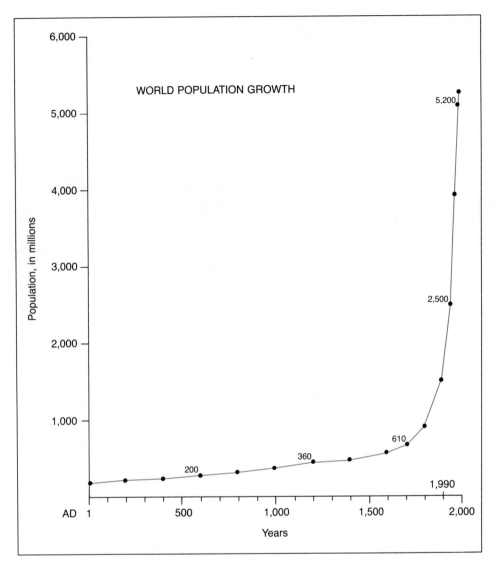

FIGURE 19.4

World Population Explosion. The world's population increased steadily from 1700 to 1900, then began to increase rapidly. Since about 1950 it has "exploded" with a very rapid increase rate.

began to sprawl and large suburban developments were constructed. More recently, similar movement has been occurring in the less developed nations of the world, but the affluent suburbs are missing.

Many rural people of the Third World have migrated to the cities seeking employment, but work is usually difficult to find. The people are then forced to live in slum and squatter settlements around the cities. The proportion of city populations in slum and squatter settlements exceeds 80 percent in many cities throughout the world. Most of these cities are located in the tropical, underdeveloped parts of the world, and some of the worst

Focus Box

International Refugees

In the past, refugees have sought asylum primarily in Europe, the United States, and other developed areas, but today the majority of the world's approximately 10 million refugees are from developing countries seeking asylum in other developing countries. Refugee groups are concentrated in four world regions: Central America, Southwest Asia, Southeast Asia, and sub-Saharan Africa. The 10 million refugees include the Palestinians but not the millions of displaced persons within countries and the international migrants who are economically motivated.

The close association that refugee flows have with political insurgencies in developing nations makes such flows particularly dangerous. They place great strains on impoverished economies and fragile political systems, such as in the Sudan. Long-time refugee groups that are not integrated or repatriated sometimes have generated successive alienated generations and have engaged in such activities as terrorism to sustain themselves or to gain attention for their cause. Thus, the regions of asylum (see Table 19.A) are termed "crisis areas" by the United Nations high commissioner for refugees.

TABLE 19.A
Main Groups of Unsettled World Refugees by Country of Asylum in Mid-1987

COUNTRY OF ASYLUM	NO. OF REFUGEES	COUNTRY OF ORIGIN
Africa (2,174,500 total) 2,238,800		
Angola (92,000)	70,000	Namibia
	13,000	Zaire
	9,000	South Africa
Benin	3,600	Chad
Botswana	3,600	Zimbabwe
Ethiopia	72,000	Sudan
Lesotho	11,500	South Africa
Nigeria	4,000	Chad
Senegal	5,000	Guinea-Bissau
Somalia	700,000	Ethiopia
Sudan (690,000)	484,000	Ethiopia
	200,000	Uganda
	5,000	Zaire
	1,000	Chad
Swaziland	6,600	South Africa
Tanzia (177,500)	153,000	Burundi
	15,000	Zaire
	2,500	Rwanda
	7,000	Other countries

TABLE 19.A—Continued

COUNTRY OF ASYLUM	NO. OF REFUGEES	COUNTRY OF ORIGIN
Zaire (329,000)	265,000	Angola
	20,000	Uganda
	14,000	Burundi and Rwanda
	30,000	Zambia and others
Zambia (94,000)	75,000	Angola
	9,500	Zaire
	7,000	Namibia
	2,500	South Africa
Zimbabwe	50,000	Mozambique
Southeast Asia (155,255 total)		
Hong Kong	10,734	Indochina
Indonesia	7,890	Indochina
Malaysia	9,155	Indochina
Thailand	127,476	Indochina
South and Southwest Asia (5,310,000 total)		
Jordan	800,000	Palestine (Israel)
Gaza Strip	430,000	Palestine (Isreal)
Iran	1,000,000	Afghanistan
Lebanon	300,000	Palestine (Israel)
Pakistan	2,500,000	Afghanistan
Syria	280,000	Palestine (Israel)
Latin America and the Caribbean (294,000 total)		
Costa Rica	16,000	El Salvador and Nicaragua
Guatemala	70,000	El Salvador
Honduras	38,000	El Salvador and Nicaragua
Mexico	170,000	Guatemala and El Salvador

Note: The above figures do not include the millions of migrants and refugees that have settled in the developed nations of the world.

SOURCE: George Demko, "The Wide World of International Refugees," *Geographic Notes* (U.S. Department of State, Bureau of Intelligence and Research), Issue 6 (July 1987): 1–6; Kim Crews, "Where Are the World's 10 Million Refugees?" *Population Today* 14 (March 1986): 4.

are found in Latin America, Tropical Africa, and Southeast Asia. The health, education, and welfare of these people constitute major problems for the countries involved.

Migration to seek employment also occurs across international boundaries. Employment is the major reason for the influx of Mexicans into the United States, but this movement is not unique. Thousands of Africans also move across borders seeking employment in more developed areas. South Africa would have an extreme labor shortage without all the workers from Botswana, Zimbabwe, and Mozambique. More than one million temporary laborers who are nationals of other countries live in Saudi Arabia. Ivory Coast and Ghana welcome thousands of workers from Niger and Cameroon. Probably the greatest amount of worker migration occurs in Europe, however. In France, for example, more than 2 million workers come from Portugal, Spain, Morocco, Tunisia, Algeria, Yugoslavia, and Italy. Other workers seek out the United Kingdom, Belgium, the Netherlands, and West Germany.

refugees People that flee their homeland to seek refuge elsewhere from war, political or religious persecution, or life-threatening conditions.

Refugees are the largest human movement problem facing world leaders today. These are the people who have been forced by political circumstances beyond their control to leave their homelands. More than 250,000 Vietnamese refugees have settled in the United States, as well as 15,000 Haitians and many Cubans. These people were escaping political persecution at home. Other refugee migrations in the world include (1) the

Afghan children in a refugee camp in Pakistan. Photograph by Jim Curtis.

> **terrorism** The use of terror and violence as a political weapon or policy to intimidate or subjugate other people.

Palestinian refugees from Israel, (2) the Afghans moving into Pakistan, the movement (3) out of Chile, (4) out of Laos, (5) out of Cambodia, (6) out of Iran, and (7) out of numerous countries of Africa. The region of turmoil in Africa stretches from Ethiopia southward. Besides the refugee movement between countries of eastern Africa, much of the internal populations have been displaced from their homes as well. Estimates vary, but in Africa the total number of displaced people runs well into the millions, possibly 5 or 6 million.

■ Economic Problems

Distribution of Income

The discrepancy between the rich people and the poor people of the world is not changing rapidly. In 1985, the average per capita gross national product for the world was $2,760, but the variation was from $9,510 for the developed countries to $700 for the less developed countries; the range was from $23,770 for the United Arab Emirates to $120 per person for Ethiopia. Variations in wealth within countries is also a major problem. The income of the top 5 percent of the people, compared with that for the bottom 20 percent is one measure of internal variation. In Brazil, the income of the top 5 percent is 30 times that of the bottom 20 percent and nearly as lopsided in Ecuador, Peru, Venezuela, and Honduras. Other countries of the world where the money and holdings of the rich people (the top 5 percent) are 20 times greater than that of the poor people (the bottom 20 percent) include South Africa, Botswana, Zimbabwe, Tanzania, Kenya, Senegal, Gabon, Turkey, Iraq, and Italy. These areas with major income discrepancies are potential world trouble spots.

Lack of Income Diversity

Another world economic problem is the lack of diversity in the incomes of many countries. In 1980, more than half of all export incomes for 44 countries came from only one commodity. All these countries are among the least developed, and half of them are in Africa. Many of the least diverse countries are the petroleum producers and some are the foodstuff producers, but other countries are dependent on copper, iron ore, cotton, fertilizers, and precious stones. Any country that depends on only one commodity for its income is trapped by the swings in world market prices and by the world need for the commodity.

Debtor Nations

Many of the one-product countries are the same countries that have tremendous debts. The debtor nations of the world are the less developed countries of the tropical areas. Those that owe the most money, compared with their GNPs, are Guyana, Mauritania, Somalia, and Rwanda. Many other countries of Central America, South America, and Africa have external public debts that are so large that they may never be paid. Much of the money is owed to banks in the industrial countries of the northern hemisphere. If the loans turn bad, the banks could be in trouble, possibly causing economic disruption in the banks' home countries. The result would be economic disaster worldwide.

■ Political Problems

Current Conflicts

One of the most difficult world political problems today is the rise of **terrorism.** This type of conflict started in earnest following the overthrow of the Shah in Iran, when militants held 52 American hostages for 444 days in 1980–81. Since then, other hostages have been taken, ships and airplanes have been commandeered, and suicide bomb attacks have been made. In 1985 alone, more than 700 terrorist attacks occurred worldwide, including bombings in London, Paris, Rome, and Vienna.

The next most serious problem for world political leaders is the conflict in the Middle East. Sharp divisions on economic, political, racial, and religious issues make the Middle East the most militarily unstable area of the world. The war between Iraq and Iran, and the resulting carryover of bombings into the Persian Gulf, created problems of major proportions for world shipping. Each country wants to cut the other's income from oil exports, so it tries to bomb ships headed for each other's ports. As a result, about 250 attacks occurred on shipping in the Persian Gulf during 1986 (see also Chapter 13). American warships are stationed in the Gulf today.

Other world conflicts have included the revolts in El Salvador and Nicaragua, the Soviet invasion of Afghanistan, and numerous trouble spots in Africa. Every decade of world history contains a series of sharp new political conflicts that rise out of racial or cultural differences among the world's peoples. Religious conflicts are still taking place in India, Northern Ireland, and the Middle East, and language differences create turmoil in many areas, such as Canada, Spain, Belgium, Algeria, and the Soviet Union. Other severe internal conflicts are based on differences in race. The apartheid policy

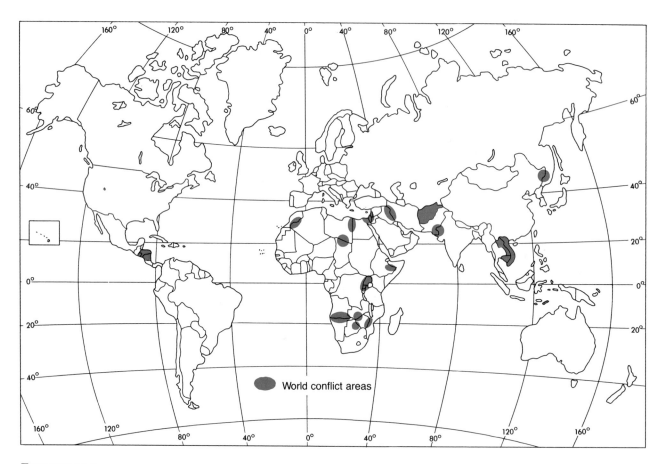

FIGURE 19.5

World Conflict Areas. The locations of wars, battles, and other conflicts since 1970 are shown on this map.

nuclear weapon Any weapon using either nuclear fission or nuclear fusion to convert mass to energy and that is used to make explosions to destroy animate or inanimate objects.

tuna wars A series of conflicts between private citizens of the United States and the Peruvian authorities over fishing rights off the shores of Peru. A major type of fish caught in the region is tuna.

in South Africa and the anti-apartheid protests and demonstrations have created constant trouble in that country for two decades. Economic sanctions imposed by the United States and 11 Western European countries against the government of South Africa are evidence of worldwide involvement with the issue.

Nuclear Weapons

The **nuclear weapons** conflict may appear to be primarily between the United States and the Soviet Union, and indeed both countries have vast storehouses of nuclear weapons. However, eight countries have manufactured and exploded nuclear weapons. In addition to the two superpowers, these are France, Great Britain, China, India, Israel, and South Africa. Furthermore, 30 other countries are now technically and industrially able to develop the capability to manufacture and explode nuclear weapons. These include most of the countries of the First World as well as such places as Brazil, Chile, Argentina, Egypt, Australia, Indonesia, Iran, South Korea, and the Philippines.

Protests against the use of nuclear weapons have taken place in Western Europe, the United States, Japan, and Australia and have become more widespread during the 1980s. The protests include strikes, occupation of facilities, sabotage, demonstrations, and serious battles with the local authorities. Many of these protests have been against the use of nuclear power for the production of energy, but the nuclear arms race remains a world problem too.

Military Governments

Suppression of human rights is an issue that affects many people of the world. The worst conditions of suppression usually occur in countries governed by military rulers. Nearly 50 countries have some type of military regime (see Figure 19.6) and about half of those were established by military coups. These countries are located in Central and South America, Middle and North Africa, and South and Southeast Asia. At least half of the 13 countries of South America have governments with substantial military participation, and half of those are total military regimes. The military also governs eight countries in Africa, and another dozen African countries are mostly controlled by the military. Besides the states controlled by the military, some countries have civilian dictators that are backed by the military. Also, in many Communist countries that are run by civilian leaders, human rights are suppressed. When these countries are all considered, it is apparent that most of the people of the world are not free to chose their own paths in life.

Territorial Expansion

Little outright takeover of other peoples' territory has occurred in recent years. The Soviets, of course, attempted a takeover of Afghanistan, and they are fighting for territory in Iran and Iraq. Other territorial disputes plague the quest for world peace, but generally no great chunks of land have been gobbled up by foreign invaders in most of the world. The great territorial expansions of the 1980s are the claims made on the oceans. Serious offshore disputes started in the 1970s with the **"tuna wars"** between the United States and Peru. Since then, most of the countries bordering oceans have extended their offshore boundaries. The traditional three-mile limit of national jurisdiction has become a 200-mile "economic zone." If all countries that border the high seas use 200 miles as their limit of jurisdiction, as much as one-third of the oceans, or 70 percent of the globe, would be claimed. This invasion of the sea has led to many demarcation disputes. More than 40 such conflicts

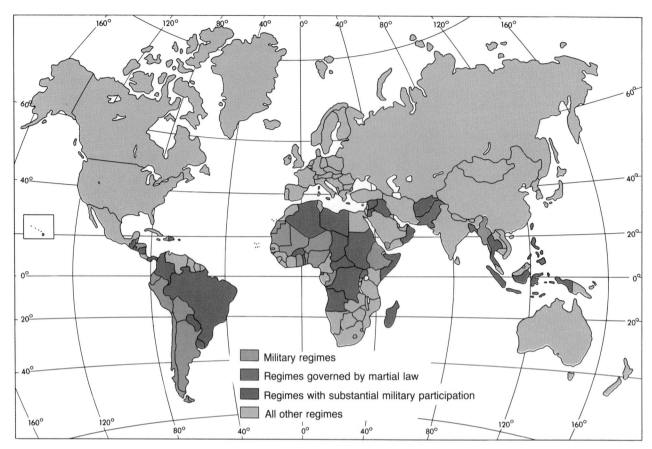

FIGURE 19.6

World Military Regimes. Many of the world's people lived under the authority of military governments or martial law. Also shown on the map are countries in which the military plays a dominant role in the government.

have occurred in the last 15 years, and the problems continue to arise.

Prospects for the Future

The pattern of world affairs may not be as gloomy as the problems mentioned above might make it appear. In many areas of the human condition there have been giant steps forward. World and regional organizations, as well as individual governments, are working on the problems that face the people of the world. With the exception of AIDS, most of the epidemic diseases have been brought under control. Governments are scrounging to find money to educate their people—not only to eliminate illiteracy but also to shed light on birth control and other health problems. Some governments have passed laws to protect the natural environment and to eliminate the destruction of land. If political problems are resolved, the nations that produce surplus food can feed the hungry of the world. Thus, much needless suffering occurs, but there is hope for a brighter future. For now, the world is our only home, and it must be preserved for our descendants.

Key Words

acid rain
air pollution
desertification
Environmental Protection Agency (EPA)
greenhouse effect
hazardous waste
nuclear weapons
population explosion
refugees
terrorism
tuna wars
water pollution
zero population growth (ZPG)

References and Readings

Boyd, Andrew. *An Atlas of World Affairs*. New York: Praeger, 1960.

James, Preston E., and Webb, Kempton E. *One World Divided: A Geographer Looks at the Modern World*. New York: Wiley, 1980.

Kidron, Michael, and Segal, Ronald. *The State of the World Atlas*. New York: Simon & Schuster, 1981.

Norris, Robert E., and Haring, L. Lloyd. *Political Geography*. Columbus, Ohio: Merrill, 1980.

Conversions for Units of Measurement

Appendix A

■ Length

1 inch (in.) = 2.54 centimeters (cm) = 25.4 millimeters (mm)
1 foot (ft) = 0.3048 meters (m)
1 meter (m) = 39.37 inches (in.) = 3.2808 feet (ft)
1 mile (mi) = 1.609 kilometers (km) = 5,280 feet (ft)
1 kilometer (km) = 0.6214 mile (mi) = 3,280.8 (ft)
1 nautical mile = 1.8533 kilometers (km) = 1.1518 miles (mi)

■ Area

1 acre = 43,650 square feet (ft^2) = 0.4047 hectare
1 hectare = 2.471 acres
1 square mile (mi^2) = 640 acres = 259 hectares = 2.59 square kilometers (km^2)
1 square kilometer (km^2) = 0.3861 square mile (mi^2)

■ Temperature

Degrees Fahrenheit = 9/5 (degrees Celsius) + 32
1° F = 0.556° C
Degrees Celsius = 5/9 (degrees Fahrenheit) − 32
1° C = 1.8° F
Number of Kelvin (k) = degrees Celsius + 273
1 Kelvin (k) = 1° C

■ Pressure

1 atmosphere (atm) = 1,013.25 millibars (mb) = 14.7 pounds per square inch = 29.92 inches of mercury (in.Hg) = 759.9 millimeters of mercury (mmHg)
1 inch of mercury (in.Hg = 25.4 millimeters of mercury (mmHg)
1 millibar (mb) = 0.02953 inch of mercury (in.Hg) = 0.75006 millimeters of mercury (mmHg)
1 millimeter of mercury (mmHg) = 0.03937 inch mercury (in.Hg) = 1.3332 millibars (mb)

Velocity

1 mile per hour (mph) = 1.467 feet per second = 0.447 meter per second = 1.61 kilometers per hour = 0.868 knot
1 meter per second = 3.6 kilometers per hour = 1,940 knots
1 knot = 0.514 meter per second = 1.854 kilometers per hour = 1.15 miles per hour

Mass and Weight

1 short ton (U.S.) = 2,000 pounds (lbs)
1 long ton (British) = 2,240 pounds (lb)
1 metric ton = 1,000 kilograms (kg) = 0.9842 long ton = 1.1032 short tons
1 kilogram = 2.20462 pounds (lb)
1 pound = 0.4536 kilograms = 453.6 grams (g)
1 pound (avoirdupois) = 7,000 grams = 16 ounces (oz)
1 pound (troy) = 5,760 grains (gr)

CLIMATIC CLASSIFICATION — *Appendix B*

Climatic classification is a tool for distinguishing regions that have different annual temperature and precipitation averages. The five basic climate groups are designated according to the Köppen system with the letters A, B, C, D, and E. The categories range in latitude from the topics (A) to the polar regions (E). The H category of climates includes most of the other categories, and is given to mountain or highland (H) areas. Subgroups of the major categories are indicated with secondary letters that refer to the seasonality of rainfall and the length of the summers. See the tables on the next two pages.

MAJOR CATEGORY	SUBGROUP (SECONDARY LETTER)	SUBGROUP (TERTIARY LETTER)	CHARACTERISTICS
A: Tropical Moist No month < 64.4° F (18° C)	Af: Tropical Wet (rainforest)	None	No dry season. Every month > 2.4 in. (6 cm) of rain.
	Aw: Savanna (grassland)	None	Winter dry season. Driest month < 2.4 in. (6 cm) of rain.
	Am: Monsoon (monsoon forest)	None	Excessive summer rain, short winter dry season.
B: Dry Evaporation exceeds precipitation	BS: Semiarid (steppe)		10 in. (25.4 cm) to 20 in. (50.8 cm) annual preciptitation.
	BW: Arid (desert)		< 10 in. (25.4 cm) annual precipitation
		BSh and BWh: Average annual temperature > 64.4° F (18° C)	Tropical and subtropical location
		BSk and BWk: Average annual temperature < 64.4° F (18° C)	Middle latitude location
C: Moist, Mild Winter Coldest month between 64.4° F (18° C) and 26.6° F (−3° C)	Cf: Humid Subtropical		No dry season; every month > 2.4 in. (6 cm) of perciptitation
	Cw: Marine		Winter dry season
	Cs: Mediterranean		West coast location, dry summers
		Cfa, Cwa, and Csa: Hot summers	Warmest month > 71.6° F (22° C)
		Cfb, Cwb, and Csb: Cool summers	Warmest month < 71.6° F (22° F)
		Cfc, Cwc, and Csc: Short, cool summers	Less than 4 months > 50° F (10° C)

MAJOR CATEGORY	SUBGROUP (SECONDARY LETTER)	SUBGROUP (TERTIARY LETTER)	CHARACTERISTICS
D: Moist, Severe Winters Coldest month, < 26.6° F (−3° C) Warmest month > 50° F (10° C)	Df: Humid Continental		No dry season; every month > 2.4 in. (6 cm) of precipitation
	Dw: Marine, Severe Winter		Winter dry season, east coast location
		Dfa and Dwa: Hot summers	Warmest month > 71.6° F (22° C)
		Dfb and Dwb: Cool summers	Warmest month < 71.6° F (22° C)
		Dfc and Dwc: Short, cool summers	Less than 4 months > 50° F (10° C)
		Dfd and Dwd: Very cold winters	Average of coldest month < −36.4° F (−38° C)
E: Polar Warmest month < 50° F (40° C)	ET: Tundra	None	Meager yearly precipitation
	EF: Ice Cap	None	Meager yearly precipitation, average annual temp. < 32° F (0° C)
H: Highlands	Many highlands, or mountain, regions are designated merely as having an H climate because the numerous climatic changes with elevation are impossible to portray on most maps, especially world maps.		

SELECTED DATA ON WORLD REGIONS

Appendix C

The following table contains selected information on the countries within the 14 regions discussed in the text. The figures on population are from the most recent estimates of the Population Reference Bureau and have been used by permission. See *World Population Data Sheet, 1988*. (Washington, D.C.: Population Reference Bureau, April, 1988).

1. **COUNTRY:** Country name.
2. **CAPITAL:** Capital City. Note: Some regions have more than one capital, and for such places as the Pacific Islands, the capital city category is blank.
3. **AREA:** The land surface area of the country given in square miles.
4. **POPULATION:** The mid-1988 population estimate for the country, given in millions of people.
5. **POPULATION DENSITY:** The population per square mile of territory for the country.
6. **CRUDE BIRTH RATE:** Annual number of births per 1,000 population.
7. **CRUDE DEATH RATE:** Annual number of deaths per 1,000 population.
8. **RATE OF NATURAL INCREASE:** Birth rate minus the death rate, implying the annual rate of population growth without regard for net migration.
9. **INFANT MORTALITY RATE:** The annual number of deaths of infants under age one year per 1,000 births.
10. **LIFE EXPECTANCY:** The average number of years a newborn infant can expect to live under current mortality levels.
11. **URBAN POPULATION:** Percentage of total population in areas termed urban by the country.
12. **PER CAPITA GNP:** The gross national product includes the value of all domestic and foreign output of the country, then the total is divided by the total population. The figures are for 1986 and are given in U.S. dollars.

Country	Capital City	Area (Sq Mi)	Population (Mid-1988 Millions)	Population Density (pop. per sq. miles)	Crude Birth Rate	Crude Death Rate	Natural Increase (%)	Infant Mortality Rate	Life Expectancy	Urban Population (%)	Per Capita GNP (1986, in U.S. $)
North America (Chapter 5)											
Canada	Ottawa	3,851,790	26.1	7	15	7	0.7	8	76	76	14,100
United States	Washington, D.C.	3,615,100	246.1	68	16	9	0.7	10	75	74	17,500
Northern Europe (Chapter 6)											
Denmark	Copenhagen	16,630	5.1	309	11	11	−0.1	8	75	84	12,640
Finland	Helsinki	130,130	4.9	38	12	10	0.3	6	74	62	12,180
Iceland	Reykjavik	39,770	0.2	6	16	7	0.9	5	77	90	13,370
Ireland	Dublin	27,140	3.5	131	17	10	0.8	9	73	56	5,080
Norway	Oslo	125,180	4.2	34	13	11	0.2	9	76	71	15,480
Sweden	Stockholm	173,730	8.4	49	12	11	0.1	6	77	83	13,170
United Kingdom	London	94,530	57.1	604	13	12	0.2	10	75	91	8,920
Western Europe (Chapter 6)											
Austria	Vienna	32,370	7.6	234	12	12	0.0	10	75	55	10,000
Belgium	Brussels	11,750	9.9	840	12	11	0.1	10	75	95	9,230
France	Paris	211,210	55.9	264	14	10	0.4	8	75	73	10,740
Germany, West	Bonn	95,980	61.2	638	10	11	−0.1	9	75	94	12,080
Luxembourg	Luxembourg	990	0.4	376	12	11	0.1	8	74	78	15,920
Netherlands	Amsterdam	14,410	14.7	1,024	13	9	0.4	8	76	89	10,050
Switzerland	Berne	15,940	6.6	413	12	9	0.3	7	77	61	17,840
Eastern Europe (Chapter 6)											
Bulgaria	Sofia	42,820	9.0	210	13	11	0.2	15	72	65	—
Czechoslovakia	Prague	49,370	15.6	317	14	12	0.2	14	71	74	—
Germany, East	East Berlin	41,830	16.6	397	13	13	0.0	9	72	77	—
Hungary	Budapest	35,920	10.6	295	12	14	−0.2	19	70	58	2,010
Poland	Warsaw	120,730	38.0	314	17	10	0.7	18	71	61	2,070
Romania	Bucharest	91,700	23.0	251	16	11	0.5	26	70	49	—
Southern Europe (Chapter 6)											
Albania	Tiranë	11,100	3.1	284	27	6	2.1	43	71	34	—
Greece	Athens	50,940	10.1	198	11	9	0.2	12	74	58	3,680

Country	Capital									
Italy	Rome	116,310	57.3	492	10	10	0.0	75	72	8,570
Malta	Valletta	2,833	0.4	120	15	8	0.7	75	85	3,470
Portugal	Lisbon	35,550	10.3	289	12	9	0.3	73	30	2,230
Spain	Madrid	194,900	39.0	200	12	8	0.4	76	91	4,840
Yugoslavia	Belgrade	98,760	23.6	239	15	9	0.6	70	47	2,300
Soviet Union (Chapter 7)										
U.S.S.R.	Moscow	8,649,500	283.0	33	20	10	1.0	69	65	7,400
Japan (Chapter 8)										
Japan	Tokyo	143,750	122.7	854	11	6	0.5	77	77	12,850
Australia and New Zealand (Chapter 9)										
Australia	Canberra	2,966,200	16.5	6	15	7	0.8	86	86	11,910
New Zealand	Wellington	103,883	3.3	32	16	8	0.8	84	84	7,110
Central America (Chapter 10)										
Belize	Belmopan	8,860	0.2	19	36	6	3.0	69	50	1,170
Costa Rica	San José	19,580	2.9	148	34	5	2.9	74	45	1,420
El Salvador	San Salvador	8,260	5.4	652	35	7	2.8	66	43	820
Guatemala	Guatemala	42,040	8.7	206	41	9	3.2	61	33	930
Honduras	Tuguciagalpa	43,280	4.8	111	39	8	3.1	63	40	740
Mexico	Mexico City	761,600	83.5	110	30	6	2.4	66	70	1,850
Nicaragua	Managua	50,190	3.6	72	43	8	3.5	62	57	790
Panama	Panama City	29,760	2.3	78	27	5	2.2	71	51	2,330
Caribbean Islands (Chapter 10)										
Antigua & Barbuda	St. Johns	170	0.1	494	15	5	1.0	72	31	2,380
Bahamas	Nassau	5,380	0.2	45	24	5	1.9	70	75	7,190
Barbados	Bridgetown	170	0.3	1,548	16	8	0.8	74	32	5,140
Cuba	Havana	42,800	10.4	242	16	6	1.0	74	71	—
Dominica	Roseau	290	0.1	338	21	5	1.6	75	—	1,210
Dominican Republic	Santo Domingo	18,810	6.9	365	32	8	2.4	65	52	710
Grenada	St. George's	130	0.1	641	26	7	1.9	72	—	1,240
Guadeloupe	Basse-Terre	690	0.3	493	20	7	1.3	72	46	—
Haiti	Port-au-Prince	10,710	6.3	588	41	13	2.8	54	25	330
Jamaica	Kingston	4,240	2.5	579	23	6	1.7	73	54	880
Martinique	Fort-de-France	420	0.3	777	18	6	1.2	74	71	—
Netherlands Antilles	Willemstad	300	0.2	607	19	6	1.4	76	—	—
Puerto Rico	San Juan	3,440	3.4	978	19	7	1.2	75	67	5,190

Country	Capital City	Area (Sq Mi)	Population (Mid-1988 Millions)	Population Density (pop. per sq. miles)	Crude Birth Rate	Crude Death Rate	Natural Increase (%)	Infant Mortality Rate	Life Expectancy	Urban Population (%)	Per Capita GNP (1986, in U.S. $)
St. Kitts-Nevis	Basse-terre	140	0.04	266	24	11	1.4	28	67	45	1,700
Saint Lucia	Castries	240	0.1	572	28	6	2.2	22	69	40	1,320
St. Vincent-Grenadines	Kingstown	150	0.1	716	26	7	2.0	27	69	—	960
Trinidad-Tobago	Port-of-Spain	1,980	1.3	646	29	7	2.2	13	70	34	5,120
South America (Chapter 11)											
Argentina	Buenos Aires	1,068,300	32.0	30	24	9	1.5	35	70	85	2,350
Bolivia	La Paz	424,160	6.9	16	40	14	2.6	110	53	49	540
Brazil	Brasilia	3,286,470	144.4	44	28	8	2.0	63	65	71	1,810
Chile	Santiago	292,260	12.6	43	22	6	1.6	20	71	82	1,320
Colombia	Bogotá	439,730	30.6	70	28	7	2.1	48	64	65	1,230
Ecuador	Quito	109,480	10.2	93	36	8	2.8	66	65	52	1,160
Guyana	Georgetown	83,000	0.8	9	26	6	2.0	36	68	32	500
Paraguay	Asunción	157,050	4.4	28	36	7	2.9	45	66	43	880
Peru	Lima	496,220	21.3	43	34	9	2.5	88	61	69	1,130
Suriname	Paramaribo	63,000	0.4	6	27	7	2.1	33	69	66	2,510
Uruguay	Montevideo	68,040	3.0	44	18	10	0.8	30	71	84	1,860
Venezuela	Caracas	352,140	18.8	53	29	5	2.4	36	70	82	2,930
Northern Africa (Chapter 12)											
Algeria	Algiers	919,590	24.2	26	42	10	3.2	81	61	43	2,570
Egypt	Cairo	386,660	53.3	138	38	9	2.8	93	59	45	760
Libya	Tripoli	679,360	4.0	6	39	8	3.1	74	65	76	—
Morocco	Rabat	172,410	25.0	145	36	10	2.6	90	61	43	590
Sudan	Khartoum	967,490	24.0	25	45	16	2.8	112	49	20	320
Tunisia	Tunis	63,170	7.7	122	31	9	2.2	71	63	53	1,140
Western Sahara	—	103,000	0.2	2	49	26	2.5	—	—	—	—
Western Africa (Chapter 12)											
Benin	Porto-Novo	43,480	4.5	103	51	20	3.0	115	45	39	270
Burkina Faso	Ouagadougou	105,870	8.5	80	48	19	2.8	145	46	8	150
Cape Verde	Praia	1,560	0.3	224	35	8	2.6	77	60	27	460
Gambia	Banjul	4,360	0.8	179	49	28	2.1	169	36	21	230
Ghana	Accra	92,100	14.4	156	42	11	3.1	72	58	31	390
Guinea	Conakry	94,930	6.9	73	47	23	2.4	153	41	22	—
Guinea-Bissau	Bissau	13,950	0.9	68	44	20	2.4	132	45	27	170

Country	Capital								
Ivory Coast	Abidjan	124,500	11.2	90	46	15	3.1	105	52	43	740
Liberia	Monrovia	43,000	2.5	57	48	16	3.1	127	50	42	450
Mali	Bamako	478,760	8.7	18	50	22	2.9	175	43	18	170
Mauritania	Nouakchott	397,950	2.1	5	50	20	3.0	132	45	35	440
Niger	Niamey	489,190	7.2	15	51	22	2.9	141	44	16	260
Nigeria	Lagos	356,670	111.9	314	46	17	2.9	122	47	28	640
Senegal	Dakar	75,750	7.0	92	46	20	2.6	137	44	36	420
Sierra Leone	Freetown	27,700	4.0	143	47	29	1.8	175	35	28	310
Togo	Lomé	21,930	3.3	152	47	14	3.3	117	54	22	250
Eastern Africa (Chapter 12)											
Burundi	Bujumbura	10,750	5.2	480	47	18	2.9	119	48	5	240
Comoros	Moroni	690	0.4	618	47	14	3.4	96	55	23	280
Djibouti	Djibouti	8,490	0.3	38	43	18	2.5	132	47	74	—
Ethiopia	Addis Ababa	471,780	48.3	102	46	15	3.0	118	50	10	120
Kenya	Nairobi	224,960	23.3	104	54	13	4.1	76	54	19	300
Madagascar	Antananarivo	226,660	10.9	48	44	16	2.8	63	51	22	230
Malawi	Lilongwe	45,750	7.7	168	53	21	3.2	157	46	12	160
Mauritius	Port Louis	790	1.1	1,392	19	7	1.2	27	68	42	1,200
Mozambique	Maputo	309,490	15.1	49	45	19	2.6	147	46	19	210
Reunion	St. Denis	970	0.6	575	24	6	1.8	12	70	60	3,940
Rwanda	Kigali	10,170	7.1	694	53	16	3.7	122	49	6	290
Seychelles	Victoria	170	0.1	404	26	8	1.9	17	70	37	—
Somalia	Mogadishu	246,200	8.0	32	48	17	3.1	147	41	34	280
Tanzania	Dar-es-Salaam	364,900	24.3	67	50	15	3.6	111	52	18	240
Uganda	Kampala	91,140	16.4	180	50	16	3.4	108	50	10	—
Zambia	Lusaka	290,580	7.5	26	50	13	3.7	87	55	43	300
Zimbabwe	Harare	150,800	9.7	65	47	12	3.5	76	57	24	620
Middle Africa (Chapter 12)											
Angola	Luanda	481,350	8.2	17	47	21	2.6	143	43	25	—
Cameroon	Yaoundé	183,570	10.5	57	43	16	2.6	126	50	42	910
Central African Rep.	Bangui	240,530	2.8	12	44	19	2.5	148	45	35	290
Chad	Ndjaména	495,750	4.8	10	43	23	2.0	143	39	27	—
Congo	Brazzaville	132,050	2.2	16	47	13	3.4	112	55	48	1,040
Equatorial Guinea	Malabo	10,830	0.3	32	38	20	1.9	130	45	60	—
Gabon	Libreville	103,350	1.3	12	34	18	1.6	112	49	41	3,020
São Tomé & Principe	São Tomé	370	0.1	316	36	9	2.7	62	65	35	340
Zaire	Kinshasa	905,560	33.3	37	45	15	3.0	103	51	34	160

Country	Capital City	Area (Sq Mi)	Population (Mid-1988 Millions)	Population Density (pop. per sq. miles)	Crude Birth Rate	Crude Death Rate	Natural Increase (%)	Infant Mortality Rate	Life Expectancy	Urban Population (%)	Per Capita GNP (1986, in U.S. $)
Southern Africa (Chapter 12)											
Botswana	Gaborone	231,800	1.3	5	48	14	3.4	67	58	22	840
Lesotho	Maseru	11,720	1.6	140	41	16	2.6	106	50	17	410
Namibia	Windhoek	318,260	1.7	5	45	17	2.8	111	49	51	1,020
South Africa	Pretoria	471,440	35.1	74	32	10	2.3	66	60	56	1,800
Swaziland	Mbabane	6,700	0.7	106	47	17	3.1	124	50	26	600
The Middle East (Chapter 13)											
Bahrain	Manama	240	0.5	2,105	32	5	2.8	32	67	81	8,530
Cyprus	Nicosia	3,570	0.7	192	20	8	1.1	12	75	62	4,360
Iraq	Baghdad	167,920	17.6	105	45	10	3.5	86	62	68	—
Israel	Jerusalem	8,020	4.4	552	23	7	1.6	11	75	89	6,210
Jordan	Amman	37,740	3.8	102	42	6	3.6	56	69	59	1,540
Kuwait	Kuwait	6,880	2.0	285	32	3	2.9	18	72	80	13,890
Lebanon	Beirut	4,020	3.3	831	28	7	2.0	51	67	80	—
Oman	Muscat	82,030	1.4	17	47	14	3.3	117	52	9	4,990
Qatar	Doha	4,250	0.4	98	30	2	2.7	42	69	86	12,520
Saudi Arabia	Riyadh	830,000	14.2	17	42	9	3.3	85	61	72	6,930
Syria	Damascus	71,500	11.3	159	47	9	3.8	59	64	49	1,560
Turkey	Ankara	301,380	52.9	175	31	9	2.2	95	63	53	1,110
United Arab Emirates	Abu Dhabi	32,280	1.5	46	30	4	2.6	38	68	81	14,410
Yemen, North	San'a	75,290	6.7	89	55	22	3.3	175	45	15	550
Yemen, South	Aden	128,560	2.4	19	48	15	3.3	116	51	40	480
Southern Asia (Chapter 14)											
Afghanistan	Kabul	250,000	14.5	58	48	24	2.4	183	39	15	—
Bangladesh	Dacca	55,600	109.5	1,969	43	17	2.7	135	50	16	160
Bhutan	Thimphu	18,150	1.5	83	38	18	2.0	139	46	5	160
India	New Delhi	1,269,340	816.8	644	33	13	2.0	104	54	25	270
Iran	Tehran	636,290	51.9	82	45	13	3.2	113	57	51	—
Maldives	Malé	120	0.2	1,753	47	10	3.7	83	60	26	310
Nepal	Kathmandu	54,360	18.3	336	42	17	2.5	112	52	7	160
Pakistan	Islamabad	310,400	107.5	346	43	15	2.9	121	54	28	350
Sri Lanka	Colombo	25,330	16.6	657	24	6	1.8	31	70	22	400

Southeast Asia (Chapter 15)									
Brunei	Bandar Seri Begawan	2,230	0.3	116	31	3	2.7	11	71	64	15,400
Myanma	Rangoon	261,220	41.1	157	34	13	2.1	103	53	24	200
Indonesia	Jakarta	735,360	117.4	241	27	10	1.7	88	58	22	500
Cambodia	Phnom Penh	69,900	6.7	96	40	17	2.3	134	48	11	—
Laos	Vientiane	91,430	3.8	42	41	16	2.5	122	50	16	—
Malaysia	Kuala Lumpur	127,320	17.0	133	31	7	2.4	30	67	35	1,850
Philippines	Manila	115,830	63.2	546	35	7	2.8	51	66	41	570
Singapore	Singapore	220	2.6	11,789	15	5	1.0	9	73	100	7,410
Thailand	Bangkok	198,460	54.7	276	29	8	2.1	52	64	17	810
Vietnam	Hanoi	127,240	65.2	512	34	8	2.6	53	63	19	—
East Asia (Chapter 16)									
China	Beijing	3,705,390	1,087.0	293	21	7	1.4	44	66	41	300
Hong Kong	Victoria	400	5.7	14,185	13	5	0.8	8	76	93	6,720
Korea, North	Pyongyang	46,540	21.9	471	31	6	2.5	33	68	64	—
Korea, South	Seoul	38,020	42.6	1,121	19	6	1.3	30	68	65	2,370
Macao	Macao	10	0.4	55,167	23	6	1.7	12	68	97	—
Mongolia	Ulan Bator	604,250	2.0	3	37	11	2.6	53	62	52	—
Taiwan	Taipei	13,890	19.8	1,422	16	5	1.1	7	73	67	—
Pacific Islands (Chapter 17)									
Fiji	Suva	7,050	0.7	105	28	5	2.3	21	67	37	1,810
French Polynesia	Papeete	1,540	0.2	124	28	4	2.4	23	71	57	—
New Caledonia	Nouméa	7,360	0.2	21	24	6	1.8	36	68	60	—
The Pacific Islands[a]	—	690	0.2	233	34	5	2.9	33	71	29	—
Papua–New Guinea	Port Moresby	178,260	3.7	20	36	12	2.4	100	54	13	690
Soloman Islands	Honiara	10,980	0.3	28	41	6	3.6	42	68	9	530
Vanuatu	Vila	5,700	0.2	27	38	5	3.3	38	69	18	—
Western Samoa	Apia	1,100	0.2	153	37	8	2.9	50	65	21	680
World		5,128		28	10	1.7	77	63	45	3,010

[a]Includes Federated States of Micronesia, Palau, and the Marshall and North Mariana islands.

BIBLIOGRAPHY

Abler, Ronald, et al. *Spatial Organization: The Geographer's View of the World.* Englewood Cliffs, N.J.: Prentice-Hall, 1971.

Ablin, David A., and Hood, Marlowe (eds.). *The Cambodian Agony.* Armonk, N.Y.: Sharpe, 1987.

Abu-Lughod, J. *Rabat: Urban Apartheid in Morocco.* Princeton: Princeton University Press, 1980.

Adalenro, I. A. *Marketplaces in a Developing Country: The Case of Western Nigeria.* Ann Arbor: University of Michigan Press, 1981.

Adams, John Q. "Settlements of the Northwestern Canadian Arctic." *Geographical Review* 27 (1941): 112–126.

Adams, Robert M. *Land Behind Baghdad: A History of Settlement on the Diyala Plains.* Chicago: University of Chicago Press, 1965.

Adler, Ron K., et al. (eds.). *Atlas of Israel.* New York: Macmillan, 1985.

Africa at a Glance. Pretoria: Africa Institute of South Africa, 1978.

Aguilar, Luis E. *Latin America, 1984.* Washington, D.C.: Stryker-Post, 1984.

Aharoni, Yohanan. *The Land of the Bible: A Historical Geography.* Translated from the Hebrew by A. F. Rainey. Philadelphia: Westminster, 1967.

Ahmad, Kazi S. U. *A Geography of Pakistan.* Pakistan Branch: Oxford University Press, 1964.

Ajayi, J. F., and Crowder, Michael (eds.). *Historical Atlas of Africa.* New York: Cambridge University Press, 1985.

Akaha, Tsuneo. *Japan in Global Ocean Politics.* Honolulu: University of Hawaii Press, 1985.

Al-Farsy, Fouad. *Saudi Arabia: A Case Study in Development.* New York: Methuen, 1986.

Allen, G. C. *A Short Economic History of Modern Japan.* New York: St. Martin's, 1981.

Allen, James P., and Turner, Eugene J. *We the People: An Atlas of America's Ethnic Diversity.* New York: Macmillan, 1987.

Alley, Roderic (ed.). *New Zealand and the Pacific.* Boulder, Colo.: Westview, 1983.

Allworth, Edward (ed.). *Ethnic Russia in the U.S.S.R.: The Dilemma of Dominance.* Elmsford, N.Y.: Pergamon, 1980.

Altschul, D. R. "Transportation in African Development," *Journal of Geography* 79 (1980): 44–56.

Al-Wohaibi, Mohammed A. "Nomads in Al-Hejaz Province: A Geographic Study of Nomads near the City of Taif." Master's thesis, Oklahoma State University, 1979.

Ambler, John (ed.). *Soviet and East European Transport Problems.* New York: St. Martin's, 1985.

Amirahmadi, Hooshang, and Parvin, Manoucher (eds.). *Post-Revolutionary Iran.* Boulder, Colo.: Westview, 1987.

Anderson, Thomas D. *Geopolitics of the Caribbean: Ministates in a Wider World.* Westport, Conn.: Praeger, 1984.

Andrusz, Gregory D. *Housing and Urban Development in the U.S.S.R.* Albany: State University of New York Press, 1985.

Anthony, John. *Arab States of the Lower Gulf: People, Politics, and Petroleum.* Washington, D.C.: Middle East Institute, 1975.

Armstrong, Warwick, and McGee, Terence G. *Theatres of Accumulation: Studies in Asian and Latin American Urbanization.* New York: Methuen, 1985.

Asad, Talal, and Owen, Roger (eds.). *The Middle East.* London: Macmillan, 1983.

Atlas of Hawaii. Honolulu: Hawaii University Press, 1983.

Atlas of North America: Space-Age Portrait of a Continent. Washington, D.C.: National Geographic Society, 1985.

Atlas of the United States: A Thematic and Comparative Approach. New York: Macmillan, 1987.

Attenborough, David. *People or Paradise.* New York: Harper, 1961.

Baird, Patrick D. *The Polar World.* New York: Wiley, 1964.

Bamford, C. G. *Geography of the EEC.* Plymouth, Eng.: MacDonald & Evans, 1983.

Baransky, N. N. *Economic Geography of the U.S.S.R.* Translated from the Russian by S. Belsky. Moscow: Foreign Language Publishing House, 1956.

Barbour, Kenneth M. *The Republic of the Sudan: A Regional Geography.* London: University of London Press, 1961.

Barbour, Nevill (ed.). *A Survey of North-West Africa: The Maghrib.* London: Oxford University Press, 1962.

Barker, Randolph, et al. *The Rice Economy of Asia.* Washington, D.C.: Resources of the Future, 1985.

Barnes, James N. *Let's Save Antarctica.* Richmond, Aust.: Greenhouse, 1982.

Barnet, A. Doak, and Clough, Ralph N. (eds.). *Modernizing China: Post-Mao Reform and Development.* Boulder, Colo.: Westview, 1985.

Barrett, Raes D., and Ford, Roslyn A. *Patterns in the Human Geography of Australia.* Melbourne: Macmillan, 1987.

Bater, James H. *The Soviet City: Ideal and Reality.* Beverly Hills, Calif.: Sage, 1980.

Bater, James H., and French, Richard A. (eds.). *Studies in Russian Historical Geography.* New York: Academic Press, 1983.

Baum, Richard D. (ed.). *China's Four Modernizations: The New Technological Revolution.* Boulder, Colo.: Westview, 1980.

Baxter, Craig. *Bangladesh: A New Nation in an Old Setting.* Boulder, Colo.: Westview, 1984.

Bayliss-Smith, Tim P., and Wanmali, Sudhir (eds.). *Understanding Green Revolution: Agrarian Change and Development Planning in South Asia.* New York: Cambridge University Press, 1984.

Beaumont, Peter, and McLachlan, Keith (eds.). *Agricultural Development in the Middle East.* New York: Wiley, 1986.

Beaumont, Peter, et al. *The Middle East: A Geographical Study.* New York: Wiley, 1976.

Beazley, Mitchell (ed.). *The Rand McNally Atlas of the Oceans.* New York: Rand McNally, 1977.

Becker, Elizabeth. *When the War Was Over: The Voices of Cambodia's Revolution and Its People.* New York: Simon & Schuster, 1986.

Bell, Morag. *Contemporary Africa: Development, Culture, and the State.* White Plains, N.Y.: Longman, 1986.

Benjamin, Roger, and Kudrle, Robert T. (eds.). *The Industrial Future of the Pacific Basin.* Boulder, Colo.: Westview, 1984.

Benjamin, Thomas, and McNellie, William (eds.). *Other Mexicos: Essays on Regional Mexican History.* Albuquerque: University of New Mexico Press, 1984.

Bennett, C. F., Jr. *Conservation and Management of Natural Resources in the United States.* New York: Wiley, 1983.

Bentley, C. R. "Glacial and Subglacial Topography of West Antarctica." *Journal of Glaciology* 3, no. 29 (1961): 882–911.

Benton, G. *The Hong Kong Crisis.* Dover, N.H.: Pluto, 1980.

Berg, Robert J., and Whitaker, Jennifer S. *Strategies for African Development.* Berkeley, Calif.: University of California Press, 1986.

Berg, V. den L., et al. *Urban Europe: A Study in Growth and Decline.* Oxford: Pergamon, 1982.

Bergson, Abram, and Levine, Herbert S. (eds.). *The Soviet Economy: Toward the Year 2000.* Winchester, Mass.: Allen & Unwin, 1983.

Bernardi, Debra (ed.). *Fodor's Australia, New Zealand, and the South Pacific.* New York: Fodor's, 1985.

Berry, Brian J. L. *Essays on Commodity Flows and the Spatial Structure of the Indian Economy.* University of Chicago, Department of Geography, Research Papers, no. 111. Chicago: University of Chicago, 1966.

Bertram, G. C. L. *Arctic and Antarctic: A Prospect of the Polar Regions.* Cambridge: Heffer & Sons, 1958.

Beyer, J. L. "Africa." In *World Systems of Traditional Resource Management,* ed. G. A. Klee. London: Arnold, 1980.

Bialer, Seweryn. *The Soviet Paradox: External Expansion. Internal Decline.* New York: Knopf, 1986.

Birdsall, Stephen S., and Florin, John W. *Regional Landscapes of the United States and Canada.* New York: Wiley, 1985.

Bishop, Elizabeth. *Brazil.* New York: Time Inc., 1967.

Blackbourn, Anthony, and Putnam, Robert G. *The Industrial Geography of Canada.* New York: St. Martin's, 1984.

Blainey, Geoffrey. *Across a Red World.* New York: St. Martin's, 1968.

Blake, Gerald H., and Lawless, Richard I. (eds.). *The Changing Middle Eastern City.* Totowa, N.J.: Barnes & Noble, 1980.

Blakemore, Harold, and Smith, Clifford T. (eds.). *Latin America: Geographical Perspectives.* New York: Methuen, 1983.

Blakemore, Harold, et al. (eds.). *The Cambridge Encyclopedia of Latin America and the Caribbean.* London: Cambridge University Press, 1985.

Blouet, Brian W., and Blouet, Olwyn M. (eds.). *Latin America: An Introductory Survey.* New York: Wiley, 1982.

Blumefeld, Jesmond (ed.). *South Africa in Crisis.* Beckenham, Eng.: Croom Helm, 1987.

Boateng, Ernest A. *A Geography of Ghana.* Cambridge: Cambridge University Press, 1966.

Boehm, Richard G., and Visser, Sent (eds.). *Latin America: Case Studies.* Dubuque, Iowa: Kendall/Hunt, 1984.

Bogardus, James Furnas. *Europe: A Geographical Survey.* New York: Harper, 1934.

Bolland, O. Nigel. *Belize: A New Nation in Central America.* Boulder, Colo.: Westview, 1986.

Bolton, G. *Spoils and Spoilers: Australians Make Their Environment, 1788–1980.* Boston: Allen & Unwin, 1981.

Borchert, John R. *America's Northern Heartland.* Minneapolis: University of Minnesota Press, 1987.

Boyd, Andrew. *An Atlas of World Affairs.* New York: Praeger, 1960.

Brander, Bruce. *Australia.* Washington, D.C.: National Geographic Society, 1968.

Braveboy-Wagner, Jacqueline A. *The Venezuela-Guyana Border Dispute: Britain's Colonial Legacy in Latin America.* Boulder, Colo.: Westview, 1984.

Brewster, Barney. *Antarctica: Wilderness at Risk.* San Francisco: Friends of the Earth, 1982.

Brice, William C. *South-West Asia: Systematic Regional Geography 8.* London: University of London Press, 1967.

Bromley, Rosemary D. F., and Bromely, R. *South American Development: A Geographical Introduction*. New York: Cambridge University Press, 1982.

Brookfield, Harold C., and Hart, Doreen. *Melanesia: A Geographical Interpretation of an Island World*. New York: Barnes & Noble, 1971.

Brown, Archie, et al. (eds.). *Cambridge Encyclopedia of Russia and the Soviet Union*. New York: Cambridge University Press, 1982.

Brown, DeSoto. *Hawaii Recalls: Nostalgic Images of the Hawaiian Islands, 1910–1950*. Boston: Routledge & Kegan Paul, 1986.

Brown, Joe David. *India*. New York: Time Inc., 1964.

Brown, Paula. *Highland Peoples of New Guinea*. New York: Cambridge University Press, 1978.

Brown, Peter G., and Shue, Henry (eds.). *The Border That Joins: Mexican Migrants and U.S. Responsibility*. Totowa, N.J.: Rowman & Littlefield, 1983.

Brown, R. N. R. "Plant Life in the Antarctic." *Discovery* 4, no. 42 (1923): 149–153.

Bruemmer, Fred. *The Arctic World*. San Francisco: Sierra Club Books, 1985.

Brunn, Stanley D., and Wheeler, James O. (eds.). *The American Metropolitan System: Present and Future*. New York: Halsted, 1980.

Buchanan, Keith, et al. *China: The Land and People*. New York: Crown, 1981.

Bunge, Frederica (ed.). *North Korea: A Country Study*. Washington, D.C.: Government Printing Office, 1981.

———. *South Korea: A Country Study*. Washington, D.C.: Government Printing Office, 1982.

Burki, Shahid J. *Pakistan: A Nation in the Making*. Boulder, Colo.: Westview, 1986.

Burks, Ardath W. *Japan: A Postindustrial Power*. Boulder, Colo.: Westview, 1984.

Burley, T. M. *The Philippines: An Economic and Social Geography*. London: Bell, 1973.

Burling, Robbins. *Hill Farms and Padi Fields: Life in Mainland Southeast Asia*. Englewood Cliffs, N.J.: Prentice-Hall, 1965.

Burnley, Ian H. *Population, Society, and Evironment in Australia*. Melbourne: Shillington House, 1982.

Burnley, Ian H., and Forrest, James (eds.). *Living in Cities: Urbanism and Society in Metropolitan Australia*. Winchester, Mass.: Allen & Unwin, 1985.

Burtenshaw, David, et al. *The City in West Europe*. New York: Wiley, 1981.

Butland, Gilbert J. *Latin America: A Regional Geography*. 3d ed. New York: Wiley, 1972.

Calvert, Peter. *Guatemala: A Nation in Turmoil*. Boulder, Colo.: Westview, 1986.

Cameron, Ian. *Antarctica: The Last Continent*. Boston: Little, Brown, 1974.

Carim, Enver (ed.). *Latin America and the Caribbean, 1983*. Essex, Eng.: World Information, 1983.

Carlson, Fred A. *Geography of Latin America*. New York: Prentice-Hall, 1943.

Carter, Gwendolen M. *Which Way Is South Africa Going?* Bloomington: Indiana University Press, 1980.

Carter, Jeff. *Outback in Focus*. London: Angus & Robertson, 1968.

Central Intelligence Agency. *The World Factbook*. Washington, D.C.: Government Printing Office, 1985.

Chang, Kuei-sheng. "The Changing Railroad Pattern in Mainland China." *Geographical Review* 51 (1961): 534–548.

Chang, S-d. "Modernization and China's Urban Development." *Annals of the Association of American Geographers* 71 (1981): 202–219.

Chapman, Murray, and Prothero, R. Mansell (eds.). *Circulation in Population Movement: Substance and Concept from the Melanesian Case*. Boston: Routledge & Kegan Paul, 1985.

Child, Jack. *Geopolitics and Conflict in South America: Quarrels Among Neighbors*. New York: Praeger, 1985.

Chiu, T. N., and So, C. L. (eds.). *A Geography of Hong Kong*. London: Oxford University Press, 1983.

Christian, Shirley. *Nicaragua: Revolution in the Family*. New York: Random House, 1985.

Christopher, Anthony J. *Colonial Africa: An Historical Geography*. Totowa, N.J.: Barnes & Noble, 1984.

———. *South Africa*. New York: Longman, 1982.

Christopher, Robert C. *The Japanese Mind: The Goliath Explained*. New York: Linden, 1983.

Chudacoff, Howard P. *The Evolution of American Urban Society*. Englewood Cliffs, N.J.: Prentice-Hall, 1981.

Clark, Audrey N. (ed.). *Longman Dictionary of Geography: Human and Physical*. White Plains, N.Y.: Longman, 1985.

Clark, David. *Post-Industrial America: Geographical Perspectives*. New York: Methuen, 1985.

Clark, John I., and Bowen-Jones, Howard (eds.). *Change and Development in the Middle East*. New York: Methuen, 1981.

Clark, John I., and Kosinski, Leszek A. (eds.). *Redistribution of Population in Africa*. Exeter, N.H.: Heinemann, 1982.

Clark, John I., et al. (eds.). *Population and Development Projects in Africa*. New York: Cambridge University Press, 1985.

Clark, M. G. *Economics of Soviet Steel*. Cambridge, Mass.: Harvard University Press, 1956.

Cloudsley-Thompson, J. L. (ed.). *Sahara Desert*. Oxford: Pergamon, 1984.

Clout, Hugh D. *Regional Variations in the European Community*. Cambridge: Cambridge University Press, 1986.

Clout, Hugh D., et al. *Western Europe: Geographical Perspectives*. White Plains, N.Y.: Longman, 1985.

Clout, Hugh D., (ed.). *Regional Development in Western Europe*. New York: Wiley, 1981.

Cohen, Saul B. *The Geopolitics of Israel's Border Question.* Boulder, Colo.: Westview, 1987.

Cole, John P. *Development and Underdevelopment: A Profile of the Third World.* New York: Methuen, 1987.

——— . *Geography of the Soviet Union.* London: Butterworths, 1984.

——— . *Latin America: An Economic and Social Geography.* London: Butterworths, 1965.

Cole, Monica M. *South Africa.* London: Methuen, 1966.

Coon, Carleton S. *Caravan: The Story of the Middle East.* New York: Holt, 1958.

Costello, Michael A., et al. *Mobility and Employment in Urban Southeast Asia: Examples from Indonesia and the Philippines.* Boulder, Colo.: Westview, 1987.

Cottrell, Alvin J. (ed.). *The Persian Gulf States: A General Survey.* Baltimore: Johns Hopkins University Press, 1980.

Couper, Alastair D. (ed.). *The Times Atlas of the Oceans.* New York: Van Nostrand Rinehold, 1983.

Courtney, Percy P. *Northern Australia: Patterns and Problems of Tropical Development in an Advanced Country.* New York: Longman, 1983.

Craig, Robert D., and King, Frank P. (ed.). *Historical Dictionary of Oceania.* Westport, Conn.: Greenwood, 1981.

Crankshaw, Edward. *Russia and the Russians.* New York: Viking, 1948.

Crassweller, Robert D. *The Caribbean Community: Changing Societies and U.S. Policy.* New York: Praeger, 1972.

Cressey, George B. *Asia's Lands and Peoples.* New York: McGraw-Hill, 1951.

——— . *Crossroads: Land and Life in Southwest Asia.* Chicago: Lippincott, 1960.

——— . *How Strong Is Russia? A Geographical Appraisal.* Syracuse, N.Y.: Syracuse University Press, 1954.

——— . *Land of the 500 Million.* New York: McGraw-Hill, 1955.

Crow, Ben, and Thomas, Allen (eds.). *Third World Atlas.* Philadelphia: Taylor & Francis, 1984.

Cumberland, Charles C. *Mexico: The Struggle of Modernity.* New York: Oxford University Press, 1968.

Cumberland, Kenneth B. *Southwest Pacific.* New York: McGraw-Hill, 1956.

Daly, M. *Sydney Boom, Sydney Bust.* Sydney, Aust.: Allen & Unwin, 1982.

Davenport, T. R. H. *South Africa: A Modern History.* Toronto: University of Toronto Press, 1982.

Davidson, Basil. *Africa: History of a Continent.* New York: Macmillan, 1966.

Davidson, William V., and Parsons, James J. (eds.). *Historical Geography of Latin America.* Baton Rouge: Louisiana State University Press, 1980.

Davis, Kingsley. *The Population of India and Pakistan.* New York: Russell & Russell, 1968.

Dawson, Owen L. *Communist China's Agriculture.* New York: Praeger, 1970.

Deacon, George E. R. *The Antarctic Circumpolar Ocean.* New York: Cambridge University Press, 1984.

Debenham, Frank. *Antarctica: The Story of a Continent.* New York: Macmillan, 1961.

de Blij, Harm J. *Africa South,* Evanston, Ill.: Northwestern University Press, 1962.

——— . *Wine: A Geographic Appreciation.* Totowa, N.J.: Rowman & Allanheld, 1983.

——— . *Wine Regions of the Southern Hemisphere.* Totowa, N.J.: Rowman & Allanheld, 1985.

de Blij, Harm J. and Martin, E. (eds.). *African Perspectives: An Exchange of Essays on the Economic Geography of Nine African States.* New York: Methuen, 1981.

Decrespigny, Rafe. *China: The Land and Its People.* New York: St. Martin's, 1974.

DeEast, William Gordon, and Spate, Oskar H. K. (eds). *The Changing Map of Asia: A Political Geography.* 4th ed. New York: Barnes & Noble, 1961.

Deeb, Marius, and Deeb, Mary Jane. *Libya Since the Revolution: Aspects of Social and Political Development.* Westport, Conn.: Praeger, 1982.

Delaplane, Stanton. *Pacific Pathways.* New York: McGraw-Hill, 1963.

Demko, George J. (ed.). *Regional Development Problems and Policies in Eastern and Western Europe.* New York: St. Martin's, 1984.

DeSilva, K. M. *Managing Ethnic Tensions in Multi-Ethnic Societies: Sri Lanka, 1880–1985.* Lanham, Md.: University Press of America, 1986.

Deutsch, Robert. *The Food Revolution in the Soviet Union and Eastern Europe.* Boulder, Colo.: Westview, 1986.

Devlin, John F. *Syria: Modern State in an Ancient Land.* Boulder, Colo.: Westview, 1983.

Dewdney, John C. *A Geography of the Soviet Union.* Elmsford, N.Y.: Pergamon, 1979.

——— . *U.S.S.R. in Maps.* New York: Holmes & Meier, 1982.

Diagram Group. *Atlas of Central America and the Caribbean.* New York: Macmillan, 1985.

Dicken, P., and Lloyd, P. E. *Modern Western Society: A Geographical Perspective on Work, Home, and Well-Being.* New York: Harper & Row, 1981.

Dickenson, John P. *Brazil.* New York: Longman, 1983.

Dickenson, John P., et al. *Geography of the Third World.* New York: Methuen, 1983.

Diem, Aubrey. *Western Europe: A Geographical Analysis.* New York: Wiley, 1979.

Ding, C. "The Economic Development of China." *Scientific American.* 243 (1980): 152–165.

Dobby, Ernest H. G. *Southeast Asia.* 10th ed. London: University of London Press, 1967.

Dohrs, Mary Ellen. *Sketches of the Russian People.* Detroit: Garelick's Gallery, 1959.

Dominguez, Jorge I. (ed.). *Cuba: Internal and International Affairs.* Beverly Hills, Calif.: Sage, 1982.

Dozier, Craig L. *Nicaragua's Mosquito Shore: The Years of British and American Presence.* Tuscaloosa: University of Alabama Press, 1985.

Drysdale, Alasdair, and Blake, Gerald H. *The Middle East and North Africa: A Political Geography.* New York: Oxford University Press, 1985.

Dutt, Ashok K. "Cities of South Asia." In *Cities of the World: World Regional Urban Development,* ed. Stanley D. Brunn and Jack F. Williams. New York: Harper & Row, 1983.

Dutt, Ashok K. (ed.). *Contemporary Perspectives on the Medical Geography of South and Southeast Asia.* Elmsford, N.Y.: Pergamon, 1981.

Dutt, Ashok K. (ed.). *Southeast Asia: Realm of Contrasts.* Boulder, Colo.: Westview, 1985.

Dutt, Ashok K., and Geib, Margaret. *An Atlas of South Asia.* Boulder, Colo.: Westview, 1987.

East, William Gordon, and Spate, Oskar H. K. (eds.). *The Changing Map of Asia: A Political Geography.* 4th ed. New York: Barnes & Noble, 1961.

Eckstein, Alexander (ed.). *Economic Trends in Communist China.* Chicago: Aldine, 1968.

Eliade, Mircea. *Australian Religions.* Ithaca, N.Y.: Cornell University Press, 1973.

El Mallakh, R. *The Economic Development of the United Arab Emirates.* New York: St. Martin's, 1981.

Embleton, Clifford (ed.). *Geomorphology of Europe.* New York: Wiley, 1984.

Enders, Thomas O., and Mattione, Richard P. *Latin America: The Crisis of Debt and Growth.* Washington, D.C.: Brookings Institute, 1984.

Falk, Pamela S. *Petroleum and Mexico's Future.* Boulder, Colo.: Westview, 1986.

Farington, Hugh. *Confrontation: The Strategic Geography of NATO and the Warsaw Pact.* Boston: Routledge & Kegan Paul, 1986.

Farmer, Bertram H. *An Introduction to South Asia.* New York: Methuen, 1984.

Fearnside, Philip M. *Human Carrying Capacity of the Brazilian Rainforest.* New York: Columbia University Press, 1986.

Feinberg, Richard E. (ed.). *Central America: International Dimensions of the Crisis.* New York: Holmes & Meier, 1982.

Ferguson, J. Halcro. *The River Plate Republics: Argentina, Paraguay, Uruguay.* New York: Time Inc., 1968.

Fernando, Tissa. *Sri Lanka: Portrait of an Island Republic.* Boulder, Colo.: Westview, 1987.

Fessler, Loren. *China.* New York: Time Inc., 1963.

Fetter, Bruce. *Colonial Rule and Regional Imbalance in Central Africa.* Boulder, Colo.: Westview, 1983.

Firth, Raymond W. *Malay Fishermen: Their Peasant Economy.* Hamden, Conn.: Shoe String Press, 1966.

Fisher, Charles A. *Southeast Asia: A Social, Economic, and Political Geography.* New York: Dutton, 1966.

Fisher, William B. *The Middle East: A Physical, Social, and Regional Geography.* Cambridge: Cambridge University Press, 1978.

Flagg, John Edwin. *Cuba, Haiti, and the Dominican Republic* Englewood Cliffs, N.J.: Prentice-Hall, 1965.

Fodor's Guide to Europe. New York: McKay, 1959.

Forbes, Dean K. *The Geography of Underdevelopment.* Baltimore: Johns Hopkins University Press, 1984.

Foster, C. R. (ed.). *Nations Without a State: Ethnic Minorities in Western Europe.* New York: Praeger, 1980.

Frank, Richard, and Chasin, Barbara H. *Seeds of Famine: Ecological Destruction and the Development Dilemma in the West African Sahel.* Montclair, N.J.: Allanheld-Osmun, 1980.

Freeman, Edward A. *The Historical Geography of Europe.* New York: Longman, Green, 1920.

Freeman, Otis W. (ed.). *Geography of the Pacific.* New York: Wiley, 1951.

Friis, Herman R. *The Pacific Basin: A History of Its Geographical Exploration.* New York: American Geographical Society, 1967.

Fuchs, Roland J. (ed.). *Urbanization and Urban Policies in Pacific Asia.* Boulder, Colo.: Westview, 1987.

Ganguly, Sumit. *The Origins of War in South Asia: Indo-Pakistani Conflicts Since 1947.* Boulder, Colo.: Westview, 1986.

Garbell, Maurice A. *Tropical and Equatorial Meteorology.* New York: Pitman, 1947.

Garreau, Joel. *The Nine Nations of North America.* Boston: Houghton Mifflin, 1981.

Gauhar, Atlaf (ed.). *Regional Integration: The Latin American Experience.* Boulder, Colo.: Westview, 1986.

Gautier, Emile F. *Sahara: The Great Desert.* Translated by Dorothy F. Mayhew. New York: Columbia University Press, 1935.

Geelan, P. J. M., and Twitchett, D. C. (eds.). *The Times Atlas of China.* 2d ed. New York: Van Nostrand Reinhold, 1984.

Gentilli, J. (ed.). *Climates of Australia and New Zealand,* New York: Elsevier, 1971.

Gibbs, James L. (ed.). *Peoples of Africa.* New York: Holt, Rinehart & Winston, 1965.

Gibson, Lay James, and Renteria, Alfonso Corona (eds.). *The U.S. and Mexico: Borderland Development and the National Economies.* Boulder, Colo.: Westview, 1985.

Gilbert, A. et al. (eds.). *Urbanization in Contemporary Latin America.* New York: Wiley, 1982.

Ginsburg, Norton, et. al. *The Pattern of Asia.* Englewood Cliffs, N.J.: Prentice-Hall, 1960.

Ginsburg, Norton S., and Lalor, Bernard A. (eds.). *China: The 80s Era.* Boulder, Colo.: Westview, 1984.

Girardet, Edward R. *Afghanistan: The Soviet War.* New York: St. Martin's, 1985.

Goldstein, Steven M., and Sears, Kathrin (eds.). *The People's Republic of China: A Basic Handbook.* Croton-on-Hudson, N.Y.: Council on Public Affairs, 1984.

Good, Charles M. *Ethnomedical Systems in Africa: Patterns of Traditional Medicine in Rural and Urban Kenya.* New York: Guilford, 1987.

Goodall, George (ed.). *Soviet Union in Maps.* Chicago: Denoyer-Geppert, 1947.

Gopalakhrishnan, Chennat (ed.). *The Emerging Marine Economy of the Pacific.* Boston: Butterworth, 1984.

Gordon, David C. *The Republic of Lebanon: Nation in Jeopardy.* Boulder, Colo.: Westview, 1983.

Gottman, Jean. *A Geography of Europe.* New York: Holt, Rinehart & Winston, 1969.

Goudie, Andrew S. *The Human Impact on the Natural Environment.* Cambridge, Mass.: MIT Press, 1986.

Goudie, Andrew S., et al. (eds.). *The Encyclopedic Dictionary of Physical Geography.* New York: Basil Blackwell, 1985.

Gradus, Y. "The Role of Politics in Regional Inequality: The Israeli Case." *Annals of the Association of American Geographers* 73 (1983): 388–403.

Green, L. P., and Thomas J. Denis. *Development in Africa: A Study in Regional Analysis with Special Reference to Southern Africa.* Johannesburg: Witwatersrand University Press, 1962.

Greenbie, Sydney. *The Central Five: Guatemala, Honduras, El Salvador, Nicaragua, and Costa Rica.* New York: Row, Peterson, 1943.

Griffin, Ernst C., and Ford, Larry R. "A Model of Latin American City Structure." *Geographical Review* 70 (1980): 397–422.

Griffiths, Ieuan L. L. (ed.). *An Atlas of African Affairs.* New York: Methuen, 1984.

Grossman, Lawrence S. *Peasants, Subsistence Ecology, and Development in the Highlands of Papua New Guinea.* Princeton: Princeton University Press, 1984.

Grove, A. T. *Africa South of the Sahara.* London: Oxford University Press, 1970.

Guinness, Paul, and Bradshaw, Michael. *North America: A Human Geography.* Totowa, N.J.: Barnes & Noble, 1985.

Gulliver, Philip H. *The Family Herds: A Study of Two Pastoral Tribes in East Africa, the Jie and Turkana.* London: Routledge & Kegan Paul, 1955.

Gunther, John. *Inside South America.* New York: Harper & Row, 1967.

Gupte, Pranay. *Vengeance: India After the Assassination of Indira Gandhi.* New York: Norton, 1985.

Gwynne, Robert N. *Industrialization and Urbanization in Latin America.* Baltimore: Johns Hopkins University Press, 1986.

Haggett, Peter. *Geography: A Modern Synthesis.* New York: Harper & Row, 1972.

Hainsworth, Geoffrey B. (ed.). *Village-Level Modernization in Southeast Asia: The Political Economy of Rice and Water.* Vancouver, Can.: University of British Columbia Press, 1982.

Hall, Carolyn. *Costa Rica: A Geographical Interpretation in Historical Perspective.* Boulder, Colo.: Westview, 1985.

Hall, Peter. "Tokyo." In *The World of Cities,* pp. 179–197. New York: St. Martin's, 1984.

———. *The World Cities.* New York: St. Martin's, 1984.

Hall, Ray, and Ogden, Philip. *Europe's Population in the 1970s and 1980s.* Cambridge: Cambridge University Press, 1985.

Hall, Robert B. *Japanese Geography.* Ann Arbor: University of Michigan Press, 1956.

———. *Japan: Industrial Power of Asia.* Princeton: Van Nostrand, 1973.

Hames, Raymond B., and Vickers, William T. (eds.). *Adaptive Responses of Native Amazonians.* New York: Academic Press, 1983.

Hamilton, Nora, and Harding, T. F. (eds.). *Modern Mexico: State, Economy, and the Social Conflict.* Beverly Hills, Calif.: Sage, 1986.

Hammond, Thomas T. *Red Flag over Afghanistan.* Boulder, Colo.: Westview, 1983.

Hance, William A. *African Economic Development.* New York: Praeger, 1967.

———. *The Geography of Modern Africa.* New York: Columbia University Press, 1975.

Hane, Mikiso. *Modern Japan: A Historical Survey.* Boulder, Colo.: Westview, 1986.

Hanley, Wayne, and Cooper, Malcolm (eds.). *Man and the Australian Environment: Current Issues and Viewpoints.* Sydney, Aust.: McGraw-Hill, 1982.

Hansen, Gary E. (ed.). *Agricultural and Rural Development in Indonesia.* Boulder, Colo.: Westview, 1981.

Hardgrave, R. L., Jr. *India: Government and Politics in a Developing Nation.* New York: Harcourt Brace Jovanovich, 1980.

Haring, Douglas G. *The Land of Gods and Earthquakes.* New York: Columbia University Press, 1929.

Harman, Carter. *The West Indies.* New York: Time Inc., 1966.

Harris, Chauncy D. "The Urban and Industrial Transformation of Japan." *Geographical Review* 72 (1982): 50–89.

Harris, George S. *Turkey: Coping with Crisis.* Boulder, Colo.: Westview, 1985.

Harris, Lillian C. *Libya: Qadhafi's Revolution and the Modern State.* Boulder, Colo.: Westview, 1986.

Harris, Stuart A. *The Permafrost Environment.* Totowa, N.J.: Rowman & Littlefield, 1986.

Harrison, Paul. *The Greening of Africa.* New York: Viking, 1987.

Harrison-Church, Ronald J. *West Africa: A Study of the Environment and Man's Use of It.* London: Longman, 1980.

Hatherton, Trevor (ed.). *Antarctica.* New York: Praeger, 1965.

Hawthorne, David. *Islands of the Pacific.* New York: Putnam, 1943.

Hedlund, Stefan. *The Crisis in Soviet Agriculture.* New York: St. Martin's, 1984.

Helms, Christian M. *Iraq: Eastern Flank of the Arab World.* Washington, D.C.: Brookings Institute, 1984.

Herman, Theodore. *The Geography of China.* New York: State Education Department, 1967.

Hibbert, Christopher. *Africa Explored: Europeans in the Dark Continent, 1769–1889.* London: Lane, 1982.

Hidore, John, and Albokhair, Y. "Sand Encroachment in Al-Hasa Oasis, Saudi Arabia." *Geographical Review* 72 (1982): 350–356.

Hill, Ronald D. (ed.). *South-East Asia: A Systematic Geography.* New York: Oxford University Press, 1979.

Hiro, Dilip. *Iran Under the Ayatollahs.* Boston: Routledge & Kegan Paul, 1985.

Hoffman, George (ed.) *Eastern Europe: Essays in Geographical Problems.* New York: Praeger, 1977.

———. *A Geography of Europe: Problems and Prospects.* 5th ed. New York: Wiley, 1983.

Hofheinz, Roy, and Calder, K. *The East Asia Edge.* New York: Basic Books, 1982.

Holderness, Mary. *New Russia.* New York: Arno, 1970.

Hole, Francis D., and Campbell, James B. *Soil Landscape Analysis.* Totowa, N.J.: Rowman & Allanheld, 1985.

Hook, Brian (ed.). *The Cambridge Encyclopedia of China.* New York: Cambridge University Press, 1982.

Hooson, David J. M. *The Soviet Union: Peoples and Regions.* Belmont, Calif.: Wadsworth, 1966.

Hopkirk, Peter. *Setting the East Ablaze: Lenin's Dream of an Empire in Asia.* New York: Norton, 1985.

Howe, G. Melvin (ed.). *The Soviet Union: A Geographical Survey.* Plymouth, Eng.: MacDonald & Evans, 1983.

Howells, William W. *The Pacific Islanders.* New York: Scribners, 1973.

Hoyle, Brian S. *Seaports and Development: The Experiences of Kenya and Tanzania.* New York: Gordon & Breach, 1983.

Hsieh, Chiao-min. *Taiwan-Ilha Formosa: A Geography in Perspective.* Washington, D.C.: Butterworth, 1964.

Hudson, Ray, and Lewis, Jim (eds.). *Uneven Development in Southern Europe.* New York: Methuen, 1985.

Huke, Robert E. *Shadows on the Land: An Economic Geography of the Philippines.* Manila: Bookmark, 1963.

Ilbery, Brian W. *Western Europe: A Systematic Human Geography.* New York: Oxford University Press, 1986.

International Financial Statistics. New York: International Monetary Fund, 1983.

Isenberg, Irwin. *Japan: Asian Power.* New York: Wilson, 1971.

Jackson, John B. *Discovering the Vernacular Landscape.* New Haven: Yale University Press, 1984.

Jackson, William A. Douglas. *The Russo-Chinese Borderlands: Zone of Peaceful Contact or Potential Conflict?* Princeton: Van Nostrand, 1968.

Jakle, John A. *The American Small Town: Twentieth-Century Place Images.* Hamden, Conn.: Shoe String Press, 1982.

James, Preston E., and Minkel, Clarence W. *Latin America.* 5th ed. New York: Wiley, 1986.

James, Preston E., and Webb, Kempton E. *One World Divided: A Geographer Looks at the Modern World.* New York: Wiley, 1980.

Jao, Y. C., and Leung, Chi-keung (eds.). *China's Special Economic Zones: Policies, Problems, and Prospects.* New York: Oxford University Press, 1986.

Japan: A Regional Geography of an Island Nation. Tokyo: Teikoku-Shoin, 1985.

Japan: The Official Guide. Tokyo: Japan National Tourist Organization, 1964.

Jeans, Dennis N. (ed.). *Australia: A Geography.* New York: St. Martin's, 1978.

Jensen, Robert G., et al. (eds.). *Soviet Natural Resources in the World Economy.* Chicago: University of Chicago Press, 1983.

John, Brian S. *Scandinavia: A New Geography.* New York: Longman, 1984.

Johnson, Basil L. C. *Bangladesh.* Totowa, N.J.: Barnes & Noble, 1982.

———. *Development in South Asia.* New York: Viking, 1983.

Johnson, William Weber. *The Andean Republics: Bolivia, Chile, Ecuador, Peru.* New York: Time Inc., 1965.

———. *Mexico.* New York: Time Inc., 1966.

Johnston, Ron J., et al. (eds.). *The Dictionary of Human Geography.* New York: Free Press, 1981.

Jones, E. L. *The European Miracle: Environments, Economies, and Geopolitics in the History of Europe and East Asia.* Cambridge: Cambridge University Press, 1981.

Jordon, Terry. *The European Culture Area.* New York: Harper & Row, 1973.

Karan, Pradyumna P. *Bhutan: A Physical and Cultural Geography.* Lexington: University of Kentucky Press, 1967.

Karklins, Rasma. *Ethnic Relations in the U.S.S.R.: The Perspective from Below.* Winchester, Mass.: Allen & Unwin, 1986.

Karnow, Stanley. *Southeast Asia.* New York: Time Inc., 1967.

———. *Vietnam: A History.* New York: Viking, 1983.

Kelly, Ian. *Hong Kong: A Political-Geographic Analysis.* Honolulu: Hawaii University Press, 1987.

Kent, George, and Valencia, Mark J. (ed.). *Marine Policy in Southeast Asia.* Berkeley, Calif.: University of California Press, 1985.

Kidron, Michael, and Segal, Ronald. *The State of the World Atlas.* New York: Simon & Schuster, 1981.

Kimble, George H. T., *Tropical Africa.* New York: Twentieth-Century Fund, 1960.

Kimble, George H. T., and Good, D. (eds.). *Geography of the Northlands.* New York: Wiley, 1956.

King, David, and Ranck, Stephen (eds.). *Atlas of Papua New Guinea: A Nation in Transition.* Bathurst, Aust.: Brown, 1981.

King, Lester C. *South African Scenery: A Textbook of Geomorphology.* New York: Hafner, 1963.

Kirby, Richard J. R. *Urbanization in China: Town and Country in a Developing Economy, 1949–2000 A.D.* New York: Columbia University Press, 1985.

Knapp, Ronald G. (ed.). *China's Island Frontier: Studies in the Historical Geography of Taiwan.* Honolulu: Hawaii University Press, 1980.

Knight, C. G., and Newman, J. (eds.). *Contemporary Africa: Geography and Change.* Englewood Cliffs, N.J.: Prentice-Hall, 1976.

Knox, Paul L. *The Geography of Western Europe.* Totowa, N.J.: Barnes & Noble, 1984.

Kolb, Albert. *East Asia: China, Japan, and Korea.* London: Methuen, 1977.

Kornhauser, David H. *Japan: Geographical Background to Urban-Industrial Development.* New York: Longman, 1982.

Kosinski, Leszek A., and Elahi, K. Maudood (eds.). *Population Redistribution and Development in South Asia.* Hingham, Mass.: Reidel, 1985.

Kumar, Raj. *The Forest Resources of Malaysia: Their Economics and Development.* Singapore: Oxford University Press, 1986.

Lall, Arthur S. *The Emergence of Modern India.* New York: Columbia University Press, 1981.

Lamb, David. *The Africans.* New York: Random House, 1983.

———. *The Arabs: Journeys Beyond the Mirage.* New York: Random House, 1987.

Lamb, H. H., and Johnson, A. I. "Climatic Variation and Observed Changes in General Circulation." *Geografiska Annalar* 41, nos. 2 and 3 (1959): 94–134.

Lapierre, Dominique. *The City of Joy (Calcutta).* Translated by Kathryn Spink. Garden City, N.Y.: Doubleday, 1985.

Latin America. New York: Americana, 1943.

Lattimore, Owen. *Inner Asian Frontiers of China.* Boston: Beacon, 1962.

Lavine, Harold. *Central America.* New York: Time Inc., 1964.

Leach, Graham. *South Africa: No Easy Path to Peace.* New York: Methuen, 1987.

Leeming, Frank. *Rural China Today.* White Plains, N.Y.: Longman, 1985.

Legum, Colin (ed.). *Africa: A Handbook to the Continent.* New York: Praeger, 1966.

Leinbach, Thomas R., and Ulack, Richard. "Cities of Southeast Asia." In *Cities of the World: World Regional Urban Development,* ed. Stanley D. Brunn and Jack F. Williams. New York: Harper & Row, 1983.

Levine, Irving R. *Travel Guide to Russia.* Garden City, N.Y. Doubleday, 1960.

Levinson, M., Ward, R., and Webb, J. *The Settlement of Polynesia.* Minneapolis: University of Minnesota Press, 1977.

Levy, Daniel C., and Szekely, Gabriel. *Mexico: Paradoxes of Stability and Change.* Boulder, Colo.: Westview, 1987.

Ley, Willy. *The Poles.* New York: Time Inc., 1962.

Linz, Susan J. (ed.). *The Impact of World War II on the Soviet Union.* Totowa, N.J.: Rowman & Allanheld, 1985.

Liversidge, D. *White Horizon.* London: Oldhams, 1951.

Lodrick, D. O. *Sacred Cows, Sacred Places.* Berkeley: University of California Press, 1981.

Lombardi, John V. *Venezuela: The Search for Order, the Dream of Progress.* New York: Oxford University Press, 1982.

Longrigg, Stephen H. *The Middle East: A Social Geography.* London: Duckworth, 1963.

———. *Oil in the Middle East: Its Discovery and Development.* New York: Oxford University Press, 1967.

Lovering, J., and Prescott, J. R. V. *Last of Lands . . . Antarctica.* Melbourne, Aust.: Melbourne University Press, 1979.

Lowery, J. H. *Europe and the Soviet Union.* London: Arnold, 1966.

Lukacs, John R. (ed.). *The People of South Asia: The Biological Anthropology of India, Pakistan, and Nepal.* New York: Plenum, 1984.

Lyde, Lionel William. *The Continent of Europe.* London: Macmillan, 1930.

Lydolph, Paul E. *Geography of the U.S.S.R.* New York: Wiley, 1977.

Ma, Lawrence J. C., and Hanten, Edward W. (eds.). *Urban Development in Modern China.* Boulder, Colo.: Westview, 1981.

Ma, Lawrence J. C., and Noble, Allen G. (eds.). *Chinese and American Perspectives on the Environment.* New York: Methuen, 1981.

Mabogunje, Akin L. *Urbanization in Nigeria.* London: University of London Press, 1968.

MacDonald, Donald. *A Geography of Modern Japan.* Ashford, Kent, Eng.: Norbury, 1985.

MacDonald, R. St. J. (ed.). *The Arctic Frontier.* Toronto: University of Toronto Press, 1966.

MacEoin, Gary. *Colombia and Venezuela and the Guianas.* New York: Time Inc., 1965.

MacInnes, Colin. *Australia and New Zealand.* New York: Time Inc., 1966.

Mackey, Sandra. *The Saudis: Inside the Desert Kingdom.* Boston: Houghton Mifflin, 1987.

MacPherson, John. *Caribbean Lands.* New York: Longman, 1980.

Malcolm, Andrew H. *The Canadians.* New York: Times Books, 1985.

Mallory, Walter H. *China: Land of Famine.* AGS Special Publication 6. New York: American Geographical Society 1926.

Mamoria, C.B. *Geography of India.* Agra: Shiva Lal Agarwala, 1975.

Mandy, Nigel. *A City Divided: Johannesburg and Soweto.* New York: St. Martin's, 1985.

Manners, I. R. "The Middle East." In *World Systems of Traditional Resource Management,* ed. G. A. Klee. London: Arnold, 1980.

McCann, Lawrence D. (ed.). *Heartland and Hinterland: A Geography of Can-*

ada. Scarborough, Canada: Prentice-Hall, 1982.

McCloud, Donald G. *System and Process in Southeast Asia: The Evolution of a Region*. Boulder, Colo.: Westview, 1986.

McGee, T. G. *The Southeast Asian City: A Social Geography of the Primate Cities of Southeast Asia*. New York: Praeger, 1967.

McKee, Jesse O. (ed.). *Ethnicity in Contemporary America: A Geographical Appraisal*. Dubuque, Iowa: Kendall/Hunt, 1985.

McKnight, Thomas L. *Australia's Corner of the World*. Englewood Cliffs, N.J.: Prentice-Hall, 1970.

_____. *Physical Geography: A Landscape Appreciation*. Englewood Cliffs, N.J.: Prentice-Hall, 1987.

McLeod, Alan L. *The Pattern of Australian Culture*. Ithaca, N.Y.: Cornell University Press, 1963.

Mehnert, Klaus. *Soviet Man and His World*. New York: Praeger, 1962.

Mellor, Roy E.H. *Geography of the U.S.S.R.* New York: St. Martin's, 1964.

_____. *The Soviet Union and Its Geographical Problems*. London: Macmillan, 1982.

Meredith, Martin. *The First Dance of Freedom: Black Africa in the Postwar Era*. New York: Harper & Row, 1985.

Mertz, John D., and Meyers, David J. (eds.). *Venezuela: The Democratic Experience*. New York: Praeger, 1986.

Meso-Lago, Carmelo. *The Economy of Socialist Cuba*. Albuquerque: University of New Mexico Press, 1981.

Meyer, Alfred H., and Strietelmeier, John H. *Geography in World Society: A Conceptual Approach*. New York: Lippincott, 1963.

The Middle East and North Africa. London: Europa, 1984.

Mikhailov, Nikolai N. *Soviet Russia: The Land and Its People*. New York: Sheridan House, 1948.

Miles, J.A.R. (ed.). *Public Health Progress in the Pacific*. Hingham, Mass.: Reidel, 1984.

Miller, E. Willard, and Miller, Ruby M. *The Far East: A Bibliography on the Third World*. Monticello, Ill.: Vance, 1982.

_____. *The Middle East (Southwest Asia): A Bibliography on the Third World*. Monticello, Ill.: Vance, 1982.

_____. *Northern and Western Africa: A Bibliography on the Third World*. Monticello, Ill.: Vance, 1981.

_____. *South America: A Bibliography on the Third World*. Monticello, Ill.: Vance, 1982.

_____. *Southeast Asia: A Bibliography on the Third World*. Monticello, Ill.: Vance, 1982.

_____. *Tropical, Eastern, and Southern Africa: A Bibliography on the Third World*. Monticello, Ill.: Vance, 1981.

Miller, Tom. *The Panama Hat Trail: A Journey from South America*. New York: Morrow, 1986.

Mills, Hugh. *Kingdom of the Ice Bear: A Portrait of the Arctic*. Austin: University of Texas Press, 1985.

Milne, Robert S., and Mawzy, Diane K. *Malaysia: Tradition, Modernity, and Islam*. Boulder, Colo.: Westview, 1986.

Minns, W. J. *A Geography of Africa*. London: Macmillan, 1984.

Minshull, G. N. *The New Europe: An Economic Geography of the EEC*. 2d ed. London: Hodder & Stoughton, 1980.

Mintz, Sidney W., and Price, Sally (eds.). *Caribbean Contours*. Baltimore: Johns Hopkins University Press, 1985.

Mirov, Nicholas T. *Geography of Russia*. New York: Wiley, 1951.

Mitchell, Robert D., and Groves, Paul A. (eds.). *North America: The Historical Geography of a Changing Continent*. Totowa, N.J.: Rowman & Littlefield, 1987.

Mitra, Sujata. "The Diffusion of Cholera in India: A Markov Chain Analysis." Master's thesis, Oklahoma State University, 1976.

Monsouri, Mehdi. "Spatial Characteristics of Population in Iran: Growth, Distribution, and Density." Doctoral dissertation, Oklahoma State University, 1981.

Moran, Emilio F. *Developing the Amazon*. Bloomington: Indiana University Press, 1981.

_____. (ed.). *The Dilemma of Amazonian Development*. Boulder, Colo.: Westview, 1983.

Morgan, Joseph R. (ed.). *Hawaii*. Boulder, Colo.: Westview, 1983.

Morgan, Joseph R., and Valencia, Mark J. (eds.). *Atlas for Marine Policy in Southeast Asian Seas*. Berkeley: University of California Press, 1983.

Morgan, William T. W. *Nigeria*. White Plains, N.Y.: Longman, 1985.

Morner, Magnus. *The Andean Past: Land, Societies, and Conflict*. New York: Columbia University Press, 1985.

Morris, Arthur S. *Latin America: Economic Development and Regional Differentiation*. Totowa, N.J.: Barnes & Noble, 1981.

_____. *South America*. Totowa, N.J.: Barnes & Noble, 1987.

Morris, Jan. *The Matter of Wales: Epic Views of a Small Country*. New York: Oxford University Press, 1985.

Morrison, Donald, et al. *Black Africa: A Comparative Handbook*. New York: Free Press, 1972.

Morton, Henry W., and Stuart, Robert C. (eds.). *The Contemporary Soviet City*. Armonk, N.Y.: Sharpe, 1984.

Mottahedeh, Roy P. *The Mantle of the Prophet: Religion and Politics in Iran*. New York: Simon & Schuster, 1985.

Mountjoy, Alan, and Hilling, David. *Africa: Geography and Development*. Totowa, N.J.: Barnes & Noble, 1987.

Muehrcke, Philip C. *Map Use: Reading—Analysis—Interpretation*. Madison, Wisc.: JP Publications, 1986.

Muller, Fritz B. *The Living Arctic*. New York: Methuen, 1981.

Murata, Kiyogi, and Ota, Isamu (eds.). *An Industrial Geography of Japan*. New York: St. Martin's, 1980.

Murdock, George P. *Africa: Its People and Their Culture and History*. New York: McGraw-Hill, 1959.

Murphy, Rhoads. *The Fading of the Maoist Vision: City and Country in Chinese Development.* New York: Methuen, 1980.

Murray, Jocelyn (ed.). *Cultural Atlas of Africa.* Oxford: Phaidon, 1981.

Musto, Stefan A., and Pinkle, Carl F. (eds.). *Europe at the Crossroads: Agendas of the Crisis.* Westport, Conn.: Praeger, 1985.

Naff, Thomas, and Matson, Ruth C. *Water in the Middle East: Conflict or Cooperation?* Boulder, Colo.: Westview, 1984.

Nairn, A. E. M. *The Ocean Basins and Margins.* New York: Plenum, 1973.

Nazaroff, Alexander I. *The Land of the Russian People,* New York: Lippincott, 1944.

Needler, Martin C., ed. *Political Systems of Latin America.* New York: Van Nostrand Reinhold, 1970.

Ness, Gayl D., and Ando, Hirofumi. *The Land Is Shrinking: Population Planning in Asia.* Baltimore: Johns Hopkins University Press, 1984.

Newman, James L., and Matzke, Gordon E. *Population: Patterns, Dynamics, and Prospects.* Englewood Cliffs, N.J.: Prentice-Hall, 1984.

1986 World Population Data Sheet. Washington, D.C.: Population Reference Bureau, 1986.

Noble, Allen G., and Dutt, Ashok K. (eds.). *India: Cultural Patterns and Processes.* Boulder, Colo.: Westview, 1982.

Nordenskjold, Otto. *The Geography of the North Polar Regions.* New York: American Geographical Society, 1928.

Norris, Robert E., et al. *Geography: An Introductory Perspective.* Columbus, Ohio: Merrill, 1982.

Norris, Robert E., and Haring, L. Lloyd. *Political Geography.* Columbus, Ohio: Merrill, 1980.

Norton, James H. K. *The Third World: South Asia.* Guilford, Conn.: Dushkin, 1984.

Nove, Alec. *The Soviet Economic System.* Winchester, Mass.: Allen & Unwin, 1980.

O'Connor, Anthony M. *The African City.* New York: Holmes & Meier, 1983.

Odell, Peter R., and Preston, David A. *Economies and Societies in Latin America: A Geographical Interpretation.* New York: Wiley, 1973.

O'Donnell, Charles P. *Bangladesh: Biography of a Muslim Nation.* Boulder, Colo.: Westview, 1984.

Oliver, Douglas L. *The Pacific Islands.* Cambridge, Mass.: Harvard University Press, 1951.

Oliver, Roland, and Crowder, Michael (eds.). *The Cambridge Encyclopedia of Africa.* Cambridge: Cambridge University Press, 1981.

Orvig, S. *Climates of the Polar Regions.* New York: Elsevier, 1970.

Osborn, Fairfield. *The Pacific World.* New York: Norton, 1944.

Osei-Kwame, P. A. *A New Conceptual Model of Political Integregation in Africa.* Laham, Md.: University Press of America, 1980.

Osleeb, Jeffery P., and ZumBrennan, Craig. *The Soviet Iron and Steel Industry.* Totowa, N.J.: Rowman & Littlefield, 1986.

Palm, Risa. *The Geography of American Cities.* New York: Oxford University Press, 1981.

Pannell, Clifton W. (ed.). *East Asia: Geographical and Historical Approaches to Foreign Area Studies.* Dubuque, Iowa: Kendall/Hunt, 1983.

Pannell, Clifton W., and Ma, Lawrence J. C. *China: The Geography of Development and Modernization.* New York: Halsted, 1983.

Parker, Franklin. *The Central American Republics.* London: Oxford University Press, 1964.

Parker, Geoffrey. *The Logic of Unity: A Geography of the European Economic Community.* 3d ed. New York: Longman, 1981.

Parker, Richard B. *North Africa: Regional Tensions and Strategic Concerns.* Westport, Conn.: Praeger, 1984.

Parker, William H. *The Soviet Union.* New York: Longman, 1983.

Parkes, Don (ed.). *Northern Australia: The Arenas of Life and Ecosystems on Half a Continent.* Orlando, Fla.: Academic Press, 1984.

Pataki-Schweizer, K. J. *A New Guinea Landscape: Community, Space, and Time in the Eastern Highlands.* Seattle: University of Washington Press, 1980.

Paterson, John H. *North America: A Geography of Canada and the United States.* 7th ed. New York: Oxford University Press, 1984.

Payne, Donald G. *Antartica: The Last Continent.* Boston: Little, Brown, 1974.

Peck, Malcolm C. *The United Arab Emirates: A Venture in Unity.* Boulder, Colo.: Westview, 1986.

Peckenham, Nancy, and Street, Annie. *Honduras: Portrait of a Captive Nation.* Westport, Conn.: Praeger, 1985.

Pederson, Alwin. *Polar Animals.* New York: Taplinger, 1966.

Pereira, Luiz B. *Development and Crises in Brazil, 1930–1983.* Boulder, Colo.: Westview, 1984.

Perkins, Dwight H. *Agricultural Development in China, 1368–1968.* Chicago: Aldine, 1969.

Perlmutter, Amos. *Israel: The Partitioned State.* New York: Scribners, 1985.

Peters, Joan. *From Time Immemorial: The Origins of the Arab-Israeli Conflict over Palestine.* New York: Harper & Row, 1984.

Pinchemel, Philippe, et al. *France: A Geographical, Social, and Economic Survey.* Cambridge: Cambridge University Press, 1987.

Pipes, Richard. *The Formation of the Soviet Union: Communism and Nationalism, 1917–1923.* New York: Atheneum, 1968.

Pitcher, Harvey J. *Understanding the Russians.* London: Allen & Unwin, 1964.

Pollock, Norman C., and Swanzie, Agnew. *An Historical Geography of South Africa.* London: Longmans, 1963.

Poltholm, Christian, and Dale, Richard (eds.). *Southern Africa in Perspective.* New York: Free Press, 1972.

Polunin, G. E. *Introduction to Plant Geography.* London: Longmans, Green, 1960.

Porsild, A. E. "Plant Life in the Arctic." *Canadian Geographical Review* 42 (1951): 122–145.

Porter, Eliot. *Antarctica.* New York: Dutton, 1978.

Pounds, Norman J. G. *Eastern Europe.* Chicago: Aldine, 1969.

_____. *Europe and the Soviet Union.* New York: McGraw-Hill, 1966.

Pred, Allen. *Urban Growth and City Systems in the United States.* Cambridge, Mass.: Harvard University Press, 1980.

Price, R. J., and Sugden, D. E. *Polar Geomorphology.* London: Institute of British Geographers, 1972.

Priestley, Sir Raymond; Adie, Raymond J.; and Robin, G. D. Q. (eds.). *Antarctic Research.* London: Butterworths, 1964.

Pritchard, John M. *Africa: A Study Geography for Advanced Students.* London: Longmans, 1979.

Prothero, R. Mansell (ed.). *A Geography of Africa.* New York: Praeger, 1969.

Pryor, Robin J. (ed.). *Migration and Development in Southeast Asia: A Demographic Perspective.* New York: Oxford University Press, 1979.

Purcell, Victor W. W. S. *The Chinese in Southeast Asia.* New York: Oxford University Press, 1965.

Raitz, Karl B., et al. *Appalachia: A Regional Geography.* Boulder, Colo.: Westview, 1984.

Rawson, R. R. *The Monsoon Lands of Asia.* Chicago: Aldine, 1963.

Rees, Henry. *Australasia: Australia, New Zealand, and the Pacific Islands.* London: Macdonald & Evans, 1962.

Reifenberg, Adolf. *The Struggle Between the Desert and the Sown: Rise and Fall of Agriculture in the Levant.* Jerusalem: Jewish Agency, 1955.

Reischauer, E. O. *Japan: The Story of a Nation.* New York: Knopf, 1981.

Rich, David C. *The Industrial Geography of Australia.* North Ryde, Aust.: Methuen, 1985.

Riding, Alan. *Distant Neighbors: A Portrait of the Mexicans.* New York: Knopf, 1985.

Roberts, Dorothy E. *A Scholar's Guide to Japan.* Honolulu: East-West Center, 1966.

Roberts, John E. (ed.). *Bold Atlas of Australia.* Sydney, Aust.: Ashton Scholastic, 1983.

Robinson, Arthur H., et al. *Elements of Cartography.* New York: Wiley, 1984.

Robinson, Harry. *Monsoon Asia: A Geographical Survey.* New York: Praeger, 1967.

Robinson, J. Lewis. *Concepts and Themes in the Regional Geography of Canada.* Vancouver, Canada: Talon, 1983.

Rogers, Rosemarie (ed.). *Guests Come to Stay: The Effects of European Labor Migration on Sending and Receiving Countries.* Boulder, Colo.: Westview, 1985.

Rogge, John R. *Too Many, Too Long: Sudan's Twenty-Year Refugee Dilemma.* Totowa, N.J: Rowman & Allenheld, 1985.

Ro'i, Yaacov (ed.). *The U.S.S.R. and the Muslim World: Issues in Domestic and Foreign Policy.* Winchester, Mass.: Allen & Unwin, 1984.

Rooney, John F., Jr., et al. (eds.). *This Remarkable Continent: An Atlas of United States and Canadian Society and Culture.* College Station: Texas A&M University Press, 1982.

Rotkin, Charles E. *Europe: An Aerial Close-up.* Philadelphia: Lippincott, 1962.

Rugg, Dean S. *Eastern Europe.* White Plains, N.Y.: Longman, 1986.

Saddler, H. *Energy in Australia: Politics and Economics.* Sydney, Aust.: Allen & Unwin, 1981.

Sagers, Matthew J., and Green, Milford B. *The Transportation of Soviet Energy Resources.* Totowa, N.J.: Rowman & Littlefield, 1986.

Sahlins, Marshall D. *Islands of History.* Chicago: University of Chicago Press, 1985.

St. John, Robert. *Israel.* New York: Time Inc., 1965.

Samuels, Marwyn S. *Contest for the South China Sea.* New York: Methuen, 1982.

Sanchez, James J. *A Bibliography for Soviet Geography.* Chicago: Council of Planning Librarians, 1985.

Sanders, Sol. *A Sense of Asia.* New York: Scribners, 1969.

Sanderson, Susan W. *Land Reform in Mexico, 1910–1980.* Orlando, Fla.: Academic Press, 1984.

Saqqaf, Abdulaziz Y. (ed.). *The Middle East City: Ancient Traditions Confront a Modern World.* New York: Paragon House, 1987.

Sater, John E. *The Arctic Basin.* Centreville, Md.: Tidewater, 1963.

Scott, Ian. *Urban and Spatial Development in Mexico.* Baltimore: Johns Hopkins University Press, 1982.

Sealey, Neil. *Tourism in the Caribbean.* London: Hodder & Stoughton, 1982.

Sears, Dudley, and Ostrom, Kjell (eds.). *The Crises of the European Regions.* New York: St. Martin's, 1983.

Seidensticker, Edward. *Japan.* New York: Time Inc., 1965.

Senior, Michael, and Okunrotifa, P. *A Regional Geography of Africa.* New York: Longman, 1983.

Settles, William F. *Life Under Communism.* Chicago: Adams, 1969.

Shabad, Theodore. *China's Changing Map: A Political and Economic Geography of the Chinese People's Republic.* New York: Praeger, 1956.

_____. *China's Changing Map: National and Regional Development, 1949–1971.* New York: Praeger, 1972.

_____. *Geography of the U.S.S.R.: A Regional Survey.* New York: Columbia University Press, 1951.

Shackleton, Margaret R. *Europe: A Personal Geography.* New York: Longmans, Green, 1936.

Shadbolt, Maurice. *Isles of the South Pacific.* Washington, D.C.: National Geographic Society, 1968.

Short, John R., and Kirby, Andrew M. *The Human Geography of Contemporary Britain.* New York: St. Martin's, 1984.

Siemans, A. H. "Wetland Agriculture in Pre-Hispanic Mesoamerica." *Geographical Review* 73 (1983): 166–181.

Simpson, Colin. *The New Australia.* Sydney, Aust.: Angus & Robertson, 1971.

Singleton, F. Seth, and Shingler, John. *Africa in Perspective.* New York: Hayden, 1967.

Sit, Victor F. S. (ed.). *Chinese Cities: The Growth of the Metropolis Since 1949.* New York: Oxford University Press, 1985.

Smil, Vaclav. *The Bad Earth: Environmental Degradation in China.* Armonk, N.Y.: Sharpe, 1984.

Smiley, Terah L., and Zumberge, James H. *Polar Deserts and Modern Man.* Tucson: University of Arizona Press, 1974.

Smith, David M. *Apartheid in South Africa.* Cambridge: Cambridge University Press, 1987.

———. (ed.). *Living Under Apartheid.* Winchester, Mass.: Allen & Unwin, 1982.

Smith, George A. *The Historical Geography of the Holy Land, Especially in Relation to the History of Israel and of the Early Church.* London: Hodder & Stoughton, 1894.

Smith, Hedrick. *The Russians.* New York: Times Books, 1983.

Smith, Michael, et al. *Asia's New Industrial World.* New York: Methuen, 1985.

Smith, Nigel J. H. *Rainforest Corridors: The Transamazonian Colonization Scheme.* Berkeley, Calif.: University of California Press, 1982.

Soja, Edward W. *The Geography of Modernization in Kenya: A Spatial Analysis of Social, Economic, and Political Change.* Syracuse, N.Y.: Syracuse University Press, 1968.

Sombitsuri, Virawan. "Migration to Bangkok: A Geographical Analysis." Doctoral dissertation, Oklahoma State University, 1977.

Songqiao, Zhao. *Physical Geography of China.* New York: Wiley, 1986.

Sopher, David E. (ed.). *An Exploration of India: Geographical Perspectives on Society and Culture.* Ithaca, N.Y.: Cornell University Press, 1980.

Spate, Oskar H. K. *Australia.* New York: Praeger, 1968.

———. *The Pacific Since Magellan: Monopolis and Freebooters.* Minneapolis: University of Minnesota Press, 1983.

———. *The Spanish Lake.* Minneapolis: University of Minnesota Press, 1979.

Spate, O. H. K., and Learmonth, A. T. A. *India and Pakistan: A General and Regional Geography.* New York: Barnes & Noble, 1967.

Spencer, Joseph E. *Shifting Cultivation in Southeast Asia.* Berkeley: University of California Press, 1976.

———. "Southeast Asia." *Progress in Human Geography* 6 (1982): 265–269.

Spencer, Joseph E., and Thomas, William L. *Asia, East by South: A Cultural Geography.* New York: Wiley, 1971.

Stamp, Lawrence Dudley. *Africa: A Study in Tropical Development.* New York: Wiley, 1959.

———. *Asia: A Regional and Economic Geography.* London: Methuen, 1967.

Steinberg, David J. *The Philippines: A Singular and Plural Place.* Boulder, Colo.: Westview, 1982.

Stevenhagen, Rudolfo (ed.). *Agrarian Problems and Peasant Movements in Latin America.* Garden City, N.Y.: Anchor Books, 1970.

Steward, Julian H., and Faron, Louis C. (eds.). *Native Peoples of South America.* New York: McGraw-Hill, 1959.

Stewart, Desmond. *The Arab World.* New York: Time Inc., 1968.

———. *Turkey.* New York: Time Inc., 1965.

Stewart, I. *Indonesia: Portrait from an Archipelago.* London: Kegan Paul, 1984.

Stolarski, Richard C. "The Antarctic Ozone Hole." *Scientific American* 258 (January 1988): 30–36.

Strahler, Arthur N., and Strahler, Alan H. *Modern Physical Geography.* New York: Wiley, 1987.

Stuart, Robert C. (ed.). *The Soviet Rural Economy.* Totowa, N. J.: Rowman & Littlefield, 1984.

Sugden, David E. *Arctic and Antarctic: A Modern Geographical Synthesis.* Totowa, N. J.: Barnes & Noble, 1982.

Sundaram, K. V. *Geography of Underdevelopment.* New Delhi: Concepts, 1983.

Super, John C., and Wright, Thomas C. (eds.). *Food, Politics, and Society in Latin America.* Lincoln: University of Nebraska Press, 1985.

Symons, Leslie. *Russian Agriculture: A Geographic Survey.* New York: Wiley, 1972.

Symons, Leslie, et al. *The Soviet Union: A Systematic Geography.* Totowa, N. J.: Barnes & Noble, 1983.

Taaffe, Edward J. (ed.). *Geography.* Englewood Cliffs, N.J.: Prentice-Hall, 1970.

Taafe, Robert N. *An Atlas of Soviet Affairs.* New York: Praeger, 1965.

Tata, Robert J. *Haiti: Land of Poverty.* Washington, D.C.: University Press of America, 1982.

Tawny, Richard H. *Land and Labour in China.* New York: Octagon, 1964.

Taylor, Alice (ed.). *Focus on Southeast Asia.* New York: Praeger, 1972.

Taylor, M. J., and Thrift, Nigel. *The Geography of Australian Corporate Power.* Beckenham, Eng.: Croom Helm, 1984.

Tayyeb, Ali. *Pakistan: A Political Geography.* London: Oxford University Press, 1966.

Tedrow, J. C. F. *Soils of the Polar Landscape.* New Brunswick, N.J.: Rutgers University Press, 1977.

Thayer, Charles W. *Russia.* New York: Time Inc., 1965.

Thernstrom, Stephen (ed.). *Harvard Encyclopedia of American Ethnic Groups.* Cambridge, Mass.: Harvard University Press, 1980.

Thesiger, Wilfred. *Arabian Sands.* New York: Dutton, 1959.

———. *The Marsh Arabs.* New York: Dutton, 1964.

Thompson, Virgina, and Adloff, Richard. *The Western Saharans: Back-*

ground to Conflict. Totowa, N.J.: Barnes & Noble, 1980.

Thorp, James. *Geography of the Soils of China*. Nanking: National Geological Survey of China, 1936.

Thrift, Nigel, and Forbes, Dean K. *The Price of War: Urbanization in Vietnam, 1954–1985*. Winchester, Mass.: Allen & Unwin, 1986.

Tregear, Thomas R. A. *China: A Geographical Survey*. New York: Wiley, 1980.

_____. *A Geography of China*. Chicago: Aldine, 1965.

Trewartha, Glen T. *Japan: A Physical, Cultural, and Regional Geography*. Madison: University of Wisconsin Press, 1945.

Trimmingham, J. Spencer. *The Influences of Islam upon Africa*. New York: Longman, 1980.

Trollope, Anthony. *Australia*. St. Lucia, Brisbane: University of Queensland Press, 1967.

Udo, Reuben K. *The Human Geography of Tropical Africa*. Exeter, N.H.: Heinemann, 1982.

Ungar, Sanford J. *Africa: The People and Politics of an Emerging Continent*. New York: Simon & Schuster, 1985.

U.N. Statistical Yearbook. New York: United Nations, 1985.

U.S.S.R. Academy of Sciences. *The Physical Geography of China*. New York: Praeger, 1969.

U.S.S.R. Energy Atlas. Washington, D.C.: U.S. Central Intelligence Agency, 1985.

Uys, Errol L. *Brazil*. New York: Simon & Schuster, 1986.

Vale, T. R. *Plants and People: Vegetation Change in North America*. Washington, D.C.: Association of American Geographers, 1982.

Vayda, Andrew P. (ed.). *Peoples and Cultures of the Pacific*. New York: Natural History Press, 1968.

Veliz, Claudio (ed.). *Latin America and the Caribbean*. New York: Praeger, 1968.

Wagret, P. (ed.). *U.S.S.R.* Geneva: Nagel, 1969.

Walker, D. F. *Canada's Industrial Space-Economy*. Toronto: Wiley, 1980.

Walker, Thomas W. *Nicaragua: The Land of the Sandino*. Boulder, Colo.: Westview, 1982.

_____. (ed.). *Nicaragua: The First Five Years*. Westport, Conn.: Praeger, 1985.

Ward, R. Gerard (ed.). *Man in the Pacific Islands: Essays on Geographical Change in the Pacific*. New York: Oxford University Press, 1972.

Webb, Kempton. *Geography of Latin America*. Englewood Cliffs, N.J.: Prentice-Hall, 1972.

Weekes, Richard V. (ed.). *Muslim Peoples: A World Ethnographic Survey*. Westport, Conn.: Greenwood, 1984.

Weil, Connie H. (ed.). *Medical Geographic Research in Latin America*. Elmsford, N.Y.: Pergamon, 1982.

Weinbaum, M. G. *Food, Development, and Politics in the Middle East*. Boulder, Colo.: Westview, 1982.

Weinstein, Bernard L., et al. *Regional Growth and Decline in the United States*. 2d ed. Westport, Conn.: Praeger, 1985.

Wiens, Herold J. *Pacific Island Bastions of the United States,* Princeton: Van Nostrand, 1962.

Wellington, John H. *South West Africa and Its Human Issue*. Oxford: Clarendon, 1967.

Wernstedt, Frederick L., and Spencer, Joseph E. *The Philippine Island World: A Physical, Cultural, and Regional Geography*. Berkeley: University of California Press, 1967.

West, Robert, and Augelli, John. *Middle America: Its Lands and People*. Englewood Cliffs, N.J.: Prentice-Hall, 1966.

White, C. Langdon, et al. *Regional Geography of Anglo-America*. 6th ed. Englewood Cliffs, N.J.: Prentice-Hall, 1985.

White, Paul. *The West European City: A Social Geography*. New York: Longman, 1984.

White, Richard. *Inventing Australia*. Boston: Allen & Unwin, 1981.

Whitmore, Timothy C. *Tropical Rain Forest of the Far East*. New York: Oxford University Press, 1984.

Whittlesey, Derwent. "The Regional Concept and the Regional Method." In *American Geography: Inventory and Prospect,* ed. Preston E. James and Clarence F. Jones. Washington, D.C.: Association of American Geographers, 1959.

Whyte, Martin King, and Parish, William L. *Urban Life in Contemporary China*. Chicago: University of Chicago Press, 1984.

Wild, M. Trevor. *West Germany: A Geography of Its People*. New York: Barnes & Noble, 1980.

Wilhelm, D. *Emerging Indonesia*. Totowa, N.J.: Barnes & Noble, 1980.

Wilkie, Richard W. *Latin American Population and Urbanization Analysis: Maps and Statistics, 1950–1982*. Los Angeles: UCLA Latin American Center, 1984.

Williams, L. E. *The Future of the Overseas Chinese in Southeast Asia*. New York: McGraw-Hill, 1966.

Wilson, David. *The Demand for Energy in the Soviet Union*. Totowa, N.J.: Rowman & Allanheld, 1983.

Wint, Guy (ed.). *Asia: A Handbook*. New York: Praeger, 1966.

Wixman, Ronald. *The Peoples of the U.S.S.R.: An Ethnographic Handbook*. Armonk, N.Y.: Sharpe, 1984.

Wood, Gordon. *The Pacific Basin: A Human and Economic Geography*. Melbourne: Oxford University Press, 1950.

Woodell, Stanley R. J. (ed.). *The English Landscape: Past, Present, and Future*. New York: Oxford University Press, 1985.

Wright, Robin B. *Sacred Rage: The Crusade of Modern Islam*. New York: Simon & Schuster, 1985.

Wubneh, Mulatu, and Abate, Yohannis. *Ethiopia: Transition and Development in the Horn of Africa*. Boulder, Colo.: Westview, 1987.

Wurm, Stephen A., and Hittori, S. (eds.). *Language Atlas of the Pacific*

Area. Canberra: Australian Academy of the Humanities, 1982.

Yawata, I., and Sinoto, Y. H. *Prehistoric Culture in Oceania*. Honolulu: Bishop Museum Press, 1968.

Yeager, Rodger. *Tanzania: An African Experiment*. Boulder, Colo.: Westview, 1983.

Yeates, Maurice H. *North American Urban Patterns*. New York: Halsted, 1980.

Young, Graham (ed.). *China: Dilemmas of Modernization*. Beckenham, Eng.: Croom Helm, 1985.

Yu-ti, Jen. *A Concise Geography of China*. Peking: Foreign Languages Press, 1964.

Zegarelli, Philip E. "Antarctica." *Focus*, September–October 1978, pp. 1–9.

GLOSSARY

Aborigines The earliest known inhabitants of Australia. These small, black people, called Australoids, live in wandering bands and have become remarkably adapted to the heat and cold of their desert homeland.

acculturation The process of becoming adapted to a new culture by one person, or the adoption by a society of the cultural traits of another society through contact between the groups.

acid rain Precipitation (in the form of rain and snow) that carries dioxides, such as those from sulfur and nitrogen, that have been emitted into the atmosphere from the burning of fossil fuels and that is dangerous to plants and animals, including humans.

Adirondack Mountains A mountain group located in northeastern New York noted for its fine scenery, numerous lakes, and many resorts, especially for winter sports. The tallest mountain in the group is Mount Marcy at 5,344 feet (1,629 m).

aeolian erosion The movement of loose particles of soil by the wind.

Afrikaners Another name for the Boers of South Africa.

age structure The relationship between the percentage of people in various age categories in a country. For example, in countries with high fertility rates, children make up a large proportion of the population, producing a pyramid shape to the age structure graph, while in low-fertility nations each generation is a smaller size, giving the population graph a boxier shape.

agglomerated Refers to a spatial distribution where items of concern are gathered together in a mass or separated only by short distances.

Ainu A member of the primitive, light-skinned race of people in Japan, now living mostly in Karafuto and Hokkaido. Also their language.

air pollution Contamination of the air through vehicle exhaust fumes, factory smoke, and other human activities that make the air impure or dirty.

Allah The name followers of Islam, the Muslim religion, have given to God.

Alliance for Progress A largely defunct organization of Latin American states that once attempted to establish free trade among its members.

animism The belief that natural phenomena, such as trees, rocks, and wind, are alive and have souls and that spirits and demons exist.

Antarctic Drift Sometimes called the West Wind Drift, it is the ocean current that extends around the world off the coast of the Antarctic continent. Cold water flows toward the east without interference from other continents, and the water movement depends on prevailing winds and the earth's rotation.

apartheid Afrikaans for "apartness". Refers to the strict racial segregation practiced in South Africa.

aphelion The point in a planet's orbit when the planet is farthest from the sun (as opposed to perihelion). Usually occurs, for the earth, around July 8.

Appalachian Plateau The upland region in the northeastern part of the United States surrounded by the mountains of the Appalachian system, which includes the following ranges: White, Green, Catskills, Alleghenies, Blue Ridge, and Cumberland.

Aquitaine Basin Historical region of southwestern France located between the Pyrenees Mountains and the Garonne River noted for its many vineyards and the production of wine.

area map A map that portrays numerous physical and human factors of a defined region, as opposed to a thematic map, which portrays only one element.

Armenian Knot Also known as the Armenian Plateau. The region immediately west of the Caspian Sea in the Armenian state of the Soviet Union. The region contains many faults with folded ridges, but many volcanic intrusions are also evident. The region is called a "knot" because numerous mountain ranges radiate from it.

Association of Southeast Asian Nations (ASEAN) An organization formed in 1967 to promote political and economic cooperation among the noncommunist states of Southeast Asia. Members include Brunei, Indonesia, Malaysia, the Philippines, Singapore, and Thailand.

autumnal equinox The time in the autumn (fall) in the northern hemisphere when the direct rays of the sun are located at the equator and the days and nights are of equal length. Usually occurs on September 22 or 23.

Avianca The national airline of Columbia, which began flying regular passenger routes in 1919. The oldest airline in the western hemisphere.

Babylon The ancient city, now in ruins, that grew up on the Euphrates River about 55 miles (89 km) south of Baghdad, Iraq, near the modern city of Hilla. The city existed from about 2200 B.C. to about 280 B.C. and was the home of the famous Babylonian kings Hammurabi and Nebuchadnezzar. It is also the place where Alexander the Great died in 323 B.C.

Baltic Lake Plain The part of the North European Plain adjacent to the Baltic Sea, stretching between Hamburg (Germany) and Leningrad (U.S.S.R.).

Bantu The large family of Negroid tribes occupying equatorial and Southern Africa.

Basques A distinct race of people whose original home was in the region of the western Pyrenees near the Bay of Biscay in Spain and France.

Bavarian Plateau An upland region of South Central Europe between Bavaria (Germany) and Tirol (Austria) containing the Bavarian Alps Mountains. The highest point is Zugspitze, at 9,719 feet (2,962 m).

Bedouins The nomadic Arabs of the Arabian and North African deserts.

Benelux The organization based on treaties among the countries of Belgium, the Netherlands, and Luxembourg, forerunners of the EEC, that allowed the lowering of trade restrictions among the three countries so that local resources (especially iron ore and coal) could be better utilized.

Berbers The Hamites of Northern Africa west of Tripoli who are closely related to Southern Europeans, Egyptians, and Ethiopians and who vary from blue-eyed blonds in the Atlas Mountains region to black-skinned inhabitants of the interior Saharan oases.

"big bang" theory The idea that the universe started with a tremendous explosion (a big bang) that threw particles outward in all directions from a central mass or core. The galaxies are currently moving away from each other at great speeds, which lends support to the theory.

Blue Ridge Mountains The eastern and southeasternmost range of the Appalachian group, located in West Virginia, North Carolina, and Georgia.

boat people After the Vietnam War, when U.S. troops were withdrawn, many South Vietnamese escaped the rule of the Communist North Vietnamese by taking to the open waters of the South China Sea in small and sometimes overloaded boats. These refugees are called "boat people."

Boers Rural descendants of the early Dutch settlers of South Africa.

Bohemian Massif A highland region (sometimes called the Bohemian-Moravian Highlands) located along the border between southeastern Bohemia and western Moravia, the two western divisions of Czechoslovakia.

bolson A flat-floored desert valley that drains into a playa. A playa is the bottom of an undrained desert basin that becomes at times a shallow lake, which when it evaporates leaves deposits of salt or gypsum.

Brownson depression Sometimes called the Brownson deep or the Puerto Rican Trench, it is the area north of Puerto Rico where the Atlantic Ocean reaches its greatest depth, 27,972 feet (8,526 m).

Buddhism A religion and philosophical system of Central and East Asia, founded in India in the sixth century B.C. by Gautama Siddhartha (Buddha) and consisting of the Great Enlightenment, which lists the causes of suffering and how salvation can be obtained through suffering.

Bushmen A race of nomadic hunters of Southern Africa, chiefly confined to the region of the Kalahari Desert.

calcification Literally the deposition of calcium salts, but used by physical geographers to describe the soil-forming process that occurs in grassland areas. The grass roots collect calcium from the soil, causing it to appear as though deposition of the salts had occurred.

campos A level, grassy plain in South America, with scattered xerophytic vegetation and a few small trees.

Canaan The old and native name of the part of Palestine between Jordan and the Mediterranean Sea, but sometimes vaguely used as the equivalent of all Palestine. The region taken over in 1200 B.C. by the Jews returning from Egypt and called the "land of milk and honey," or Promised Land.

Central America The part of Middle America located on the North American continent, but not including Mexico. The countries of Central America are Belize, Costa Rica, El Salvador, Guatemala, Honduras, Nicaragua, and Panama.

Central American Common Market (CACM) An organization of Central American countries started in 1960 to promote local industrialization and trade among the countries.

Central German Upland Uplifted area of central Germany that includes the Thuringian Forest (Thüringer Wald) and other wooded mountain ranges.

Central Massif Plateau region (also Massif Central) of the central part of southeastern France and the source area of many streams, including the Loire River. The highest point on the plateau is Puy de Sancy at 6,185 feet (1,885 m).

central place model A model designed in the 1930s by Walter Christaller to describe the location of market towns using marketing principles and distance.

Central Plateau (of Mexico) The uplifted region of central Mexico running from Mexico City to the northern border and bounded on the east by the Sierra Madre Oriental and on the west by the Sierra Madre Occidental.

cephalic index A number obtained by dividing the width of the cranium by its length and multiplying by 100. The number 80 is usually used as a base measurement to distinguish between long narrow heads (<80), and heads that are more round (>80).

Cerro Aconcagua *Cerro* in Spanish means "highland" or mountain region, so the massive Aconcagua (a-kon-*ka*-gwa) Mountain is called Cerro Aconcagua to distinguish it from the river and province in Chile with the same name. It is the highest peak in the western hemisphere.

Cevennes Pronounced ca-*ven*. The southern edge of the Central Massif, located between Montpellier and Valance in southern France.

China's Sorrow A phrase used to describe the Huang He (Hwang Ho, or Yellow River) in China, because its periodic flooding destroys crops and kills people.

Chinese dynasties Dynasties are successions of rulers who are members of the same family. In China the dynastic period lasted from the beginning of the Chou Dynasty in 1122 B.C. until Emperor Pu Yi of the Manchu Dynasty abdicated the throne on Feburary 12, 1912. The Chinese Empire was ruled by "barbarian" dynasties for much of its history, including the Mongol (Yuan, 1279–1368) and the Manchu (Ch'ing, 1644–1912). The Manchu Empire included not only China proper but also Tibet, Chinese Turkestan (Sinkiang), Mongolia, and Manchuria.

choropleth map A type of thematic map that portrays various quantities, aggregated by unit areas, through shades of one color or through various colors.

cohorts Individual age categories on a population pyramid.

coking coal A type of bituminous coal that has been heated to remove the gases and used for industrial fuel.

Coloreds Strictly meaning "having color," this word refers to the people of South Africa that are Caucasian-Negroid mixtures.

communism A doctrine and program based on socialism which calls for regulation of all social, cultural, political, and economic activities through a single authoritarian party.

Confucianism The ethical teachings of Confucius introduced into the Chinese religion, emphasizing devotion to parents, family, and friends, ancestor worship, and the maintenance of law and order.

continental climate The climate designated by Köppen as D, located where extreme continental influences create hot summers and very cold winters. No such climate exists in the southern hemisphere because there are no large land masses to retard the stabilizing influence of the oceans.

continental shelf The submerged shelf of land that slopes gradually from the shoreline to the point where the steep descent to the ocean bottom begins.

conurbation A region containing a large aggregation of urban communities and usually dominated by one large center, such as the Tokyo conurbation.

coolies From the hindi word *quli,* meaning hired servant but used to describe an unskilled native laborer, especially in East Asia.

core area The place of origin and usually where the current heaviest concentration of certain types of people live. Could be either a region or a small area within a city.

Council for Mutual Economic Assistance (CMEA) The organization of the Communist countries of Eastern Europe based on the idea of the EEC and designed to lower trade restrictions among member countries.

crude birth rate The annual number of births for a country for every 1,000 people.

crude death rate The annual number of deaths for a country for every 1,000 people.

crust The outermost layer of the lithosphere, solidified from the molten material inside the earth.

cultural diffusion The geographic dispersion of a culture or cultural trait. (See also *spatial diffusion.*)

culture The ideas, habits, skills, arts, instruments, institutions, and so forth of a given group of people for a given time.

cylindrical projection A type of map projection where the earth's curved surface is projected onto a cylinder so that the resulting image can be drawn on a flat surface. Distortion increases with distance away from the area where the earth's surface and the cylinder are tangent.

demographic transition A four-stage change process, over time, in birth and death rates that typically occurs when less developed nations become more developed.

density One characteristic of all spatial distributions. Defined as the number of items (geographic facts) per unit area.

desertification Literally, making a desert out of land that was not a desert. This occurs through human misuse of the land, such as allowing animals to overgraze the grasslands adjacent to the deserts, and through atmospheric heat balance changes, such as that caused by the greenhouse effect.

diastrophism All processes that change the shape of the earth's surface through crustal movements.

direction arrow A symbol used on some maps to show the orientation of the map by indicating the direction of north.

dispersion One characteristic of all spatial distributions. Defined as the degree of spread between the individual items (geographic facts).

dissected Erosion of the landscape that creates a complex of gullies and canyons, such as the borders of Mexico's Central Plateau.

distance decay The normal decrease in human activities with increased distance away from a central place. Could be measured in terms of traffic, telephone calls, migrants, and so forth.

dot map A map that portrays a spatial distribution by using a dot to represent either each object or a group of objects. Various sizes of dots also can be used to represent varying quantities of objects.

double-cropping A method of planting another crop immediately after one has been harvested, so that two crops can be grown during the same season.

Dravidian (1) A member of a group of intermixed races of people living in southern India and southern Sri Lanka. (2) The family of non–Indo-European languages spoken by those people, including Kanarese, Kurukh, Malayalam, Tamil, Telugu, and others.

drumlin An oval-shaped hill of sand and gravel deposited under glacial ice, then exposed when the glacier melts.

Environmental Protection Agency (EPA) An independent agency of the U.S. government that oversees the protection of the environment by setting standards, patrolling abusers, and setting fines for abusers.

erosion The wearing away of soil and rocks by the action of running water, wind, or moving ice (glaciers).

esker A narrow, sometimes winding ridge of sand and gravel deposited by a stream flowing under glacial ice, then exposed after the glacier melts.

estancia Spanish for "farm" or "ranch."

European Economic Community (EEC) A group of 12 Western European countries that have abolished customs, duties, and other trade restrictions in order to promote the

diurnal range Diurnal means "daily," so "diurnal range" refers to the high and low readings of such things as temperature, barometric pressure, and wind velocities for a 24-hour period.

Donbas The manufacturing region of the Ukraine area of the Soviet Union where high-grade coal, the basis for the local industry, is mined. The region is centered in the Donets Basin located northeast of the Sea of Azov (Azoskoye More) and east of the city of Donetsk.

exchange of goods among themselves. The organization started in 1957, when the Benelux countries (Belgium, the Netherlands, and Luxembourg) joined with France, Germany, and Italy in signing treaties expanding the 1951 agreement on the free movement of coal and iron ore. Denmark, Ireland, and the United Kingdom joined the organization in 1973, Greece became the tenth member in 1981, and Spain and Portugal joined in 1986. The EEC has created a large market that competes with the United States and Japan, but EEC members have been plagued in recent years by unemployment. Workers also are free to move among the member countries, so some areas have too many people looking for jobs.

expansion diffusion A type of spatial diffusion where the idea, disease, or whatever is spreading moves through an existing population.

fast ice Ice that develops in polar oceans by growing outward from the land as the water freezes, creating an ice shelf off the land.

fault block mountains Mountains created by the tilting of a large block of the earth's crust, as opposed to mountains created by diastrophism or vulcanism.

fault lines A fault is a fracture in the earth's crust along which movement has occurred, causing displacement of one side in relation to the other. A fault line, then, is the line followed by a fault.

fell fields Open areas in the Arctic where the plant association is primarily grass and small shrubs found in scattered clumps.

fiords Large U-shaped valleys along high-latitude coasts, created by past glaciers and now filled with sea water.

Five-Year Plan An economic planning period in the Soviet Union.

fluvial erosion Movement of soil and rock materials by running water.

French Indochina Indochina, or "Farther India," included the southeastern peninsula of Asia and all the countries from Myanma to Vietnam. The eastern part of the region (Vietnam) was controlled by France for many years and was therefore called French Indochina.

front The plane where two air masses converge but do not mix, causing uplifting of the warmer air mass.

functional region An area of the earth's surface demarcated for special purposes (functions), such as administration, distribution, collection, or other political or economic activities.

Garonne Basin The basin of the Garonne River in southwestern France that runs from the Pyrenees northwestward past Toulouse and Bordeaux.

geographic isolation The isolation of a species of plant or animal so that mixing of inherited traits cannot occur.

geomorphology The study of the earth's landform features.

glacial erosion The movement of soil and rock materials by moving ice (glaciers).

gley soil Soils created in poorly drained areas under cool or cold climatic conditions and thus usually found in subarctic or tundra areas.

graben A place where the earth's crust has been lowered due to a diastrophic movement, creating a deep valley. The opposite of a horst.

grasslands Any of the semiarid regions of the world where precipitation amounts are too low to support trees. Precipitation determines whether the grass is tall (savanna), medium (prairie), or short (steppe).

gravity models Social models based on Newton's gravity concept, where mass and distance are the important elements. Bodies of large mass (large cities) have more attraction for people than bodies of less mass (small towns), so more people visit and move to large places.

Great Dividing Range Sometimes called the Eastern Highlands of Australia, the "Great Divide" is a complex region of uplifted blocks and folds with intervening lowlands. The highland region parallels the eastern and southeastern coasts for nearly 2,500 miles (4,023 km). Mount Kosciusko (7,328 ft, or 2,234 m), in the southeastern corner of the continent, is the highest point in Australia.

Greater Antilles The large east-west trending islands between the Gulf of Mexico and the Caribbean Sea, composed of Cuba, Hispaniola, Jamaica, Puerto Rico, and the Virgin Islands, which comprise 90 percent of the land area of the West Indies.

greenhouse effect An atmospheric condition where the sun's heat energy is trapped by layers of carbon dioxide in the atmosphere, similar to the way heat energy is trapped inside a greenhouse by the panels of glass.

Hamites Members of the native races of North Africa, which include the Berber peoples north of the Sahara Desert as well as the Fula , the Tibbu, and the Tuaregs of the Sudan, the ancient Egyptians and their ancestors, the tribes of Ethiopia, and the Galles and Somalies to the south.

Hamito-Semitic Languages of the people who live in North Africa and the Middle East. The Hamitic languages include Berber, Tuareg, and Kyshit, and the Semitic languages include Arabic, Hebrew, and Amharic.

hazardous waste Any of the 403 toxic chemicals that have been identified by the Environmental Protection Agency as being harmful to humans, animals, plants, or other parts of the environment and that have been stored or dumped and are of no further use.

hierarchical diffusion A type of spatial diffusion characterized by the movement of the idea or disease, or whatever is moving, from one large place (such as a city) to another, then down the hierarchy to smaller places in a hopscotch manner.

high-yield rice hybrids Hybrid plants are produced by crossbreeding two plants of a different variety or species. Plant geneticists strive to grow plants that will produce a higher yield per unit area of land than either of the parent plants. This process has been successful with the rice plant.

Hinayana The Mahayana (Greater Vehicle) form of Buddhism carries a profusion of relics and liturgies and is a dramatic, colorful form of the religion. It spread northward from India into Tibet, China, and Japan. The Hinayana (Lesser Vehicle) form of Buddhism is more fundamental and subtle than the Mahayana form, and it spread eastward into Southeast Asia.

Hindi The group of Indo-European languages associated with northern India, including Assamese, Bengali, Marathi, Punjabi, Hindustani, and others.

Hinduism The religion and social system of the Hindus, developed from Brahmanism with elements from Buddhism.

hiragana The cursive and more widely used of the two Japanese forms of writing. (See also *katakana*)

Homo sapiens The various races of humans regarded as a single species. *Homo* is from the Greek word *homo*, meaning "the same," and *sapiens* is from the Latin word *sapientia,* meaning "wisdom" or "knowledge," a main factor that separates humans from other animals.

horst A raised portion of the earth's crust caused by diastrophism. The opposite of a graben.

Hottentots A race of people of Southern Africa thought to be a mixture of Bushmen and Bantu with yellowish-brown complexions. Noted for the clicks in their speech that make them appear to stutter.

Huk Communists In the Philippines during World War II, organized guerrilla forces fought the Japanese. They were called the Hukbong Bayan Laban Sa Hapon, or People's Anti-Japanese Resistance Army and were known as Huks. These fighters, as well as similar guerrilla forces in other parts of Southeast Asia, later became Communists and fought for national liberation, so they became known as Huk Communists.

human geography The branch of geography dealing with the place-to-place differences in human activities, such as economics, politics, and cultures (including languages and religions).

hurricanes Large rotating tropical storms generating surface wind speeds of at least 74 mph, originating in the central portion of the Atlantic Ocean and moving northwestward across the West Indies. Called "typhoons" in the Pacific region.

ice sheet A thick layer of ice covering a large area for a long time, such as during the Ice Age. Sometimes called "ice caps" or "continental glaciers" (as opposed to mountain glaciers) and found today only on Greenland and on the continent of Antarctica.

import substitution A method common in Third World countries, of substituting expensive imported finished products, such as automobiles, with locally produced "substitutes" made in assembly plants utilizing inexpensive local labor and imported parts.

Inca Empire The Inca state had reached and passed its zenith of development before the arrival of the Europeans, but at one time it extended along the west coast of South America from central Ecuador, through Peru, and as far south as the present location of Santiago, Chile. The nucleus of the empire was centered in the Cuzco Basin in southern Peru. The empire was formed through the conquest and assimilation of numerous separate Indian groups and eventually had political control over a large land area.

Indo-European A group of ten language families related through a common origin. The group includes the Germanic (German, English, and some Scandinavian), Romance, Celtic, Slavic, Baltic, Greek, Albanian, Iranian, Armenian, and Indo-Aryan.

inner core The innermost portion of the earth, assumed to be a rigid mass about 1,200 miles (1,920 km) in diameter and very hot (8,000 to 10,000° F).

intertropical convergence zone The tropical region of the world wind system, where the trade winds converge, causing uplifting of warm, moist air resulting in nearly continuous precipitation.

irredenta Literally, "unredeemed." A region containing people (irredentists) who advocate policies that seek to reincorporate the region into the country of their national origin. The term was applied originally to Italians living within Austrian territory. *Italia irredenta* means "unredeemed Italy."

Islam The Muslim religion, a monotheistic religion whose supreme deity is Allah and whose chief prophet and founder was Muhammad.

isohyet A type of isoline that connects points having the same amount of precipitation for a specified period, such as a year.

isoline A line on a map that connects points of equal value, such as elevation, temperature, precipitation, or other measurements.

isostasy Refers to the universal equilibrium in the earth's crust caused by the gravitational yielding of the magma beneath the earth's surface.

isotherm An isoline drawn on a map that connects points where actual or average temperatures for a specified time are the same.

Japanese Alps A name sometimes used for the line of mountains that stretch through the Japanese Islands from Hokkaido to Kyushu.

Japanese-Korean Related languages of the Altaic group.

Java Man A type of primitive human known from fossil remains first found in Java in 1891. Also called *Pithecanthropus erectus*.

Judaism Observance of Jewish customs, ceremonies, and rules, but especially the observance of the Jewish religion.

junks From the Spanish and Portuguese word *junco* and referring to a Chinese flat-bottomed ship with one accordion-like sail.

Jura Mountains Mountain range extending about 200 miles (322 km) along the boundary between France and Switzerland. The highest peak is Reculet, at 5,642 feet (1,720 m).

jute Either of two East Indian plants whose strong, glossy fibers are used in making burlap, sacks, mats, rope, and other items. (Sometimes the fiber itself is called "jute".)

Kaaba From the Arabic word *ka-'bah*, meaning "square building". The sacred shrine of Islam at Mecca in Saudi Arabia, the building contains a small black stone (a meteorite) supposedly given to Abraham by the angel Gabriel. The Kaaba is located in the courtyard of a large mosque toward which believers turn when praying.

kanji The Chinese ideographs used in Japanese writing.

kata-kana The more angular and less commonly used of the two forms of Japanese writing. (See also *hiragana*).

kolkhoz A collective farm of the Soviet Union.

Koran The sacred book of Islam, written in Arabic. The book's contents are reportedly revelations made to Muhammad by Allah.

Krakatoa The Indonesian island volcano in the center of the Soenda Strait between Sumatra and Java, noted for its violently explosive eruptions. The eruption on August 26, 1883, was the greatest ever recorded and was heard as far away as Turkey and Japan. The resulting tidal wave killed about 36,000 people in western Java.

Kuzbas An area of heavy manufacturing in the Soviet Union, located in the Kuznetsk Basin about 1,200 miles (2,000 km) east of the Ural Mountains in Central Asia. Both coal and iron ore have been found and utilized in the Kuzbas.

labor-intensive Describes any economic activity that requires or is done by hand labor with a noted absence of machines and that requires many work-hours to complete.

landlocked Refers to areas that do not have a border along an open ocean and thus are "locked in" by neighboring countries.

large-scale Map scale is the relationship between the size of an area shown on a map and the size of the actual area on the surface of the earth. The representative fraction for a large-scale map is large (1:63, 360, for example), so only small areas of the earth can be shown on a large-scale map. (See also *small scale*.)

lateritic soils Relatively infertile soils formed by the laterization process in tropical areas. Soils are dark red and yellow in color from the abundance of iron and aluminum, but humus is notably lacking. Soils created by laterization are called lateritic soils, or "latosols."

Latin America All the countries of the western hemisphere located south of the U.S.-Mexican boundary.

Latin American Free Trade Association (LAFTA) An organization of the countries of South America and Mexico begun in 1960 to promote trade among the countries, instead of having each country trade commodities on its own with North America and Europe.

latitude Angular measurement from the center of the earth that measures distances north and south of the equator. By rotating the angle, parallel lines that run east-west around the world are created (the lines are called "parallels" or "latitude lines").

latosols See *lateritic soils*.

Lesser Antilles A north-south trending arc of islands that separates the Caribbean Sea from the western portion of the North Atlantic Ocean and extends from the island of Sombrero on the north to Grenada on the south.

lichens A large group of mosslike plants consisting of algae and fungi and growing in patches on rocks and tree trunks.

lithosphere The part of the earth composed primarily of rock, as opposed to the hydrosphere (water) and atmosphere (air).

llano In Spanish, a plain, especially one that is very flat.

logistic curve A curve that rises slowly at first from left to right, then more quickly, and finally slows down again and flattens out at the upper bounds so the curve has an S-shape. Sometimes called an S-curve, Lorenz curve, or sigmoid curve.

longitude Angular measurement from the center of the earth used to measure distances east and west of the prime meridian (0° longitude). When rotated, the angles form lines that run from pole to pole. The lines are called "meridians" or "longitude lines."

Lowland of Brabant The lowland area of the southern Netherlands and central and northern Belgium.

magma Molten, fluid rock occurring naturally within the earth. Igneous rocks are formed through the cooling and solidification of magma.

Malayo-Polynesian The Austronesian languages based on Malay, including Indonesian, Polynesian, and Melanesian.

Malaysian Federation The 13 states of Malaysia located on the southeastern tip of Asia and the north coast of Borneo. The country, created on September 16, 1963, after being under British rule since 1867, is a federal parliamentary democracy with a constitutional monarch.

mantle The portion of the earth's interior located between the outer shell (the crust) and the outer core. The mantle is thought to be somewhat solid, but fluid enough so that convectional currents move material laterally between the core and the crust. The mantle is about 1,800 miles (2,900 km) thick.

Maoris The New Zealand aboriginal people, primarily Polynesian with some Melanesian admixture. They are tall in stature and were originally very warlike, but now are citizens of New Zealand.

map scale The relationship between distances on a map and the actual distances they represent on the surface of the earth.

market city A city that serves as the single market for all goods produced in an imaginary landscape. Used for modeling land-use patterns in the von Thünen model.

marsupials An order of species (found mainly in Australia) comprising the lowest mammals and having a pouch for carrying their young. It contains kangaroos, wombats, bandicoots, opossums, and others.

mass wasting All processes where soil and rock debris move downslope.

Mediterranean climate Climates classified by Köppen as Csa and found typically in the Mediterranean region and in the semitropical regions on the west coasts of all continents (except the Antarctic). Climates are noted for long, hot, dry summers and mild, wet winters.

merino sheep A hardy breed of fine-wooled white sheep originating in Spain. Their wool excels all others in weight and quality, but the breed does not rank high as a mutton producer.

mestizo The racial mixture in Latin America composed of Indian and European ancestry.

Middle America All the land area in the western hemisphere between the U.S.-Mexican boundary on the north and the Panama-Colombia boundary on the south.

milpa A type of primitive agriculture carried on in the tropical rainforests. Trees are cut and burned, and the land is farmed for a few years until the soil gives out. Then the land is allowed to lie fallow and return to its original forest state. (See also *roza* and *purma*.)

Moho The mohorovičić discontinuity. The boundary between the molten rock of the earth's mantle and the solid rock of the earth's crust. Named after the person who proved its existence.

Mongols The native tribes of Mongolia that historically were nomadic tent-dwellers, herdsmen by occupation, and followed their own sect of Buddhism. In the twelfth and thirteenth centuries, they conquered most of Asia and Eastern Europe.

monsoon forests Forests of the tropical areas of Southeast Asia and Africa where the monsoon climate produces a marked wet and dry seasonality, causing the trees to shed their leaves at the start of the dry season and giving the forest the aspect of a deciduous forest of temperate regions.

monsoons Seasonal wind of the Indian Ocean and South-Asia blowing from the southwest from April to October and from the northeast during the rest of the year. The winds coming from the southwest carry moisture off the ocean, creating heavy rains when driven over the land.

moraine An accumulation of soil, rocks, and rock debris left after a glacier has melted and retreated.

mortality The annual number of deaths in a country for every 1,000 people. Usually "infant mortality," which means the annual number of deaths of children under the age of one for every 1,000 births.

Moscow Region The manufacturing region surrounding the city of Moscow in the Soviet Union, including satellite cities as far away as Gorky and Smolensk.

mosses Very small green plants that grow in velvety clusters on rocks, trees, and moist ground.

mulatto The racial mixture in Latin America composed of Black African and white European ancestry.

multifactor region An area of the earth's surface demarcated by numerous factors, such as both the physical environment and human activities.

An example could be the American South, distinguished by its climate as well as the "southern accent."

mutation A sudden change in an inheritable characteristic of a plant or animal which reappears in all following generations.

National Health Scheme Australia's socialized medical program, which covers the health costs of all citizens.

natural selection The survival of certain forms of plants and animals that have adjusted best to the conditions under which they live.

nodal region An area demarcated by lines of movement from a "node" to peripheral points, or from outlying places to a point (node). For example, the distribution area for a newspaper is a nodal region (point-to-area movement), as is a garbage collection area (area-to-point movement).

nomadic pastoralism "Pastoralism" pertains to caring for livestock (shepherding), and "nomadic" means roaming around for pastures, so nomadic pastoralism is the economic activity carried on by nomadic herders.

normal lapse rate The constant decrease in temperature with increased elevation above the surface of the earth, usually given as 3.6° F for every 1,000 feet of elevation.

nortenos Literally translated from Spanish as "northerner" and used by Mexicans to refer to people from the United States.

North Sea oil Petroleum produced by using drilling platforms stationed offshore in the North Sea.

nuclear weapon Any weapon that uses either nuclear fission or nuclear fusion to convert mass to energy which is used for explosions for destruction.

nunataks Individual mountain peaks in the polar regions that protrude above glacial ice and remain dry from the ice and snow. They look like islands in a sea of ice.

oblate spheroid A spheroid is a solid sphere, and "oblate" means flat on two sides. Describes the shape of the earth: the polar regions are flat because of the equatorial bulge caused by rotation.

ocean trench Sometimes called an "ocean deep." A place in the ocean where the water is very deep. The underwater trenches are similar to canyons on the surface of the earth. The deepest known point of any ocean is the Philippine or Mindanao Trench of the Philippine Sea in the western Pacific.

Organization of Petroleum Exporting Countries (OPEC) An organization created on November 14, 1960, with the aim of determining world oil prices and advancing members' interests in trade and dealings on development with the industrialized oil-consuming nations. Venezuela led the initiative. Member nations today include Algeria, Ecuador, Gabon, Indonesia, Iran, Iraq, Kuwait, Libya, Nigeria, Qatar, Saudi Arabia, United Arab Emirates, and Venezuela.

Orient From Latin meaning "east," as opposed to "occident," meaning "west." Thus the term refers to Asia, especially East Asia.

orographic precipitation "Orography" refers to mountains, so orographic precipitation is produced when mountain barriers deflect moisture-laden air upward, causing cooling and condensation, which in turn causes the moisture to fall to the ground.

outer core The portion of the earth located between the mantle and the inner core. Assumed to be molten or in a liquid state with a radius of about 1,500 miles (2,400 km).

overseas Chinese Chinese nationals (citizens of China) that live in countries of Southeast Asia other than China. They are noted for their industrious work habits and ability to obtain and wisely use their money, sometimes sending large sums back to relatives in China.

pack ice Ice that has been broken into chunks and refrozen to create a jumbled mass over polar waters, especially in the Arctic region. The ice pack also drifts with the oceanic circulation, causing further breakage and refreezing.

paddy From the Malay *padi* which originally referred to unmilled or rough rice, whether growing or cut. Now, the field where the rice is grown.

Pangaea The name given to the supercontinent assumed to have existed on earth about 200 million years ago and to have separated into several large sections that became the current continents.

papiamento The jargon spoken by the black people of Curaçao, the island off the north coast of South America, with a vocabulary of Spanish, Portuguese, Dutch, English, Carib, and native African.

paramos A mountain grassland in Latin America consisting of tall bunch grasses and shrubs. Includes some areas above the snowline.

parent material The unconsolidated rock material from which soils are formed.

Paris Basin A bowl-shaped depression with Paris, France, located near the center. The depression is rimmed by high ground formed by cuestas, and the Seine River cuts through the middle of it.

pattern The geometric arrangement of the individual items (geographic facts) within a spatial distribution. An essential characteristic of all spatial distributions. (See also *density* and *dispersion*.)

perihelion The point in a planet's orbit when the planet is nearest the sun (as opposed to aphelion). Usually occurs, for the earth, around January 8.

permafrost Subsoil that is frozen throughout the year.

physical geography The branch of geography that focuses on the physical environment, including landforms, weather, climate, vegetation, and soils.

physiographic province A region of the earth's surface that is somewhat uniform in terms of the underlying rock and surface configuration.

Piedmont region An upland belt, part of the Atlantic Coastal Plain of the eastern United States, lying east of the Blue Ridge and Appalachian mountains and extending from the Hudson River to central Alabama. Also called Piedmont Plateau.

pixels Originally, the dots that make up a picture on a television screen. Used in remote sensing to describe the small unit areas (62 × 85 yards) that make up the image recorded by satellite cameras.

plane of the ecliptic The imaginary plane that contains the sun and all its planets (except Pluto). The solar system, then, is shaped like a plane, and the most direct rays of the sun that reach the earth extend outward along the plane.

plane projection A map projection where the earth's curved surface is projected onto a plane surface so that a two-dimensional image can be produced. Distortion of the image increases with distance away from the point of tangency between the circle (the earth) and the plane.

plate tectonics The theory that the earth's crust consists of a mosaic of rigid plates that move and generate landforms (especially mountains) where the plates meet.

podzol soil Soil developed through the podzolization process, which occurs only where climates are cold enough to inhibit bacterial action but have sufficient moisture to permit the growth of large green plants. Podzolization is associated with cool marine climates of west coasts and the interior continental climates with forests of coniferous trees such as spruce, fir, and pine.

polar ice The stationary ice in the Arctic Ocean about which the pack ice rotates. Polar ice is about 10 feet (3 m) thick and located directly over the North Pole.

polder A region of low land reclaimed from the sea by using dikes or dams to hold back the sea water.

pole of inaccessibility Describes the south polar region, because it was inaccessible to the early explorers for many years. Not only are the temperatures very low at the South Pole, but the elevation is very high (13,000 feet/3,962 m), making the lack of oxygen a major problem to explorers.

polyconic projection A type of map projection where the curved surface of the earth is projected onto numerous ("poly") cones, so that the image can be drawn onto a two-dimensional surface. The most popular of the various map projections.

population explosion The very rapid growth rate of the world's population during the last 50 years.

population pyramid A graph used to depict the age structure of the population of a country. The percentage of males and females in categories of five-year increments are shown. Although the graphs now often do not resemble pyramids, the traditional name of the graphs has been retained.

Povolzhye region The heavily utilized region in the Soviet Union that extends along the Volga River. A major function is the transportation of foodstuffs and raw materials on the river, but the region is also the country's major source of petroleum and natural gas. Water transportation extends from the Volga to the Don River and Black Sea through a series of canals, and pipelines for the movement of petroleum products extend outward to Moscow and the Ukraine.

prefecture A subdivision of a province for local administration under a governor, assembly, and council. Japan, for example, has 47 prefectures.

prices The amounts of money offered for goods or services and set by the threshold and range of the goods or services (associated with the Christaller model).

primate city Many Third World countries have only one large city, as opposed to industrial nations that have many large cities. These primate cities are usually the capitals and the economic centers drawing migrants from all parts of the countries.

profit The amount of income left after the costs of production and transportation have been deducted (associated with the von Thünen model of land use).

projections Methods used for changing the curved surface of the earth so it can be represented on the flat surface of a map.

Protestant Reformation The religious reform movement in Western Europe beginning in the early sixteenth century which resulted in the formation of the various Protestant denominations. Martin Luther led the movement by attacking the Roman Catholic concepts of paying indulgences for church favors, the veneration of the Virgin Mary and the saints, and the practice of clerical celibacy.

purma A type of tropical agriculture where the trees of the rainforest are cut and burned and the plots of land are farmed until the soil is depleted. (See also *milpa* and *roza*).

Pygmies The dwarf people of Central Africa, generally under 5 feet (1.5m) in height and dark-skinned, but lighter in color than the true Negroid.

Qing Zang Gaoyuan Also Plateau of Tibet. A broad, somewhat oval-shaped plateau in South-Central Asia that contains the highest mountains in the world, the Himalayas.

rainforest A tropical woodland characterized by an immense variety of lofty evergreen trees, lianas, and woody epiphytes and found in any considerable quantity only in the Amazon Valley of South America, Tropical West Africa, parts of India, and Southeast Asia.

range The average distance people are willing to travel in order to purchase a particular type of good or service.

rate of natural increase (RNI) A country's annual rate of population growth without regard for net migration. Calculated as the birth rate minus the death rate.

refugees People that flee their homeland to seek refuge elsewhere from war, political or religious persecution, or life-threatening conditions.

region An area of the earth's surface, usually with some internal homogeneity of physical or human factors and usually different from other areas.

regional geography An approach to the study of geography in which certain demarcated areas (regions) are important. The focus is on the human and physical aspects of the region.

relocation diffusion A type of spatial diffusion where the spreading process occurs through relocation, as in the building of an alluvial fan or the migration of people into a previously unsettled area. The opposite of expansion diffusion.

remote sensing Gathering information (data) by means of remotely located instruments, but usually refers to gathering data from the earth's surface by means of satellite photography.

representative fraction The fraction indicating the relationship between distances on a map and the actual distances on the surface of the earth. Usually given as something between 1:1,000 and 1:10,000,000, or where one inch on the map represents 1,000 or 10,000,000 inches on the earth. A common representative fraction is 1:36,360, where one inch on the map equals 36,360 inches (one mile) on the earth.

rift valley A large linear depression created by a lowering of part of the earth's crust through diastrophism, similar to a graben but on a larger scale. The lakes of eastern Africa appear in the north-south trending rift valley of that region.

"ring of fire" The line of active volcanoes that circles the Pacific Ocean from the southern tip of South America northward through North America, and west and south through East Asia and Southeast Asia. The ring contains about 80 percent of the world's active volcanoes.

Romanic languages (Also Romance languages.) The Indo-European languages originating from the countries that succeeded the Roman Empire, especially Italy, Spain, Portugal, France, and Romania.

roza A type of primitive agriculture used in the tropical rainforests. The trees are cut and burned, the land is farmed for a few years until the soil gives out, then the land is allowed to lie fallow and return to its original forest state. (See also *milpa* and *purma*.)

Russian Orthodox church A branch of the dominant Christian body of Eastern Europe, the Orthodox church. Church members are in union with the "ecumenical patriarch" (the Orthodox patriarch of Turkey) and do not recognize the leadership of the pope in Rome. Otherwise, the rituals are similar to those of the Roman Catholic church.

salinization The soil-forming process associated with desert regions where severe soil-moisture deficiency occurs. Sulfates and chlorides of calcium and sodium are common salts found in abundance in such soils.

Sanskrit The classical Old Indic literary language that developed in India from the third century B.C. onward. It is still used in the ritual of the Northern Buddhist church.

savanna A tropical or semitropical grassland containing scattered trees and xerophytic plants, such as the *campos* of Brazil or similar areas in eastern Africa. Usually contains the tallest of the grass types. Associated with the Köppen Aw-type climate.

scrublands Regions too dry for forests but wetter than grasslands. Plants are stunted trees and shrubs, or plants intermediate between trees and grass.

seismic waves A group of elastic waves generated within the earth whenever a sudden displacement of rock material occurs. Also describes long-period ocean waves generated by a seismic disturbance.

selection Any natural or artificial process that prevents certain individuals or groups of organisms from surviving and allows others to do so. The result is that certain traits of those that survive become pronounced, as in "natural selection" from Darwin.

sensible temperature A comfort (or discomfort) index that is the combination of air temperature and humidity but is given in terms of temperature. For example, a moist, hot day is more uncomfortable than a dry, hot day, so a temperature of 100° F might be given as 105° F in terms of sensible temperature.

Shinto The ethnic cult and religion of the Japanese, in which spirits of ancestors and some deities of nature are revered. Shinto is not regarded as a separate religion, so its adherents can be members of all faiths, including Buddhists and Christians.

sial The upper layer of the earth's crust that is lighter in color and less dense than the sima (the lower layer). The word is a combination of "si" and "al" from the first two letters of the most prominent minerals in the rocks: silica and aluminum.

Siberian High A large, stable high-pressure system that builds up over eastern Asia (Siberia) during the winter months because of the coldness caused by the continental location.

Sierra Madre del Sur The mountains of southern Mexico that connect the Sierra Madre Occidental with those of the Sierra Madre Oriental and fringe the southern portion of the Central Plateau.

Sierra Madre Occidental The range of mountains in Mexico running parallel to the Pacific coast and bordering the Central Plateau on the west.

Sierra Madre Oriental The range of mountains in Mexico running parallel to the Gulf of Mexico coast and bordering the Central Plateau on the east.

sima The lower layer of the earth's crust that is dense, dark in color, and made primarily of basaltic rocks. The word is a combination of "si" and "ma" from the first two letters of the most prominent minerals in the rocks: silica and magnesium.

single-factor region An area of the earth's surface demarcated by only one factor, such as where a particular crop is grown or where certain types of languages are spoken.

Sino-Tibetan A group of similar languages including Chinese, Tibetan, Burmese, and Thai.

siroccos A hot, steady, oppressive wind blowing from the deserts of the Middle East and North Africa across the wetter lands along the coast and sometimes continuing into Southern Europe.

Slavic languages A group of related languages spoken by the Slavic people, with an alphabet based on Greek, as opposed to the alphabet of the Romance languages, which is based on Latin. The Slavic people include the Russians, Poles, Bohemians, Moravians, Bulgarians, Serbians, Croatians, Wends, Slovaks, and other smaller groups.

small scale A map scale with a small representative fraction, such as 1:5,000,000. Maps with small scales portray large areas of the earth's surface or the entire earth.

solifluction Any mass movement of earth materials (including soil and rock) caused by alternate freezing and thawing.

South America The continent in the western hemisphere south of the boundary between Panama and Colombia.

Southern Alps The chain of mountains mostly on South Island of New Zealand but extending onto North Island. About half the surface of South Island is mountainous, as opposed to only about one-tenth of North Island. The Southern Alps, with their serrated crests, are among the world's more spectacular mountain ranges.

sovkhoz A farm, or agricultural estate, owned by the government of the Soviet Union.

Spanish Meseta The large plateau or tableland with underlying ancient crystalline masses that makes up most of the country of Spain.

spatial diffusion The dispersal of items by moving them through space, creating a new spatial distribution. (See also *cultural diffusion*.)

spatial distribution The distribution of things in space. Can be three-dimensional, but usually refers to the geometric arrangement of geographic phenomena on the surface of the earth.

spatial interaction Movement (interaction) through space, between sets of spatial distributions.

spatial models Methods that help us understand why things are located in particular places. Models can be verbal (descriptions), visual (maps and graphs), or symbolic (mathematical expressions).

Spice Islands Usually refers to the Molucca Islands, a group in the Malay archipelago between the islands of Celebes and New Guinea noted for producing spices, especially nutmeg, mace, and cloves. Sometimes also used to refer to all the Southeast Asia islands.

spring wheat Wheat planted in the spring, as opposed to winter wheat (planted in the fall), in the higher latitudes, where the ground freezes during the winter.

state farms Large, single-unit farms run by cooperative farmers who share the workload. The state owns all the land and equipment. Found primarily in the Soviet Union but also in other Communist countries.

statistical map Any thematic map based on information (statistics) for a single phenomenon, such as rainfall distribution, crop production, or population.

steppe Usually refers to one of the grassy, treeless plains of southeastern Europe and Asia, but also used to describe semiarid climate regions anywhere in the world.

Strait of Malacca The channel between the southern part of the Malay Peninsula and the island of Sumatra, connecting the Indian Ocean with the South China Sea. It is about 500 miles (805 km) long and varies in width between 35 miles (56 km) and 185 miles (298 km).

subduction The lowering of sections of the earth's crust at plate boundaries so that the rock melts and combines with the magma of the mantle.

subtropical forest Woodlands of the subtropical areas, such as the southeastern part of the United States, containing both evergreen and deciduous trees.

summer maximum rainfall Seasonal precipitation that comes mostly in the three months of the high sun period (summer), such as the Köppen Am climate areas.

summer solstice The longest day of the year, or when the direct rays of the sun are located at the Tropic of Cancer or Capricorn. Usually occurs on June 21 in the northern hemisphere and on December 21 in the southern hemisphere.

taiga The coniferous forests of the far northern regions of Asia, Europe, and North America.

Taoism The Chinese word "tao" means "the way," and Taoism is the Chinese religion and philosophy based on the doctrines of Lao-tse of the sixth century B.C., who advocated simplicity and selflessness.

Tasmania An island separated from the Australian continent by the Bass Strait, which is about 150 miles (241 km) in width. The island is essentially a southern extension of the Eastern Highlands of Australia.

temperate mixed forest Woodlands of the mid-latitudes containing both evergreen and deciduous trees.

terrorism The use of terror and violence as a political weapon or policy to intimidate or subjugate other people.

thematic map A map that contains information on one factor (or theme) only, such as precipitation amounts or population density.

threshold The minimum price and quantity of goods needed to bring a market into existence and to keep it functioning. Usually measured in terms of potential buyers (people) in an area surrounding the market location.

tierra caliente The Spanish term *tierra* means earth or ground, and *caliente* means hot or fiery. Thus, in the mountains of Central and South America the "hot ground" is a hot, humid zone found at the lowest elevations (0–3,000 feet), where bananas and sugar cane are grown.

tierra fria The Spanish term *tierra* means earth or ground, and *fria* means cold. Thus, in the mountains of Central and South America the "cold ground" lies in zones at the higher elevations, or between about 7,000 and 10,000 feet. The phrase generally refers to the vegetation types found in the zone.

tierra templada The Spanish term *tierra* means earth or ground, and *templada* means temperate or moderate. Thus, in the mountains of Central and South America the "temperate ground" is a zone of moderate temperatures lying between about 3,000 and 7,000 feet in elevation.

topographic maps Maps that portray the surface features of the earth, as well as human-made features, such as roads, bridges, and towns. These maps are noted for indicating accurate elevations above sea level through special isolines known as contour lines.

topography The surface features of the earth, such as mountains, plains, and lowlands.

transhumance The seasonal movement of livestock to and from mountain regions to utilize the mountain grasses for summer grazing and to protect the animals by bringing them to lower elevations during the winter.

transmigration Because the island of Java in Indonesia is so crowded, the Indonesian government encourages people to migrate to some of the less densely populated islands. Free passage and free land are used as inducements. The process is called "transmigration."

Treaty of Tordesillas After Columbus's first voyage of discovery, and to prevent a clash between Portugal and Spain, the pope drew an imaginary line from north to south through the Atlantic Ocean. He awarded Portugal the lands on the African side of the line, and gave to Spain the lands on the western side. The line was drawn 100 leagues (one league is about 3 miles) west of the Azores and the Cape Verde Islands. In 1494 the line was redrawn during the Treaty of Tordesillas at 370 leagues west of the Cape Verde Islands. Portugal retained rights to the land east of the line, and Spain those to the west. Spain and Portugal maintained their obligations faithfully, and no serious conflicts arose between them. The Protestant countries of northern Europe, however, did not recognize the pope's right to make presents of the New World lands, so British and Dutch explorers ignored the "line of demarcation."

tuna wars A series of conflicts between private citizens of the United States and the Peruvian authorities over fishing rights off the shores of Peru. A major type of fish caught in the region is tuna.

tundra The vast, nearly level, treeless plains of the Arctic regions.

Turkic languages Subfamilies of the Ural-Altaic languages spoken by people who live in the vast region ranging from the Adriatic Sea to the Okhotsk. The people are of similar Mongol ancestry but racially mixed. The numerous languages, however, are remarkably similar and are noted for agglutination and vowel harmony.

typhoons Intense tropical storms of the western Pacific Ocean (called hurricanes in the Atlantic Ocean) that produce wind speeds of over 100 mph and torrential rains.

unbounded plain An imaginary plain that is not constrained by boundaries, used to model the location of market towns.

uniform distribution A spatial distribution where all the items (geographic facts) are equally spaced or have equal distances between them. A completely uniform distribution forms a hexagonal pattern.

uniform plain An imaginary plain used for modeling spatial distributions that has no variation in the physical or human environments.

uniform transportation network A road or railroad network that is completely connected, so that movement in all directions and between all points is equally easy.

United Fruit Company A North American company founded in 1899 to develop banana production in Costa Rica.

Ural-Altaic Languages of Asia, including the Uralian, such as Finnish and Samoyed, and the Altaic, such as Turkic, Mongolic, Manchu, and Korean.

Ural-Volga region Two linear regions of manufacturing in the Soviet Union that run along the eastern and western slopes of the Ural Mountains. The regions are noted for their oil production, mining, and transportation connections with the Moscow area, as well as for manufacturing.

Urdu A language used by the followers of Islam living in India. It developed from Hindustani, but with Arabic characters.

Valley and Ridge region A physiographic region of the eastern United States (especially Pennsylvania) that contains valley and ridge topography that resembles a washboard.

variation In an organism, divergence in structural or physiological character from characteristics typical of the group to which it belongs.

vernal equinox The time of the year (spring in the northern hemisphere) when the sun's most direct rays are located at the equator, making the days and nights of equal length throughout the world.

vulcanism The releasing of magma from the mantle either into vents in the earth's crust (intrusive) or out onto the earth's surface (extrusive).

water pollution Contamination of the earth's water, especially the rivers and oceans near port cities, with garbage, factory waste, and other refuse.

Watusi A branch of the Bahima people of Central Africa living near Uganda who are Hamite-Bantu mixtures.

weathering The breakdown of rock material by exposure to the atmosphere, creating small, easily moved (eroded) fragments from the larger rocks.

West Indies The island portion of Middle America including the Greater and Lesser Antilles.

windchill Both low temperature and wind cause heat loss from body surfaces, and the greater the velocity of the wind, the lower the temperature feels, up to about 45 mph wind; after that there is no additional chilling effect.

winter maximum rainfall Seasonal precipitation that comes primarily in the three low-sun period months (winter), such as with climates designated by Köppen as Cs.

winter solstice The time of the year when the most direct rays of the sun are located at the Tropics of Cancer or Capricorn opposite the hemisphere of concern. For example, it occurs on December 21 in the northern hemisphere, when the sun's direct rays are at the Tropic of Capricorn (in the southern hemisphere). That day is also the shortest day of the year for northern hemisphere locations.

winter wheat Wheat that is planted in the fall, starts to grow, lies dormant over the winter, then continues to grow the following spring, giving it a head start on the growing season. This method is used at lower latitudes, where the ground does not freeze.

world regions Large areas of the earth's surface noted for their internal homogeneity or designated for some functional reason.

xerophytic Describes all vegetation that has adapted to arid conditions (*xero* means dry, and *phytic* refers to plants).

Yamato One of the Japanese proper, as distinguished from the Ainus and the naturalized foreigners. The Yamatos are of ancient origin and probably entered Japan from the Chinese mainland during prehistoric times.

zambo The racial mixture in Latin America composed of Black African and native Indian ancestry.

zero population growth (ZPG) Occurs when a country's rate of natural increase is zero percent. Can be achieved if no family has more than two children and immigration does not exceed outmigration.

zooplankton Zooplankton, which consists principally of algae and diatoms, is to the life of the sea what pastures are to the life of the land. It is more abundant in cold water than in warm water, because of the smaller amounts of denitrifying bacteria in the colder water.

Zulu A branch of the Bantu tribes of Central Africa.

Zulu wars The conflicts between the Boers and later the British with the great Bantu nations of the Natal region of South Africa.

INDEX

ABC countries, 324
Aborigines, 241, 242, 246
Abortion, 211
Abraham, 412
Acacia trees, 239
Acapulco, Mexico, 265, 292, 296
Acculturation, 38, 437
Acetone, 322
Acid, 561
Acid rain, 558
Acid soil, 25
Aconcagua, Cerro (Mountain), 305
Acorns, 149
Acquired Immunodeficiency Syndrome (AIDS), 370, 371, 570
Adams, John Quincy, 77
Adelaide, Australia, 235, 244, 248, 249, 252
Aden, Gulf of, 339, 395-97, 407
Adirondack Mountains, 83, 110
Adriatic Sea, 183
Aegean Sea, 398
Aeolian erosion, 17
Aerosol sprays, 549
Afghanistan, 170, 396, 404, 416, 424, 426, 427, 430, 431, 435-41, 444, 445, 448, 566-69
Africa, 12, 36, 42, 43, 46, 49, 131, 147, 154, 245, 275, 279, 283, 288, 303, 304, 339-93, 396, 398, 403, 407, 413, 463, 467, 509, 540, 564, 566-69
  agriculture, 354, 372, 385
Afrikaan language, 385
Afrikaners, 385, 387, 390
Age structure, 32, 246
Agglomerated distribution, 58
Agricultural machinery, 114
Agriculture, shifting, 460
Agriculture, slash and burn, 320, 321, 561
Agung, Mount, 457
Ainu, 214
Air conditioners, 549
Air Mali, 358
Air photo interpretation, 77
Air photograph, 58, 78, 526
Air pollution, 558
Aircraft manufacturing, 115, 328
Al-Hijaz Azir Mountains, 398
Alabama, 110
Alaska, 87-89, 93, 110, 120, 166, 231, 297, 305, 538, 540, 547, 551-53

Alaskan pollack, 220
Albanese, 145
Albania, 138, 142, 145, 147, 154
Albany, Australia, 248
Albany, New York, 116
Albatross Plateau, 516
Albert, Lake, 362
Alberta, Canada, 93, 107, 110
Alcoholism, 187
Aleutian Islands, 88, 527
Alexander the Great, 399, 435
Alexandria, Egypt, 349, 356
Alfalfa, 107, 324
Algae, 539, 545
Algeria, 147, 346-48, 355, 356, 409, 414, 566, 568
Algiers, Algeria, 346, 349
Alice Springs, Australia, 252
Allah, 407
Alliance for Progress, 289
Alluvial fans, 60
Alluvium, 205
Alpine System of Europe, 130, 131
Alps Mountains, 131
Alsace-Loraine, 148, 149
Altai (Altay) Mountains, 170, 484
Altaic languages, 214
Altyn Tagh Mountains, 484
Aluminum (See also Bauxite), 13, 190, 328, 375, 474, 561
Amazon River, 308, 309, 313, 322, 331, 473
American Civil War, 77
American Indians, 95, 102, 103, 275
American Somoa, 526, 530, 532, 533
Americans, 167, 351, 404, 557
Amundsen-Scott station, 550
Amur River, 170, 487
Andaman Sea, 454
Andean Pact, 329
Andes Mountains, 304, 305, 308-13, 317, 319, 322, 325-27, 331, 334, 518
Andesite, 518, 519
  line, 518
Andorra, 125, 131
Anglican church, 101, 144, 217, 246, 384, 391, 529
Anglo-America, 89-91, 95, 100, 102, 103, 113-15, 120, 133, 137, 151, 186, 197, 203, 205, 211, 227, 261
Angola, 379

Animism, 46, 215, 353, 371
Ankara, Turkey, 403, 404
Antananarivo, Madagascar, 383
Antarctic Circle, 515, 537, 544
Antarctic Drift, 238
Antarctic Ocean, 222
Antarctic Treaty, 553
Antarctica, 538-40, 542-45, 548-50, 552, 559
Anthracite (See also Coal), 49, 190, 238
Anthropologist, 35, 36
Anti-Atlas Mountains, 347
Anticyclone, 207
Antimony, 288, 388, 411, 476, 562
Apartheid, 378, 390, 391, 568
Apennine Mountains, 131
Aphelion, 10, 11
Appalachian Highlands, 82, 83, 110, 113
Appalachian Mountains, 103, 108
Appalachian Plateau, 82
Apples, 151, 220, 249, 501
Aqaba, Gulf of, 395, 397, 407
Aquitaine Basin, 129
Arab World, 4, 353, 398, 418
Arabia (See also Saudi Arabia), 343
Arabian Desert, 399, 405, 414, 419
Arabian Peninsula, 42, 339, 343, 396, 398, 399, 407, 411
Arabian Sea, 395, 398
Arabic language, 42, 43, 346, 353, 354, 358, 406, 440
Arabs, 47, 343, 346, 360, 383, 409, 412, 418, 419, 475, 526, 557
Arafura Sea, 454
Aral Sea, 170, 171
Arapaho, 103
Arctic, 539, 546, 547, 552
Arctic Circle, 42, 170, 171, 180, 537, 545, 552, 553
Arctic Coastal Plain, 83
Arctic fox, 547
Arctic Ocean, 128, 130, 167, 170, 537-40, 545, 547, 553
Arctic poppy, 546
Arctic willow, 546
Ardennes, 131
Area map, 69, 76
Arequipa, Peru, 315
Argentina, 304, 305, 308, 310, 313, 319, 324-29, 331, 503, 538, 550, 569
Argon, 548

I-1

Arizona, 18, 19, 37, 88, 93, 103, 106, 113, 548
Arkansas, 86
Armenian Knot, 398
Armenian Plateau, 397
Armenian Republic, U.S.S.R., 175, 176, 189, 398
Aroostock Plain, 83
Arsenic, 288
Artesian wells, 235, 247
Aruba, 273, 279, 294
Arum, 525
Asbestos, 388, 411, 476
Asia, 22, 28, 43, 49, 95, 123, 167, 203, 207, 208, 211, 214, 228, 255, 265, 276, 317, 318, 339, 340, 341, 343, 353, 383, 395, 396, 398, 403, 413, 423, 427, 445, 474, 484, 491, 518, 526, 564, 569
Asians, 245, 343, 391
Assam, India, 441, 477
Assamese language, 440
Association of Southeast Asian Nations (ASEAN), 480
Astronomical unit (AU), 10
Asuncion, Paraguay, 309
Aswan High Dam, 347, 356
Atacama Desert, 312, 327, 328, 382
Atafu Islands, 529
Athens, Greece, 138, 140
Atlanta, Georgia, 103, 116
  International Airport, 118
Atlantic Ocean, 12, 28, 83, 101, 106, 126, 129, 133, 172, 220, 262, 272, 274, 304, 308, 309, 324, 339, 361, 362, 380, 382, 459
Atlas Mountains, 343, 347
Atolls, 519, 522
Atomic energy, 505
Auckland, New Zealand, 244, 246
Australia, 41, 43, 49, 189, 223, 224, 231–56, 374, 454, 503, 518, 523, 532, 538, 540, 553, 562, 569
  agriculture, 247
Australian aborigines, 525
Australian Federation, 243
Australoids, 241, 528
Austria, 103, 131, 140, 144, 148, 151, 152, 413
Austrians, 147
Austro-Hungarian Empire, 147
Autobahns, 156
Automobile, 226, 228, 250, 355, 503, 507
  manufacturing, 115, 297, 328, 329, 476
  racing, 98
Autumnal equinox, 11
Avars, 147
Avianca Airlines, 333
Ayrshire cattle, 151
Azerbaijan Republic, U.S.S.R., 188
Azizia, Libya, 349
Azores, 304
Azov Sea, 190
Aztec civilization, 296

Bab al-Mandab, 339
Baby boom, 34, 282
Babylon, 399
Badminton, 99
Baffin Island, 540
Baghdad, Iraq, 399, 403, 404, 416
Bahama Islands, 262, 271, 279, 295
Bahasia language, 471
Bahia Blanca, Argentina, 310
Bahima, 343, 397, 417, 418
Bahrain, 398, 409, 416
Baikal, Lake, 170, 190, 207
Baker, Lorenzo, 287
Baku, U.S.S.R., 191, 198
Bali, 457, 458
Balkan countries, 147, 155, 395, 413
Balkan Mountains, 131, 147
Balkanization, 147
Ballet, 199
Balsam, 286
Baltic Lake Plain, 130
Baltic Sea, 130, 158, 170, 172, 175, 177
Baltimore, Maryland, 103, 114
Baluchi language, 406
Bamboo, 209, 430, 431
Bamileke tribe, 368
Banana Republics, 284
Bananas, 50, 209, 249, 261, 266, 269, 271, 280, 283–87, 290, 293, 321, 331, 359, 409, 441, 529
Banda Sea, 454, 458
Banderas Bay, Mexico, 265
Bandicoots, 239
Bandung, Indonesia, 467
Banghazi, Libya, 346, 349
Bangkok, Thailand, 456, 467, 476
Bangladesh, 28, 424–27, 429, 431, 432, 435, 439–41, 444, 445, 447, 448
Bantu language, 339, 342, 372
Bantus, 343, 359, 368, 371, 378, 379, 383–85, 390, 391
Bantustans, 391
Banyan, 431
Baptist religion, 43, 101, 182, 217
Barbados, 273, 276, 278
Barbed wire, 419
Barcelona, Spain, 161
Barents Sea, 180
Barley, 50, 107, 148, 149, 220, 222, 249, 354, 411, 441, 501, 503
Barracuda, 233
Barranquilla, Colombia, 310, 326
Barrier reefs, 519
Basalt, 518
Baseball, 98, 99, 218, 219
Basin and Range Province, 19, 88, 89, 107
Basketball, 98, 99, 219
Basque language, 145, 146
Basques, 131, 145, 147
Bass Strait, 233
Bats, 524
Battle of the Bulge, 129
Bauxite, 154, 190, 222, 250, 289, 291, 328, 374, 442, 476, 562

Bavarian Plateau, 130, 131
Bay of Bengal, 429, 454, 456
Beans (*See also* Soybeans), 283, 320, 325, 354, 473, 503
Beaumont, Texas, 108
Bedouins, 343, 414, 416, 419
Beech trees, 209, 241
Beef (*See also* Cattle), 325, 331, 388, 529, 530
Beijing (Peking), China, 492, 493, 495, 498, 507, 510
Beirut, Lebanon, 416
Belgium, 129, 131, 135, 136, 138, 145, 148, 151, 152, 157, 343, 566, 568
Belize, 261, 277, 279, 280, 286, 296
Belukha, Mount, 170
Benelux, 152
Bengali language, 43, 440
Benguela Current, 382
Benin, 368, 373
Benzine, 407
Berber language, 354
Berbers, 343, 346, 354, 358, 371
Berezniki, U.S.S.R., 191
Bering Strait, 87, 166, 171, 515, 550
Berkshire Hills, 83
Berlin, East Germany, 142
Berlin, West Germany, 138, 140, 413
Berlin-to-Baghdad railroad, 416
Berries, 241, 539
Bethlehem, Israel, 407
Bhutan, 424, 427, 431, 433, 436, 439, 440, 444, 445, 446
Bible, 401, 405, 407, 410, 412
Bicycles, 375, 443, 503, 507
Big bang theory, 7
Birch trees, 209, 430, 539
Birmingham, Alabama, 82, 103, 110, 209
Birth control, 211, 352, 493, 570
Birth rate, crude, 30, 32
Bituminous coal (*See also* Coal), 153, 190, 251
Black Africa, 42
Black Hills, 87
Black Hole of Calcutta, 434
Black Sea, 130, 131, 135, 158, 166, 167, 170, 177, 186, 191, 196, 395, 398, 403
Blacks, American, 103
Blacks, West Indies, 147
Bligh, William, 245
Blizzards, 545
Block-fault mountains, 264, 272
Blue gas, 548
Blue Nile River, 347
Blue Ridge Mountains, 82
Boat people, 246, 465
Bobo tribe, 367
Boer War, 379
Boers, 379, 385
Bogota, Colombia, 316, 332
Bohemian Basin, 131
Bohemian Massif, 130, 131
Bohemians, 145, 183
Boke, Guinea, 374

Bolivia, 304, 305, 309, 310, 311, 313, 318, 319, 322, 327–29, 333, 334
Bolivian Plateau, 311
Bolsons, 264, 265
Bombay, India, 430, 434, 441, 445
Bonaire, 279
Bophuthatswana, 391
Borate, 411
Borneo, 455
Bosporus, 403, 404
Boston Mountains, 86
Boston, Massachusetts, 114, 116, 287
Botanists, 211, 473
Botswana, 378, 379, 382–85, 388, 389, 566, 567
Bowling, 98, 99
Boxing, 98, 219
Brahmaputra River, 427, 441
Brandy, 129
Brazil, 42, 46, 223, 303, 304, 309, 310, 313, 314, 317–19, 321, 322, 324–29, 331, 333, 473, 567, 569
  Plateau of, 308, 321
Brazilian Highlands, 309, 310
Bread, 66
Breton language, 145
Bridge on the River Kwai, 464
Bridgetown, Barbados, 276
Brisbane, Australia, 244, 249
British Columbia, 89, 107
British Empire, 120, 425
British Imperial system, 77
British India, 434
British Isles, 133, 145, 151
British Mandate of Israel, 406, 411
British, 226, 242, 246, 280, 304, 343, 379, 386, 395, 406, 435, 445, 448, 453, 462, 463, 473, 483, 553
Britons, 245
Brittany, 128, 145
Broadleaf decidous trees, 91, 135, 172, 209, 211
Broadleaf evergreen trees, 271, 309, 312
Bronze, 526
Bronze Age, 526
Brooklyn, New York, 287
Brooks Mountains, 540
Brown Swiss cattle, 151
Brownson depression, 272
Brownsville, Texas, 83
Brunei, 424, 456, 473, 475, 480
Buddha (Siddhartha Gautama), 216, 436, 439
Buddhism, 46, 181, 215, 216, 220, 436, 439, 470, 471, 493, 496–98
Buddhist monks, 434, 439, 496
Bude, England, 161
Buenos Aires, Argentina, 29, 309, 314, 316, 324, 325, 332
Buffalo, New York, 114–16
Bukavu, Zaire, 376
Bulgaria, 142, 145, 147, 154, 396
Bulgarians, 145, 183
Bullfinch Bay, Antarctica, 546

Bunker Hill, 83
Burkina Faso (Upper Volta), 359, 367, 368, 374
Burlap, 441
Burma (*See also* Myanma), 445
Burma Road, 477
Burmese language, 471
Burundi, 359, 363, 368, 370–72, 374, 376
Bushmen, 343, 359, 378, 379, 384
Buthelezi, Mangosuthu, 391
Butter, 49, 107, 248, 252, 388
Byelorussian language, 182
Byelorussian S.S.R., 174, 176, 177
Byrrange Mountains, 540
Byzantines, 147, 182,

Cabbage, 220
Cabinda, 361
Cacoa, 288, 321, 325, 372
Cacti, 22
Caicos Islands, 271
Cairns, Australia, 249
Cairo, Egypt, 349, 351, 364
Cajamarca Department, Peru, 322
Calais, France, 156
Calcification of soils, 24, 95
Calculators, 224
Calcutta, India, 430, 434, 445
California, 22, 89, 93, 101–3, 106–8, 110–15, 135, 261, 265, 312, 388, 560
  Gulf of, 267
Calvinistic religion, 379
Camaguey, Cuba, 278
Cambay, Gulf of, 441
Cambodia (Kampuchea), 424, 454, 456, 465, 467, 471, 473, 476, 480, 567
Cambridge University, 28
Camels, 354, 358, 411, 507,
Cameras, 224–26, 228
Cameroon, 343, 363, 368, 372, 378, 566
Cameroon, Mount, 362
Camphor, 209
Campos, 310, 324
Canaan, 406, 412
Canada, 81–121, 170, 175, 223, 224, 226, 251, 279, 291, 317, 328, 503, 538, 553, 562, 568
Canadian Shield, 83, 85, 113, 540
Canberra, Australia, 244
Cancer, Tropic of, 260, 427
Cancun, Mexico, 293, 296
Cannibals, 525
Canoe, dugout, 297
Cantabrian Mountains, 130
Cantaloupes, 107
Canton (*See* Guangzhou)
Cape Coloreds (*See also* Coloreds), 382
Cape Town, South Africa, 339, 382, 387, 391
Cape Verde Islands, 303, 304,
Cape York Peninsula, 233, 239, 252
Capricorn, Tropic of, 239, 321
Caracas, Venezuela, 332

Carbon dioxide, 548, 559, 560
Carbon monoxide, 559
Carib language, 279
Caribbean Islands, (*See also* West Indies), 259, 262, 267, 271, 274, 275, 279, 286, 297
Caribbean Sea, 262, 263, 268, 269, 271, 273, 276, 283, 294, 295, 305, 308
Caribou, 548, 550
Caroline Islands, 526, 532
Carpathian Mountains, 131, 167, 188
Carpentaria, Gulf of, 234
Carrots, 107
Cartago, Costa Rica, 47
Cartesian Coordinate System, 70
Carthaginian ruins, 355
Carthaginians, 346
Cartographers, 74
Casablanca, Morocco, 346, 347, 351, 356
Casagrande, Louis, 291
Cascade Mountains, 89, 93
Casein, 248
Caspian Sea, 167, 170, 175, 191, 362, 395, 398, 411
Cassava (manioc), 320, 373, 529
Caste system, 436
Castille, Plateau of, 130
Castro, Fidel, 286
Cat Island, 271
Catalonia, Spain, 147
Cataracts, 548
Catherine the Great, 174
Catholic religion (*See* Roman Catholic)
Cattle, 49, 106, 107, 149, 151, 220, 235, 243, 247, 248, 252, 271, 288, 324, 354, 374, 385, 441, 529
Caucasoids, 214, 276
Caucasus Mountains, 130, 166, 167, 191
Caviar, 187, 190, 411
Cayman Islands, 279
Cedar trees, 211
Celebes, 462
Celebes Sea, 454
Celtic language, 145, 146
Celtics, 128
Cement industry, 223, 290, 294, 411, 443, 530
Centipedes, 524
Central African Republic, 359, 374
Central America, 259-300, 322, 329, 564, 567, 569
Central American Common Market (CACM), 329
Central German Upland, 130, 131
Central Lowlands, Australia, 233-35
Central Lowlands, Europe, 126, 129, 130
Central Massif, 130
Central place model, 65, 66
Central Plateau of Europe, 126, 130
Central Plateau of Mexico, 264, 265, 268
Central Valley of California, 89, 90, 93, 107, 108
Cephalic index, 36
Ceram Sea, 454

Ceramics, 225
Cereal grains, 409
Cevennes, 130
Chad, 346, 348, 349, 351–56
Chang Jiang, 484–86, 488, 501, 503, 505, 507
Chao Phraya River, 456
Chapare region, Bolivia, 322
Chaparral, 93, 135, 312
Chatham Island, 232
Cheese, 49, 65, 107, 151, 248, 388
Chelyabinsk, U.S.S.R., 177, 186
Chemical industry, 250, 290, 294
Chemical warfare, 561
Chemical weathering, 17
Chengdu Plain (Red Basin), 485, 488
Chengdu, China, 495
Chernozem soils, 188
Cherrapunji, India, 429
Chesapeake Bay, Virginia, 309
Chevrolet, 197
Chiapas Highlands, 265
Chic Choc Mountains, 83
Chicago, Illinois, 63, 68, 95, 103, 114–18
Chickens, 106, 151, 461, 525, 529
Child-care centers, 142
Chile, 304, 305, 308, 312, 313, 318, 319, 324, 327–29, 334, 519, 538, 567, 569
China's Sorrow, 484
China, 28, 46, 166, 170, 189, 214, 216, 225, 424, 426, 427, 445, 446, 453, 456, 461, 468, 470, 483, 484, 487–90, 491–94, 497–99, 503–5, 507–9, 511, 569
Chindwin River, 456
Chinese, 103, 245, 246, 276, 468, 526
Chinese Dynasties, 492
Chinese Exclusion Act, 103
Chinese food, 103
Chinese language, 101, 214, 215
Chinese tribes, 461
Chins, 470
Chlorofluorocarbons, 549
Chocolate, 324, 372
Cholera, 432
Chongquig (Chungking), China, 485, 486, 488
Choropleth map, 72
Chou Dynasty, China, 492, 496
Choukoutien, China, 491
Christaller, Walter, 65–67
Christchurch, New Zealand, 232, 244
Christian religion, 529
Christianity, 43, 45, 47, 100, 142, 181, 182, 215, 216, 318, 353, 368, 371, 384, 396, 399, 405, 412, 413, 415, 497, 498
Christians, 318, 406, 407, 415, 498
Christmas, 236
Chrome, 388, 411, 476, 562
Chronium ore, 113, 154, 190, 222, 289, 529
Chrysler Mfg. Co., 116, 329
Chungking (See Chongquig)

Church of Jesus Christ of Latter-Day Saints (Mormons), 101, 217
Churchill, Winston, 404
Cigarettes, 322, 530
Cinchona, 473
Cincinnati, Ohio, 115
Circulation of the atmosphere, 20, 21
Cirey-sur-Vezouze, France, 149
Ciskei, 391
Citrus fruit, 131, 134, 149, 249, 354, 387, 409, 415, 473, 503
City of David (See also Jerusalem), 405
Ciudad Bolivar, Venezuela, 327, 328
Clay, 113
Cleveland, Ohio, 103
Climate regions, world, 20
Climatic classification, 18
Climatologists, 559
Clitorectomy, 370
Cloves, 373, 462
Club Mex, 292
Coal (See also Anthracite and Bituminous) 50, 51, 108, 114, 118, 129, 135, 152, 190, 220, 223, 249, 288, 290, 328, 388, 411, 442, 476, 504–7, 530, 540, 552, 558
Coastal Plains, 83, 86,
Cobalt, 113, 355, 375, 529, 552
Coca, 322
Coca–Cola, 511
Cocaine, 322, 324
Cocoa, 269, 324, 372
Coconut oil, 529
Coconuts, 269, 288, 385, 387, 441, 473, 525, 529
Cocos Islands, 516
Codfish, 128
Coffee, 50, 120, 261, 267, 268, 283, 290, 319, 321, 325, 330, 372, 441, 473
Cohorts, age, 33, 34,
Collective farms, 385, 501
Colombia, 259, 269, 297, 304, 305, 308–10, 313, 319, 321, 322, 326–29, 331, 333, 334
Colombo, Sri Lanka, 435, 445
Colon, Panama, 287
Colonialism, 464, 480, 532, 533
Color blindness, 36
Colorado, 88
Colorado Plateau, 88
Colorado River, 18, 88
Coloreds, 382, 383, 391
Columbia Plateau, 88
Columbia River, 88
Columbium, 113, 375
Columbus, Christopher, 259, 304, 462, 463
Commercial farming, 48–50, 325
Commercial herding, 48, 49
Communism, 286, 295, 297, 386, 414, 417–19, 448, 465, 480, 497, 498, 501, 502, 509, 511
Communism Peak, 170
Communist China (See also China), 483, 493

Communist Party, 147, 165, 170, 174, 177, 180, 182, 187, 199
Commuter airlines, 118
Computer cartography, 76
Computer maps, 76
Computers, 224
Concentric zonation, 64, 65
Concepcion, Chile, 312, 328
Confucianism, 46, 471, 496–98, 501
Confucius, 492, 496, 501
Congo, 363, 372
Congolese language, 42
Congressional townships, 69
Conic projection, 75, 76
Coniferous trees, 131, 133, 209, 211, 241
Connecticut River Valley, 114
Constantinople, Turkey (See also Istanbul), 413
Continental climate, 20, 170, 236
Continental glaciers, 538
Continental shelf, 83
Conurbation, 205, 211
Cook Islands, 529, 532
Cook Strait, 232
Cook, Captain James, 242
Cook, Mount, 233, 235
Coolies, 486
Copper, 113, 190, 222, 288, 289, 327, 330, 355, 375, 388, 411, 476, 506, 507, 526, 529, 552, 567
Copra, 385, 476, 529
Coral, 233, 516, 518
Coral atolls, 520
Coral islands, 518
Coral reefs, 518
Coral Sea, 232
Core area, 103, 147
Cork oaks, 149
Corn Belt, 106
Corn, 4, 49, 50, 65, 107, 188, 249, 283, 319, 320, 324, 325, 331, 354, 388, 441, 444, 473, 501
Cornwall, 128
Costa Rica, 37, 261, 268, 275, 276, 280, 283, 284, 297, 329
Cottage cheese, 49
Cotton Belt, 106
Cotton, 106, 107, 115, 189, 225, 284, 286, 321, 331, 354, 385, 409, 411, 441, 444, 473, 501, 503, 505, 567
Council for Mutual Economic Assistance, 135
Cozumel, Mexico, 288
Creeping willow trees, 539
Crimea, 167
Crimean Mountains, 167
Croatians, 145, 148, 183
Cross-country, 99
Crude-oil (See also Oil and Petroleum), 289, 325
Crusades, 405
Crust of the earth, 14, 16
Crustaceans, 220
Cuba, 263, 271–73, 276, 278, 279, 286, 288, 290, 293–97, 368

Cubans, 102, 103, 566
Cuernavaca, Mexico, 291
Cuestas, 83
Cultural diffusion, 103
Cultural traits, 38
Curacao, 279, 294
Curling, 99
Cuzco Basin, 304
Cuzco, Peru, 304, 318, 322, 327
Cyanide, 561
Cyclonic storms, 429, 491, 522, 538, 552
Cylindrical projection, 74, 75
Cyprus, 398, 403, 406, 411, 415, 417–19
Czechoslovakia, 131, 144, 145, 152, 158, 413

da Gama, Vasco, 462
Dairy Belt, 107
Dairy products, 65, 220, 243, 247
Dakar, Senegal, 376
Dallas, Texas, 63, 116
Dallas-Ft. Worth International Airport, 118
Damavand, Mount, 397
Danish language, 41, 145
Danube River, 131, 135, 158, 170
Dari Persian language, 440
Darien, Gulf of, 269
Darling River, 235
Darwin, Charles, 520
Daryal Pass, 167
Dates, 131, 134, 354, 411
David, King of Jews, 405
DeAzuero Peninsula, Panama, 269
DeNicoya Peninsula, Costa Rica 269
de Palma, Emilio, 550
Dead Sea, 399, 401, 405
Dead Sea Scrolls, 405
Death rate, crude, 30
Decathlon, 99
Deccan Plateau, 427, 429
Deception Island, 553
Delhi, India, 431, 434
Demographic transition, 31, 32
Demons, 371
Denmark, 115, 135, 140, 142, 151, 153, 553
Density, 57, 58
Denver, Colorado, 21, 63, 383, 495
  Stapleton International Airport, 118
Deodorants, 549
Dependency ratio, 33
Desalinization plants, 409
Desert soils, 23
Desertification, 559
Detroit, Michigan, 103, 114
Developed World, 558, 561, 562
Developing World, 260, 334
Devil's Island, French Guiana, 313
Dewey, George, 463
Dezhneva Cape, 166
Diabetes, 37
Diameter of the earth, 9
Diamonds, 374, 388, 415, 562

Diastrophism, 16
Diatoms, 545
Dictators, 569
Diffusion of an innovation, 60
Dikes, 484, 485
DiMaggio, Joe, 100
Dinaric Alps Mountains, 131
Dingo, 239
Direction arrow, 59, 73
Dismal Swamp, 83
Dispersion, 57, 58
Distance decay, 68, 69
Distance, earth to sun, 9
Divehi language, 440
Djibouti, 374
DNA molecules, 548
Dnieper River, 170, 182, 190
Dniester River, 170
Doldrums, 520, 522
Dome of the Rock, 405
Dominica Island, 276
Dominican Republic, 263, 272, 276, 278, 279, 296
Domino effect, 464
Don River, 170, 191
Donbas, 186, 190, 191, 196
Donets Basin, 186, 190
Donetsk, U.S.S.R., 190
Dong Hai (East China Sea), 484, 485, 498
Donkeys, 354, 357, 374, 411, 417
Dortmund, West Germany, 152
Dot map, 71, 72
Double-cropping, 209
Drake Passage, 304
Drakensberg Mountains, 380–83
Dravidian language, 39, 42
Drug trafficking, 291
Drugs, 250
Drumlins, 540
Druze Christians, 406
Duckbill platypus, 239
Ducks, 151
Dugout canoes, 526
Duisburg, West Germany, 152
Dunedin, New Zealand, 244
Durango, Colorado, 88
Durban, South Africa, 383, 387
Dutch, 145, 226, 304, 343, 379, 385, 456, 463
Dutch language, 41, 145, 279
Dutch Reformed Church, 384
Dutch Shell Oil Company, 294
Dutch United East India Company, 463
Dwarfs, 546
Dwarf shrubs, 547
Dwarf trees, 211
Dzongkha language, 440

Early majority, 63
Earthquakes, 14, 204, 205, 265, 273, 519
Earthsat, 78
East Asia, 265, 423, 424, 483–512
  agriculture, 501–3
East China Sea (See Dong Hai)

East European Plain, 167, 173, 196
East Germany, 140, 152
East India Trading Company, 425, 434
East Indies, 463
East London, South Africa, 383
East Pakistan, 439
East Siberian Highlands, 167
Easter Island, 516
Easter Highlands, Australia, 233
Eastern Orthodox Religion, 43, 142, 181, 182, 217
Ebony, 372
Economic zone, off shore, 569
Ecuador, 304, 305, 309, 311, 313, 317, 318, 321, 329, 333, 334, 409, 567
Edward, Lake, 362
Edwards Plateau, 87
Eggs, 107, 529
Equatorial Guinea, 368
Egypt, 34, 35, 346, 347, 349, 352–56, 358, 359, 363, 364, 398, 412, 414, 569
Egyptians, 343, 352, 410
Egyptian Pharaohs, 412
Eiffel Tower, Paris, 29
El Djem, Tunisia, 344
El Jebel Plateau, 411
El Paso, Texas, 115
El Salvador, 261, 283, 297, 329, 568
Elbe River, 130, 131, 140, 158
Elbrus, Mount, 167
Elburz Mountains, 398, 404
Ellesmere Island, 540
Ellsworth Mountains, 544
Emeralds, 411
Emi Koussi, Mount, 348
Empire State Building, 29
Emu, 239, 241
Encounter Bay, 235
England, 50, 101, 129, 144, 148, 157, 158, 161, 242, 243, 249, 296, 413, 529
English Channel, 128, 129, 136, 158
English language, 41, 43, 101, 145, 246, 279, 317, 354, 371, 385, 406, 434, 440, 471, 498, 501
Ensilage, 107
Environmental Protection Agency (EPA), 560, 561
Epicanthic fold, 36
Epidemiologists, 60
Epiphytes, 241
Episcopal church, 217
Equator, 303, 308, 520
Equatorial Guinea, 372
Equinox, 309, 363
Erebus, Mount, 544
Erg, 348
Erie Canal, 116
Erosion, 15, 17
Erratics, 85
Erythroxylon coca, 322
Erythroxylon novogranatense, 322
Eskers, 540
Eskimos, 35, 37, 183, 550
Essen, West Germany, 152
Estancias, 324

I-5

Estonian Republic, U.S.S.R., 176
Estuaries, 205
Ether, 322
Ethiopia, 343, 344, 363, 369, 371, 372, 374, 567
Ethiopians, 343
Eucalyptus trees, 239
Euphrates River, 398, 399
Europe, 28, 32, 36, 41–43, 49, 123–63, 166, 167, 186, 197, 225, 245, 251, 275, 276, 283, 294, 297, 303, 312, 321, 322, 324, 329, 339, 340, 355, 358, 374, 395, 396, 401, 403, 407, 408, 412–14, 425, 433, 442, 453, 462, 464, 476, 491, 492, 526, 550, 552, 560, 564, 566, 569
  agriculture, 148, 150
European Common Market, 560
European Economic Community (EEC), 135, 251, 329, 377, 391, 479
European Inquisition, 410
Everest, Mount, 427, 516
Everglades, 83
Evergreen forests, 172
Exclaves, 145
Expansion diffusion, 60, 62
Eyre, Lake, 235

Falkland Islands, 553
Fall line, 82
Fang tribe, 368
Far East (See also East Asia), 395, 483
Farsi language, 42, 406
Fast ice, 539
Fault block mountains, 89, 234, 264, 341
Faulting, 16, 204, 269, 305, 398
Federal Aid Highway Act, 116
Fell Fields, 546
Fencing, 99
Fertility, 31
Fertilizer, 319, 328, 443, 476, 506, 560, 567
  chemical, 440
  commercial, 501
Fiat, 329
Field hockey, 99
Figs, 354
Fiji, 528, 529, 532
Filipino language, 471
Filipinos, 103, 463
Film (movies) industry, 355
Finland, 42, 126, 135, 144, 145, 154
Finland, Gulf of, 177
Finnish language, 146, 183
Finno-Ugrians, 550
Fiordland, New Zealand, 235
Fiords (fjords), 126, 128, 308, 540
Fir trees, 211, 430
Fire ants, 60
First order (goods and towns), 66
First World, 529, 562, 569
Fishing, 128, 189, 196, 207, 220, 312, 375, 411, 444, 476, 507, 525, 530, 537, 545, 547, 548, 550, 569
Five-Year Plans, 187

Fjords (See Fiords)
Flanders Plain, 129, 148
Flax, 188
Fleming, Alexander, 32
Flemish language, 41, 145, 148
Flies, 524
Florence, Italy, 136
Flores Sea, 454
Florida, 83, 113, 262, 271, 275, 295, 430
Fluvial erosion, 17
Folding, 16, 204, 269, 272, 305, 398
Fontana, California, 110
Football, 98, 99, 219
Ford Motor Company, 116, 329
Formosa (See Taiwan)
Fossil fuels, 50, 374, 375, 558, 561
Foveaux Strait, 232
France, 29, 76, 120, 126, 128–30, 135, 136, 142, 144, 145, 147–49, 151, 152, 154, 157, 259, 296, 297, 304, 313, 386, 343, 413, 463, 476, 532, 538, 566, 569
Frank, Carl, 287
Frank, Otto, 287
Frankfurt, Germany, 413
Frankincense, 411
Free market economy, 106
Freeport, Bahamas, 294
French, 318, 347, 351, 367, 386, 471
French Canadians, 103
French Guiana, 304, 313, 317, 318
French Indochina, 463, 464
French language, 39, 42, 43, 101, 102, 279, 317, 346, 354, 371, 385, 471
French Polynesia, 532
French Revolution, 148
Friction of distance, 68
Front, 21
Fuji, Mount, 205
Fukuoka, Japan, 212
Fula, 343
Fulani tribe, 367
Fungus, 370, 539
Furniture, 388

Gabon, 363, 368, 369, 372, 375, 409, 567
Gabriel, 407
Gaelic language, 128, 145
Galapagos Islands, 516
Galilee, Sea of, 399, 401, 407, 409
Galles, 343
Galveston, Texas, 108
Gambia, The, 344, 368, 371, 372, 376
Gambling, 294
Ganges plain, 45
Ganges River, 425, 427, 430, 431, 433, 441
Garonne Basin, 129
Gasoline, 66, 407, 557
Gaya, India, 436
Geese, 151
Gem stones, 567
General Motors Mfg. Corp., 116, 329
General Sherman tree, 22
Genes, 35, 37

Geneticists, 473
Genocide, 413
Geographers, 91
Geographic fact, 57
Geographic isolation, 37
Geography:
  economic, 3, 48, 557
  human, 3, 27, 53, 557
  physical, 3, 557
  political, 3, 53, 557
  regional, 4
Geologists, 7, 8
Geomorphologists, 14, 60, 81
Geomorphology, 14
Geopolitics, 419
George, Lake, 362
Georgia, 82, 83, 115
Georgia Republic, U.S.S.R., 177, 189
German Catholics, 144
German Jews, 411
German language, 41, 43, 145, 148, 287, 414
German Lowlands, 167
German Protestants, 144
Germanic languages, 41, 42, 145
Germans, 59, 145, 147, 180, 191, 390, 413, 456
Germany, 29, 64, 103, 120, 129–31, 135, 144, 151, 160, 228, 245, 252, 304, 343, 419, 476, 566
Gerona, Spain, 149
Geysers, 272
Ghana, 372, 374, 566
Giant clams, 233
Gibraltar, Strait of, 339, 346
Gibson Desert, 235
Gilbert Islands, 526
Glacial erosion, 17
Glaciers, 235, 240, 253, 267, 308, 490, 523, 539, 540, 543, 544
Glasnost, 182
Gleization soils, 24, 95
Goa, 446
Goanna, 239
Goats, 135, 149, 354, 374, 418, 529
Gobi Shamo (Gobi Desert), 490, 498
God, 412
Gold, 113, 191, 243, 245, 249, 288, 374, 388, 461, 476, 505–7, 529, 552
Golf, 98, 99, 218, 219
Goliath of Gath, 405
Good Hope, Cape of, 379
Goodyear, Charles, 473
Gora Chen, Mount, 167
Gorbachev, Mikhail, 182
Gorky, U.S.S.R., 177, 186, 191
Goruma tribe, 367
Goths, 147
Graben, 131
Gran Chaco, Paraguay, 310
Grand Canyon, 18–19
Grapes, 107, 129, 130, 134, 148, 149, 151, 189, 249, 354, 409
Graphite, 288, 506
Gravity model, 68, 69
Grease, 407

Great Artesian Basin, 235
Great Australian Bight, 234
Great Barrier Reef, 233
Great Basin, 95
Great Britain, 29, 77, 102, 103, 126, 128, 144, 152, 246, 251, 343, 395, 418, 425, 441, 442, 463, 569
Great Depression, 34
Great Dividing Range, 233, 234, 238, 239, 243, 244, 252, 257
Great Enlightenment, 216
Great Lakes, 85, 86, 101, 110, 114, 115, 118
Great Plains, 86, 93
Great Russians, 182
Great Sandy Desert, 235
Great Trek, 379
Great Victoria Desert, 235
Great Wall of China, 489, 492
Greater Antilles, 262–64, 271, 272, 278
Greece, 131, 135, 137, 142, 145, 154, 161, 396
Greek language, 37, 145, 146, 183, 406
Greek Orthodox Church (*See also* Eastern Orthodox), 406
Greeks, 36, 419
Green Mountains, 83
Greenhouse effect, 559, 560
Greenland, 36, 538, 540, 550, 553, 559
Greenwich, England, 70
Grenada Island, 262
Greytown, Nicaragua, 270
Gross National Product (GNP), 225
Groundnuts, 386
Guadalajara, Mexico, 277, 287
Guadeloupe Island, 273, 276, 279
Guam, 515, 529, 530, 532, 533
Guangzhou (Canton), China, 495, 503
Guantanomo Bay, Cuba, 263
Guard hairs, 548
Guatemala, 261, 265, 276, 277, 280, 283, 290, 329
Guernsey cattle, 151
Guiana Highlands, 308, 327
Guianas, 313, 317, 321
Guilders, 463
Guinea, 362, 374, 376
Guinea, Gulf of, 362, 363
Guinea-Bissau, 372, 373, 376
Gulf and Atlantic Coastal Plain, 83
Gulf Coast oilfields, 118
Gulf Coastal Plain, 106
Gulf Stream, 126, 128, 133
Gum arabic, 355
Gum trees, 239
Gutenfels Castle, West Germany, 160
Guyana, 304, 313, 317, 318, 321, 328, 567
Gymnastics, 99, 219
Gypsum, 265, 411

Haifa, Israel, 416
Haileybury College, 28
Hair sprays, 549
Hair texture, 35

Haiti, 263, 272, 278, 279, 283, 286, 296, 328
Haitians, 566
Halley Bay, Antarctica, 548
Hamburg, West Germany, 140, 413
Hamites, 343
Hamito-Semitic languages, 39, 42, 353, 372
Hammurabi, 399
Han Chinese, 461
Han Dynasty, China, 496
Hanoi, Vietnam, 456, 467
Harare, Zimbabwe, 383
Hare (*See also* Rabbit), 548
Havana, Cuba, 278
Hawaii, 42, 103, 231, 516–19, 526, 527, 529–33
Hay, 107, 152
Hazardous waste, 561
Heavy industry, 50, 191, 196, 249, 286, 290, 291, 328, 356, 375, 411, 443, 476, 504–6, 530
Hebei, China, 505
Hebrew language, 42, 353, 406, 414
Hebrews (*See also* Jews), 410, 412
Helium, 548
Hellenic languages, 146
Hells Canyon, 88
Hemlock trees, 211
Hemp, 189
Henequen, 286, 290
Henry the Navigator, 462
Heredity, 35
Hermon, Mount, 401
Herring, 128
Hess, Harry, 13
Hevea brasiliensis (rubber tree), 473
Hexagonal patterns, 66
Hides, 49, 247, 252, 374, 411, 503
Hierarchical diffusion, 62, 63
High Plains, 87
High Veld, 379
High-technology, 50
High-yield seeds, 440
Hill people, 468
Hilla, Iraq, 399
Himalaya Mountains, 22, 28, 46, 341, 425–27, 430, 434, 456, 484, 518
Hinayana (Lesser Vehicle), 470
Hindi language, 42, 43, 406, 440, 486
Hindu Kush Mountains, 425, 426, 437
Hinduism, 45, 46, 318, 435, 436, 439, 441
Hindus, 318, 384, 434
Hindustani language, 440
Hinterlands, 65
Hira-gana, 215
Hispaniola, 263, 272, 278
Ho Chi Minh City (Saigon), Vietnam, 467, 468, 470, 476
Hockey, 98
Hoggar Mountains, 348
Hogs, 106, 149, 151, 188, 189, 247, 374, 461, 501, 525, 529
Hokkaido, 204, 205, 209, 211, 212, 220
Holland (*See also* The Netherlands), 413
Holmes, Arthur, 13

Holstein cattle, 151
Holy Land, 399, 406, 407, 410, 415
Homelands (South Africa), 391
Homesteading, 326
Homo sapiens, 37, 343
Honda, 116, 226
Honduras, 261, 269, 279, 284, 296, 329, 567
Hong Kong (Xianggang), 28, 95, 103, 115, 424, 483, 489, 491, 494, 495, 499, 503, 507, 508
Honolulu, Hawaii, 226
Honshu, 204, 205, 209, 211
Hooghly River, 434
Hoosac Mountains, 83
Hope Bay Colony, 550
Horn, Cape of, 303
Hornets, 524
Horse latitudes, 520
Horseracing, 98, 294
Horses, 324, 374, 411, 507
Hostages, 404, 408, 417, 567
Hot springs, 205
Hottentots, 378, 379, 384
House of Hanover, 362
Houston, Texas, 108
Huallaga Valley, Peru, 322
Huang Hai (Yellow Sea), 484
Huang He (Hwang Ho, or Yellow River), 484, 485, 488–91, 498, 501, 503–7
Hudson Bay, 83
Huk Communists, 464
Humus, 23
Hungarian Basin, 131, 135
Hungarian language, 145, 146
Hungary, 42, 103, 140, 144, 145, 414
Huns, 135, 147
Huron, Lake, 118
Hurricanes, 209, 274, 275, 277, 279, 459
Hutu tribe, 368
Hybrid seeds, 319
Hydrochloric acid, 322
Hydroelectricity, 152, 205, 234, 251, 374, 444
Hydrogen, 548
Hydroids, 519
Hyundai Excel, 502

Ibadan, Nigeria, 364
Ibak, Sultan of Delhi, 439
Iberian Peninsula, 130, 155, 303, 317
Ice Age, 167, 538
Ice caps, 538, 559
Ice cream, 49
Ice hockey, 99
Ice sheets, 126, 129, 538–40, 542, 543, 549
Icebergs, 543
Iceland, 133, 137, 154, 155
Icelandic language, 145
Idaho, 88
Illinois, 106, 108, 110, 115, 116
Illyrian languages, 145, 146
Immigration and Naturalization Act, 103

I-7

Immune deficiencies, 548
Import substitution, 329
Inca civilization, 304, 313, 318
India, 28, 42, 45, 46, 147, 166, 215, 216, 223, 259, 276, 313, 317, 353, 404, 406, 407, 423–27, 429–37, 439–43, 445, 446, 448, 453, 456, 461, 463, 470, 496, 509, 568, 569
Indian languages:
  Middle America, 280
  South America, 317
Indian Ocean, 42, 220, 238, 339, 380, 381, 395, 427, 429, 454, 455
Indiana, 106
Indians, North America, 147, 276, 280, 283, 286
Indians, South America, 305, 313, 319, 322, 324
Indians, India, 462
Indic languages, 42
Indo–Aryan language, 317
Indo–European language, 39, 41, 42, 145, 146, 182
Indonesia, 29, 353, 386, 409, 424, 455, 456, 462, 464–67, 470, 471, 473–77, 480, 515, 516, 525, 528, 529, 569
Indus River, 407, 425, 427
Industrial Revolution, 148, 152, 559
Infant mortality, 467
Infrared light, 560
Innovators, 63
Insects, 460, 524
Inter–American Highway, 297
Intercoastal Waterway, 116
Interior Lowlands, 86
Interior Plains, 86, 93, 103
Intermontane basins, 269
International Court of Justice, 390
International Geophysical Year, 553
Interstate Highway System, 116
Intertropical convergence zone (intertropical front), 21, 270, 458, 459, 522
Inventions, 38
Iodine, 328
Iowa, 86, 103, 106
Iquitos, Peru, 331
Iran, 43, 120, 398, 399, 401, 406, 408, 409, 411, 415, 416, 418, 419, 437, 567–69
Iranian Desert, 399
Iranic languages, 42
Iraq, 43, 398, 399, 401, 404, 406–9, 411, 415, 416, 418, 567–69
Irazu, Mount, 271
Ireland, 103, 126, 135, 149, 154
Irish, 246
Irish language, 128, 145
Irish potatoes, 149
Irkutsk, U.S.S.R., 186, 190, 197, 507
Iron (*See also* iron ore), 8, 9, 50, 51, 113, 114, 152, 288–90, 325, 561
Iron Age, 50
Iron Gate, 131
Iron industry, 290

Iron ore (*See also* Iron), 108, 110, 118, 129, 135, 152, 190, 222, 223, 225, 249, 250, 288, 289, 327, 331, 355, 375, 388, 411, 442, 476, 504–6, 530, 567
Iron smelters, 115
Iron Springs, Utah, 110
Irrawaddy River, 456
Irredentas, 148
Irredentists, 148
Irtysh River, 167, 188
Islam religion, 43, 45, 47, 181, 353, 358, 367, 368, 371, 396, 407, 412, 419, 435, 439, 461–63, 470, 497, 498
Islamic World, 4
Isohyet, 310
Isoline, 19, 72, 310
Isostasy, 542
Isotherm, 19, 107
Israel, 120, 398, 406, 408, 409, 411, 413–15, 417–19, 567, 569
Israel Museum, Jerusalem, 405
Israelites, 405
Istanbul, Turkey, 156, 182, 403, 404, 411
Italian food, 103
Italian language, 42
Italians, 148
Italy, 29, 103, 135, 137, 144, 145, 148, 149, 151, 152, 154, 155, 245, 246, 259, 304, 343, 413, 462, 566, 567
Ivan the Great, 174
Ivan the Terrible, 174
Ivanovo, U.S.S.R., 191
Ivory Coast, 368, 371, 566
Ixtapa, Mexico, 293

J Hadur Shu–ayb, Mount, 398
Jacksonville, Florida, 116
Jade, 476
Jakarta, Indonesia, 467
Jamaica, 263, 272, 278–80, 286, 287, 289, 291, 293, 296, 328
Jamshedpur, India, 443
Janesville, Wisconsin, 115
Japan, 28, 30, 36, 46, 120, 135, 203–29, 232, 252, 296, 297, 327, 329, 330, 391, 408, 423, 442, 445, 457, 470, 471, 476, 483, 484, 488, 492, 493, 495, 498, 502, 504, 507, 511, 519, 563, 569
  agriculture, 217
Japan, Sea of, 180, 207
Japanese, 464
Japanese Alps, 205
Japanese food, 103
Japanese language, 39, 42, 43, 214, 215, 471
Jari River, 473
Java, 42, 456–58, 466, 467, 477
Java Man (Pithecanthropus erectus), 461
Java Sea, 454, 458
Java Trench, 458
Jebusites, 405
Jefferson, Thomas, 77

Jenner, Edward, 32
Jersey cattle, 151
Jersey City, New Jersey, 287
Jerusalem, Israel, 43, 405, 407, 412
Jesus, 405, 407, 412
Jet airplanes, 297, 416, 507
Jet fuel, 407
Jews (*See also* Judiasm), 36, 43, 45, 181, 405, 413, 414, 418
Johannesburg, South Africa, 383, 384, 388
Jordan, 398, 406, 409, 411, 415, 419
Jordan River, 399, 401, 407
Juan Fernandez Islands, 516
Juarez, Mexico, 115
Judaism, 43, 45, 396, 406
Judea (Judaea), 406, 412
Jumbuck, 246
Jungle, 276, 430, 524
Junks (junco), 461, 477, 486
Jura Mountains, 130, 131
Jute, 434, 441, 444, 503

Kaaba, 407
Kabul, Afghanistan, 435–37
Kachins, 470
Kagoshima, Japan, 209
Kai-shek, Chiang, 492, 496
Kaiser Aluminum Company, 291
Kalahari Desert, 379, 382, 385
Kalinin, U.S.S.R., 191
Kangaroos, 239, 241
Kanji, 215
Kano, Nigeria, 364
Kansas, 107, 108
Kansas City, Missouri, 116
Kansi Plain, Japan, 205, 213
Kanuri language, 354
Karachi, Pakistan, 435, 439, 445, 446
Karaganda, U.S.S.R., 186, 190, 197
Karakoram Mountains, 426, 427, 430, 487
Karate, 219
Karens, 470
Kariba Dam, 377, 381
Kariba Gorge, 381
Kariba Lake, 377, 381
Kashmir, 446
Kata-kana, 215
Katanga region, Zaire, 375, 380
Katmandu, Nepal, 434
Kawasaki, 211
Kazakh Republic, U.S.S.R., 175, 186, 190
Keith, Henry, 287
Keith, Minor, 287
Kentucky, 108, 110
Kenya, 30, 359, 362, 366, 368, 370–72, 376, 567
Kerosene, 322, 407
Key West, Florida, 90
Khalkha Mongol, 499
Khan, Genghis, 492, 495
Khan, Kublai, 492
Kharkov, U.S.S.R., 177, 186, 190
Khartoum, Sudan, 347, 351, 356

I-8

Khasi Hills, 423, 425, 429
Khmer language, 471
Khomeini, Ayatollah, 419
Khrebet Cherskogo Mountains, 170
Khrushchev, Nikita, 182
Khyber Pass, 425, 437
Kibo, Mount, 362
Kiev, U.S.S.R., 170, 174, 177, 182, 186, 190
Kievan Rus', 182
Kilimanjaro, Mount, 362
Killer bees, 60
Kimberley, South Africa, 388
Kingston, Jamaica, 278
Kinshasa, Zaire, 364
Kirgiz Republic, U.S.S.R., 175, 186
Kiribati Islands, 529
Kiro Shio current, 208
Kiruna iron mine, 553
Kitakyushu, Japan, 212, 223
Klamath Mountains, 89
Koala bears, 239
Kobe, Japan, 212
Kolkhozes, 187, 188
Kolomna, U.S.S.R., 191
Komsa Stone Age, 552
Köppen, Wladimir, 18, 19, 22
Koran, 407
Korea, (See also North and South Korea) 28, 46, 95, 204, 213, 214, 407, 499, 507
Korean language, 39, 42, 214
Korean Peninsula, 484, 488, 493, 503, 505, 507
Korean War, 255, 487, 493
Koreans, 103
Kosciusko, Mount, 233, 235
Krakatoa, 457, 458, 519
Krasnoyarsk, U.S.S.R., 197
Kremlin, Moscow, U.S.S.R., 177, 199
Krypton, 548
Kuibyshev, U.S.S.R., 177, 186, 199
Kunlun Shan Mountains, 484, 485
Kunming, China, 477
Kurdish Hills, 411
Kuwait, 396, 407, 409, 411
Kuzbas, 186, 190, 191, 196
Kuznetsk Basin, 177, 190, 196
Kuznetsk, U.S.S.R., 186
Kwajalein Island, 519
Kyoto, Japan, 46, 215
Kyshit language, 354
Kyushu, 204, 205, 212

La Paz, Bolivia, 331, 332
Labor-intensive, 217
Labrador, Canada, 83, 110, 430
Lacrosse, 99
Laggards, 63
Lagoons, 519
Lagos, Nigeria, 364
Lake Charles, Louisiana, 108
Lamb meat, 189, 247

Lambert glacier, 539
Landlocked countries, 484
Landsat, 78
Landslide, 17, 205
Lao language, 471
Lao-tse, 496
Laos, 424, 454, 456, 465, 467, 470–74, 476, 480, 567
Lapp language, 145, 146
Lapps, 550, 552
Large-scale map, 73
Lashio, Myanma, 477
Last Supper, 405
Late majority, 63
Lateritic soils, 24, 93
Latin America, 259–61, 266, 267, 275–78, 284, 296, 304, 566
Latin American Free Trade Association (LAFTA), 329
Latin language, 37, 42, 145, 166, 183
Latitude lines, 20, 70, 71, 73
Latvian Republic, U.S.S.R., 176
Launceton, England, 133
Lava, 265, 458, 516, 527
Lawrence, T. E. (Lawrence of Arabia), 416
Laxatives, 324
Lead, 476, 505, 506, 552, 559
Leather products, 355
Leather tanning, 505
Lebanese, 276
Lebanon, 398, 406, 409, 415, 417, 419
Ledo Road, 477
Leeward Islands, 263, 272, 279
Leipzig, East Germany, 142
Lemming, 548
Lena River, 170, 191
Lenin, Vladimir, 174, 182
Leningrad, U.S.S.R., 177
Leprosy, 432
Lesotho, 370, 378, 382–85, 387
Lesser Antilles, 262, 264, 271, 272, 279
Lettuce, 107
Levees, 484, 485
Lianas, 241
Liao He, 488
Liberia, 363, 371, 374–76
Libya, 344, 346, 351, 355, 356, 359, 368, 409
Lichens, 545–47
Liechtenstein, 125
Light industry, 50, 290, 375, 476
Light year, 10
Lignite (See also Coal), 152, 190, 251
Lille, France, 39
Lima, Peru, 304, 316, 332
Lime, 113, 519
Limes, 411
Limestone, 114, 222, 223, 265, 271, 272, 274, 290, 405, 411, 519, 527
Limon, Costa Rica, 279, 287, 296
Line of Demarcation (See also Treaty of Tordesillas), 304
Lippe River, 152
Lisbon, Portugal, 462

Lithosphere, 13
Lithuanian Republic, U.S.S.R., 144, 176
Liverpool, England, 136
Llano Estacado, 87
Llanos, 310, 322
Lochs, 128
Loess soil, 129, 167, 501
Logistic curve, 63
Loire River, 129
London, England, 70, 138–40, 568
Long Beach, California, 108
Longitude lines, 70, 71, 73
Lorenz curve, 63
Lorraine district, 152
Los Angeles, California, 63, 95, 101–3, 108, 115–16, 226, 362
  International Airport, 118
Louisiana, 101, 102, 108
Lourenco Marques, Mozambique (See Maputo)
Loveland Pass, 21
Lowland of Brabant, 129
Luanda, Angola, 382
Lugansk, U.S.S.R., 190
Lusaka, Zambia, 370
Luther, Martin, 142–43
Lutherans, 101, 144
Luxembourg, 129, 135, 140, 152
Luzon, Philippines, 456, 471
Lyon, France, 156, 160

Macao, 28, 424, 483, 489, 499, 507
Macarthur, John, 242, 245
Macchia, 135
Mace, 462
Machine guns, 203
Machu Pichu, Peru, 309
Mackerel, 220
Macquarie, Lachlan, 245
Madagascar, 42, 344, 378, 380, 382, 384–87, 529
Madras, India, 445
Madrid, Spain, 138
Magdalena River, 308, 331
Magellan, Strait of, 305
Magellan, Ferdinand, 463
Magma, 13, 16, 518, 541
Magnesite, 506
Magnesium, 9, 13, 113
Magyars, 147
Mah-jongg, 506
Mahayana (Greater Vehicle), 470
Mahogany, 372
Maine, 83, 116
Maize (See also Corn), 49, 283, 319, 373
Major league baseball, 100
Makis, 135
Malacca, Strait of, 454
Malagasy (See also Madagascar), 386
Malagasy language, 385, 386
Malan, Daniel, 391
Malaria, 36, 287, 473
Malawi, 368

I-9

Malay culture, 464, 471, 473, 525, 528
Malay language, 471
Malay Peninsula, 454, 455, 459, 476, 515
Malaya (See also Malaysia), 464
Malayans, 214, 383, 460–62, 468
Malayo-Polynesian language, 39, 42, 43
Malaysia, 424, 427, 440, 441, 454, 463, 464, 470, 471, 473, 474, 476, 480
  Eastern, 455, 456
  Western, 455, 459
Malaysian Federation, 455
Mali, 346, 349, 351–56, 362, 367
Malta, 125, 136, 137
Malthus, Thomas, 28
Mamison Pass, 167
Manadnock, Mount, 83
Manaus, Brazil, 331
Manchester, England, 136
Manchu Dynasty, China, 492
Manchu language, 183
Manchuria, 488, 490–92, 501, 504, 507
Manchus, 492, 493
Mandalay, Myanma, 456
Mandarin Chinese language, 43, 471, 498, 499
Mandarin oranges, 220
Mandela, Nelson, 391
Manganese, 113, 154, 190, 222, 288, 289, 355, 375, 388, 411, 442, 476, 529, 562
Mango, 431
Manhattan Island, 101
Manila Bay, 463
Manioc, 283, 320
Manitoba, Canada, 107
Mantle, 9, 14, 16
Manufacturing, 52, 196
Manzanillo, Mexico, 293
Maoris, 242, 243, 246, 526
Map projections, 73
Map scale, 73
Maple, 430
Maputo, Mozambique, 383
Maquiladora, 115, 290
Maquis, 135
Maracaibo, Lake, 326
Maranon River, Peru, 322
Marathi language, 440
Marcos, Ferdinand, 464
Mare's milk, 503
Mariana Islands, 209, 532
Maritime Atlas Mountains, 347
Market city, 64
Markham, Mount, 544
Marmara, Sea of, 403, 404
Marrakech, Morocco, 347
Marseille, France, 139
Marshall Islands, 519, 526, 532
Marsupials, 239, 523
Martinique Island, 273, 276, 279
Marx, Karl, 180
Marxism, 496
Mashhad, Iran, 403, 404, 416
Mass wasting, 17
Massachusetts, 83
Matamata, Tunisia, 341
Mauna Kea, 516

Mauna Loa, 516
Mauritania, 346, 351–55, 567
Mayan civilization, 296
Mayflower, 126
Mazatlán, Mexico, 287
Mazda, 226
Mbabane, Swaziland, 383
McKinley, Mount, 89, 305
McMurdo Sound, 544, 549
Meat (See also Cattle, Hogs, and Sheep), 49, 183, 355, 385
Mecca, Saudi Arabia, 45, 405, 407
Medan, Indonesia, 467
Medina, Saudi Arabia, 407, 416
Mediterranean climate, 249, 312
Mediterranean Riviera, 131
Mediterranean Sea, 130, 131, 134–36, 148, 149, 151, 339, 346, 348, 349, 353, 355, 361, 395, 398, 399, 403, 404, 406, 407, 409, 412, 415, 416, 560
Mekong River, 456
Melanesia, 515, 525, 528, 529
Melanesian, 242
Melbourne, Australia, 233, 243, 244, 252
Melons, 409
Memphis, Tennessee, 103, 116
Mendoza, Argentina, 305
Mercedes Benz, 329
Mercury, 288, 552
Mesabi iron ore range, 110
Meso-America (See also Middle America), 259
Mesopotamia, 399, 412, 416
Mestizo, 275, 276, 283, 313, 325
Meteorite, 407
Methodists, 101, 246, 384, 496, 529
Metric Conversion Act, 77
Metric system, 498
Metromex, 291, 292
Mexamerica, 291
Mexican Americans, 103
Mexican food, 103
Mexicans, 95, 103, 566
Mexico, 42, 49, 95, 175, 236, 259–300, 320, 329, 503
Mexico City, Mexico, 29, 262–65, 268, 277, 283, 290, 291, 297
Mexico, Gulf of, 83, 263, 265, 277
Miami, Florida, 83, 102, 103, 116, 118, 271, 295, 303
Mica, 190, 388, 442
Michigan, 107, 114, 115
Michoacian, Mexico, 265
Microchips, 203
Micronesia, 515, 526, 528, 529
Mid-continent oil field, 108
Middle Ages, 148, 187, 414, 439
Middle America, 259–303, 309, 310 312, 317, 328
  agriculture, 283–88
Middle Atlas Mountains, 347
Middle East, 4, 43, 45, 154, 191, 250, 294, 329, 330, 353, 395–421, 423, 568,
  agriculture, 409–11
Middle Rocky Mountains, 87

MIG fighter planes, 507
Migration, 30, 59, 60, 291, 314, 317, 328, 406, 414, 437, 438, 562, 566
Milan, Italy, 152
Military governments, 569
Milk, 49, 64, 66, 107, 151, 189, 220, 248
Millet, 49, 373, 411, 441, 501, 503, 526
Milpa, 320
Milwaukee, Wisconsin, 29, 114
Mimosa trees, 239, 355
Minas Gerias, Brazil, 327
Mindanao, 456, 515
Ming Dynasty, China, 492
Mining, 196
Minneapolis, Minnesota, 116, 166
Minnesota, 85, 106, 110, 113
Minsk, U.S.S.R., 177
Missionaries, 371
Mississippi, 103
Mississippi River, 86, 116
Mississippi River Valley, 83, 93
Missouri, 86, 106
Missouri River, 86, 116
Mitchell, Mount, 82
Mites, 524
Mobile, Alabama, 260
Mocha coffee, 411
Mogul tribes, 437
Mohair, 387
Mohorovicic discontinuity (Moho), 9
Moist, mild winter climate, 19
Moist, severe winter climate, 19
Moldavia, 176
Moles, 239
Molodezhnaya station, 550
Molucca Islands, 462
Molucca, Sea of, 454
Molybdenum, 288
Monaco, 125
Monadnocks, 83
Mongolia, 46, 170, 424, 483, 484, 488, 490, 492–95, 498–501, 503, 507, 508, 511
Mongolian language, 183, 499
Mongoloids, 35, 528, 550
Mongols, 174, 183, 214, 425, 461, 493, 509
Monsoon, 207, 270, 429, 430, 461, 489
Monsoon forest, 431
Montana, 107, 204
Monterrey, Mexico, 277, 286, 290, 291, 297
Montreal, Canada, 95, 114
Moon, 8
Moors, 131
Moraine, 83, 167
Moravians, 145, 183
Morocco, 346, 347, 349, 355, 358, 414, 566
Mortality rate, 31
Moscow Region, 191
Moscow, U.S.S.R., 135, 165, 166, 170, 174, 176, 177, 180, 181, 188, 191, 196, 197, 511
Moses, 406, 412
Mosque of the Prophet, 407
Mosques, 405

Mosquito Coast, Nicaragua, 297
Mosquitoes, 287
Moss, 539, 545–47
Mossi tribe, 367
Mount Cook Air Service Ltd., 253
Mountbatten, Earl (Lord), 453
Mozambique, 378, 380, 381, 383, 385, 389, 390, 566
Muese River, 131
Mufumbiro (Virunga) Mountains, 362
Muhammad, The Prophet, 43, 353, 405, 407
Mulatto, 275, 276, 283, 312, 313
Mulberry trees, 220, 503
Mules (couriers), 324
Mummies, 318
Munich, Germany, 131
Murmansk, U.S.S.R., 180, 537, 552
Murray Basin, 235
Murray River 235, 243, 248, 249
Muscat, Oman, 405
Muscovy, 174
Musk ox, 548
Muslim World, 4
Muslims 43, 142, 181, 318, 353, 358, 359, 371, 384, 404, 405, 419, 434, 435, 439, 497
Mustard, 151
Mutation, 35
Mutton, 242, 243, 325, 388
Myanma (Burma), 424, 454, 456, 463, 470, 471, 473–75

Nagoya, Japan, 211, 223
Nairobi, Kenya, 366, 376
Namib Desert, 382, 385
Namibia, 378–80, 382, 385, 389, 390
Nan Hai (South China Sea), 454, 456, 465, 484, 498
Nan Shan Mountains, 484
Nanjing (Nanking), China, 485
Naphtha, 407
Napoleon, 174
Naradnaya, Mount, 167
Narcotic, 322
Nasser, Lake, 347
Natal region, 379, 387
National Association of Stock Car Auto Racing, 98
National Baseball League, 218
National Council of Churches, 102
National Health Scheme, Australia, 247
Natural gas, 108, 118, 191, 250, 251, 325, 375, 552
Natural selection, 35
Nauru, 527, 529
Navaho Indians, 37
Nawab of Bengal, 434
Nazis, 413
Ndebele tribe, 383
Near East (See also Middle East), 395
Nebraska, 87, 106, 125
Nebuchadnezzar, 399
Needleleaf evergreen trees, 91, 135, 211
Neon, 548

Nepal, 424, 427, 433–36, 439, 440, 445, 446
Nepali language, 440
Netherlands Antilles, 279, 294
Netherlands, The, 129, 130, 135–38, 145, 151, 157, 245, 566
New Brunswick, 83
New Caledonia, 524, 525, 527, 529
New England, 101, 114, 115
New England Petroleum Company, 294
New Granada, 304
New Guinea, 453, 456, 460, 462, 477
New Hampshire, 83
New Jersey, 83, 101, 107
New Mexico, 87, 88, 103
New Orleans, Louisiana, 103, 108, 116
New South Wales, 245, 247, 249
New Spain, 291, 292
New York, 83, 101, 103, 107, 115, 125
New York, New York, 29, 45, 63, 68, 95, 101, 115–18, 165, 226, 360, 466
New York State Barge and Canal Line, 116
New Zealand, 41, 43, 231–56, 532, 538, 553
    agriculture, 247
Newark, New Jersey, 103
Newcastle, Australia, 244
Newfoundland Territory, 82, 553
Newspapers, 198, 417
Newton, Issac, 68
Niagara Falls, 380
Nicaragua, 261, 269, 270, 276, 278, 279, 284, 297, 329, 568
Nicholas II, 174
Nickel, 8, 113, 154, 190, 250, 288, 289, 388, 476, 529, 552, 562
Nicolle, Charles J.H., 32
Niger, 346, 351–56, 362, 367, 566
Niger River, 362
Nigeria, 288, 294, 362–64, 371, 375, 376, 409
Night soil, 501
Nile River, 346, 347, 349, 354, 356, 361, 412
Nirvana, 439
Nissan, 116, 226, 329
Nitrate, 328
Nitric acid, 328
Nitrogen, 546, 548
Nitrogen dioxide, 558
Niue Islands, 532
Nobel Prize, 375, 391
Nobi Plain, Japan, 205
Nomadic herders, 241, 354, 406, 411, 424, 461, 499, 503
Nomadic hunters, 241
Norfolk Islands, 532
Norilsk, U.S.S.R., 552
Normal curve, 63
Normal lapse rate, 90
Norteamericanos, 295
Nortenos, 291
North America, 28, 29, 35, 81 90, 91, 95, 110, 118, 130, 166, 259, 265, 269, 279, 284, 303, 305, 321, 325, 329, 341, 374, 550, 552, 553

North Atlantic Drift, 126
North Carolina, 29, 82, 83, 103, 209
North European Plain, 129, 136
North Island, New Zealand, 232, 233, 235, 236, 244
North Korea, 424, 483, 487, 491, 493, 495, 498, 501, 502, 505–8, 511
North Pole, 11, 70, 539, 545, 547, 554
North Sea, 128–30, 148, 153, 158
    oil, 153
North Slope of Alaska (Arctic Slope), 83, 540, 551
North Vietnam, 464
North Yemen, 396, 409, 411, 417
Northeast trade winds, 269, 270, 520
Northern Africa, 344–59, 397, 399, 401, 407, 560
Northern Ireland, 568
Northern Rocky Mountains, 87, 89
Northern Territory, 253
Northwest Territory, 83, 553
Northwestern Uplands of Europe, 126–28
Norway, 126, 128, 133, 152, 153, 538, 552–54
Norwegian language, 41, 145
Notre Dame Mountains, 83
Nouakchott, Mauritania, 351
Novosibirsk, U.S.S.R., 177, 186, 190
Nuclear energy, 152, 533
Nuclear weapons, 255, 569
Nukunono Islands, 529
Nunataks, 543, 544
Nutmeg, 462

O'Hare International Airport, 118
Oak, 209, 430
Oakland, California, 103
Oats, 50, 149, 222, 249, 354
Ob' River, 167, 170, 177, 191
Obadhi Muslims, 406
Oblate spheroid, 9
Observational balloons, 76
Ocean trench, 458
Oceania, 246
Octopus, 286, 287
Oder River, 130, 131
Odessa, U.S.S.R., 166, 177
Offshore fishing zones, 220, 222
Ogbomosho, Nigeria, 364
Ohio River, 86, 93, 114, 116
Ohio, 106, 115
Oil (See also Crude Oil and Petroleum), 250, 288, 294, 326, 375, 396, 407, 411, 417, 456, 505, 545, 557
Oil shale, 540
Oil tankers, 203, 224, 407
Okhotsk, 183
Okhotsk, Sea of, 170
Oklahoma, 86, 103, 107, 108
Oklahoma State University, 99
Olduvai Gorge, 359
Olive oil, 149, 354
Olives, 131, 134, 148, 149, 354, 409, 441
Olympic Games, 100
Omaha, Nebraska, 103, 116

I-11

Oman, 396, 401, 405, 406, 409, 411, 416, 419
Oman, Gulf of, 395, 404
Onions, 107, 267
Opal, 288
Opium, 473
Opossums, 239
Orange Free State, 379, 387
Orange River, 380–82, 387
Oranges, 107
Oregon, 88, 89, 93, 103
Organization of Petroleum Exporting Countries (OPEC), 408, 409
Orient, 483
Orient Express, 156
Orinoco River, 308, 309, 327, 328, 331
Orizaba, Mount, 265
Orographic precipitation, 274, 363
Orohena, Mount, 522
Osaka, Japan, 211
Osaka-Kobe conurbation, 212, 223
Otago region, New Zealand, 235
Ottawa, Canada, 553
Ottoman Empire, 404, 406, 413
Ouagadougou, Burkina Faso, 367
Outback, 247, 255
Outer core, 8
Outrigger canoes, 526
Overseas Chinese, 468, 480
Oxygen, 9, 542, 548, 559, 561
Oymyakon, U.S.S.R., 172
Oysters, 220
Ozark Plateau, 86
Ozone, 548, 549, 559
Ozone depletion, 548
Ozonosphere, 559

Pacific Islands, 515–34
    agriculture, 529–30
Pacific Mountains and Valleys, 88
Pacific Ocean, 13, 42, 89, 204, 207–9, 220, 228, 231, 240, 242, 265, 268, 269, 275, 297, 304, 306, 454, 456, 458, 459, 463, 484, 515, 516, 521–23, 526
Pack ice, 539, 547
Padre Island, 83
Pago Pago, 530, 533
Pakistan, 28, 30, 37, 353, 396, 424, 426, 427, 430, 431–33, 435, 437, 439–41, 444–46, 448, 566, 567
Pakistanis, 147
Palestine, 406, 408, 412, 413, 418
Palestinians, 564
Palm oil, 288, 373, 530
Palm trees, 209, 372, 431, 524, 530
Palmer Peninsula, 544, 545, 549, 553
Palo Duro Canyon, 87
Pamir Knot, 426
Pamir Mountains, 170, 429
Pampas of Argentina, 313, 324, 325, 331
Pan American Airlines, 333
Pan American Highway, 332
Panama, 259, 261, 269, 278, 279, 284, 296, 297, 304, 515

Panama Canal, 295, 296
Pangaea, 12, 13
Pantheistic, 529
Papiamento, 279
Papua New Guinea, 456, 527–29, 532
Paraffin, 407
Paraguay, 304, 309, 313, 317, 318, 334
Paraguay River, 309, 331
Paraguay-Parana-Plata River system, 309
Paramos, 267
Parana River, 309, 331
Parasitic infections, 370
Parent material, 23
Paricutin volcano, 265
Paris Basin, 129, 152
Paris, France, 29, 129, 138, 140, 152, 156, 160, 304, 568
Pashto language, 440
Pashtun tribes, 437, 438
Passion fruit, 529
Pasteur, Louis, 32
Patagonia, 310
Pattern of spatial distribution, 57, 58
Peaches, 107, 249
Peanuts, 356, 372, 373, 388, 441
Pearl Harbor, Hawaii, 100
Pears, 107, 220, 249
Peas, 503
Peking (See Beijing)
Peking man, 491
Pelee, Mount, 273, 276
Penghu Islands, 494
Penguins, 545, 548
Pennsylvania, 82, 107, 108, 110, 115
Pentecostalists, 182
People's Republic of China (See also China), 483
Pepper, 462, 473
Perestroika, 199
Perihelion, 10, 11
Permafrost, 83, 171, 540, 546
Persia (See also Iran), 399, 412
Persian carpets, 415
Persian Gulf, 395, 398, 399, 406, 407, 416, 417, 568
Persian languages, 406
Persians, 526
Perth, Australia, 235, 239, 244, 248, 249, 252
Peru, 304, 305, 309, 311–13, 318, 322, 327–29, 333, 334, 395, 567, 569
Peshawar, Pakistan, 437
Pesticides, 561
Peter the Great, 174, 177, 182
Petrochemical industries, 407
Petroleum (See also Oil, Crude oil, and Natural gas), 50, 108, 115, 118, 154, 190, 191, 198, 220, 222, 225, 290, 294, 326, 330, 358, 388, 396, 407, 411, 419, 475, 552, 567
Petroleum jelly, 407
Peugeot-Citroen, 329
Pharmaceuticals, 561
Philadelphia, Pennsylvania, 103
Philippine Sea, 458
Philippine Trench (Mindanao Trench), 458

Philippines, 43, 95, 405, 424, 456, 460, 463, 464, 467, 473, 475, 476, 480, 525, 526, 529, 569
Philistines, 405
Phillip, Arthur, 245
Phnom Pehn, Cambodia, 456, 467
Phoenicia, 405
Phoenician language, 353
Phoenix, Arizona, 63, 226
Phosphate, 113, 191, 355, 411, 506, 529
Photogrammetrist, 78
Photogrammetry, 77
Physiographic provinces, 81, 82
Physiological incompatibility, 35
Picadilly Circus, 139
Pickup trucks, 226
Pico Duarte, 272
Piedmont, 82, 115
Pig iron (See also Iron, Iron ore and Iron smelting), 152
Pigs (See hogs)
Pilcomaya River, 309
Pimar Knot, 484
Pin-yin, 498
Pindus Mountains, 131
Pine trees, 211, 430
Pineapples, 50, 249, 288, 529
Ping-pong, 219
Pipelines, 407
Pitcairn Islands, 529, 532
Pittsburgh, Pennsylvania, 116
Pixel, 78
Plague, 370
Plane of the ecliptic, 8, 11
Plane projection, 75, 76
Planetesimal, 8
Planets, 8
Plankton, 545
Plantains, 283
Plastics industry, 290, 388, 474
Plata River, 309, 324
Plate tectonics, 13, 14, 16
Platinum, 113, 374, 388, 562
Playa, 265
Pleistocene, 83, 306
Ploesti, Romania, 153
Pluto, 10
Plymouth, England, 126
Plywood, 443, 473
Po River Valley, 152, 154
Podzol soils, 24, 93, 188
Point Barrow, Alaska, 90, 540
Poland, 125, 129–31, 133, 136, 144, 145, 151, 152, 155, 174, 197, 245, 413, 414
Polar bears, 547
Polar climate, 19
Polar easterlies, 545
Polar ice, 539
Polar Regions, 537–55
Polders, 130
Poles, 145, 183
Polish language, 182
Polish Lowlands, 167
Polish Silesia, 152
Pollution, 226, 227, 558, 559
Polo, Marco, 462

Polyconic projection, 76
Polyester, 106
Polynesia, 516, 528, 529
Polynesian languages, 529
Polynesians, 214, 242, 526
Polyps, 518, 519
Pomegranates, 354, 411
Pontus Mountains, 398
Pope, Roman Catholic, 181, 304
Poplar trees, 209
Poppies, 473
Population:
　birth rate, 30
　death rate, 30
　explosion, 28, 562, 563
　pyramid, 32, 33
　rate of natural increase, 30
Population Reference Bureau, 175
Porcelain, 492
Pork (See also Hogs), 189, 529
Port Arthur, Texas, 108
Port Elizabeth, South Africa, 382
Port Harcourt, Nigeria, 364
Port Phillip Bay, 244
Port Sudan, Sudan, 356
Port-of-Spain, Trinidad, 278
Portsmouth, England, 245
Portugal, 128, 135, 145, 149, 154, 155, 259, 303, 304, 324, 343, 378, 385, 390, 413, 566
Portuguese, 317, 324, 360, 434, 462, 463, 483, 486
Portuguese language, 42, 43
Potash, 411
Potassium permanganate, 322
Potassium salts, 191
Potatoes, 65, 149, 187–89, 220, 267, 269, 319
Pottery, 291
Povolzhye region, 191
Prague, Czechoslovakia, 131
Prairie Dog Town Fork, 87
Precipitation, frontal, 21
Prefectures, 211
Presbyterian church, 144, 246
Prescott, Arizona, 89
Pressure, earth's internal, 8
Preston, Andrew, 287
Pretoria, South Africa, 383, 384
Prevailing westerlies, 520, 545, 552
Primary production, 48
Primate cities, 467
Prime meridian, 70
Prostitutes, 242, 245, 370
Protestant Reformation, 143
Protestant religions, 43, 46, 142, 217, 246, 304
Prudhoe Bay, Alaska, 551
Prunes, 107
Pu Yi, Emperor of Manchu Dynasty, China, 492
Puebla, Mexico, 265
Puerto Escondido, Mexico, 293
Puerto Monte, Chile, 332
Puerto Rican Trench, 272
Puerto Ricans, 103

Puerto Rico, 263, 272, 276, 278, 279, 283, 286, 288, 290, 293, 295–97
Puerto Vallarta, Mexico, 293
Puget Sound, 89, 107
Pumice, 518, 529
Punjab plain, 45, 46
Punjab, India, 448
Punjabi language, 440
Puntarenas, Costa Rica, 270
Purma, 320
Pusan, South Korea, 496
Putorana Mountains, 540
Puzta, 135
Pygmies, 343, 359, 378
Pyrenees Mountains, 129, 131

Qantas Airlines, 253
Qatar, 396, 406, 409, 411, 415, 416, 456
Qing Zang Gaoyuan (Plateau of Tibet), 456, 484, 485, 489, 490
Quachita Mountains, 86
Quebec, Canada, 101–3, 110
Quebec Territory, 553
Queen Maud Mountains, 544
Queensland, 235, 253
Queretaro, Mexico, 291
Quili (See Coolies)
Quinine, 473
Quito, Ecuador, 332

Rabbits (See also Hare), 151, 239
Race, 35
Racial segregation, 391
Radio, 53, 198, 224, 358, 375, 376, 389, 417, 448, 477, 508, 530
　AM, 60, 446
　FM, 60, 61, 63, 446
　shortwave, 530
　transistor, 417, 477, 507
Radishes, 220
Railroads, 197, 225, 252, 253, 287, 297, 331, 332, 375, 376, 389, 416, 418, 445, 476, 477, 504, 505, 507, 530
Ramadan, fast of, 407
Range of goods, 66
Rangoon, Myanma, 456, 467
Rape, 151
Rayon, 444
Reculet, Mount, 131
Red Basin (See Chengdu Plain)
Red China (See also China), 483
Red River, Oklahoma-Texas, 87
Red River, Vietnam, 456
Red Sea, 339, 356, 395, 398, 407
Reefs, 272, 518, 519
Refugees, 414, 418, 437, 438, 448, 562, 564
Reg, 348
Region:
　functional, 4
　multi-factor, 4
　nodal, 4
　single factor, 4

Reincarnation, 371, 436
Reindeer, 548, 552
Relative location, 396
Relocation diffusion, 60, 62
Remote sensing, 76, 78
Renault, 329
Representative fraction, 73
Republic of China (See Taiwan)
Resins, 223
Reunion Islands, 368
Revolutionary War, 83, 120
Rh-negative blood, 35
Rh-positive blood, 35
Rhine River, 130, 131, 136, 152, 157, 158, 160
Rhode Island, 114
Rhodesia (See Zimbabwe)
Rice, 49, 106–8, 189, 217, 220, 354, 373, 387, 441, 443, 444, 461, 471, 473, 493, 503, 505, 526, 557
　brown, 474
　dry-land, 474
　high-yield hybrids, 473
　polished, 474
　wet-land, 474
　wild, 474
Rickshaw, 470
Riflery, 99
Rift valleys, 341, 362, 516
Ring of fire, 13, 204, 265, 458, 518
Rio de Janeiro, Brazil, 29, 316, 328
Rio de la Plata, 324
Rio Grande River, 120, 268
Riyadh, Saudi Arabia, 404, 416
Robots, 224
Rock glaciers, 60
Rocky Mountains, 4, 21, 86–88, 93, 101, 107, 116, 341
Rodents, 547
Rodeos, 98
Roman Catholic religion, 43, 47, 101, 103, 142–44, 181, 182, 216, 217, 246, 261, 280, 283, 317, 318, 371, 384, 463, 471, 497, 498, 529
Roman Empire, 135, 142, 145, 406, 412
Roman ruins, 355
Romance languages, 42, 145, 146, 183, 528
Romania, 135, 142, 145, 147, 153, 154, 170
Romans, 147, 344, 346
Rome, Italy, 125, 135, 138, 140, 142, 181, 568
Ronne Ice Shelf, 544
Rooney, John F., 98
Roosevelt, Franklin D., 404
Rose Bowl-Pasadena, California, 99
Ross Sea, 543, 544
Rostock, Germany, 64
Rostov, U.S.S.R., 186, 190
Royal Naval Observatory, 70
Roza, 320
Ruapehu, Mount, 235
Rubber, 269, 290, 356, 387, 473, 476, 480, 545
Rudolf (Turkan), Lake, 362

Rugby, 99
Ruhr district, 152
Ruhr River, 152
Rum, 286
Russia (*See also* Soviet Union), 165
Russian language, 42, 43, 182, 499
Russian Orthodox Church (*See also* Eastern Orthodox), 180, 182
Russian Plain, 167
Russian Republic, U.S.S.R., 176, 188
Russians, 145, 147, 183, 553
Rutabagas, 149
Rwanda, 359, 363, 368, 370, 371, 374–76, 567
Rye, 149, 187
Rzhev, U.S.S.R., 191

S-curve, 63
Saar district, 152
Sabah, 455
Sacramento, California, 90
Sahara Desert, 42, 340, 341, 343, 344, 346–49, 351, 354, 356, 360, 367, 371, 382, 399, 545, 559
Saharan Atlas Mountains, 347
Sahel, 367, 368
Saint Francois Mountains, 86
St. Lawrence River, 29, 110, 118
St. Lawrence Seaway, 118
St. Lawrence Valley, 107
St. Louis, Missouri, 29, 103, 114, 116
Saint Lucia, Guadeloupe, 279
Saint Petersburg, U.S.S.R., 177
Saint Pierre, Martinique, 273
Sakhalin Island, 204
Salay Gomez Islands, 516
Salinization of soils, 24, 95
Salisbury, England, 46
Salisbury, Zimbabwe (*See* Harare)
Salt, 114, 115, 218, 265, 288
Salween River, 456
Sampans, 477
San Antonio, Texas, 103
San Diego, California, 88, 116, 226, 260
San Francisco Bay, 110
San Francisco, California, 103, 226, 362, 495
San Jose, Costa Rica, 293
San Juan Parangaricutiro, Mexico, 265
San Juan, Puerto Rico, 278
San Luis Potosi, Mexico, 268
San Marino, 125, 136, 137
Sandstone, 411, 540
Sanskrit, 215, 440, 471
Santa Cruz, Bolivia, 322
Santa Marta, Colombia, 330
Santiago de Cuba, 278
Santiago, Chile, 304, 316, 331, 332
Sao Francisco River, 331
Sao Paulo, Brazil, 314
Sao Tome and Principe, 363, 372
Sapphires, 476
Sapporo, Japan, 212
Sarawak, 455

Sardines, 128, 220
Sardinia, 154
Saskatchewan, Canada, 93, 107, 110
Satellite transmission of television, 530
Saudi Arabia 250, 396, 398, 401, 407, 409, 411, 415, 416, 419, 566
Saul, King of Jews, 405
Sault Sainte Marie Canals, 118
Savanna grassland, 22, 270, 309, 343, 360, 363, 460, 524, 561
Savu Sea, 454
Scandinavia, 126, 130, 133, 137, 144, 145, 151, 155, 167, 540, 550
Schweitzer, Albert, 375
Scorpions, 524
Scotch language, 128
Scotland, 128, 144, 151
Scrub forests, 135
Sea lions, 548
Sea-floor spreading, 16
Seals, 545, 547, 548
Seattle, Washington, 115, 116, 226
Seaweed, 547
Second-order towns, 66
Second Temple, Jerusalem, 405
Sediment, 16
Sedimentary rock, 235
Seine River, 129, 140
Seismic waves, 8, 519
Selva, 363
Semiarid climates, 490
Semitic languages, 353, 371, 372, 406
Senegal, 368, 371, 372, 376, 567
Sensible temperature, 430
Seoul, South Korea, 496
Sequoia, 22
Serbians, 145, 148, 183
Serengeti Plain, 362
Serpukhov, U.S.S.R., 191
Sesame, 411
Seventh-Day Adventists, 529
Sewage, 560
Seward Peninsula, 166
Sewing machines, 443
Seychelles, 359, 363, 368
Sfax, Tunisia, 354
Shah of Iran, 120, 419, 567
Shale, 290, 540
Shang Dynasty, China, 491, 492
Shanghai, China, 485, 486, 495, 498, 503, 505, 506
Shans, 470
Shanxi, China, 505
Sharks, 233
Sheba, Queen of, 411
Sheep, 49, 107, 128, 135, 149, 151, 189, 233, 235, 243, 246, 247, 249, 252, 324, 325, 354, 374, 379, 385, 388, 411, 444
Sheep, merino, 242, 245, 247
Shenyang, China, 495
Shetland Islands, 544, 553
Shia Muslims (Shiite), 43, 404, 406
Shiite Muslims (*See* Shia Muslims)
Shikoku, 204, 205

Shinto, 46, 215, 216
Shinto shrines, 210
Shipbuilding, 207, 507
Shoes, 114, 290
Shona tribe, 383
Sial, 13, 518
Siberia, 35, 174, 183, 190, 191, 198, 491, 545, 552, 553
Siberian High, 173, 207
Sicily, 145
Sickle-cell anemia, 36
Sierra de Guadarrama, 130
Sierra Leone, 344, 362, 363, 368, 371, 374–76
Sierra Madre del Sur, 265
Sierra Madre Occidental, 264, 265
Sierra Madre Oriental, 265
Sierra Nevada Mountains, Spain, 131
Sierra Nevada Mountains, United States, 22, 89, 93
Sigmoid curve, 63
Silica, 13
Silicate, 9
Silicon, 8, 9
Silk, 189, 220, 224, 225, 415, 444, 503, 505
Silkworms, 220, 503
Silver, 288, 374, 476, 506, 529, 552, 562
Sima, 13
Singapore, 424, 454, 455, 466–68, 471, 473, 475, 476, 480
Sinhala language, 440
Sinkiang (*See* Xinjiang)
Sinkiang Uighur Autonomous Region, China, 494
Sinkiang, China, 505
Sino-Tibetan language, 39, 42
Siroccos, 401
Sisal, 286, 385
siSwati language, 385
Sitka, Alaska, 166
Skiing, 99, 219
Skin cancer, 548
Slavery, 103, 360
Slavic languages, 42, 145, 146, 182, 183
Slavic people, 145, 147, 182
Sleeping sickness, 370
Slovaks, 145, 148, 183
Slums, 563
Slurry, 120
Small-scale map, 73
Smallpox, 370
Smog, 297, 357
Snowfields, 235
Soap, 476
Soccer, 99, 199, 327
Social Security, New Zealand, 247
Social selection, 35
Socialism, 419
Society Islands, 522, 526
Sodium carbonate, 322
Sodium nitrate, 328
Soende Strait, 457
Softball, 99
Soil, 17, 24

Soil conservation, 560
Solar system, 8
Solder, 475
Solifluction, 540
Solo River, Java, 461
Solomon Islands, 527–29, 532
Solomon, King of Jews, 405, 411
Solstices, 309
Solvents, 561
Somalia, 343, 363, 371, 374, 567
Sombrero Island, 262
Sonora, Mexico, 265
Soochow (*See* Suzhou)
Sorcerism, 371, 529
Sorghum, 249
Sotho language, 385
Sotho tribe, 383
Soufriere, Mount, 273, 276
Sousse, Tunisia, 47, 357
South Africa, Republic of, 113, 120, 378, 379, 380, 382–85, 387–91, 562, 567–69
South America, 12, 35, 87, 149, 245, 259, 260, 264, 265, 269, 271–73, 279, 287, 297, 303–36, 344, 374, 382, 473, 509, 516, 540, 544, 567, 569
  agriculture, 318–25
South Asia, 423–50, 467, 484
  agriculture, 440–42
South China Sea (*See* Nan Hai)
South Dakota, 87, 103, 106, 107
South Georgia Islands, 553
South Island, New Zealand, 232, 233, 235, 236, 253
South Korea, 115, 224, 424, 483, 489, 493, 495, 498, 501, 502, 504–8, 511, 569
South Orkney Islands, 553
South Pacific Commission (SPC), 532
South Pacific Forum (SPF), 532
South Pole, 70, 538, 539, 542, 545, 549, 550
South Sandwich Islands, 553
South Vietnam, 464, 465
South Yemen, 396, 401, 411, 417
Southeast Asia, 46, 49, 224, 231, 265, 270, 327, 423, 424, 439, 453–81, 484, 489, 503, 516, 523, 525, 566, 567
  agriculture, 471–74
Southeast trade winds, 459, 520
Southern Africa, 378–93
Southern Alps, 233, 235, 240
Southern Rocky Mountains, 87, 88
Soviet Union, 29, 42, 125, 129, 130, 135, 144, 145, 154, 162, 165–201, 203–5, 223, 252, 285, 286, 289, 294, 347, 353, 368, 386, 396, 403, 404, 407, 414, 417, 418, 423, 426, 435, 437, 438, 445, 448, 484, 487, 493, 507, 509, 537, 538, 540, 545, 550, 552, 554, 562, 568, 569
  agriculture, 187
Sovkhozes, 187, 188
Soya Strait, 204
Soybeans, 50, 106, 107, 188, 220, 321, 501

Spain, 131, 133, 135, 144, 145, 147, 149, 151, 152, 155, 242, 246, 259, 287, 296, 303, 304, 324, 343, 413, 463, 566, 568
Spanish, 276, 279, 303, 313, 316, 317, 324, 471, 486
Spanish colonialism, 316
Spanish language, 42, 43, 101, 103, 261, 279, 280, 305, 312, 471
Spanish Meseta, 130, 135, 148, 151
Spatial diffusion, 60, 406
Spatial distribution, 57, 58
Spatial interaction, 59, 60, 67
Spatial model, 5, 63, 67
Spice Islands, 462
Spices, 462, 463
Sports, American, 98
Sports, Japanese, 218, 219
Spring wheat (*See also* Wheat), 107, 249
Squatter settlements, 317, 563
Squid, 220
Squirrels, 523
Sri Lanka, 424, 426, 427, 436, 439–41, 443, 448, 470
Stalin Peak, 170
Stalin, Joseph, 174, 175, 177, 180, 182, 404
Standard Oil Company of California, 294
Staple Food Management Law, 217
State farms, 187
Stations for sheep and cattle, 247, 252, 253
Statistical map, 69, 71
Steel, 50, 51, 108, 110, 113, 152, 155, 190, 222–25, 249, 251, 290, 327, 328, 375, 388, 415, 442, 443, 504, 545
Steepe grass, 135, 171, 188, 235, 238, 239, 310, 343, 363, 490, 491
Stereos, 224
Stereoscope, 78
Stewart Island, 232
Stillwater, Oklahoma, 99
Stone Age, 525
Strandline, 524
Strawberries, 107
Sturgeon, 190, 411
Stuttgart, Germany, 136
Subduction, 13, 14
Subsistence farming, 48, 49, 148, 325
Subtropical forests, 209
Sudan, 343, 346, 347, 351, 354–56, 564
Sudanese language, 372
Sudirman Mountains, 456
Suez Canal, 296, 339, 343
Sugar beets, 107, 149, 188, 501
Sugar cane, 50, 103, 131, 249, 252, 266, 284, 286, 288, 290–94, 321, 324, 331, 354, 385, 387, 388, 441, 444, 473, 503, 529
Sukarno Mountain, 456
Sukarno, President of Indonesia, 465
Sulfate haze, 558
Sulfur dioxide, 558, 559
Sulfur, 8, 113, 115, 222, 288, 505
Sulfuric acid, 322

Sulphuric acid, 559
Sulu Sea, 454
Sumatra, 454, 456–58, 466, 467
Summer maximum precipitation, 239
Summer monsoon, 209, 460
Summer solstice, 11
Sumo wrestling, 218
Sunflowers, 151, 188
Sunni Muslims, 43, 406
Superior, Lake, 118, 362
Superior, Arizona, 113
Supermarkets, 66
Supersonic airplanes, 549
Supertankers, 294
Surabaja, Indonesia, 467
Surinam, 304, 313, 318, 328
Surrey, England, 28
Suzhou (Soochow), China, 485
Svalbard Islands, 552–54
Sverdlovsk, U.S.S.R., 177, 186, 190, 197
Swahili language, 42, 371
Swazi tribe, 383
Swaziland, 378, 383–85, 387–89
Sweden, 140, 142, 144, 152, 154, 553
Swedes, 174
Swedish language, 41, 145
Sweet potatoes, 220, 473, 503, 529
Swimming, 219
Switzerland, 144, 151, 152, 224
Sydney, Australia, 233, 243–45, 248, 252
Syncline, 540
Synthetic fibers, 223
Syria, 276, 398, 399, 401, 403, 406, 415, 417
Szechwan Basin, 503

Tabriz, Iran, 416
Tabuk, Saudi Arabia, 416
Taca Airlines, 297
Tadzhik Republic, U.S.S.R., 175, 186
Taegu, South Korea, 496
Tagalog, 471
Tahat, Mount, 348
Tahiti, 522, 525, 526, 531
Taiga, 133, 135, 172, 552
Taipei, Taiwan, 496
Taiwan, 28, 224, 424, 483, 484, 487, 489, 492, 494, 498, 499, 503, 504, 507, 508, 511, 525
Taklimakan Shamo (Takla Makan Desert), 489, 490
Talbot, William H. F., 77
Tamboro, 458
Tamil language, 440
Tampico, Mexico, 268
Tanganyika, Lake, 343, 361, 362
Tanning, 375
Tantalum, 113
Tanzania, 359, 362, 370, 372, 374, 567
Tao, 496
Taoism, 46, 471, 496–98
Tape recorders, 224
Tapioca, 387, 473
Tar, 560

I-15

Tarim Basin, 487, 490
Tarim He (Tarim River), 487, 498
Tarn River, 130
Taro, 525, 529
Tartars, 147, 174, 493
Tashkent, U.S.S.R., 177, 186
Tasman Glacier, 233
Tasman Sea, 232, 244
Tasman, Abel, 242
Tasmania, 233, 239, 242
Tasmanian devil, 239
Taurus Mountains, 398
Tbilisi, U.S.S.R., 177, 186
Tea, 50, 189, 220, 225, 372, 441, 443, 473, 503
Teak, 372, 431, 473
Tehran, Iran, 398, 403, 404, 416
Tehuantepec, Isthmus of, 265, 268
Tel Aviv, Israel, 416
Telegraph, 358, 446
Telephones, 53, 358, 377, 389, 443, 446, 477, 508
Television, 53, 60, 198, 218, 224–28, 250, 358, 369, 376, 389, 417, 440, 446, 477, 503, 508, 530
Temperate mixed forests, 209, 241
Tennessee Valley Authority, 442
Tennis, 98, 99
Terrorism, 415, 417, 567
Teutonic languages, 145, 146
Texas, 87, 93, 101–3, 106–8, 113, 115, 118, 166, 261, 348
Texcoco, Lake, 291
Textiles, 114, 115, 189, 191, 224, 250, 286, 290–94, 321, 328, 375, 385, 388, 411, 443, 444, 476, 503, 505, 507
Thabana Ntlenyana, Mount, 380
Thai language, 471
Thailand, 49, 424, 454, 456, 463, 468, 470, 471, 473–76, 480
Thames River, 140, 152
Thana, India, 434
Thar Dessert, 431
Thematic map, 69, 76
Thimbu, Bhutan, 434
Third World, 50, 187, 329, 418, 432, 467, 476, 509, 563
Thirty Years' War, 143
Three-mile limit, 569
Threshold population, 66
Tianjin, China, 495
Tibbu, 343
Tibesti Mountains, 348
Tibet (See Xizang)
Tibet, Plateau of, (See Qing Zang Gaoyuan)
Tibetan language, 499
Tidal waves, 458
Tien Shan Mountains, 170, 177
Tierra caliente, 266, 271, 284, 312
Tierra del Fuego, 310
Tierra fria, 267, 269, 271, 276, 312
Tierra templada, 266, 268, 269, 271, 312, 313
Tigers, 523

Tigris River, 398, 399, 404
Timber, 220, 224, 234
Tin, 317, 327, 328, 388, 474, 475, 480, 507, 562
Tinfoil, 475
Titanium, 355
Titicaca, Lake, 304, 308
Tobacco, 106, 151, 189, 288, 290, 321, 328, 354, 385, 409, 411, 473, 503
Tofo, Chile, 327
Togas, 386
Tokelau Islands, 529, 532
Tokyo Bay, Japan, 213
Tokyo, Japan, 29, 205, 209, 211–13, 216, 218
Tokyo-Yokohama conurbation, 205, 211, 223
Toledo, Ohio, 68
Toltec civilization, 296
Tomatoes, 107, 409
Tonga Islands, 529
Topographic map, 72
Topsoil, 560
Toronto, Canada, 95
Toubkal, Mount, 347
Tourism, 263, 293–97, 355, 358, 375, 388, 405, 415, 435, 444, 476, 507, 517, 530, 533
Tower of Babel, 146
Townsville, 249
Toxic chemicals, 561
Toyota, 226
Toys, 225
Track and field, 99
Trade winds, 21, 273, 520, 522
Train à Grande Vitesse, 160
Trans-Siberian Railroad (See also Railroads), 177, 180, 183, 197
Transhumance, 151
Transkei, 391
Transmigration, 467
Transvaal, 379, 388
Transylvanian Basin, 131
Treaty of Tordesillas, 303, 304
Tree ferns, 241
Trinidad and Tobago, 276, 278, 279, 286, 288, 296, 308
Tripoli, Libya, 343, 349, 356
Tropical Africa, 359–78
Tropical cyclones, 274
Tropical diseases, 297
Tropical fish, 233
Tropical moist climate, 19
Tropical monsoon climate, 22
Tropical rainforest, 22, 239, 269, 270, 284, 309, 310, 317, 320, 321, 325, 333, 343, 360, 363, 373, 378, 424, 430, 473, 524, 561
Troposphere, 560
Trotsky, Leon, 174
Trucks, 507
Truffles, 149
Tse-tung, Mao, 492, 496
Tsuhima, Strait of, 204
Tsunamis, 519

Tuaregs, 343
Tuberculosis, 432
Tucson, Arizona, 103
Tula, U.S.S.R., 191
Tulsa, Oklahoma, 68, 69, 71
Tuna war, 569
Tundra, 93, 171, 172, 539, 540, 546, 547, 552
Tungsten, 154, 190, 375, 476, 506, 507
Tungus-Manchu, 550
Tunis, Tunisia, 346, 349, 351
Tunisia, 346, 351, 355, 356, 414, 566
Turan Lowland, 167
Turfan Pendi (Turfan Depression), 490
Turin, Italy, 152
Turk Islands, 271
Turkestan, 404
Turkey, 42, 398, 401, 406, 409, 411, 413, 415, 417, 419, 457, 567
Turkeys, 107
Turkish chieftains, 439
Turkish language, 147, 183, 406, 499
Turkistan, 171, 183
Turkmen Republic, U.S.S.R., 175, 189
Turks, 36, 147, 182, 346
Turnips, 149
Turquoise, 411
Tutsi tribe, 368
Tutu, Desmond, 391
Tuvalu Islands, 529
Twa (pygmy), 368
Tyasa, Lake, 362
Type-setting metals, 475
Typewriters, 443
Typhoons, 52, 209, 275, 459, 460
Typhus, 370

Ubangi River, 361
Uganda, 49, 343, 359, 362, 370–72
Ukraine, 170, 174, 176, 177, 188, 190, 191, 196
Ukrainian Republic, U.S.S.R., 170, 174, 188
Ukrainian language, 182
Ukrainians, 182
Ulan Bator, Mongolia, 499, 507
Ultraviolet light, 548, 560
Unbounded plain, 65
Uniform distribution, 65
Uniform plain, 64
Uniform transportation network, 65
United Arab Emirates, 396, 409, 411, 416, 456, 567
United Church of Christ, 217
United Fruit Company, 284, 287
United Kingdom, 135, 138, 142, 152–55, 211, 245, 251, 252, 255, 532, 538, 566
United Nations, 294, 319, 377, 390, 448, 455, 403, 532, 564
United States, 4, 32, 34–36, 39, 43–46, 49, 63, 69, 77, 81–121, 135, 142, 155, 162, 166, 170, 175, 188–90, 203, 204, 209, 211, 217–20, 226, 228, 231, 232, 239, 251, 252, 255, 259–64, 279, 282,

I-16

286–97, 317, 322, 324, 326–30, 333, 348, 353, 368, 404, 407, 408, 417–19, 441, 433, 440, 442, 446, 448, 463, 464, 467, 475, 480, 489, 493, 495, 501, 502, 504, 506, 507, 530, 532, 533, 538, 549, 553, 557–62, 564, 566, 569
  agriculture, 106–8
  Constitution of, 120
University of Costa Rica, 39
University of Oklahoma, 99
University of Sydney, Australia, 247
Untouchables, 439
Upper Silesia, 152
Ural Mountains, 28, 166, 167, 177, 182, 183, 188, 190, 191, 197, 540
Ural River, 170
Ural-Altaic language, 39, 42, 183
Ural-Volga region, 191
Uralic languages, 145
Uranium, 190, 250, 388, 552, 562
Uranus, 8
Urdu language, 406, 440
Uruguay, 304, 309, 310, 313, 324, 325
Uruguay River, 309
Ussher, James, 7
Ussuri River, 487
Utah, 88, 101
Uzbek language, 440
Uzbek Republic, U.S.S.R., 175, 177, 186, 189

Vaal River, 382, 387, 388
Valdez, Alaska, 120, 551
Valley and ridge topography, 82
Van Nuys, California, 118
Van Valkenburg, Samuel, 43
Vancouver Island, 110
Vandals, 346
Vanuatu Islands, 529, 532
Vatican City, 125
Veal, 220, 247
Vegetables, 106, 130, 148, 149, 188, 216, 220, 283, 354, 409, 415, 529
Venda, 391
Venetians, 147
Venezuela, 262, 273, 294, 304, 308–10, 313, 319, 321, 326–29, 409, 567
Venus, 8
Veracruz, Mexico, 265, 292
Verkhoyansk, U.S.S.R., 172, 545
Vermont, 83
Vernal equinox, 11
Vertical zonation of crops, 265
Victoria, 247, 249, 252
Victoria Falls, 380, 381
Victoria Land, 544
Victoria, Hong Kong, 495
Victoria, Lake, 343, 362
Video recorders, 224
Vienna Basin, 131
Vienna, Austria, 413, 568
Vientiane, Laos, 456, 467
Viet tribes, 461

Vietnam, 424, 454, 456, 459, 465, 467, 468, 471, 473, 476, 480, 509
Vietnam War, 255, 456, 464, 465, 480
Vietnamese, 103, 566
Vietnamese food, 103
Vinson Massif, 544
Virgin Islands, 263
Virginia, 83, 115, 309
Vistula River, 130
Vladimir, Prince, 182
Vladivostok, U.S.S.R., 165, 180, 183, 197, 198
Vodka, 187, 188
Volcanic soil, 236, 273, 283, 529
Volcanoes, 13, 14, 204, 205, 236, 264, 265, 271–73, 341, 362, 398, 456–58, 516, 519, 522
Volga automobile, 197
Volga River, 170, 177, 191, 196–98
Volgograd, U.S.S.R., 190, 191
Volkswagen, 297, 329
Volleyball, 99, 219
Volta Redonda, Brazil, 328
Volta River, 374
Von Thünen, Johann Heinrich, 64–67, 324
Voodoo, 318
Vostok Station, 545
Vulcanism (See also Volcanoes), 13, 14, 16, 305

Wai He, 498
Wailing Wall, 405
Wales, 128, 151
Wallace's Line, 523
Wallace, Alfred Russell, 523
Wallis and Futuna Islands, 529, 530, 532
Walloons, 145, 148
Walruses, 547, 548
Wapsipinicon River, 103
War of the Pacific, 328
Warping, 16
Warsaw, Poland, 177
Washington, 88, 89, 93, 103, 107
Washington, D.C., 103, 262
Watches, 114, 224
Water buffalo, 461, 474
Water pollution, 559
Water polo, 99
Watt, James, 50
Watusi, 343
Wax trees, 209
Weathering, 16, 17
Weber's Line, 523
Wegener, Alfred, 11, 13
Wellington, New Zealand, 244, 246
Welsh language, 128, 145
Wends, 145, 183
Weser River, 131
West Germany, 138, 140, 142, 152–54, 157, 296, 297
West Indies (See also Caribbean Islands), 147, 259, 262, 263, 269

West Pakistan, 439
West Siberian Lowlands, 167, 191
West Virginia, 108, 362
West Wind Drift, 238
Westerly winds (See also Prevailing westerlies), 133
Western Civilization, 123
Western hemisphere, 259, 305
Western Plateau, Australia, 233, 235
Western Sahara, 346, 356
Western Samoa, 529, 530
Whales, 222, 545, 547, 553
Wheat Belt, 4
Wheat, 49, 50, 65, 106, 120, 135, 148, 149, 151, 187, 188, 220, 222, 248, 249, 324, 354, 388, 411, 441, 444, 474, 501, 503, 505
White Australia Policy, 245
White Mountains, 83
White Nile River, 347, 361
White Russians, 182
White Sea, 171
Wichita, Kansas, 103, 115
Willamette Valley, 89, 107
Williams, Ted, 100
Willy-willy, 246
Windchill, 539, 546
Windmills, 247
Windward Islands, 263, 272
Wine, 129, 149, 249, 354, 388, 411
Winter maximum precipitation, 239
Winter monsoon, 208
Winter solstice, 11
Winter wheat (See also Wheat), 107, 249
Wisconsin, 107
Witch doctors, 370
Wittenburg, East Germany, 142
Witwatersrand, 388
Wolf, 548
Wombats, 239
Wood products, 64
Wood pulp, 120, 251
Wool, 49, 189, 242, 243, 247, 325, 387, 388, 411, 444, 503, 505
World Bank, 368
World problems, 557–71
World religions, distribution, 44
World Trade Center, New York, 101
World War I, 77, 129, 147, 174, 333, 406, 413, 415
World War II, 13, 34, 59, 77, 102, 103, 115, 129, 177, 180, 182, 191, 213, 217, 225, 232, 245, 255, 333, 404, 407, 413, 425, 445, 453, 464, 477, 488, 493, 507, 509, 552, 562
Wrestling, 99
Wuhan, China, 485, 495

Xenon, 548
Xerophytic vegetation, 22, 239, 310, 312, 348, 430
Xianggang (See Hong Kong)
Xinjiang Uygur (Sinkiang), 490

I-17

Xizang (Tibet), 46, 434, 470, 484, 496, 509
Xizang Autonomous Region, 484

Yahweh, 412
Yak butter, 444
Yaks, 434
Yakut-Turkic, 550
Yalu River, 487
Yamato, 214
Yams, 283, 373, 526, 529
Yang, 496
Yangtze River (*See* Chang Jiang)
Yaroslavl, U.S.S.R., 191
Yat-sen, Sun, 492
Yellow fever, 287, 370
Yellow River (*See* Huang He)
Yellow Sea (*See* Huang Hai)
Yenisey River, 167, 170

Yiddish language, 42, 414
Yin, 496
Yogurt, 49
Yokohama, Japan, 39, 205, 211, 213
York, Cape of, 252
Yosemite National Park, 103
Yu Shan Mountain, 484
Yuan Dynasty, China, 492
Yucatan Peninsula, 265, 268, 269, 271, 275
Yugoslavia, 131, 138, 142, 145, 147, 148, 154, 566
Yun Ho Canal, 507
Yungas region, Bolivia, 322
Yunnan Province, China, 477

Zagros Mountains, 398, 416
Zaire (Congo), 113, 343, 359, 360, 368, 370–75, 380

Zaire River, 361, 362, 380
Zambezi River, 377, 380, 381, 384
Zambia, 113, 370, 375, 377, 380
Zambo, 275, 313
Zanzibar, 373
Zaskar Mountains, 426
Zebu cattle, 386
Zero Population Growth (ZPG), 31, 562
Zhontang, Liang, 505
Zihuatanejo, Mexico, 293
Zimbabwe, 113, 377, 378, 380, 382, 384, 385, 388–90, 562, 566, 567
Zinc, 154, 190, 288, 355, 411, 476, 506, 552, 562
Zooplankton, 545
Zulu language, 42
Zulu tribe, 391
Zulu wars, 379